BIG IDEAS MATH 7®

VIRGINIA EDITION

Ron Larson
Laurie Boswell

BIG IDEAS LEARNING®

Erie, Pennsylvania
BigIdeasLearning.com

Big Ideas Learning, LLC
1762 Norcross Road
Erie, PA 16510-3838
USA

For product information and customer support, contact Big Ideas Learning at **1-877-552-7766** or visit us at ***BigIdeasLearning.com***.

Printed in the U.S.A.

ISBN 13: 978-1-60840-165-9
ISBN 10: 1-60840-165-0

4 5 6 7 8 9 10 WEB 14 13 12

AUTHORS

Ron Larson is a professor of mathematics at Penn State Erie, The Behrend College, where he has taught since receiving his Ph.D. in mathematics from the University of Colorado in 1970. Dr. Larson is well known as the lead author of a comprehensive program for mathematics that spans middle school, high school, and college courses. His high school and Advanced Placement books are published by Holt McDougal. Ron's numerous professional activities keep him in constant touch with the needs of students, teachers, and supervisors. Ron and Laurie Boswell began writing together in 1992. Since that time, they have authored over two dozen textbooks. In their collaboration, Ron is primarily responsible for the pupil edition and Laurie is primarily responsible for the teaching edition of the text.

Laurie Boswell is the Head of School and a mathematics teacher at the Riverside School in Lyndonville, Vermont. Dr. Boswell received her Ed.D. from the University of Vermont in 2010. She is a recipient of the Presidential Award for Excellence in Mathematics Teaching. Laurie has taught math to students at all levels, elementary through college. In addition, Laurie was a Tandy Technology Scholar, and served on the NCTM Board of Directors from 2002 to 2005. She currently serves on the board of NCSM, and is a popular national speaker. Along with Ron, Laurie has co-authored numerous math programs.

ABOUT THE BOOK

This book is brand new! It is not a revision of a previously published book.

When the NCTM released its new *Curriculum Focal Points* for mathematics for grades 6–8, we were delighted. The traditional mile-wide and inch-deep programs that have been followed for years have clearly not worked. Middle school students need something new . . . fewer topics with deeper coverage.

- **DEEPER** Each section is designed for 2–3 day coverage.
- **DYNAMIC** Each section begins with a full class period of active learning.
- **DOABLE** Each section is accompanied by full student and teacher support.
- **DAZZLING** How else can we say this? This book puts the dazzle back in math!

Ron Larson Laurie Boswell

TEACHER REVIEWERS

- Gail Englert
  Math Department Chairperson
  Norfolk Public Schools
  Norfolk, VA

- J. Patrick Lintner
  Mathematics Supervisor
  Harrisonburg City Public Schools
  Harrisonburg, VA

- Jamie Rosati Perkins
  Middle School Mathematics Coach
  Henrico County Public Schools
  Richmond, VA

- Dianne Schoonover
  Secondary Math Teacher Specialist
  Hampton City Schools
  Hampton, VA

- Bill Setzer
  Retired Mathematics Supervisor
  Roanoke County Schools
  Salem, VA

- Beth Swain
  Mathematics Coordinator
  Salem City Schools
  Salem, VA

- Denise Walston
  Mathematics Coordinator
  Norfolk Public Schools
  Norfolk, VA

STUDENT REVIEWERS

- Ashley Benovic
- Vanessa Bowser
- Sara Chinsky
- Kaitlyn Grimm
- Lakota Noble
- Norhan Omar
- Jack Puckett
- Abby Quinn
- Victoria Royal
- Madeline Su
- Lance Williams

CONSULTANTS

- Patsy Davis
  Educational Consultant
  Knoxville, Tennessee
- Bob Fulenwider
  Mathematics Consultant
  Bakersfield, California
- Deb Johnson
  Differentiated Instruction Consultant
  Missoula, Montana
- Mark Johnson
  Mathematics Assessment Consultant
  Raymond, New Hampshire
- Ryan Keating
  Special Education Advisor
  Gilbert, Arizona
- Michael McDowell
  Project-Based Instruction Specialist
  Scottsdale, Arizona
- Sean McKeighan
  Interdisciplinary Advisor
  Midland, Texas
- Bonnie Spence
  Differentiated Instruction Consultant
  Missoula, Montana

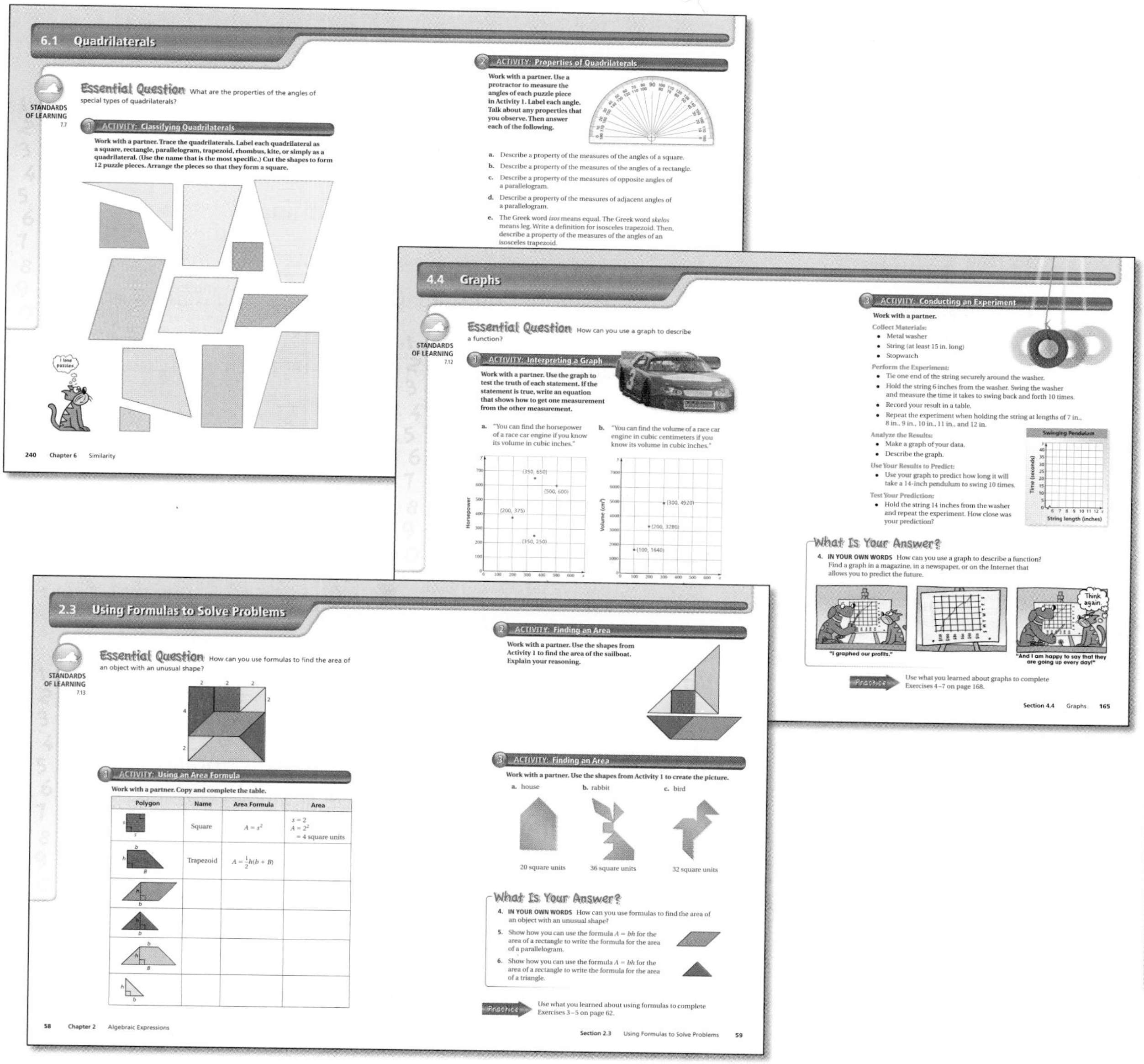

Virginia Standards of Learning for Grade 7

Chapter Coverage for Content Strands

1 - 2 - 3 - **4** - 5 - 6 - 7 - 8 - **9** - 10

Strand: Number and Number Sense

Focus: Proportional Reasoning

Standards of Learning

7.1 The student will **(a)** investigate and describe the concept of negative exponents for powers of ten; **(b)** determine scientific notation for numbers greater than zero; **(c)** compare and order fractions, decimals, percents, and numbers written in scientific notation; **(d)** determine square roots; and **(e)** identify and describe absolute value for rational numbers.

7.2 The student will describe and represent arithmetic and geometric sequences, using variable expressions.

1 - 2 - **3** - 4 - **5** - **6** - 7 - 8 - 9 - 10

Strand: Computation and Estimation

Focus: Integer Operations and Proportional Reasoning

Standards of Learning

7.3 The student will **(a)** model addition, subtraction, multiplication, and division of integers; and **(b)** add, subtract, multiply, and divide integers.

7.4 The student will solve single-step and multistep practical problems, using proportional reasoning.

1 - 2 - 3 - 4 - 5 - **6** - 7 - **8** - 9 - 10

Strand: Measurement

Focus: Proportional Reasoning

Standards of Learning

7.5 The student will **(a)** describe volume and surface area of cylinders; **(b)** solve practical problems involving the volume and surface area of rectangular prisms and cylinders; and **(c)** describe how changing one measured attribute of a rectangular prism affects its volume and surface area.

7.6 The student will determine whether plane figures—quadrilaterals and triangles—are similar and write proportions to express the relationships between corresponding sides of similar figures.

1 - 2 - 3 - 4 - 5 - 6 - 7 - 8 - 9 - 10

Strand ## Geometry

Focus: Relationships between Figures

Standards of Learning

7.7 The student will compare and contrast the following quadrilaterals based on properties: parallelogram, rectangle, square, rhombus, and trapezoid.

7.8 The student, given a polygon in the coordinate plane, will represent transformations (reflections, dilations, rotations, and translations) by graphing in the coordinate plane.

1 - 2 - 3 - 4 - 5 - 6 - 7 - 8 - 9 - 10

Strand ## Probability and Statistics

Focus: Applications of Statistics and Probability

Standards of Learning

7.9 The student will investigate and describe the difference between the experimental probability and theoretical probability of an event.

7.10 The student will determine the probability of compound events, using the Fundamental (Basic) Counting Principle.

7.11 The student, given data for a practical situation, will **(a)** construct and analyze histograms; and **(b)** compare and contrast histograms with other types of graphs presenting information from the same data set.

1 - 2 - 3 - 4 - 5 - 6 - 7 - 8 - 9 - 10

Strand ## Patterns, Functions, and Algebra

Focus: Linear Equations

Standards of Learning

7.12 The student will represent relationships with tables, graphs, rules, and words.

7.13 The student will **(a)** write verbal expressions as algebraic expressions and sentences as equations and vice versa; and **(b)** evaluate algebraic expressions for given replacement values of the variables.

7.14 The student will **(a)** solve one- and two-step linear equations in one variable; and **(b)** solve practical problems requiring the solution of one- and two-step linear equations.

7.15 The student will **(a)** solve one-step inequalities in one variable; and **(b)** graph solutions to inequalities on the number line.

7.16 The student will apply the following properties of operations with real numbers: **(a)** the commutative and associative properties for addition and multiplication; **(b)** the distributive property; **(c)** the additive and multiplicative identity properties; **(d)** the additive and multiplicative inverse properties; and **(e)** the multiplicative property of zero.

Operations with Integers

"I love my math book. It has so many interesting examples and homework problems. I have always liked math, but I didn't know how it could be used. Now I have lots of ideas."

Algebraic Expressions

"I like starting each new lesson with a partner activity. I just moved to this school and the activities helped me make friends."

Rational Numbers, Equations, and Inequalities

"I like having the book on the Internet. The online tutorials help me with my homework when I get stuck on a problem."

4

Tables, Graphs, and Functions

"I love the cartoons. They are funny and they help me remember the math. I want to be a cartoonist some day."

Ratios and Proportions

"I like how I can click on the words in the book that is online and hear them read to me. I like to pronouce words correctly, but sometimes I don't know how to do that by just reading the words."

6 Similarity

"I really liked the projects at the end of the book. The history project on ancient Egypt was my favorite. Someday I would like to visit Egypt and go to the pyramids."

Transformations

"I like how the glossary in the book is part of the index. When I couldn't remember how a vocabulary word was defined, I could go to the index and find where the word was defined in the book."

Surface Area and Volume

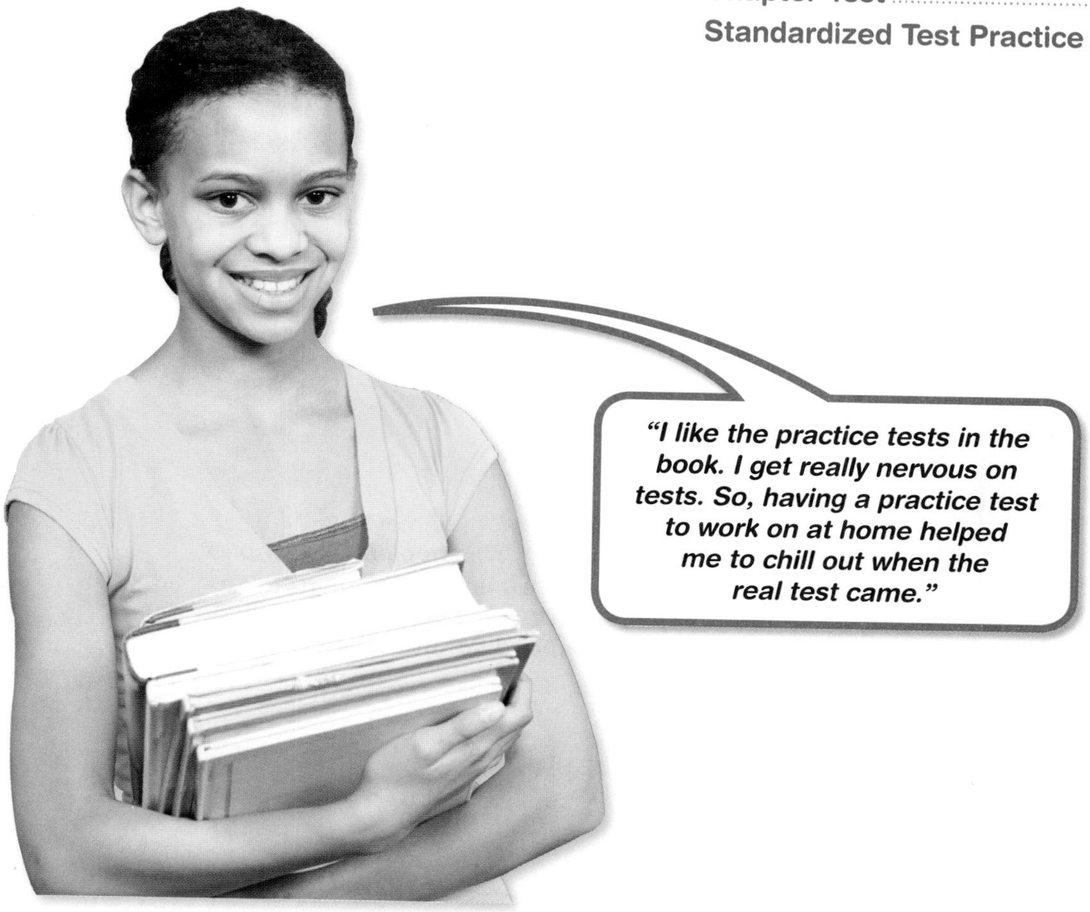

"I like the practice tests in the book. I get really nervous on tests. So, having a practice test to work on at home helped me to chill out when the real test came."

Exponents and Scientific Notation

"I like the review at the beginning of each chapter. This book has examples to help me remember things from last year. I don't like it when the review is just a list of questions."

Data Analysis and Probability

"I like the workbook (Record and Practice Journal). It saved me a lot of work to not have to copy all the questions and graphs."

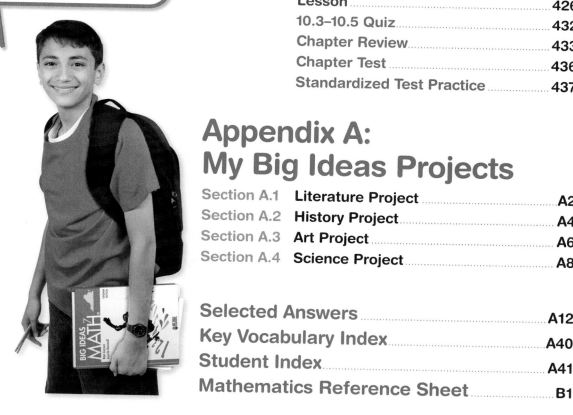

Appendix A: My Big Ideas Projects

How to Use Your Math Book

- Read the **Essential Question** in the activity.

  Work with a partner to decide **What Is Your Answer?**

  Now you are ready to do the **Practice** problems.

- Find the **Key Vocabulary** words, **highlighted in yellow**.

  Read their definitions. Study the concepts in each **Key Idea**.
  If you forget a definition, you can look it up online in the

  Multi-Language Glossary at BigIdeasMath.com.

- After you study each **EXAMPLE**, do the exercises in the **On Your Own**.

  Now You're Ready to do the exercises that correspond to the example.

  As you study, look for a **Study Tip** or a **Common Error**.

- The exercises are divided into 3 parts.

  Vocabulary and Concept Check

  Practice and Problem Solving

  Fair Game Review

  If an exercise has a 1 next to it, look back at Example 1 for help with that exercise.

  More help is available at **Check It Out Lesson Tutorials BigIdeasMath.com**.

- To help study for your test, use the following.

  Quiz **Study Help**

  Chapter Review **Chapter Test**

SCAVENGER HUNT

Use this *Scavenger Hunt* to find where things are in **Chapter 1**.

Start

Newton

Find an Essential Question.

Find a Practice.

Find the definition of **opposites**.

Find the Absolute Value.

Use the index in the back of the book.

Find an **EXAMPLE** of subtracting integers.

Find an On Your Own about |10| 11.

Find Now You're Ready Exercises 32–37.

Find a Remember about using number lines.

Find a Vocabulary question about adding integers.

Find a Problem Solving question about a **TUNNEL**.

Find the answer in the back of the book.

Which examples ? help with Exercise 13 on page 12?

Find 2 Quizzes.

How many questions are on the Chapter Test?

You're Ready

Descartes

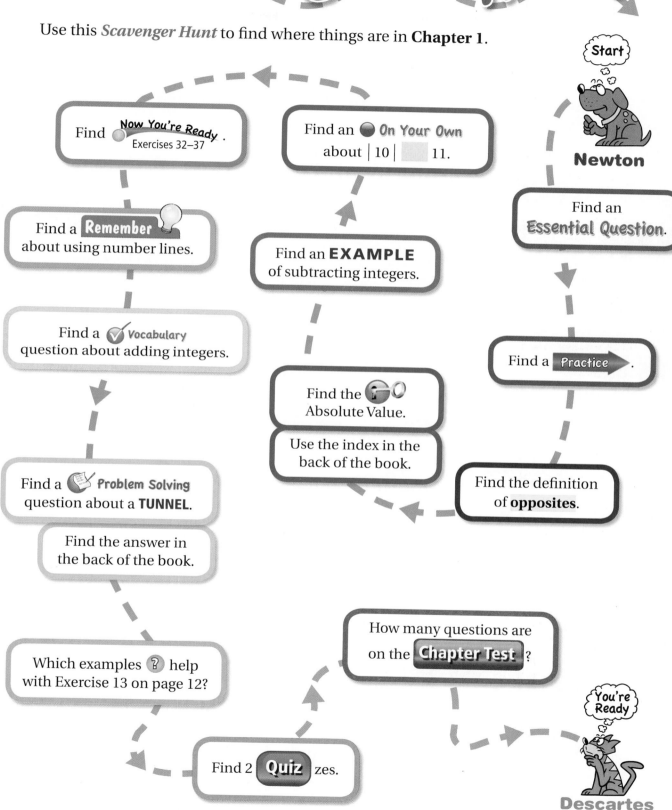

1 Operations with Integers

"Look, subtraction is not that difficult. Imagine that you have five squeaky mouse toys."
5 + (-1) = 4
"After your friend Fluffy comes over for a visit, you notice that one of the squeaky toys is missing."
4 - (-1) = 5
I'm going to count my toys
"Now, you go over to Fluffy's and retrieve the missing squeaky mouse toy. It's easy."

"Dear Sir: You asked me to 'find' the opposite of −1."
That's not the only thing that is missing.
"I didn't know it was missing."

What You Learned Before

"I liked it because it is the opposite of the freezing point on the Fahrenheit temperature scale."

Writing Integers

Example 1 **Write a positive or negative integer to represent the temperature 28 degrees below zero.**

"Below" indicates a negative integer.

∴ So, −28 represents 28 degrees below zero.

Try It Yourself

Write a positive or negative integer that represents the situation.

1. A savings account earns $34 in interest.

2. You go down 4 floors in an elevator.

3. A football team gains 56 yards.

Ordering Integers

Example 2 **Use a number line to order 0, −1, 2, 5, and −6.**

∴

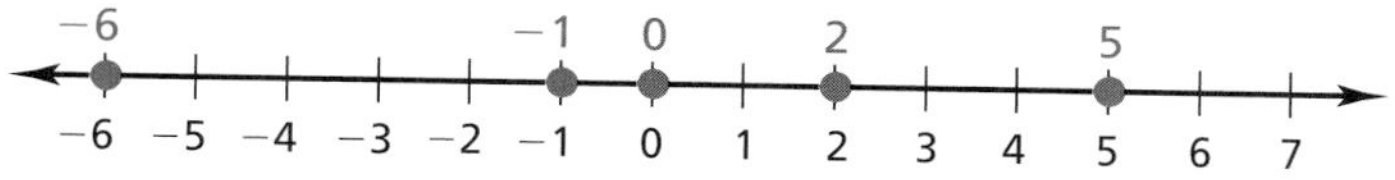

Try It Yourself

Use a number line to order the integers.

4. −10, 15, 4, −2, −12

5. 7, −5, 3, −3, 1

Using Order of Operations

Example 3 **Evaluate $6^2 \div 4 - 2(9 - 5)$.**

$6^2 \div 4 - 2(9 - 5)$	$= 6^2 \div 4 - 2 \cdot 4$	Perform operation in parentheses.
	$= 36 \div 4 - 2 \cdot 4$	Evaluate 6^2.
	$= 9 - 8$	Divide 36 by 4 and multiply 2 and 4.
	$= 1$	Subtract 8 from 9.

Try It Yourself

Evaluate the expression.

6. $15\left(\frac{8}{4}\right) + 2^2 - 3 \cdot 7$ = 13

7. $5^2 \cdot 2 \div 10 + 3 \cdot 2 - 1$ = 10

8. $3^2 - 1 + 2(4(3 + 2))$

1.1 Integers and Absolute Value

Essential Question How are velocity and speed related?

STANDARDS OF LEARNING
7.1

On these two pages, you will investigate vertical motion (up or down).

- Speed tells how fast an object is moving, but does not tell the direction.
- Velocity tells how fast an object is moving and also tells the direction.
  If velocity is positive, the object is moving up.
  If velocity is negative, the object is moving down.

1 EXAMPLE: Falling Parachute

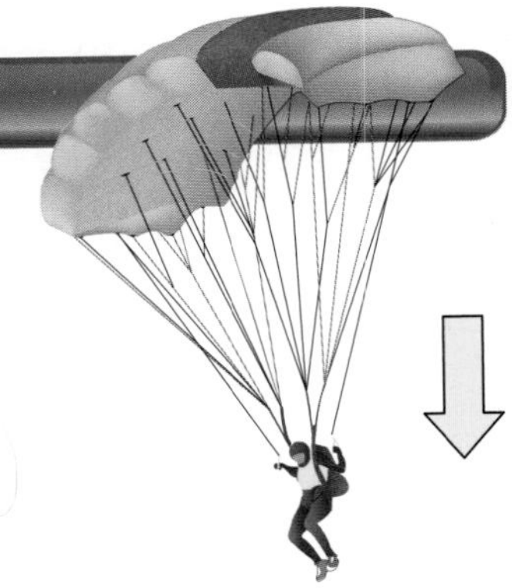

You are gliding to the ground wearing a parachute. The table shows your height at different times.

Time (seconds)	0	1	2	3
Height (feet)	45	30	15	0

a. How many feet do you move each second?
b. What is your speed? Give the units.
c. Is your velocity positive or negative?
d. What is your velocity? Give the units.

a. For each 1 second of time, your height is 15 feet less.
b. You are moving at 15 feet per second.
c. Because you are moving down, your velocity is negative.
d. Your velocity is -15 feet per second. This can be written as -15 ft/sec.

2 ACTIVITY: Rising Balloons

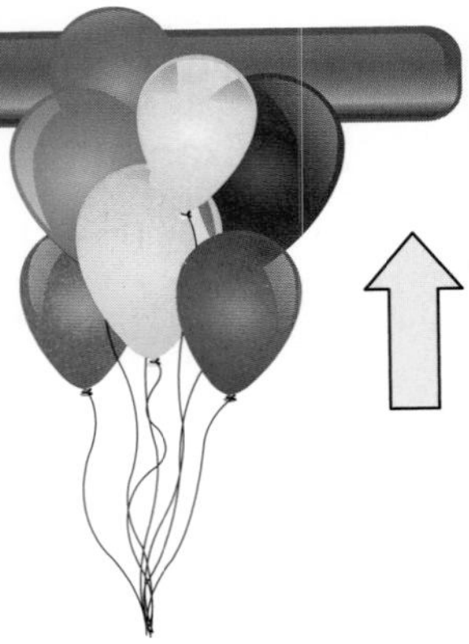

Work with a partner. The table shows the height of a group of balloons.

Time (seconds)	0	1	2	3
Height (feet)	0	4	8	12

a. How many feet do the balloons move each second?
b. What is the speed of the balloons? Give the units.
c. Is the velocity positive or negative?
d. What is the velocity? Give the units.

3 ACTIVITY: Finding Speed and Velocity

Work with a partner. The table shows the height of a firework's parachute.

Time (seconds)	Height (feet)
0	480
1	360
2	240
3	120
4	0

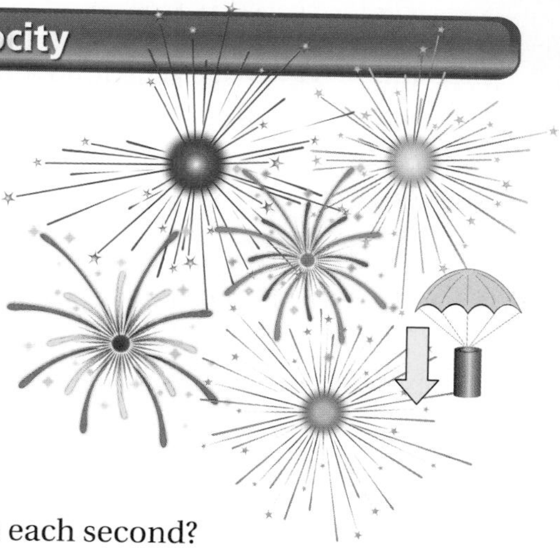

a. How many feet does the parachute move each second?

b. What is the speed of the parachute? Give the units.

c. Is the velocity positive or negative?

d. What is the velocity? Give the units.

Inductive Reasoning

4. Copy and complete the table.

Velocity (feet per second)	−14	20	−2	0	25	−15
Speed (feet per second)						

5. Find two different velocities for which the speed is 16 feet per second.

6. Which number is greater: −4 or 3? Use a number line to explain your reasoning.

7. One object has a velocity of −4 feet per second. Another object has a velocity of 3 feet per second. Which object has the greater speed? Explain your answer.

What Is Your Answer?

In this lesson, you will study **absolute value**. Here are some examples:

Absolute value of −16 = 16   Absolute value of 16 = 16

Absolute value of 0 = 0   Absolute value of −2 = 2

8. IN YOUR OWN WORDS How are velocity and speed related?

9. Which of the following is a true statement? Explain your reasoning.

a. Absolute value of velocity = speed

b. Absolute value of speed = velocity

Practice Use what you learned about absolute value to complete Exercises 4–11 on page 6.

Key Vocabulary 🔊
integer, *p. 4*
absolute value, *p. 4*

The following numbers are **integers.**

$$\ldots, -3, -2, -1, 0, 1, 2, 3, \ldots$$

Key Idea

Absolute Value

Words The **absolute value** of an integer is the distance between the number and 0 on a number line. The absolute value of a number a is written as $|a|$.

```
          4 units          4 units
    •———————————|———————————•
 -5  -4  -3  -2  -1  0  1  2  3  4  5
```

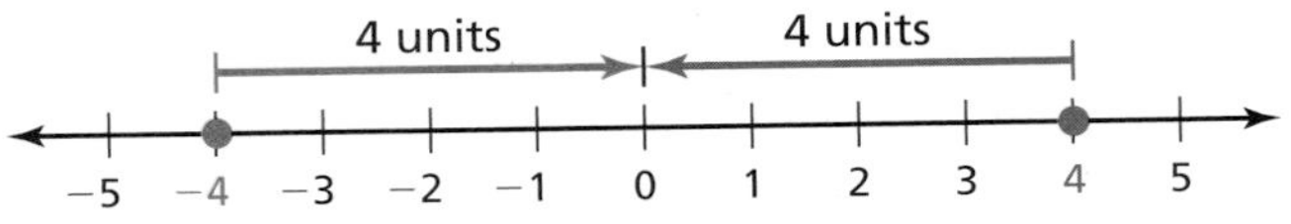

Numbers $|-4| = 4$ $|4| = 4$

EXAMPLE 1 **Finding Absolute Value**

Find the absolute value of 2.

Graph 2 on a number line.

```
    -4  -3  -2  -1  0  1  2  3  4
                       |—— 2 ——|
```

The distance between 2 and 0 is 2.

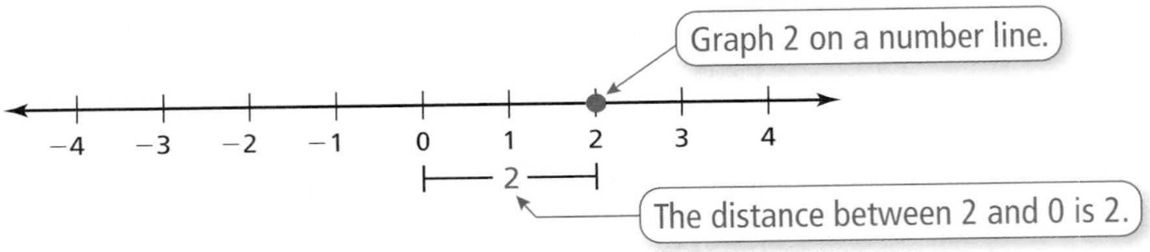

∴ So, $|2| = 2$.

EXAMPLE 2 **Finding Absolute Value**

Find the absolute value of −3.

Graph −3 on a number line.

```
    -6  -5  -4  -3  -2  -1  0  1  2
                |—— 3 ——|
```

The distance between −3 and 0 is 3.

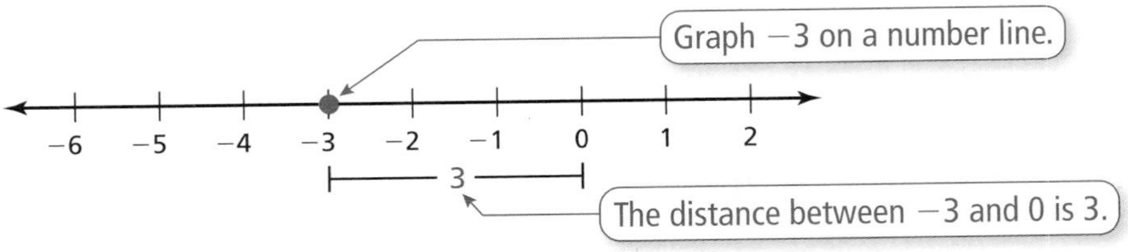

∴ So, $|-3| = 3$.

On Your Own

Now You're Ready
Exercises 4–19

Find the absolute value.

1. $|7|$ 2. $|-1|$ 3. $|-5|$ 4. $|14|$

🔊 Multi-Language Glossary at BigIdeasMath✓com.

EXAMPLE 3 Comparing Values

Remember

A number line can be used to compare and order integers. Numbers to the left are less than numbers to the right. Numbers to the right are greater than numbers to the left.

Compare 1 and $|-4|$.

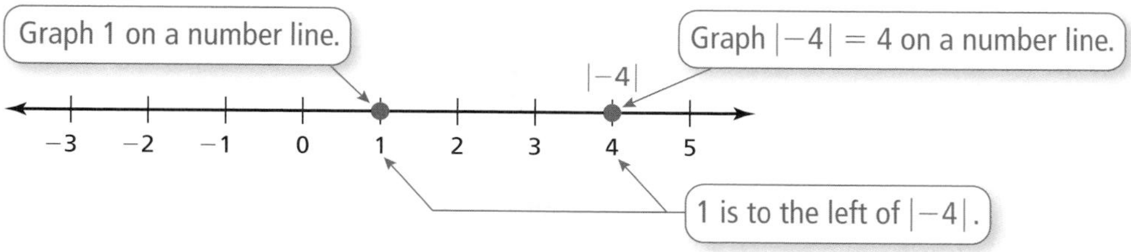

So, $1 < |-4|$.

On Your Own

Now You're Ready Exercises 20–25

Copy and complete the statement using <, >, or =.

5. $|-2|$ ☐ -1

6. -7 ☐ $|6|$

7. $|10|$ ☐ 11

8. 9 ☐ $|-9|$

EXAMPLE 4 Real-Life Application

Substance	Freezing Point (°C)
Butter	35
Airplane fuel	−53
Honey	−3
Mercury	−39
Candle wax	55

The *freezing point* is the temperature at which a liquid becomes a solid.

a. Which substance in the table has the lowest freezing point?

b. Is the freezing point of mercury or butter closer to the freezing point of water, 0°C?

a. Graph each freezing point.

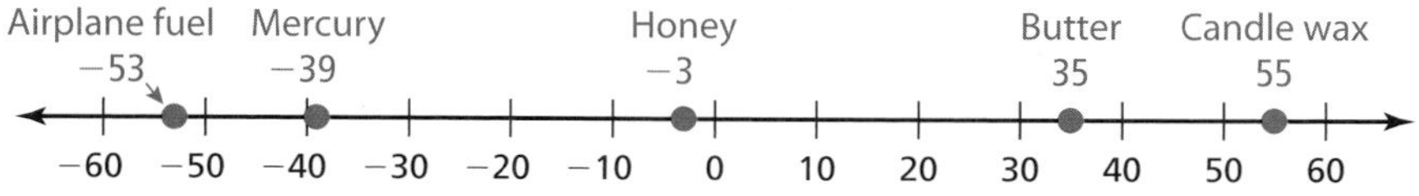

Airplane fuel has the lowest freezing point, −53°C.

b. The freezing point of water is 0°C, so you can use absolute values.

Mercury: $|-39| = 39$ **Butter:** $|35| = 35$

Because 35 is less than 39, the freezing point of butter is closer to the freezing point of water.

On Your Own

9. Is the freezing point of airplane fuel or candle wax closer to the freezing point of water? Explain your reasoning.

1.1 Exercises

Check It Out
Help with Homework
BigIdeasMath.com

Vocabulary and Concept Check

1. **VOCABULARY** Which of the following numbers are integers?

$$9, 3.2, -1, \frac{1}{2}, -0.25, 15$$

2. **VOCABULARY** What is the absolute value of an integer?

3. **WHICH ONE DOESN'T BELONG?** Which expression does *not* belong with the other three? Explain your reasoning.

$\lvert 6 \rvert$	6	-6	$\lvert -6 \rvert$

Practice and Problem Solving

Find the absolute value.

(1) (2) 4. $\lvert 9 \rvert$ 5. $\lvert -6 \rvert$ 6. $\lvert -10 \rvert$ 7. $\lvert 10 \rvert$

8. $\lvert -15 \rvert$ 9. $\lvert 13 \rvert$ 10. $\lvert -7 \rvert$ 11. $\lvert -12 \rvert$

12. $\lvert 5 \rvert$ 13. $\lvert -8 \rvert$ 14. $\lvert 0 \rvert$ 15. $\lvert 18 \rvert$

16. $\lvert -24 \rvert$ 17. $\lvert -45 \rvert$ 18. $\lvert 60 \rvert$ 19. $\lvert -125 \rvert$

Copy and complete the statement using <, >, or =.

(3) 20. $2 \;\square\; \lvert -5 \rvert$ 21. $\lvert -4 \rvert \;\square\; 7$ 22. $-5 \;\square\; \lvert -9 \rvert$

23. $\lvert -4 \rvert \;\square\; -6$ 24. $\lvert -1 \rvert \;\square\; \lvert -8 \rvert$ 25. $\lvert 5 \rvert \;\square\; \lvert -5 \rvert$

ERROR ANALYSIS Describe and correct the error.

26. ✗ $\lvert 10 \rvert = -10$

27. ✗ $\lvert -5 \rvert < 4$

28. **SAVINGS** You deposit \$50 in your savings account. One week later, you withdraw \$20. Write each amount as an integer.

29. **ELEVATOR** You go down 8 floors in an elevator. Your friend goes up 5 floors in an elevator. Write each amount as an integer.

Order the values from least to greatest.

30. $8, \lvert 3 \rvert, -5, \lvert -2 \rvert, -2$

31. $\lvert -6 \rvert, -7, 8, \lvert 5 \rvert, -6$

32. $-12, \lvert -26 \rvert, -15, \lvert -12 \rvert, \lvert 10 \rvert$

33. $\lvert -34 \rvert, 21, -17, \lvert 20 \rvert, \lvert -11 \rvert$

Simplify the expression.

34. $\lvert -30 \rvert$ 35. $-\lvert 4 \rvert$ 36. $-\lvert -15 \rvert$

37. PUZZLE Use a number line.

a. Graph and label the following points on a number line: $A = -3$, $E = 2$, $M = -6$, $T = 0$. What word do the letters spell?

b. Graph and label the absolute value of each point in part (a). What word do the letters spell now?

38. OPEN-ENDED Write a negative integer whose absolute value is greater than 3.

REASONING Determine whether $n \geq 0$ or $n \leq 0$.

39. $n + |-n| = 2n$

40. $n + |-n| = 0$

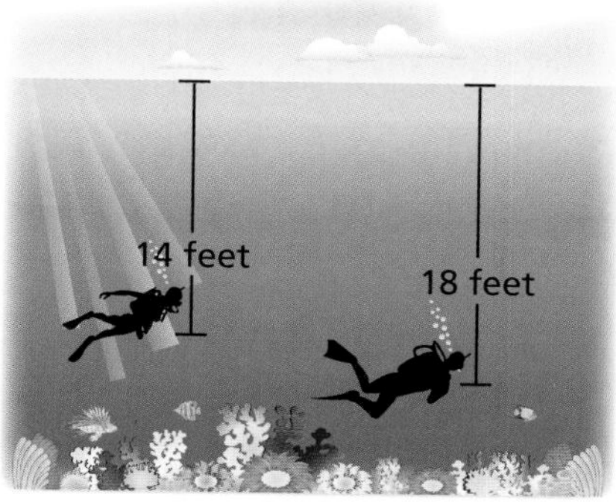

41. CORAL REEF Two scuba divers are exploring a living coral reef.

a. Write an integer for the position of each diver relative to sea level.

b. Which integer in part (a) is greater?

c. Which integer in part (a) has the greater absolute value? Compare this with the position of the diver farther from sea level.

42. VOLCANOES The *summit elevation* of a volcano is the elevation of the top of the volcano relative to sea level. The summit elevation of the volcano Kilauea in Hawaii is 1277 meters. The summit elevation of the underwater volcano Loihi in the Pacific Ocean is −969 meters. Which summit is closer to sea level?

43. MINIATURE GOLF The table shows golf scores, relative to *par*.

a. The player with the lowest score wins. Which player wins?

b. Which player is at par?

c. Which player is farthest from par?

Player	Score
1	+5
2	0
3	−4
4	−1
5	+2

True or False? **Determine whether the statement is *true* or *false*. Explain your reasoning.**

44. If $x < 0$, then $|x| = -x$.

45. The absolute value of every integer is positive.

Fair Game Review What you learned in previous grades & lessons

Add. *(Skills Review Handbook)*

46. $19 + 32$

47. $50 + 94$

48. $181 + 217$

49. $1149 + 2021$

50. MULTIPLE CHOICE Which value is *not* a whole number? *(Skills Review Handbook)*

Ⓐ −5 Ⓑ 0 Ⓒ 4 Ⓓ 113

1.2 Adding Integers

STANDARDS OF LEARNING

7.3

Essential Question Is the sum of two integers *positive*, *negative*, or *zero*? How can you tell?

1 EXAMPLE: Adding Integers with the Same Sign

Use integer counters to find $-4 + (-3)$.

Combine 4 negative counters and 3 negative counters.

Find the total number of counters.

$$-4 \quad + \quad -3 \quad = \quad -7$$

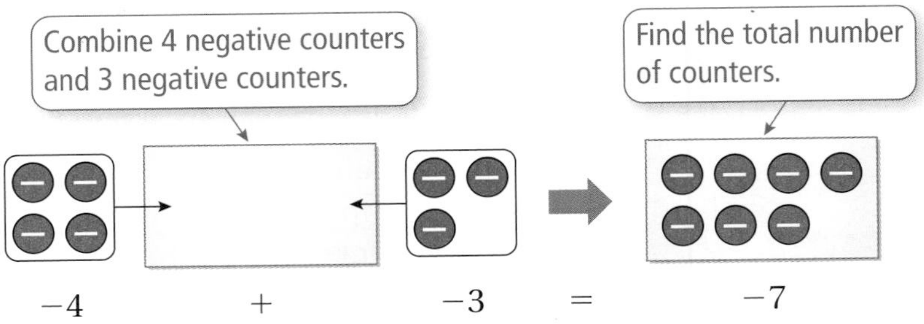

∴ So, $-4 + (-3) = -7$.

2 ACTIVITY: Adding Integers with Different Signs

Work with a partner. Use integer counters to find $-3 + 2$.

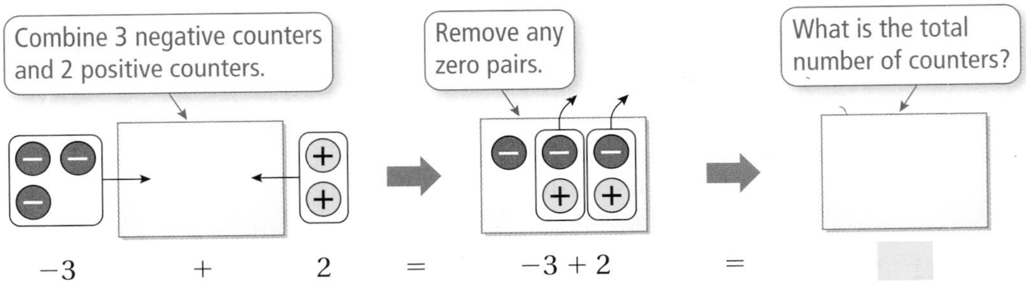

Combine 3 negative counters and 2 positive counters.

Remove any zero pairs.

What is the total number of counters?

$$-3 \quad + \quad 2 \quad = \quad -3 + 2 \quad = \quad \boxed{}$$

∴ So, $-3 + 2 = \boxed{}$.

3 EXAMPLE: Adding Integers with Different Signs

Use a number line to find $5 + (-3)$.

Start at 0. Move 5 units to the right.

Add -3.

Then move 3 units left to end at 2.

5

$$-8 \quad -6 \quad -4 \quad -2 \quad 0 \quad 2 \quad 4 \quad 6 \quad 8$$

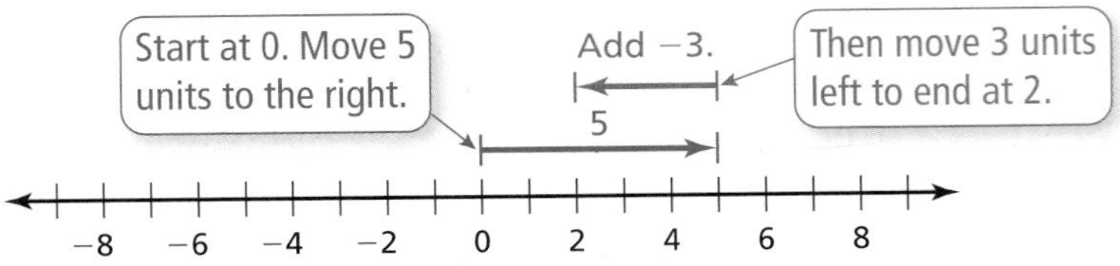

∴ So, $5 + (-3) = 2$.

4 ACTIVITY: Adding Integers with Different Signs

Work with a partner. Write the addition expression shown. Then find the sum.

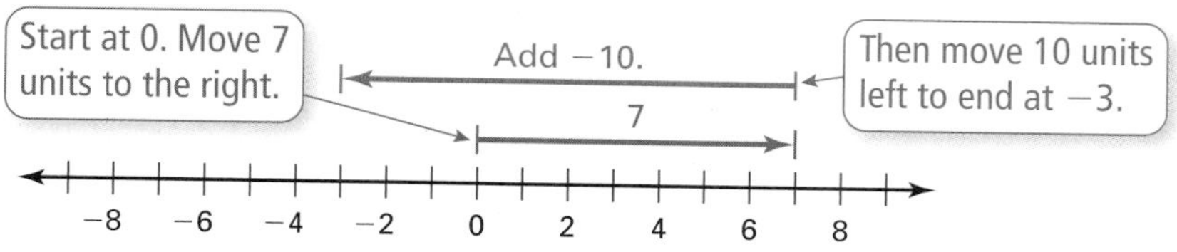

Start at 0. Move 7 units to the right.
Add −10.
7
Then move 10 units left to end at −3.
−8
−6
−4
−2
0
2
4
6
8

Inductive Reasoning

Work with a partner. Use integer counters or a number line to complete the table.

	Exercise	Type of Sum	Sum	Sum: Positive, Negative, or Zero
1	**5.** $-4 + (-3)$	Integers with the same sign	−7	Negative
2	**6.** $-3 + 2$	Integers with different signs	−1	Negative
3	**7.** $5 + (-3)$	Integers with different signs	2	Positive
4	**8.** $7 + (-10)$	Integers with different signs	−3	Negative
	9. $2 + 4$	Integers with the same signs	6	Positive
	10. $-6 + (-2)$	Integers with the same signs	−8	Negative
	11. $-5 + 9$	Integers with different sign	4	Positive
	12. $15 + (-9)$	Different signs	6	Positive
	13. $-10 + 10$	Different signs	0	Positive
	14. $-6 + (-6)$	Same signs	−12	Negative
	15. $12 + (-12)$	different signs	0	Positive

What Is Your Answer?

16. **IN YOUR OWN WORDS** Is the sum of two integers *positive*, *negative*, or *zero*? How can you tell?

17. Write general rules for adding (a) two integers with the same sign, (b) two integers with different signs, and (c) an integer and its opposite.

Practice Use what you learned about adding integers to complete Exercises 8–15 on page 12.

1.2 Lesson

Key Vocabulary
opposites, *p. 10*
additive inverse, *p. 10*

Key Idea

Adding Integers with the Same Sign

Words Add the absolute values of the integers. Then use the common sign.

Numbers $2 + 5 = 7$ $\quad -2 + (-5) = -7$

EXAMPLE 1 **Adding Integers with the Same Sign**

Find $-2 + (-4)$. Use a number line to check your answer.

$-2 + (-4) = -6$ $\quad$ Add $|-2|$ and $|-4|$.

Use the common sign.

∴ The sum is -6.

Check

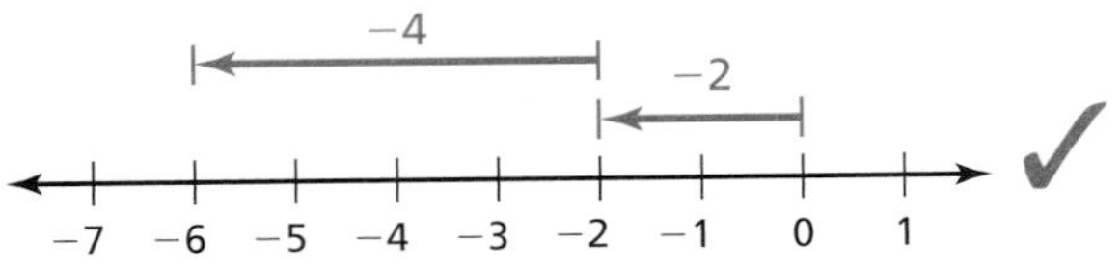

The Meaning of a Word

Opposite

When you walk across a street, you are moving to the **opposite** side of the street.

On Your Own

Add.

1. $7 + 13$ $\quad$ **2.** $-8 + (-5)$ $\quad$ **3.** $-20 + (-15)$

Two numbers that are the same distance from 0, but on opposite sides of 0, are called **opposites.** For example, -3 and 3 are opposites.

Key Ideas

Adding Integers with Different Signs

Words Subtract the lesser absolute value from the greater absolute value. Then use the sign of the integer with the greater absolute value.

Numbers $8 + (-10) = -2$ $\quad -13 + 17 = 4$

Additive Inverse Property

Words The sum of an integer and its **additive inverse,** or opposite, is 0.

Numbers $6 + (-6) = 0$ $\quad -25 + 25 = 0$ $\quad$ **Algebra** $a + (-a) = 0$

EXAMPLE 2 Adding Integers with Different Signs

a. Find $5 + (-10)$.

$5 + (-10) = -5$ — $|-10| > |5|$. So, subtract $|5|$ from $|-10|$.

Use the sign of -10.

The sum is -5.

b. Find $-3 + 7$.

$-3 + 7 = 4$ — $|7| > |-3|$. So, subtract $|-3|$ from $|7|$.

Use the sign of 7.

The sum is 4.

c. Find $-12 + 12$.

$-12 + 12 = 0$ — The sum is 0 by the Additive Inverse Property.

-12 and 12 are opposites.

The sum is 0.

On Your Own

Now You're Ready
Exercises 8–23

Add.

4. $-2 + 11$ **5.** $13 + (-8)$ **6.** $9 + (-10)$

7. $-8 + 4$ **8.** $7 + (-7)$ **9.** $-31 + 31$

EXAMPLE 3 Adding More than Two Integers

The list shows four bank account transactions in July. Find the change *C* in the account balance.

JULY TRANSACTIONS	
Deposit	**$50**
Withdrawal	**-$40**
Deposit	**$75**
Withdrawal	**-$50**

Find the sum of the four transactions.

$C = 50 + (-40) + 75 + (-50)$ — Write the sum.

$= 10 + 75 + (-50)$ — Add 50 and -40.

$= 85 + (-50)$ — Add 10 and 75.

$= 35$ — Add 85 and -50.

Because $C = 35$, the account balance increased $35 in July.

On Your Own

Now You're Ready
Exercises 28–33

10. WHAT IF? In Example 3, the deposit amounts are $30 and $55. Find the change *C* in the account balance.

1.2 Exercises

Check It Out
Help with Homework
BigIdeasMath.com

Vocabulary and Concept Check

1. **WRITING** How do you find the additive inverse of an integer?
2. **NUMBER SENSE** Is $3 + (-4)$ the same as $-4 + 3$? Explain.

Tell whether the sum is *positive*, *negative*, or *zero* without adding. Explain your reasoning.

3. $-8 + 20$
4. $50 + (-50)$
5. $-10 + (-18)$

Tell whether the statement is *true* or *false*. Explain your reasoning.

6. The sum of two negative integers is always negative.
7. An integer and its absolute value are always opposites.

Practice and Problem Solving

Add.

(1) (2)

8. $6 + 4$
9. $-4 + (-6)$
10. $-2 + (-3)$
11. $-5 + 12$
12. $5 + (-7)$
13. $8 + (-8)$
14. $9 + (-11)$
15. $-3 + 13$
16. $-4 + (-16)$
17. $-3 + (-4)$
18. $14 + (-5)$
19. $0 + (-11)$
20. $-10 + (-15)$
21. $-13 + 9$
22. $18 + (-18)$
23. $-25 + (-9)$

ERROR ANALYSIS **Describe and correct the error in finding the sum.**

24. ✗ $9 + (-6) = -3$

25. ✗ $-10 + (-10) = 0$

26. **TEMPERATURE** The temperature is −3°F at 7 A.M. During the next four hours, the temperature increases 21°F. What is the temperature at 11 A.M.?
27. **BANKING** Your bank account has a balance of −\$12. You deposit \$60. What is your new balance?

Add.

(3)

28. $13 + (-21) + 16$
29. $22 + (-14) + (-35)$
30. $-13 + 27 + (-18)$
31. $-19 + 26 + 14$
32. $-32 + (-17) + 42$
33. $-41 + (-15) + (-29)$

Tell how the Commutative and Associative Properties of Addition can help you find the sum mentally. Then find the sum.

34. $9 + 6 + (-6)$
35. $-8 + 13 + (-13)$
36. $9 + (-17) + (-9)$
37. $7 + (-12) + (-7)$
38. $-12 + 25 + (-15)$
39. $6 + (-9) + 14$

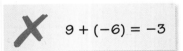

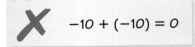

40. **OPEN-ENDED** Write two integers with different signs that have a sum of -25. Write two integers with the same sign that have a sum of -25.

MENTAL MATH **Use mental math to solve the equation.**

41. $x + 8 = 0$ **42.** $3 + a = -3$ **43.** $y + (-1) = 3$

44. $d + 12 = 2$ **45.** $b + (-2) = 0$ **46.** $-8 + m = -15$

47. **FIRST DOWN** In football, a team must gain 10 yards to get a first down. The team gains 6 yards on the first play, loses 3 yards on the second play, and gains 8 yards on the third play. Which expression can be used to decide whether the team gets a first down?

$10 + 6 - 3 + 8$ $6 + (-3) + 8$ $6 + (-3) + (-8)$

48. **DOLPHIN** Starting at point A, the path of a dolphin jumping out of the water is shown.

a. Is the dolphin deeper at point C or point E? Explain your reasoning.

b. Is the dolphin higher at point B or point D? Explain your reasoning.

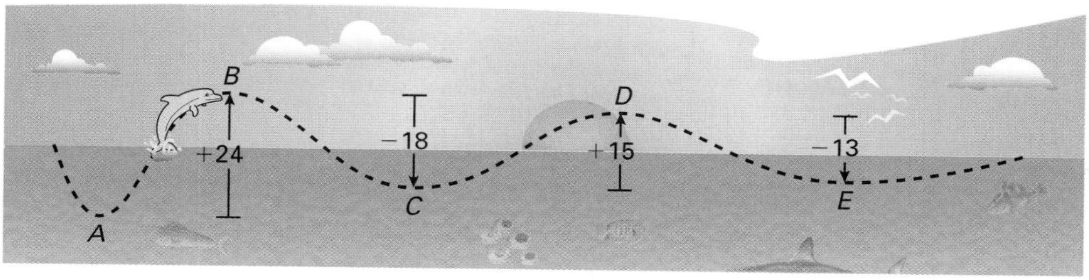

49. Puzzle According to a legend, the Chinese Emperor Yu-Huang saw a magic square on the back of a turtle. In a *magic square*, the numbers in each row and in each column have the same sum. This sum is called the magic sum.

Copy and complete the magic square so that each row and each column has a magic sum of 0. Use each integer from -4 to 4 exactly once.

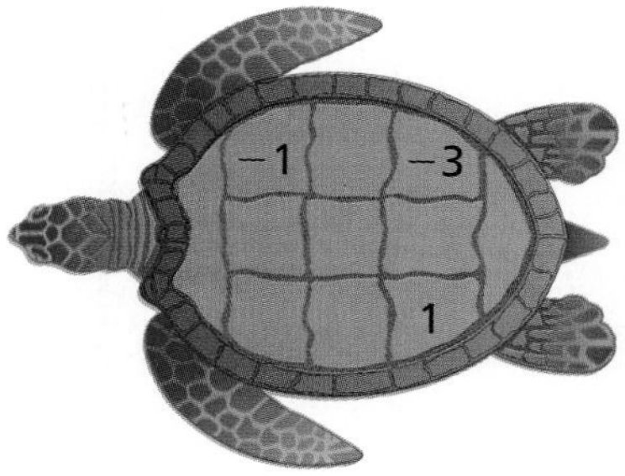

Fair Game Review What you learned in previous grades & lessons

Subtract. *(Skills Review Handbook)*

50. 69 − 38 **51.** 82 − 74 **52.** 177 − 63 **53.** 451 − 268

54. **MULTIPLE CHOICE** What is the range of the numbers below? *(Skills Review Handbook)*

12, 8, 17, 12, 15, 18, 30

Ⓐ 12 Ⓑ 15 Ⓒ 18 Ⓓ 22

1.3 Subtracting Integers

STANDARDS OF LEARNING
7.3

Essential Question How are adding integers and subtracting integers related?

1 EXAMPLE: Subtracting Integers

Use integer counters to find 4 − 2.

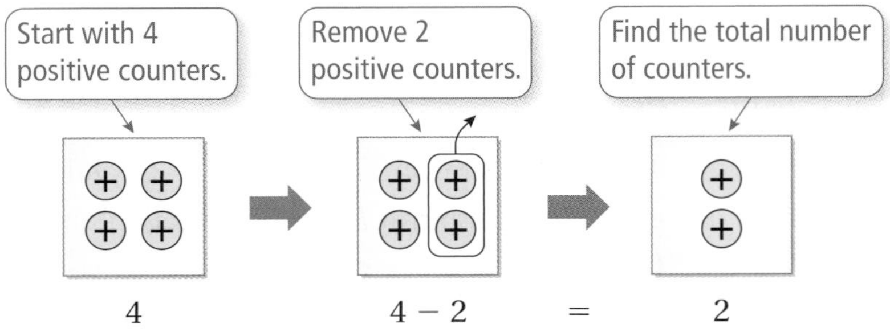

∴ So, 4 − 2 = 2.

2 ACTIVITY: Adding Integers

Work with a partner. Use integer counters to find 4 + (−2).

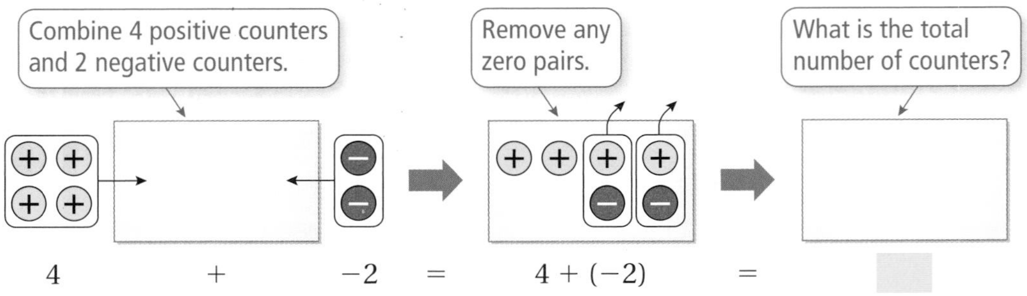

Combine 4 positive counters and 2 negative counters.
Remove any zero pairs.
What is the total number of counters?
4
+
−2
=
4 + (−2)
=

3 EXAMPLE: Subtracting Integers

Use a number line to find −3 − 1.

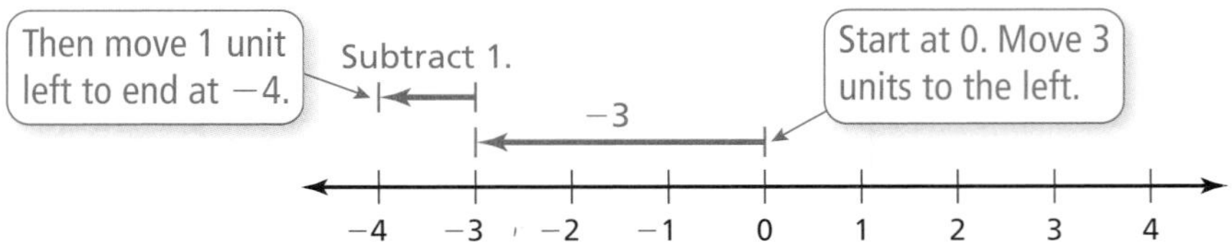

Then move 1 unit left to end at −4.

Start at 0. Move 3 units to the left.

∴ So, −3 − 1 = −4.

4 ACTIVITY: Adding Integers

Work with a partner. Write the addition expression shown. Then find the sum.

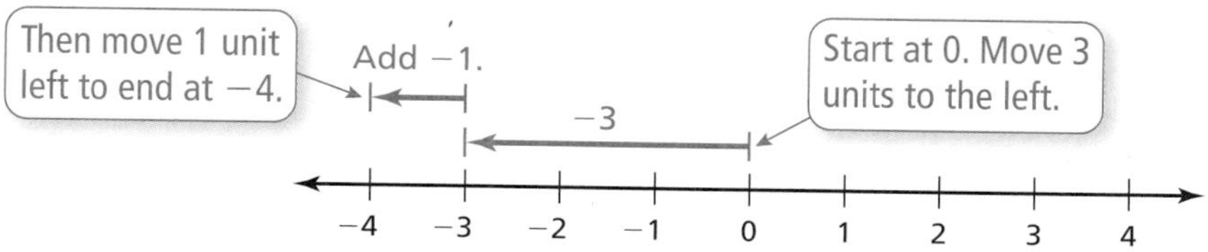

Then move 1 unit left to end at −4.
Add −1.
−3
Start at 0. Move 3 units to the left.
−4
−3
−2
−1
0
1
2
3
4

Inductive Reasoning

Work with a partner. Use integer counters or a number line to complete the table.

Exercise	Operation: Add or Subtract	Answer
1 **5.** $4 - 2$	Subtract 2	2
2 **6.** $4 + (-2)$	Add −2	2
3 **7.** $-3 - 1$	Subtract 1	−4
4 **8.** $-3 + (-1)$	Add −1	−4
9. $3 - 8$	Subtract 8	5
10. $3 + (-8)$	Add −8	−5
11. $9 - 13$	Subtract 13	4
12. $9 + (-13)$	Add −13	−4
13. $-6 - (-3)$	Subtract −3	−3
14. $-6 + (3)$	Add 3	3
15. $-5 - (-12)$	Subtract −12	7
16. $-5 + 12$	Add 12	7

What Is Your Answer?

17. **IN YOUR OWN WORDS** How are adding integers and subtracting integers related?

18. Write a general rule for subtracting integers.

Practice Use what you learned about subtracting integers to complete Exercises 8–15 on page 18.

1.3 Lesson

Check It Out
Lesson Tutorials
BigIdeasMath.com

Key Idea

Subtracting Integers

Words To subtract an integer, add its opposite.

Numbers $3 - 4 = 3 + (-4) = -1$

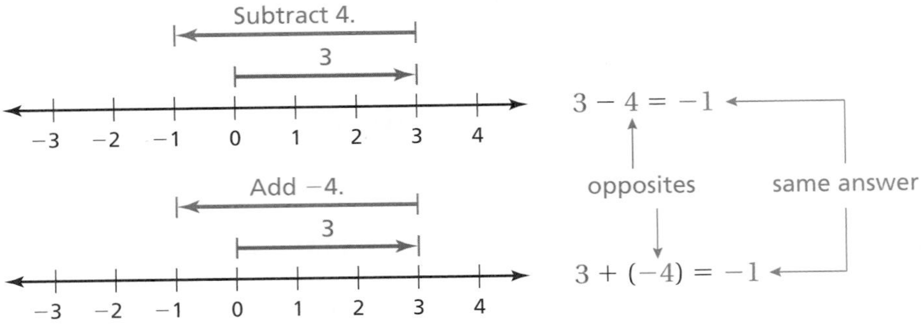

EXAMPLE 1 Subtracting Integers

a. Find 3 − 12.

$3 - 12 = 3 + (-12)$ Add the opposite of 12.

$= -9$ Add.

∴ The difference is −9.

b. Find −8 − (−13).

$-8 - (-13) = -8 + 13$ Add the opposite of −13.

$= 5$ Add.

∴ The difference is 5.

c. Find 5 − (−4).

$5 - (-4) = 5 + 4$ Add the opposite of −4.

$= 9$ Add.

∴ The difference is 9.

On Your Own

Now You're Ready
Exercises 8–23

Subtract.

1. $8 - 3$ **2.** $9 - 17$ **3.** $-3 - 3$

4. $-14 - 9$ **5.** $9 - (-8)$ **6.** $-12 - (-12)$

EXAMPLE 2 Subtracting Integers

Evaluate $-7 - (-12) - 14$.

$-7 - (-12) - 14 = -7 + 12 - 14$	Add the opposite of −12.
$= 5 - 14$	Add −7 and 12.
$= 5 + (-14)$	Add the opposite of 14.
$= -9$	Add.

∴ So, $-7 - (-12) - 14 = -9$.

On Your Own

Now You're Ready
Exercises 27–32

Evaluate the expression.

7. $-9 - 16 - 8$
8. $-4 - 20 - 9$
9. $0 - 9 - (-5)$
10. $0 - (-6) - 8$
11. $15 - (-20) - 20$
12. $13 - 18 - (-18)$

EXAMPLE 3 Real-Life Application

Which continent has the greater range of elevations?

	North America	Africa
Highest Elevation	6198 m	5895 m
Lowest Elevation	−86 m	−155 m

To find the range of elevations for each continent, subtract the lowest elevation from the highest elevation.

North America

range $= 6198 - (-86)$
$= 6198 + 86$
$= 6284$ m

Africa

range $= 5895 - (-155)$
$= 5895 + 155$
$= 6050$ m

∴ Because 6284 is greater than 6050, North America has the greater range of elevations.

On Your Own

13. The highest elevation in Mexico is 5700 meters, on Pico de Orizaba. The lowest elevation in Mexico is −10 meters, in Laguna Salada. Find the range of elevations in Mexico.

1.3 Exercises

Check It Out
Help with Homework
BigIdeasMath.com

Vocabulary and Concept Check

1. **WRITING** How do you subtract one integer from another?
2. **OPEN-ENDED** Write two integers that are opposites.
3. **DIFFERENT WORDS, SAME QUESTION** Which is different? Find "both" answers.

Find the difference of 3 and -2.	What is 3 less than -2?
How much less is -2 than 3?	Subtract -2 from 3.

MATCHING **Match the subtraction expression with the corresponding addition expression.**

4. $9 - (-5)$
5. $-9 - 5$
6. $-9 - (-5)$
7. $9 - 5$

A. $-9 + 5$ B. $9 + (-5)$ C. $-9 + (-5)$ D. $9 + 5$

Practice and Problem Solving

Subtract.

(1)
8. $4 - 7$
9. $8 - (-5)$
10. $-6 - (-7)$
11. $-2 - 3$
12. $5 - 8$
13. $-4 - 6$
14. $-8 - (-3)$
15. $10 - 7$
16. $-8 - 13$
17. $15 - (-2)$
18. $-9 - (-13)$
19. $-7 - (-8)$
20. $-6 - (-6)$
21. $-10 - 12$
22. $32 - (-6)$
23. $0 - (20)$

24. **ERROR ANALYSIS** Describe and correct the error in finding the difference $7 - (-12)$.

✗ $7 - (-12) = 7 + (-12) = -5$

25. **SWIMMING POOL** The floor of the shallow end of a swimming pool is at -3 feet. The floor of the deep end is 9 feet deeper. Which expression can be used to find the depth of the deep end?

$-3 + 9$ $-3 - 9$ $9 - 3$

26. **SHARKS** A shark is at -80 feet. It swims up and jumps out of the water to a height of 15 feet. Write a subtraction expression for the vertical distance the shark travels.

Evaluate the expression.

(2)
27. $-2 - 7 + 15$
28. $-9 + 6 - (-2)$
29. $12 - (-5) - 8$
30. $8 + 14 - (-4)$
31. $-6 - (-8) + 5$
32. $-15 - 7 - (-11)$

MENTAL MATH **Use mental math to solve the equation.**

33. $m - 5 = 9$

34. $w - (-3) = 7$

35. $6 - c = -9$

36. $d - 3 = -10$

37. $k - (-2) = -6$

38. $4 - n = -1$

39. **PLATFORM DIVING** The figure shows a diver diving from a platform. The diver reaches a depth of 4 meters. What is the change in elevation of the diver?

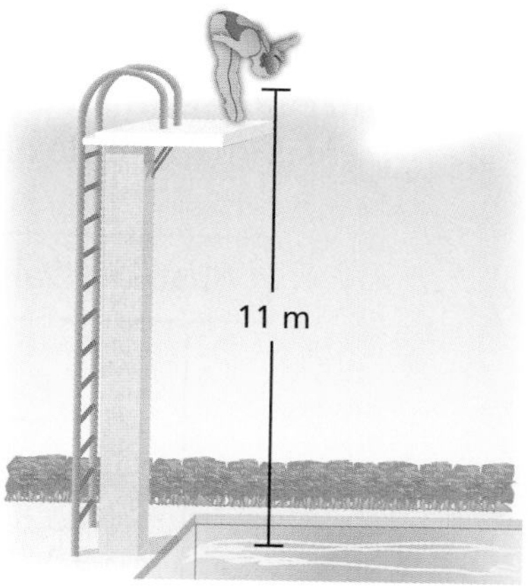

40. **OPEN-ENDED** Write two different pairs of negative integers, x and y, that make the statement $x - y = -1$ true.

41. **TEMPERATURE** The table shows the record monthly high and low temperatures in Anchorage, AK.

	Jan	Feb	Mar	Apr	May	Jun	Jul	Aug	Sep	Oct	Nov	Dec
High (°F)	56	57	56	72	82	92	84	85	73	64	62	53
Low (°F)	−35	−38	−24	−15	1	29	34	31	19	−6	−21	−36

a. Find the range of temperatures for each month.

b. What are the all-time high and all-time low temperatures?

c. What is the range of the temperatures in part (b)?

REASONING **Tell whether the difference between the two integers is *always*, *sometimes*, or *never* positive. Explain your reasoning.**

42. two positive integers

43. two negative integers

44. a positive integer and a negative integer

45. a negative integer and a positive integer

Number Sense **For what values of a and b is the statement true?**

46. $|a - b| = |b - a|$

47. $|a + b| = |a| + |b|$

48. $|a - b| = |a| - |b|$

Fair Game Review What you learned in previous grades & lessons

Add. *(Section 1.2)*

49. $-5 + (-5) + (-5) + (-5)$

50. $-9 + (-9) + (-9) + (-9) + (-9)$

Multiply. *(Skills Review Handbook)*

51. 8×5

52. 6×78

53. 36×41

54. 82×29

55. **MULTIPLE CHOICE** Which value of n makes the value of the expression $4n + 3$ a composite number? *(Skills Review Handbook)*

Ⓐ 1 Ⓑ 2 Ⓒ 3 Ⓓ 4

Check It Out
Graphic Organizer
BigIdeasMath.com

You can use an **idea and examples chart** to organize information about a concept. Here is an example of an idea and examples chart for absolute value.

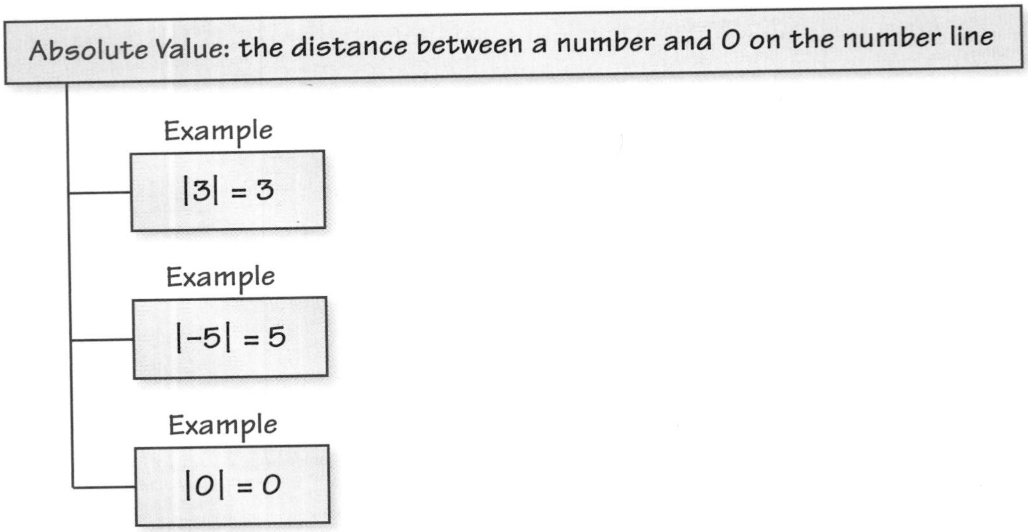
Absolute Value: the distance between a number and 0 on the number line
Example
|3| = 3
Example
|−5| = 5
Example
|0| = 0

On Your Own

Make an idea and examples chart to help you study these topics.

1. integers
2. adding integers
   a. with the same sign
   b. with different signs
3. Additive Inverse Property
4. subtracting integers

After you complete this chapter, make idea and examples charts for the following topics.

5. multiplying integers
   a. with the same sign
   b. with different signs
6. dividing integers
   a. with the same sign
   b. with different signs

"I made an idea and examples chart to give my owner ideas for my birthday next week."

1.1–1.3 Quiz

Check It Out
Progress Check
BigIdeasMath.com

Find the absolute value. *(Section 1.1)*

1. $|-13|$ **2.** $|3|$ **3.** $|0|$

Copy and complete the statement using <, >, or =. *(Section 1.1)*

4. $|-8| \;\square\; 3$ **5.** $7 \;\square\; |-7|$

6. $|-3| \;\square\; |-7|$ **7.** $|-12| \;\square\; |12|$

Order the values from least to greatest. *(Section 1.1)*

8. $-4, |-5|, |-4|, 3, -6$ **9.** $12, -8, |-15|, -10, |-9|$

Evaluate the expression. *(Section 1.2 and Section 1.3)*

10. $-3 + (-8)$ **11.** $-4 + 16$

12. $3 - 9$ **13.** $-5 - (-5)$

14. EXPLORING Two climbers explore a cave. *(Section 1.1)*

a. Write an integer for the position of each climber relative to the surface.

b. Which integer in part (a) is greater?

c. Which integer in part (a) has the greater absolute value?

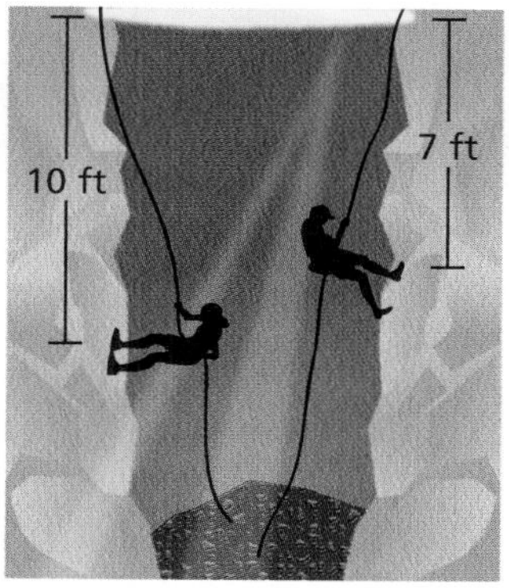

15. SCHOOL CARNIVAL The table shows the income and expenses for a school carnival. The school's goal was to raise \$1100. Did the school reach its goal? Explain. *(Section 1.2)*

Games	Concessions	Donations	Flyers	Decorations
\$650	\$530	\$52	−\$28	−\$75

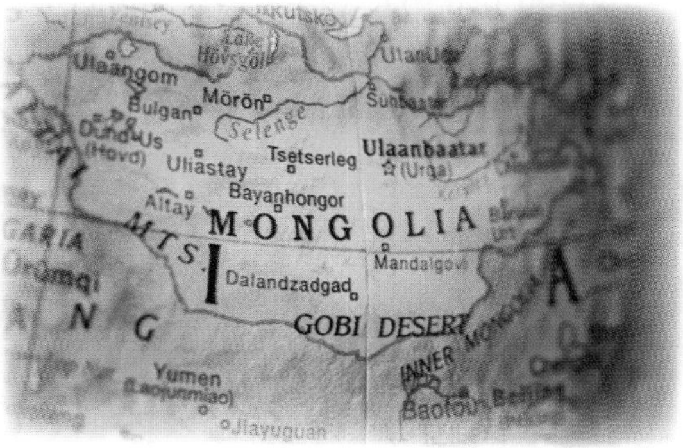

16. TEMPERATURE Temperatures in the Gobi Desert reach $-40°F$ in the winter and $90°F$ in the summer. Find the range of the temperatures. *(Section 1.3)*

1.4 Multiplying Integers

STANDARDS OF LEARNING
7.3

Essential Question Is the product of two integers *positive*, *negative*, or *zero*? How can you tell?

1 EXAMPLE: Multiplying Integers with the Same Sign

Use repeated addition to find 3 • 2.

Recall that multiplication is repeated addition. 3 • 2 means to add 3 groups of 2.

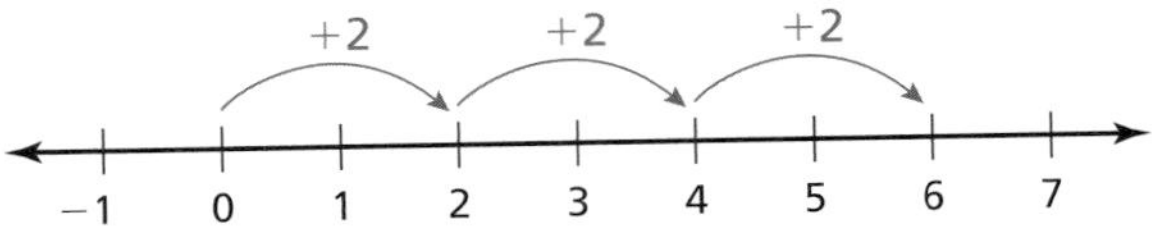

Now you can write
$3 \cdot 2 = 2 + 2 + 2 = 6.$

So, $3 \cdot 2 = 6$.

2 EXAMPLE: Multiplying Integers with Different Signs

Use repeated addition to find 3 • (−2).

$3 \cdot (-2)$ means to add 3 groups of −2.

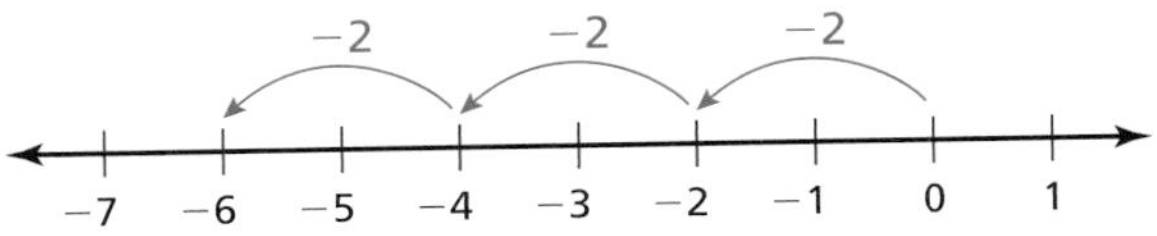

Now you can write
$3 \cdot (-2) = (-2) + (-2) + (-2)$
$= -6.$

So, $3 \cdot (-2) = -6$.

3 ACTIVITY: Multiplying Integers with Different Signs

Work with a partner. Use a table to find −3 • 2.

Describe the pattern in the table. Use the pattern to complete the table.

2	•	2	=	4
1	•	2	=	2
0	•	2	=	0
−1	•	2	=	
−2	•	2	=	
−3	•	2	=	

Notice the products decrease by 2 in each row.

So, continue the pattern.

$-1 \cdot 2$: $0 - 2 =$ ____

$-2 \cdot 2$: $-2 - 2 =$ ____

$-3 \cdot 2$: $-4 - 2 =$ ____

So, $-3 \cdot 2 =$ ____.

4 ACTIVITY: Multiplying Integers with the Same Sign

Work with a partner. Use a table to find $-3 \cdot (-2)$.

Describe the pattern in the table. Use the pattern to complete the table.

-3	$\cdot$	3	$=$	-9
-3	$\cdot$	2	$=$	-6
-3	$\cdot$	1	$=$	-3
-3	$\cdot$	0	$=$	
-3	$\cdot$	-1	$=$	
-3	$\cdot$	-2	$=$	

Notice the products increase by 3 in each row.

So, continue the pattern.

$-3 \cdot 0$: $-3 + 3 =$ ____

$-3 \cdot -1$: $0 + 3 =$ ____

$-3 \cdot -2$: $3 + 3 =$ ____

So, $-3 \cdot (-2) =$ ____.

Inductive Reasoning

Work with a partner. Complete the table.

	Exercise	Type of Product	Product	Product: Positive or Negative
1	**5.** $3 \cdot 2$	Integers with the same sign	6	Positive
2	**6.** $3 \cdot (-2)$	Integers with different signs	-6	negative
3	**7.** $-3 \cdot 2$	Integers with different signs	-6	negative
4	**8.** $-3 \cdot (-2)$	Integers with the same sign	6	Positive
	9. $6 \cdot 3$	Same sign	18	Positive
	10. $2 \cdot (-5)$	different sign	-10	negative
	11. $-6 \cdot 5$	different sign	-30	negative
	12. $-5 \cdot (-3)$	Same sign	15	Positive

13. Write two integers whose product is 0.

What Is Your Answer?

14. **IN YOUR OWN WORDS** Is the product of two integers *positive, negative,* or *zero*? How can you tell?

15. Write general rules for multiplying (a) two integers with the same sign and (b) two integers with different signs.

Practice Use what you learned about multiplying integers to complete Exercises 8–15 on page 26.

1.4 Lesson

Check It Out
Lesson Tutorials
BigIdeasMath.com

Key Ideas

Multiplying Integers with the Same Sign

Words The product of two integers with the same sign is positive.

Numbers $2 \cdot 3 = 6$ $-2 \cdot (-3) = 6$

Multiplying Integers with Different Signs

Words The product of two integers with different signs is negative.

Numbers $2 \cdot (-3) = -6$ $-2 \cdot 3 = -6$

EXAMPLE 1 Multiplying Integers with the Same Sign

Find $-5 \cdot (-6)$.

The integers have the same sign.

$$-5 \cdot (-6) = 30$$

The product is positive.

The product is 30.

EXAMPLE 2 Multiplying Integers with Different Signs

Multiply.

a. $3(-4)$

b. $-7 \cdot 4$

The integers have different signs.

$$3(-4) = -12$$

$$-7 \cdot 4 = -28$$

The product is negative.

The product is -12.

The product is -28.

On Your Own

Now You're Ready
Exercises 8–23

Multiply.

1. $5 \cdot 5$

2. $4(11)$

3. $-1(-9)$

4. $-7 \cdot (-8)$

5. $12 \cdot (-2)$

6. $4(-6)$

7. $-10(6)$

8. $-5 \cdot 7$

EXAMPLE 3 Using Exponents

Study Tip

Place parentheses around a negative number to raise it to a power.

a. Evaluate $(-2)^2$.

$(-2)^2 = (-2) \cdot (-2)$ — Write $(-2)^2$ as repeated multiplication.

$= 4$ — Multiply.

b. Evaluate -5^2.

$-5^2 = -(5 \cdot 5)$ — Write 5^2 as repeated multiplication.

$= -25$ — Multiply.

c. Evaluate $(-4)^3$.

$(-4)^3 = (-4) \cdot (-4) \cdot (-4)$ — Write $(-4)^3$ as repeated multiplication.

$= 16 \cdot (-4)$ — Multiply.

$= -64$ — Multiply.

On Your Own

Now You're Ready
Exercises 32–37

Evaluate the expression.

9. $(-3)^2$ **10.** $(-2)^3$ **11.** -7^2 **12.** -6^3

EXAMPLE 4 Real-Life Application

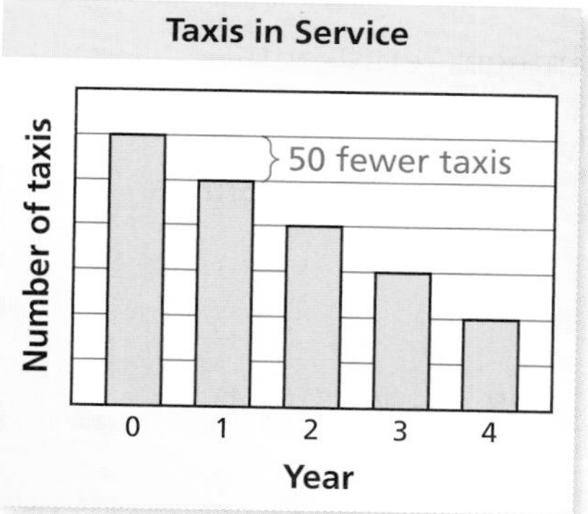

The bar graph shows the number of taxis a company has in service. The number of taxis decreases by the same amount each year for four years. Find the total change in the number of taxis.

The bar graph shows that the number of taxis in service decreases by 50 each year. Use a model to solve the problem.

Total change = Change per year $\cdot$ Number of years

$= -50 \cdot 4$

Use -50 for the change per year because the number *decreases* each year.

$= -200$

The total change in the number of taxis is -200.

On Your Own

13. A manatee population decreases by 15 manatees each year for 3 years. Find the total change in the manatee population.

1.4 Exercises

Check It Out
Help with Homework
BigIdeasMath.com

Vocabulary and Concept Check

1. **WRITING** What can you conclude about the signs of two integers whose product is (a) positive and (b) negative?
2. **OPEN-ENDED** Write two integers whose product is negative.

Tell whether the product is *positive* or *negative* without multiplying. Explain your reasoning.

3. $4(-8)$
4. $-5(-7)$
5. $-3 \cdot (12)$

Tell whether the statement is *true* or *false*. Explain your reasoning.

6. The product of three positive integers is positive.
7. The product of three negative integers is positive.

Practice and Problem Solving

Multiply.

1 2
8. $6 \cdot 4$
9. $7(-3)$
10. $-2(8)$
11. $-3(-4)$
12. $-6 \cdot 7$
13. $3 \cdot 9$
14. $8 \cdot (-5)$
15. $-1 \cdot (-12)$
16. $-5(10)$
17. $-13(0)$
18. $-9 \cdot 9$
19. $15(-2)$
20. $-10 \cdot 11$
21. $-6 \cdot (-13)$
22. $7(-14)$
23. $-11 \cdot (-11)$

24. **JOGGING** You burn 10 calories each minute you jog. What integer represents the change in your calories after you jog for 20 minutes?
25. **WETLANDS** About 60,000 acres of wetlands are lost each year in the United States. What integer represents the change in wetlands after 4 years?

Multiply.

26. $3 \cdot (-8) \cdot (-2)$
27. $6(-9)(-1)$
28. $-3(-5)(-4)$
29. $-7(-3)(-5)$
30. $-6 \cdot 3 \cdot (-6)$
31. $3 \cdot (-12) \cdot 0$

Evaluate the expression.

3
32. $(-4)^2$
33. $(-1)^3$
34. -8^2
35. -6^2
36. $-5^2 \cdot 4$
37. $-2 \cdot (-3)^3$

ERROR ANALYSIS Describe and correct the error in evaluating the expression.

38. ✗ $-2(-7) = -14$
39. ✗ $-10^2 = 100$

Evaluate the expression.

40. $-5 \cdot |-3|$ **41.** $3(-7) - 4(-2)$ **42.** $4 \cdot (-3)^2 + 4(-5)$

NUMBER SENSE **Find the next two numbers in the pattern.**

43. $-12, 60, -300, 1500, \ldots$ **44.** $7, -28, 112, -448, \ldots$

45. GYM CLASS You lose four points each time you attend gym class without sneakers. You forget your sneakers three times. What integer represents the change in your points?

46. AIRPLANE The height of an airplane during a landing is given by $22{,}000 + (-480t)$, where t is the time in minutes.

a. Copy and complete the table.

b. Estimate how many minutes it takes the plane to land. Explain your reasoning.

Time	5 min	10 min	15 min	20 min
Height				

47. INLINE SKATES In June, the price of a pair of inline skates is \$165. The price changes each of the next three months.

a. Copy and complete the table.

Month	Price of Skates
June	$165 = \$165$
July	$165 + (-12) = \$$____
August	$165 + 2(-12) = \$$____
September	$165 + 3(-12) = \$$____

b. Describe the change in the price of the inline skates for each month.

c. The table at the right shows the amount of money you save each month to buy the inline skates. Do you have enough money saved to buy the inline skates in August? September? Explain your reasoning.

Amount Saved	
June	\$35
July	\$55
August	\$45
September	\$18

48. Reasoning Two integers, a and b, have a product of 24. What is the least possible sum of a and b?

Fair Game Review What you learned in previous grades & lessons

Divide. *(Skills Review Handbook)*

49. $27 \div 9$ **50.** $48 \div 6$ **51.** $56 \div 4$ **52.** $153 \div 9$

53. MULTIPLE CHOICE What is the prime factorization of 84? *(Skills Review Handbook)*

(A) $2^2 \times 3^2$ (B) $2^3 \times 7$ (C) $3^3 \times 7$ (D) $2^2 \times 3 \times 7$

1.5 Dividing Integers

Essential Question Is the quotient of two integers *positive*, *negative*, or *zero*? How can you tell?

STANDARDS OF LEARNING
7.3

1 EXAMPLE: Dividing Integers with Different Signs

Use integer counters to find $-15 \div 3$.

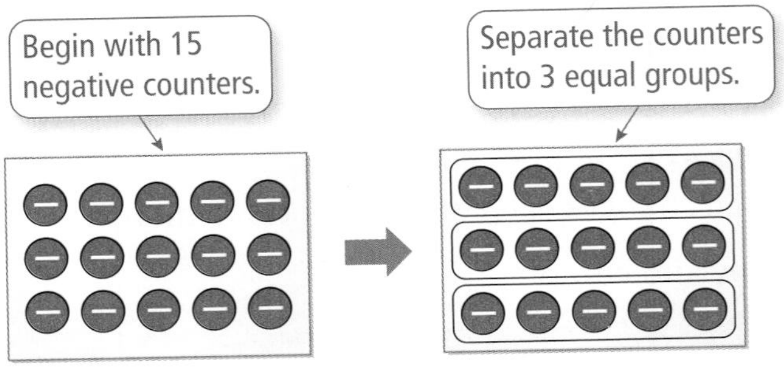

Begin with 15 negative counters.
Separate the counters into 3 equal groups.

Because there are five negative counters in each group, $-15 \div 3 = -5$.

2 ACTIVITY: Rewriting a Product as a Quotient

Work with a partner. Rewrite the product $3 \cdot 4 = 12$ as a quotient in two different ways.

First Way

12 is equal to 3 groups of 4.

So, $12 \div 3 =$ ____.

Second Way

12 is equal to 4 groups of ____.

So, $12 \div 4 =$ ____.

3 EXAMPLE: Dividing Integers with Different Signs

Rewrite the product $-3 \cdot (-4) = 12$ as a quotient in two different ways. What can you conclude?

First Way

$12 \div (-3) = -4$

Second Way

$12 \div (-4) = -3$

In each case, when you divide a positive integer by a negative integer, you get a negative integer.

4 EXAMPLE: Dividing Negative Integers

Rewrite the product $3 \cdot (-4) = -12$ as a quotient in two different ways. What can you conclude?

First Way

$-12 \div (-4) = 3$

Second Way

$-12 \div (3) = -4$

When you divide a negative integer by a negative integer, you get a positive integer. When you divide a negative integer by a positive integer, you get a negative integer.

Inductive Reasoning

Work with a partner. Complete the table.

	Exercise	Type of Quotient	Quotient	Quotient: Positive, Negative, or Zero
1	**5.** $-15 \div 3$	Integers with different signs	−5	negative
2	**6.** $12 \div 4$	Same signs	3	Positive
3	**7.** $12 \div (-3)$	different signs	−4	negative
4	**8.** $-12 \div (-4)$	Integers with the same sign	3	Positive
	9. $-6 \div 2$	different signs	−3	negative
	10. $-21 \div (-7)$	same signs	3	Positive
	11. $10 \div (-2)$	different signs	−5	negative
	12. $12 \div (-6)$	different signs	−2	negative
	13. $0 \div (-15)$	different signs	0	positive
	14. $0 \div 4$	Same signs	0	Positive

What Is Your Answer?

15. **IN YOUR OWN WORDS** Is the quotient of two integers *positive, negative,* or *zero*? How can you tell?

16. Write general rules for dividing (a) two integers with the same sign and (b) two integers with different signs.

Practice Use what you learned about dividing integers to complete Exercises 8–15 on page 32.

1.5 Lesson

Check It Out
Lesson Tutorials
BigIdeasMath.com

Key Ideas

Dividing Integers with the Same Sign

Words The quotient of two integers with the same sign is positive.

Numbers $8 \div 2 = 4$ $\qquad -8 \div (-2) = 4$

Dividing Integers with Different Signs

Words The quotient of two integers with different signs is negative.

Numbers $8 \div (-2) = -4$ $\qquad -8 \div 2 = -4$

EXAMPLE 1 Dividing Integers with the Same Sign

Find $-18 \div (-6)$.

The integers have the same sign.

$$-18 \div (-6) = 3$$

The quotient is positive.

The quotient is 3.

EXAMPLE 2 Dividing Integers with Different Signs

Divide.

a. $75 \div (-25)$ **b.** $\frac{-54}{6}$

The integers have different signs.

$$75 \div (-25) = -3 \qquad \frac{-54}{6} = -9$$

The quotient is negative.

The quotient is -3. The quotient is -9.

On Your Own

Now You're Ready
Exercises 8–23

Divide.

1. $14 \div 2$ **2.** $-32 \div (-4)$ **3.** $-40 \div (-8)$

4. $0 \div (-6)$ **5.** $\frac{-49}{7}$ **6.** $\frac{21}{-3}$

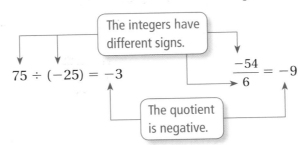

EXAMPLE 3 Using Order of Operations

Evaluate $10 - 8^2 \div (-4)$.

$10 - 8^2 \div (-4)$	$= 10 - 8 \cdot 8 \div (-4)$	Write 8^2 as repeated multiplication.
	$= 10 - 64 \div (-4)$	Multiply 8 and 8.
	$= 10 - (-16)$	Divide 64 by -4.
	$= 26$	Subtract.

On Your Own

Now You're Ready
Exercises 28–31

Evaluate the expression.

7. $-18 \div (-6) + 3$ **8.** $\dfrac{-15 + 7}{2^2}$ **9.** $\dfrac{(-10)^2}{25} - 9$

EXAMPLE 4 Real-Life Application

You measure the height of the tide using support beams of a pier. Your measurements are shown in the picture. What is the mean hourly change in the height?

Use a model to solve the problem.

$$\text{Mean hourly change} = \frac{\text{Final height} - \text{Initial height}}{\text{Elapsed Time}}$$

$= \dfrac{8 - 59}{6}$	Substitute. The elapsed time from 2 P.M. to 8 P.M. is 6 hours.
$= \dfrac{-51}{6}$	Subtract.
$= -8.5$	Divide.

∴ The mean change in the height of the tide is -8.5 inches per hour.

On Your Own

10. The height of the tide at the Bay of Fundy in New Brunswick decreases 36 feet in 6 hours. What is the mean hourly change in the height?

1.5 Exercises

Check It Out
Help with Homework
BigIdeasMath.com

Vocabulary and Concept Check

1. **WRITING** What can you tell about two integers when their quotient is positive? negative? zero?
2. **VOCABULARY** A quotient is undefined. What does this mean?
3. **OPEN-ENDED** Write two integers whose quotient is negative.
4. **WHICH ONE DOESN'T BELONG?** Which expression does *not* belong with the other three? Explain your reasoning.

$\frac{10}{-5}$ $\frac{-10}{5}$ $\frac{-10}{-5}$ $-\left(\frac{10}{5}\right)$

Tell whether the quotient is *positive* or *negative* without dividing.

5. $-12 \div 4$
6. $\frac{-6}{-2}$
7. $15 \div (-3)$

Practice and Problem Solving

Divide, if possible.

1 2
8. $4 \div (-2)$
9. $21 \div (-7)$
10. $-20 \div 4$
11. $-18 \div (-6)$
12. $\frac{-14}{7}$
13. $\frac{0}{6}$
14. $\frac{-15}{-5}$
15. $\frac{54}{-9}$
16. $-33 \div 11$
17. $-49 \div (-7)$
18. $0 \div (-2)$
19. $60 \div (-6)$
20. $\frac{-56}{14}$
21. $\frac{18}{0}$
22. $\frac{65}{-5}$
23. $\frac{-84}{-7}$

ERROR ANALYSIS **Describe and correct the error in finding the quotient.**

24. ✗ $\frac{-63}{-9} = -7$

25. ✗ $0 \div (-5) = -5$

26. **ALLIGATORS** An alligator population in a nature preserve in the Everglades decreases by 60 alligators over 5 years. What is the mean yearly change in the alligator population?
27. **READING** You read 105 pages of a novel over 7 days. What is the mean number of pages you read each day?

Evaluate the expression.

3
28. $12 \div (-2) + 5$
29. $\frac{10(-4)^2}{-20}$
30. $\left|\frac{10(-5)}{-(-2)}\right|$
31. $\frac{-9^2 + 2(9)}{-3}$

Find the mean of the integers.

32. 3, −10, −2, 13, 11

33. −26, 39, −10, −16, 12, 31

Evaluate the expression.

34. $-8 - 14 \div 2 + 5$

35. $24 \div (-4) + (-2) \cdot (-5)$

36. PATTERN Find the next two numbers in the pattern −128, 64, −32, 16, ... Explain your reasoning.

37. SNOWBOARDING A snowboarder descends a 1200-foot hill in 3 minutes. What is the mean change in elevation per minute?

38. GOLF The table shows a golfer's score for each round of a tournament.

a. What was the golfer's total score?

b. What was the golfer's mean score per round?

Scorecard	
Round 1	−2
Round 2	−6
Round 3	−7
Round 4	−3

39. TUNNEL The Detroit-Windsor Tunnel is an underwater highway that connects the cities of Detroit, Michigan, and Windsor, Ontario. How many times deeper is the roadway than the bottom of the ship?

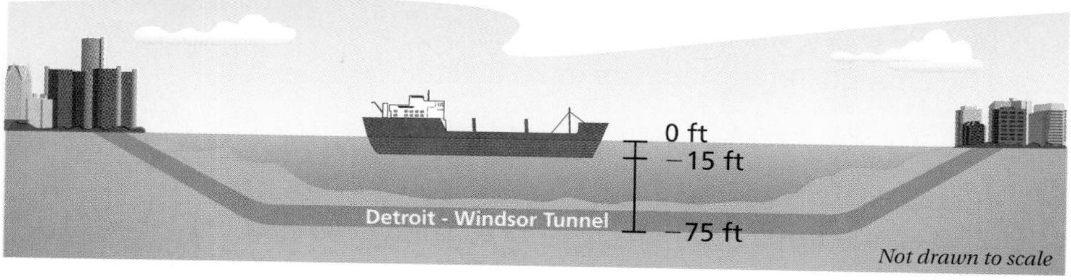

40. AMUSEMENT PARK The regular admission price for an amusement park is \$72. For a group of 15 or more, the admission price is reduced by \$25. How many people need to be in a group to save \$500?

41. Number Sense Write five different integers that have a mean of −10. Explain how you found your answer.

Fair Game Review What you learned in previous grades & lessons

Graph the values on a number line. Then order the values from least to greatest. *(Section 1.1)*

42. $-6, 4, |2|, -1, |-10|$

43. $3, |0|, |-4|, -3, -8$

44. $|5|, -2, -5, |-2|, -7$

45. MULTIPLE CHOICE What is the value of $4 \cdot 3 + (12 \div 2)^2$? *(Skills Review Handbook)*

Ⓐ 15 Ⓑ 48 Ⓒ 156 Ⓓ 324

1.4–1.5 Quiz

Check It Out
Progress Check
BigIdeasMath.com

Tell whether the value of the expression is *positive* or *negative* without evaluating. *(Section 1.4 and Section 1.5)*

1. $-6 \cdot 7$
2. $-66 \div (-11)$
3. $\dfrac{48}{-12}$
4. $9(-12)$

Evaluate the expression. *(Section 1.4 and Section 1.5)*

5. $-7(6)$
6. $-1(-9)$
7. $\dfrac{-72}{-9}$
8. $-24 \div 3$
9. $-3 \cdot 4 \cdot (-6)$
10. $(-3)^3$

11. **SPEECH** In speech class, you lose 3 points for every 30 seconds you go over the time limit. Your speech is 90 seconds over the time limit. What integer represents the change in your points? *(Section 1.4)*

12. **MOUNTAIN CLIMBING** On a mountain, the temperature decreases by 18°F every 5000 feet. What integer represents the change in temperature at 20,000 feet? *(Section 1.4)*

13. **GAMING** You play a video game for 15 minutes. You lose 165 points. What integer represents the average change in points per minute? *(Section 1.5)*

14. **DIVING** You dive 21 feet from the surface of a lake in 7 seconds. *(Section 1.4 and Section 1.5)*

    a. What is the mean change in your position in feet per second?

    b. You continue diving. What is your position relative to the surface after 5 more seconds?

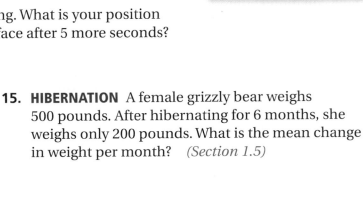

15. **HIBERNATION** A female grizzly bear weighs 500 pounds. After hibernating for 6 months, she weighs only 200 pounds. What is the mean change in weight per month? *(Section 1.5)*

1 Chapter Review

Check It Out
Vocabulary Help
BigIdeasMath.com

Review Key Vocabulary

integer, *p. 4*
absolute value, *p. 4*
opposites, *p. 10*
additive inverse, *p. 10*

Review Examples and Exercises

1.1 Integers and Absolute Value *(pp. 2–7)*

Find the absolute value of −2.

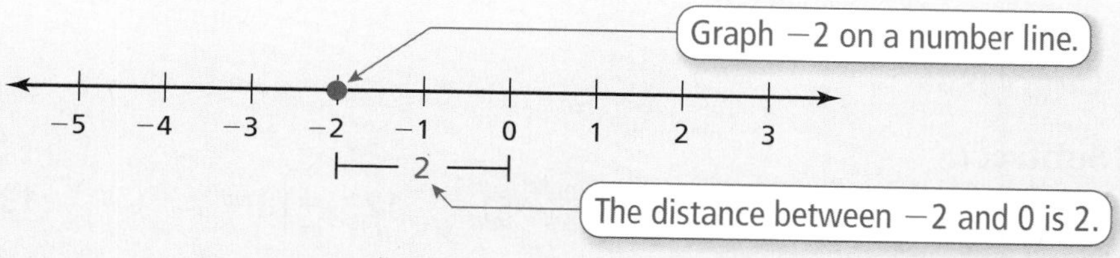

Graph −2 on a number line.
−5
−4
−3
−2
−1
0
1
2
3
2
The distance between −2 and 0 is 2.

∴ So, $|-2| = 2$.

Exercises

Find the absolute value.

1. $|3|$ **2.** $|-9|$ **3.** $|-17|$ **4.** $|8|$

5. ELEVATION The elevation of Death Valley, CA is −282 feet. The Mississippi River in Illinois has an elevation of 279 feet. Which is closer to sea level?

1.2 Adding Integers *(pp. 8–13)*

Find 6 + (−14).

$6 + (-14) = -8$

$|6|$ is less than $|-14|$. So, subtract $|6|$ from $|-14|$.

Use the sign of −14.

∴ The sum is −8.

Exercises

Add.

6. $-16 + (-11)$ **7.** $-15 + 5$ **8.** $100 + (-75)$ **9.** $-32 + (-2)$

1.3 Subtracting Integers (pp. 14–19)

Subtract.

a. $7 - 19 = 7 + (-19)$ — Add the opposite of 19.

$= -12$ — Add.

∴ The difference is -12.

b. $-6 - (-10) = -6 + 10$ — Add the opposite of -10.

$= 4$ — Add.

∴ The difference is 4.

Exercises

Subtract.

10. $8 - 18$ **11.** $-16 - (-5)$ **12.** $-18 - 7$ **13.** $-12 - (-27)$

14. GAME SHOW Your score on a game show is -300. You answer the final question incorrectly, so you lose 400 points. What is your final score?

1.4 Multiplying Integers (pp. 22–27)

a. Find $-7 \cdot (-9)$.

The integers have the same sign.

$$-7 \cdot (-9) = 63$$

The product is positive.

∴ The product is 63.

b. Find $-6(14)$.

The integers have different signs.

$$-6(14) = -84$$

The product is negative.

∴ The product is -84.

Exercises

Multiply.

15. $-8 \cdot 6$ **16.** $10(-7)$ **17.** $-3 \cdot (-6)$ **18.** $-12(5)$

1.5 Dividing Integers (pp. 28–33)

a. Find $30 \div (-10)$.

The integers have different signs.

$$30 \div (-10) = -3$$

The quotient is negative.

The quotient is -3.

b. Find $\frac{-72}{-9}$.

The integers have the same sign.

$$\frac{-72}{-9} = 8$$

The quotient is positive.

The quotient is 8.

Exercises

Divide.

19. $-18 \div 9$

20. $\frac{-42}{-6}$

21. $\frac{-30}{6}$

22. $84 \div (-7)$

Evaluate the expression.

23. $\frac{3^2 \cdot (-2)}{-6}$

24. $\frac{-8(-4)}{2} - (-5)$

25. $7 + (-39) \div 3$

Find the mean of the integers.

26. $-3, -8, 12, -15, 9$

27. $-54, -32, -70, -25, -65, -42$

28. PROFITS The table shows the weekly profits of a fruit vendor. What is the mean profit for these weeks?

Week	1	2	3	4
Profit	−\$125	−\$86	\$54	−\$35

29. RETURNS You return several shirts to a store. The receipt shows that the amount placed back on your credit card is −\$30.60. Each shirt is −\$6.12. How many shirts did you return?

1 Chapter Test

Check It Out
Test Practice
BigIdeasMath.com

Find the absolute value.

1. $|-9|$
2. $|64|$
3. $|-22|$

Copy and complete the statement using <, >, or =.

4. $4 \;\square\; |-8|$
5. $|-7| \;\square\; -12$
6. $-7 \;\square\; |3|$

Evaluate the expression.

7. $-6 + (-11)$
8. $2 - (-9)$
9. $-9 \cdot 2$
10. $-72 \div (-3)$
11. $-5 + 17 - (-4)$
12. $\dfrac{-14(21)}{-7}$

Find the mean of the integers.

13. $11, -7, -14, 10, -5$
14. $-32, -41, -39, -27, -33, -44$

15. **NASCAR** A driver receives -25 points for each rule violation. What integer represents the change in points after four rule violations?

16. **GOLF** The table shows your scores, relative to *par*, for nine holes of golf. What is your total score for the nine holes?

Hole	1	2	3	4	5	6	7	8	9	Total
Score	+1	−2	−1	0	−1	+3	−1	−3	+1	?

17. **VISITORS** In a recent 10-year period, the change in the number of visitors to U.S. National Parks was about $-11{,}150{,}000$ visitors.

    a. What was the mean yearly change in the number of visitors?

    b. During the seventh year, the change in the number of visitors was about 10,800,000. Explain how the change for the 10-year period can be negative.

1 Standardized Test Practice

Test-Taking Strategy
Solve Directly or Eliminate Choices

You ripped out $(-1)^2 + (-2)(-3)$ whiskers. How many did you rip out?
Ⓐ −5 Ⓑ 5 Ⓒ −7 Ⓓ 7
Yeow, why the biggest number?

"You can eliminate A and C. Then, solve directly to determine that the correct answer is D."

1. A football team gains 2 yards on the first play, loses 5 yards on the second play, loses 3 yards on the third play, and gains 4 yards on the fourth play. What is the team's overall gain or loss for all four plays?

A. a gain of 14 yards **C.** a loss of 2 yards

B. a gain of 2 yards **D.** a loss of 14 yards

2. Point P is plotted in the coordinate plane below.

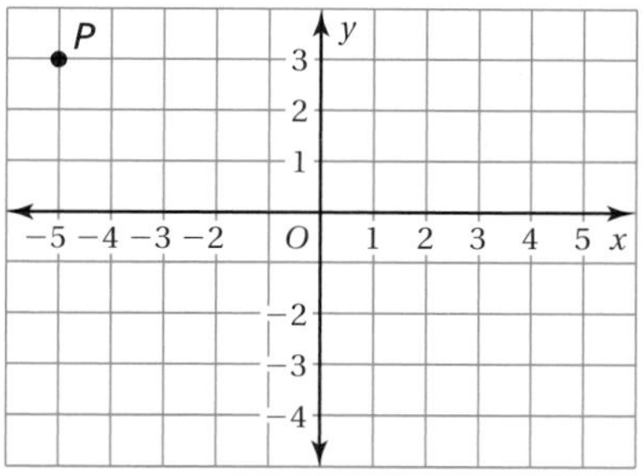

What are the coordinates of point P?

F. $(-5, -3)$ **H.** $(-3, -5)$

G. $(-5, 3)$ **I.** $(3, -5)$

3. What is the value of the expression below?

$$17 - (-8)$$

4. Sam was evaluating an expression in the box below.

$$\begin{aligned} |-8 + 6 + (-3)| &= |-8| + |6| + |-3| \\ &= 8 + 6 + 3 \\ &= 17 \end{aligned}$$

What should Sam do to correct the error that he made?

A. Find the absolute value of the sum of 8, 6, and 3 and make that the final answer.

B. Find the sum of -8, -6, and -3 and make that the final answer.

C. Find the sum of -8, 6, and -3 and make that the final answer.

D. Find the absolute value of the sum of -8, 6, and -3 and make that the final answer.

5. The expression below can be used to find the temperature in degrees Celsius when given F, the temperature in degrees Fahrenheit.

$$\frac{5}{9}(F - 32)$$

What is the temperature in degrees Celsius, to the nearest degree, when the temperature in degrees Fahrenheit is 27°?

F. $-33°$ **H.** $-5°$

G. $-17°$ **I.** $-3°$

6. What is the missing number in the sequence below?

9, ____, 39, 54, 69

7. Which equation is *not* true for all numbers n?

A. $-n + 0 = -n$ **C.** $n - 0 = -n$

B. $n \cdot (-1) = -n$ **D.** $-n \cdot 1 = -n$

8. What is the area of the semicircle below? (Use 3.14 for π.)

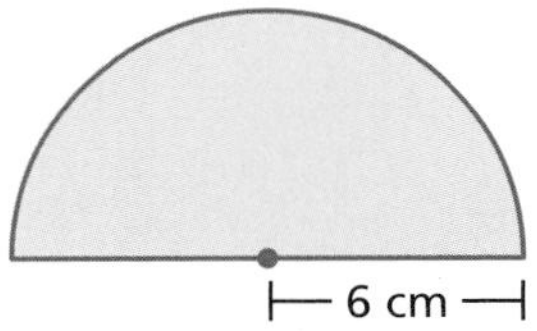

F. 18.84 cm^2 **H.** 56.52 cm^2

G. 37.68 cm^2 **I.** 226.08 cm^2

9. The campers at a summer camp held a contest in which they had to run across a field carrying buckets of water that were full at the beginning. The team who lost the least water from its bucket was the winner.

- Team A *lost* 40% of the water from its bucket.
- Team B *lost* 0.3 of the water from its bucket.
- Team C *kept* $\frac{5}{8}$ of the water in its bucket.
- Team D *kept* 67% of the water in its bucket.

Which team was the winner?

A. Team A **C.** Team C

B. Team B **D.** Team D

10. Which integer is closest to the value of the expression below?

$$-5.04 \cdot (16.89 - 20.1)$$

F. -105

G. -15

H. 15

I. 105

11. Answer the following questions in the coordinate plane.

Think Solve Explain

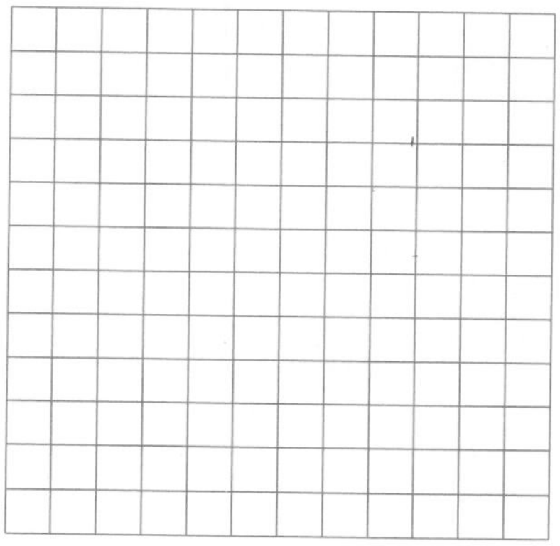

Part A Draw an x-axis and y-axis in the coordinate plane. Then plot and label the point $(2, -3)$.

Part B Plot and label *four* points that are 3 units away from $(2, -3)$.

12. What is the mean of the data set in the box below?

$$-8, -6, -2, 0, -6, -8, 4, -7, -8, 1$$

A. -8

B. -7

C. -6

D. -4

13. Jane and Manuel measured the lengths of their pet rats.

- Jane's pet rat
  - Body length: $10\frac{1}{2}$ inches
  - Tail length: $7\frac{3}{4}$ inches
- Manuel's pet rat
  - Body length: $9\frac{5}{8}$ inches
  - Tail length: $8\frac{1}{4}$ inches

The total length of each rat is determined by the sum of its body length and its tail length. Whose rat has the longer total length and by how much?

F. Jane's rat is longer by $\frac{3}{8}$ inch.

G. Jane's rat is longer by $\frac{1}{6}$ inch.

H. Manuel's rat is longer by $\frac{5}{8}$ inch.

I. Manuel's rat is longer by $\frac{1}{4}$ inch.

2 Algebraic Expressions

What You Learned Before

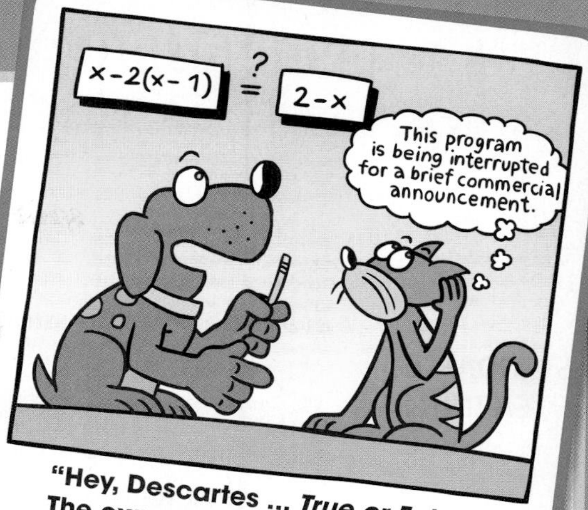

"Hey, Descartes ... *True or False:* The expressions are equivalent."

Finding Area

Example 1 **Find the area of the trapezoid.**

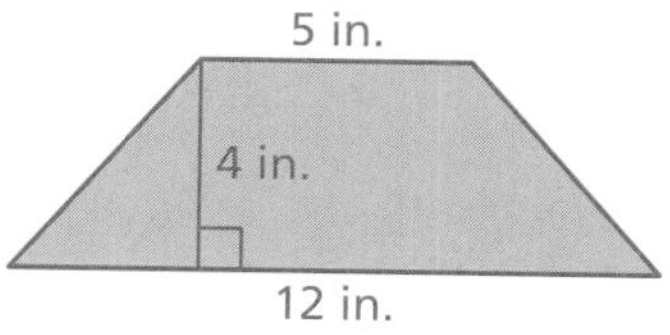

$$A = \frac{1}{2}h(b + B)$$
$$= \frac{1}{2}(4)(5 + 12)$$
$$= 34$$

∴ The area of the trapezoid is 34 square inches.

Try It Yourself

Find the area of the polygon.

1.

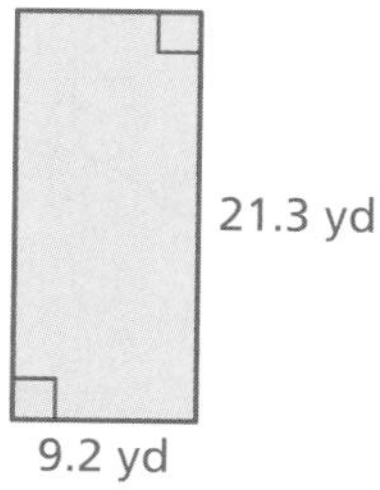

2.

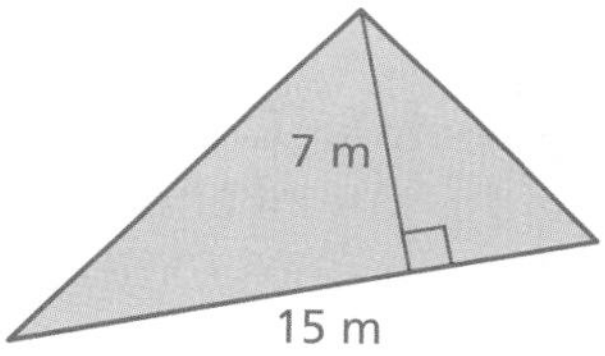

3.

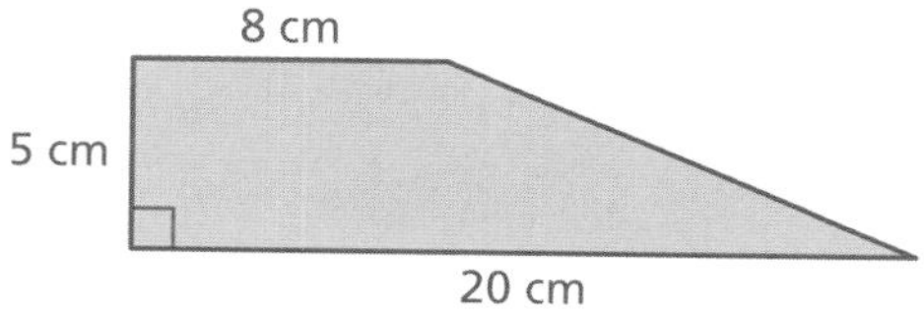

4.

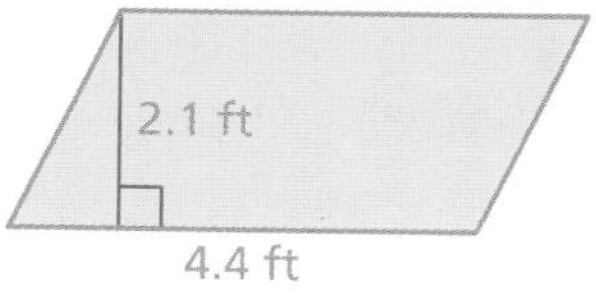

Using the Distributive Property

Example 2 **Use the Distributive Property to simplify the expression 5(*a* − 3).**

$5(a - 3)$	$= 5(a) - 5(3)$	Distributive Property
	$= 5a - 15$	Multiply.

Try It Yourself

Use the Distributive Property to simplify the expression.

5. $2(3.4 + y)$

6. $7(d - 4)$

7. $\frac{1}{3}(9 + w + 6)$

8. $12(0.5 + b - 1.2)$

2.1 Evaluating Algebraic Expressions

STANDARDS OF LEARNING
7.13

Essential Question What does it mean to *evaluate* an expression that has variables?

Share Your Work at... My.BigIdeasMath.com

1 ACTIVITY: Rate and Time Problems

Work with a partner. Many real-life uses of mathematics involve equations of the form

$$\text{Amount} = \text{Rate} \cdot \text{Time}.$$

Write and solve a real-life question for each "rate-time" problem. Illustrate your question with a drawing, diagram, or graph.

Sample: **a.** $D = r \cdot t$ Distance = Rate • Time

km km per h h

Question: Earth is moving around the Sun at the rate of about 100,000 km per hour. How far does Earth travel in a day?

Answer: Substitute $r = 100{,}000$ km/h and $t = 24$ h.

$$D = r \cdot t$$

$$= \frac{100{,}000 \text{ km}}{1 \cancel{\text{h}}} \cdot 24 \cancel{\text{h}}$$

$$= 2{,}400{,}000 \text{ km}$$

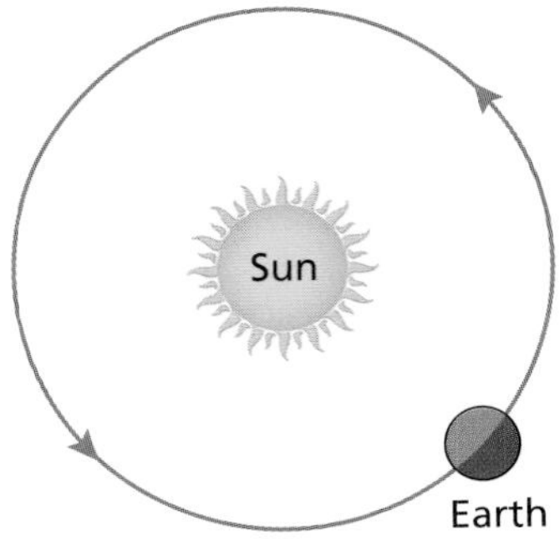

b. $D = r \cdot t$ Distance = Rate • Time

ft ft per sec sec

c. $P = w \cdot t$ Pay = Wage • Time

$ $ per h h

d. $P = r \cdot t$ Population = Rate • Time

people people per yr yr

e. $C = r \cdot t$ Rental Cost = Rate • Time

$ $ per day day

2 ACTIVITY: Solving Puzzles

Work with a partner. Complete each puzzle using each number from 1 through 9 *exactly once.* Read each row from left to right and each column from top to bottom.

Note: An arrow from one dashed circle to another indicates that the same number goes in both circles.

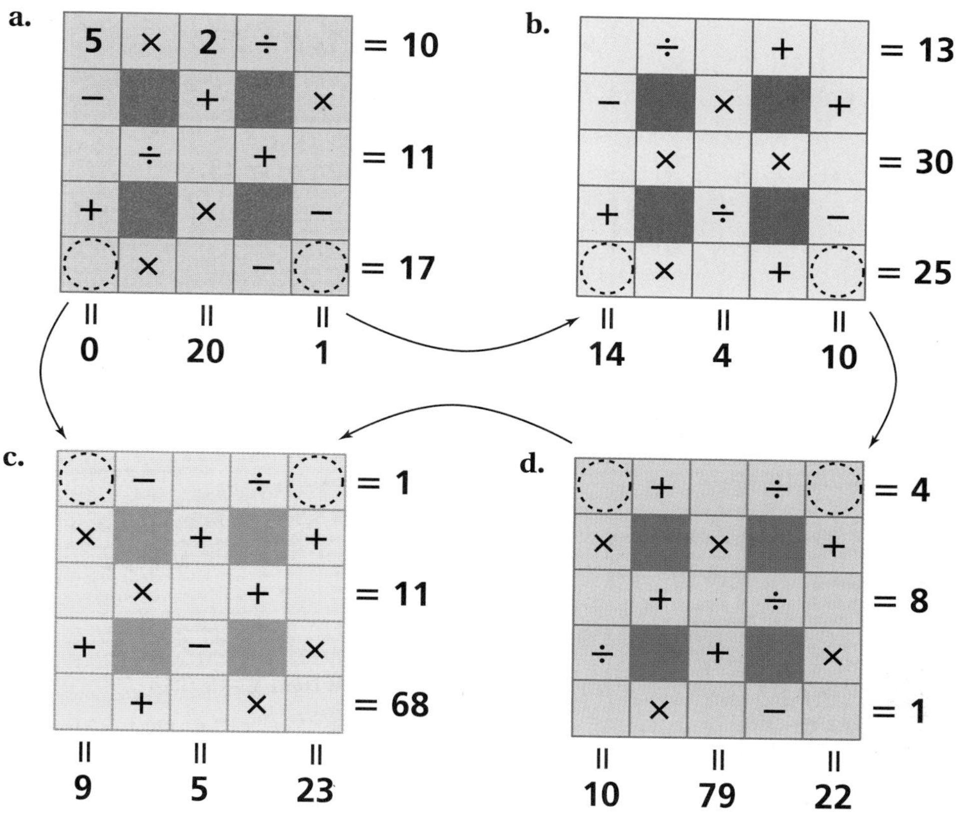

e. Make up your own puzzle. (Write the answers on a separate piece of paper.) Trade puzzles with your partner and try solving each other's puzzle.

What Is Your Answer?

3. IN YOUR OWN WORDS What does it mean to *evaluate* an expression that has variables? Give an example to illustrate your answer.

Practice Use what you learned about evaluating algebraic expressions to complete Exercise 4 on page 48.

2.1 Lesson

Check It Out
Lesson Tutorials
BigIdeasMath.com

Key Vocabulary
numerical expression, p. 46
algebraic expression, p. 46

A **numerical expression** contains only numbers and operations. An **algebraic expression** may contain numbers, operations, and one or more variables. Here are some examples.

Numerical Expression	***Algebraic Expression***
$6 \cdot 3.5 - 2$	$x \div 8 + y$

To evaluate an algebraic expression, substitute a number for each variable. Then use the order of operations to find the value of the numerical expression.

EXAMPLE 1 Evaluating Algebraic Expressions

a. Evaluate $-18 + m$ when $m = 23$.

Substitute 23 for *m*.

$$-18 + m = -18 + 23$$

$$= 5 \quad \text{Add.}$$

b. Evaluate $\frac{1}{2} \cdot y$ when $y = 48$.

Substitute 48 for *y*.

$$\frac{1}{2} \cdot y = \frac{1}{2} \cdot 48$$

$$= 24 \quad \text{Multiply.}$$

Study Tip

You can write the product of $\frac{1}{2}$ and *y* in several ways.

$\frac{1}{2} \cdot y$ $\frac{1}{2}y$ $\frac{1}{2}(y)$

On Your Own

Now You're Ready
Exercises 5–10

Evaluate the expression when $g = -20$.

1. $g + 6$
2. $15 - g$
3. $-2g$
4. $\frac{g}{10}$

EXAMPLE 2 Evaluating an Expression with Two Variables

Evaluate $a \div b$ when $a = 51$ and $b = -17$.

Substitute 51 for *a*.

$$a \div b = 51 \div (-17)$$

Substitute −17 for *b*.

$$= -3 \quad \text{Divide.}$$

On Your Own

Now You're Ready
Exercises 11–16

Evaluate the expression when $x = -24$ and $y = -8$.

5. $x \div y$
6. $y + x$
7. $x - y$
8. xy

EXAMPLE 3 Evaluating Expressions Using Order of Operations

Remember

Order of Operations

1. Perform operations in **P**arentheses.
2. Evaluate numbers with **E**xponents.
3. **M**ultiply or **D**ivide from left to right.
4. **A**dd or **S**ubtract from left to right.

a. Evaluate $3x + 10$ when $x = -4$.

$3x + 10 = 3(-4) + 10$	Substitute -4 for x.
$= -12 + 10$	Multiply.
$= -2$	Add.

b. Evaluate $c^2 - 2d$ when $c = 5$ and $d = 6.7$.

$c^2 - 2d = 5^2 - 2(6.7)$	Substitute 5 for c and 6.7 for d.
$= 25 - 2(6.7)$	Evaluate the power.
$= 25 - 13.4$	Multiply.
$= 11.6$	Subtract.

EXAMPLE 4 Real-Life Application

The expression $2.5g + 2s$ gives the cost for a group of friends to bowl g games and rent s pairs of shoes.

a. What is the cost for 12 games and 4 pairs of shoes? 9 games and 3 pairs of shoes? 16 games and 4 pairs of shoes?

Substitute the number of games for g and the number of pairs of shoes for s.

Number of Games, g, and Pairs of Shoes, s	$2.5g + 2s$	Total Cost
12 games, 4 pairs	$2.5(12) + 2(4)$	$30 + 8 = \$38$
9 games, 3 pairs	$2.5(9) + 2(3)$	$22.5 + 6 = \$28.50$
16 games, 4 pairs	$2.5(16) + 2(4)$	$40 + 8 = \$48$

b. Is \$45 enough to pay for 16 games and 4 pairs of shoes? Explain.

The cost for 16 games and 4 pairs of shoes is \$48. So, \$45 is not enough.

On Your Own

Now You're Ready
Exercises 20–25

Evaluate the expression when $y = -6$ and $z = 2.9$.

9. $7y + 2$ **10.** $24 - 12 \div y$ **11.** $y^2 - 7z$ **12.** $1.5 + y^2 - z$

13. WHAT IF? In Example 4, is \$50 enough to pay for 12 games and 6 pairs of shoes? Explain.

2.1 Exercises

Check It Out
Help with Homework
BigIdeasMath.com

Vocabulary and Concept Check

1. **WRITING** Describe how to evaluate the expression $12x - 5y$ when $x = 3$ and $y = 9$.
2. **VOCABULARY** How are numerical expressions and algebraic expressions different?
3. **DIFFERENT WORDS, SAME QUESTION** Which is different? Find "both" answers.

Evaluate $a + b$ when $a = -12$ and $b = 8$.	Find the sum of -12 and 8.
Evaluate $a + b$ when $a = 12$ and $b = -8$.	Add -12 and 8.

Practice and Problem Solving

Write and solve a real-life question for the problem. Illustrate your question with a drawing, diagram, or graph.

4. $C = r \cdot t$ Parking Cost ($) = Rate ($ per h) • Time (h)

Evaluate the expression when $x = -24$, $y = 4$, and $z = -3$.

(1) (2)

5. $-32 + z$
6. $\frac{x}{3}$
7. $y - 7$
8. $-45 - x$
9. $6.3y$
10. $24 \div z$
11. yz
12. $x \div z$
13. $x - y$
14. $z + y$
15. $\frac{x \div y}{z}$
16. $\frac{yz}{x}$

17. **ERROR ANALYSIS** Describe and correct the error in evaluating the expression when $a = -6$.

✗
$$12 - a + 4 = 12 - 6 + 4$$
$$= 6 + 4$$
$$= 10$$

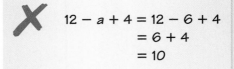

18. **TEMPERATURE** The expression $37 - 2h$ represents the temperature (in degrees Fahrenheit) after sunset in January, where h is the number of hours after sunset. What is the temperature 4 hours after sunset? 7 hours after sunset?

19. **NECKLACES** A store owner purchases n necklaces for c dollars. To determine the minimum price for a necklace, the store owner uses the expression $c \div n$. Find the minimum price when $c = 82$ and $n = 8$.

Evaluate the expression when $b = 1.2$ and $c = -8$.

20. $5b - 13$ **21.** $56 - c^2$ **22.** $-32 \div c + b$

23. $c^2 - 9b$ **24.** $-4(15b + c)$ **25.** $\dfrac{c^2 - 21.7b}{2}$

26. FOOTBALL The expression $6t + p + 3f$ gives the score of a football team after scoring t touchdowns, p extra points, and f field goals. Which combination of touchdowns, extra points, and field goals will give the home team the lead?

- 3 touchdowns, 3 extra points, and 2 field goals
- 4 touchdowns, 2 extra points, and 0 field goals
- 2 touchdowns, 1 extra point, and 5 field goals

27. STALACTITES The expression $0.13y$ represents the length (in millimeters) of a stalactite after y years. Use guess, check, and revise to find the number of years it takes for a stalactite to grow 65 millimeters.

28. Number Sense The expression $t + 2m + 4s$ is the number of points earned on a test, where t is the number of correct true-false questions, m is the number of correct multiple choice questions, and s is the number of correct short answer questions. Find several ways in which you can earn 45 points to get an A.

	Number of Questions
True-False	8
Multiple Choice	15
Short Answer	3

Fair Game Review What you learned in previous grades & lessons

Identify the property. Then simplify. *(Skills Review Handbook)*

29. $8 + (-3) + 2 = 8 + 2 + (-3)$

30. $2 \cdot (4.5 \cdot 9) = (2 \cdot 4.5) \cdot 9$

31. $\dfrac{1}{4} + \left(\dfrac{3}{4} + \dfrac{1}{8}\right) = \left(\dfrac{1}{4} + \dfrac{3}{4}\right) + \dfrac{1}{8}$

32. $\dfrac{3}{7} \cdot \dfrac{4}{5} \cdot \dfrac{14}{27} = \dfrac{3}{7} \cdot \dfrac{14}{27} \cdot \dfrac{4}{5}$

33. MULTIPLE CHOICE The regular price of a photo album is $18. You have a coupon for 15% off. How much is the discount? *(Skills Review Handbook)*

(A) $2.70 (B) $3 (C) $15 (D) $15.30

2.2 Writing Expressions and Equations

STANDARDS OF LEARNING
7.13

Essential Question How can you use math, logic, and diagrams to solve word problems?

Share Your Work at... My.BigIdeasMath.com

1 ACTIVITY: Solving Brain Teasers

Work with a partner. When you solve real-life problems that involve math, communication and clear writing are important. Here are some brain teasers and puzzles that illustrate this. Solve each brain teaser.

a. You have two U.S. coins whose value is $0.55. One of them is not a nickel. What are the coins?

b. One brick weighs one pound plus half a brick. How much, in pounds, does one brick weigh?

c. You drive 60 miles to town at a speed of 30 miles per hour. How fast do you have to drive on the return trip so that your average round-trip speed is 60 miles per hour?

d. If it were two hours later, it would be half as long until midnight as it would be if it were an hour later. What time is it now?

e. You and your friend live in different sections of town but go to the same school. You left for school ten minutes before your friend and you meet in the park. When you meet, who is closer to school?

2 ACTIVITY: Writing a Brain Teaser

Work with a partner.

Write your own brain teaser. Make it fun and interesting. Use color and a drawing to make it appealing.

When you are finished, trade brain teasers with your classmates and see who can solve the brain teasers quickest.

3 ACTIVITY: Brain Teasers in History

In about 1650 B.C., the Egyptian scribe Ahmes made a transcript of an even more ancient mathematical manuscript dating to the reign of the Pharaoh Amenemhat III. The manuscript is filled with math riddles. Here is one that is similar to a riddle that was in the manuscript. Work with your partner to solve this riddle.

One hundred measures of corn must be divided among 5 workers. The third, fourth, and fifth workers together get four times as much as the first and second workers together. The second worker gets 4 times as much as the first worker. The third and fourth workers get equal amounts. The fifth worker gets twice as much as the third worker. How many measures of corn does each worker get?

What Is Your Answer?

4. **IN YOUR OWN WORDS** How can you use math, logic, and diagrams to solve word problems?

Practice Use what you learned about math, logic, and diagrams to complete Exercises 4 and 5 on page 54.

2.2 Lesson

Check It Out
Lesson Tutorials
BigIdeasMath.com

Key Vocabulary
equation, *p. 52*

Some words imply math operations.

Operation	Addition	Subtraction	Multiplication	Division
Key Words and Phrases	added to plus sum of more than increased by total of and	subtracted from minus difference of less than decreased by fewer than take away	multiplied by times product of twice of	divided by quotient of

EXAMPLE 1 Writing Algebraic Expressions

Remember

When writing expressions involving subtraction or division, order is important.

Write the phrase as an expression.

a. negative 4 divided by a number z

$$-4 \div z, \text{ or } \frac{-4}{z}$$

The phrase "divided by" means division. So, divide -4 by z.

b. 5.5 more than the product of 2 and a number x

$$2 \cdot x + 5.5, \text{ or } 2x + 5.5$$

The phrase "product of" means multiplication. So, multiply 2 and x.

The phrase "more than" means addition. So, add 5.5 to $2x$.

On Your Own

Now You're Ready
Exercises 6–9

Write the phrase as an expression.

1. the number y times 50
2. the difference of negative 1 and a number b
3. twice a number n minus 15
4. 3 less than the quotient of 7 and a number m

An **equation** is a mathematical sentence that uses an equal sign to show that two expressions are equal.

Expressions	*Equations*
$-5 + 2$	$-5 + 2 = -3$
$x + 2$	$x + 2 = -3$

To write a word sentence as an equation, look for key words or phrases such as "is," "the same as," or "equals" to determine where to place the equal sign.

EXAMPLE 2 Writing Equations

Write the word sentence as an equation.

a. The difference of a number x and 8 equals negative 5.

The difference of a number x and 8 equals negative 5.

$$x - 8 \quad = \quad -5$$

"Difference of" means subtraction.

An equation is $x - 8 = -5$.

b. 3 multiplied by a number w is 13.75.

3 multiplied by a number w is 13.75.

$$3w \quad = \quad 13.75$$

"Multiplied by" means multiplication.

An equation is $3w = 13.75$.

On Your Own

Now You're Ready
Exercises 10–15

Write the word sentence as an equation.

5. A number p plus 100 is 250.
6. The quotient of a number q and 0.6 equals negative 10.

EXAMPLE 3 Real-Life Application

The temperature at a point on the moon is twice the temperature at Earth's North Pole minus 8°C. Write an equation that can be used to find the temperature at the North Pole.

Words The temperature on the moon | is | twice the temperature at the North Pole | minus | 8°C.

Variable Let t be the temperature at the North Pole.

Equation $-58 \quad = \quad 2t \quad - \quad 8$

An equation is $-58 = 2t - 8$.

On Your Own

7. You answer 22 questions correctly and receive 5 bonus points on a test. Each question is worth the same number of points. You score a 93 on the test. Write an equation that can be used to find the number of points each question is worth.

2.2 Exercises

Check It Out
Help with Homework
BigIdeasMath.com

Vocabulary and Concept Check

1. **VOCABULARY** Describe how writing an expression and writing an equation are different.
2. **OPEN-ENDED** Describe a real-world situation represented by the equation $m \div 5 = 6$.
3. **DIFFERENT WORDS, SAME QUESTION** Which is different? Write "both" equations.

6 less than a number x is 10.	A number x equals 6 minus 10.
A number x decreased by 6 equals 10.	The difference of x and 6 is 10.

Practice and Problem Solving

Solve the brain teaser.

4. Why are 2011 pennies worth more than 2010 pennies?
5. You buy an apple and a water with a dollar and a quarter. The apple costs a quarter less than the water. How much does the water cost?

Write the phrase as an expression.

1 6. 3 increased by a number x
7. negative 10 take away a number c
8. 4 divided by the sum of a number n and 1
9. 3 times the quotient of a number y and 4

Write the word sentence as an equation.

2 10. The sum of a number x and 8 is 22.
11. A number p subtracted from 10 equals 1.2.
12. 6.5 times a number c equals 52.
13. The quotient of a number z and 7 is 5.
14. Negative 6.8 is 2 plus one-third of a number y.
15. 9 more than the product of a number b and negative 5 is 10.

ERROR ANALYSIS **Describe and correct the error in writing the word sentence as an equation.**

16. The difference of a number d and 5 is 15.

✗ $5 - d = 15$

17. The quotient of a number w and 3 is 8.

✗ $\frac{3}{w} = 8$

18. **PIZZA** A group of friends share the cost of a large pizza equally. Write an expression for the cost per person.

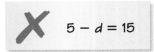

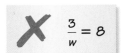

The Pizza Pantry
dine-in, take-out, or delivery
Large One-Topping Pizza
$12

GEOMETRY **Write an equation that can be used to find the value of x.**

(3) **19.** Circumference of circle: 9π in.

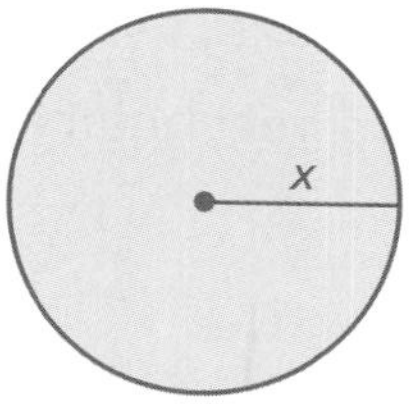

20. Volume of rectangular prism: 400 cm^3

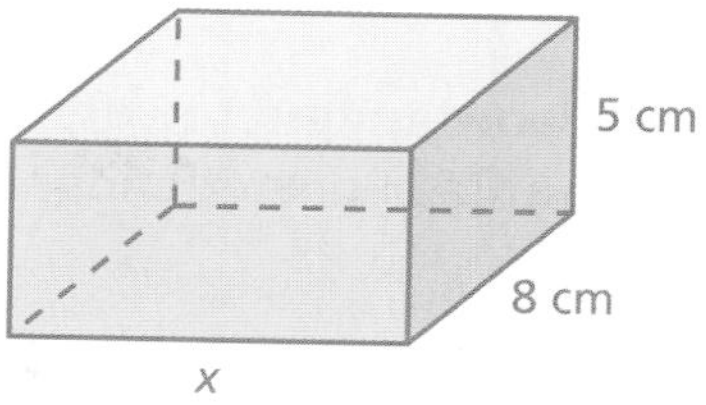

21. WRITING Use the words and operations to form two different correct equations. Write a real-life question for each equation. Use mental math to answer each question.

Total pay ($) | Time (hours) | Hourly wage ($ per hour) | + | − | × | ÷ | =

22. TICKETS A concert advertisement says the price of each ticket is $5 off. Write an expression for the cost of 3 tickets.

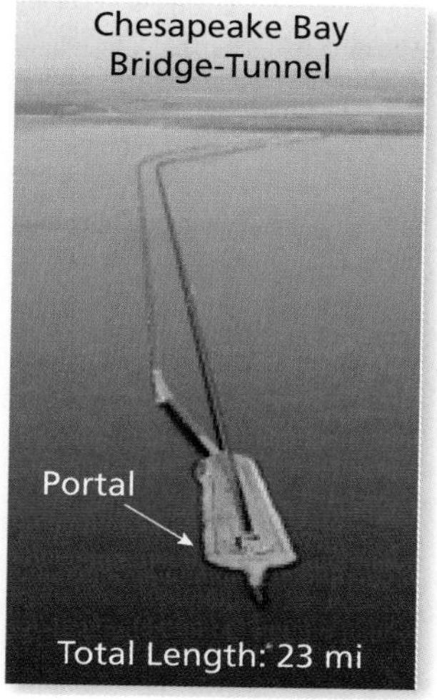

23. BRIDGE-TUNNEL The Chesapeake Bay Bridge-Tunnel is a combination of bridges and tunnels over and under two shipping channels, using four man-made islands built in the bay as portals.

a. The shore-to-shore length of the bridge-tunnel is 5.4 miles less than the total length. Write an equation that can be used to find the shore-to-shore length.

b. The minimum water depth, 25 feet, along the route is one-fourth the maximum water depth. Write an equation that can be used to find the maximum water depth.

c. RESEARCH Use the Internet or some other resource to find how long it took to construct the bridge-tunnel.

24. Geometry Two semicircles are cut from a rectangle. The perimeter of the shaded region is 43 centimeters. Write an equation that can be used to find the height h.

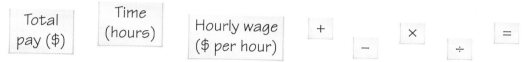

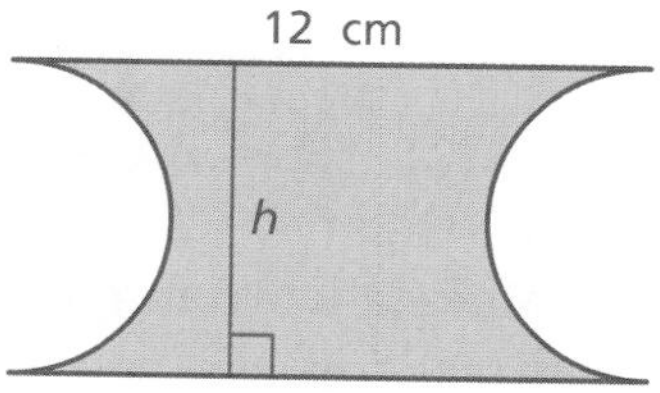

Fair Game Review What you learned in previous grades & lessons

Evaluate the expression when $x = 30$ and $y = -5$. *(Section 2.1)*

25. $12 - y$ **26.** $x \div 6$ **27.** xy **28.** $x \div y$

29. MULTIPLE CHOICE What is the surface area of a cube that has a side length of 5 feet? *(Skills Review Handbook)*

(A) 25 ft^2 (B) 75 ft^2 (C) 125 ft^2 (D) 150 ft^2

2 Study Help

Check It Out
Graphic Organizer
BigIdeasMath.com

You can use a **word magnet** to organize information associated with a vocabulary word. Here is an example of a word magnet for equation.

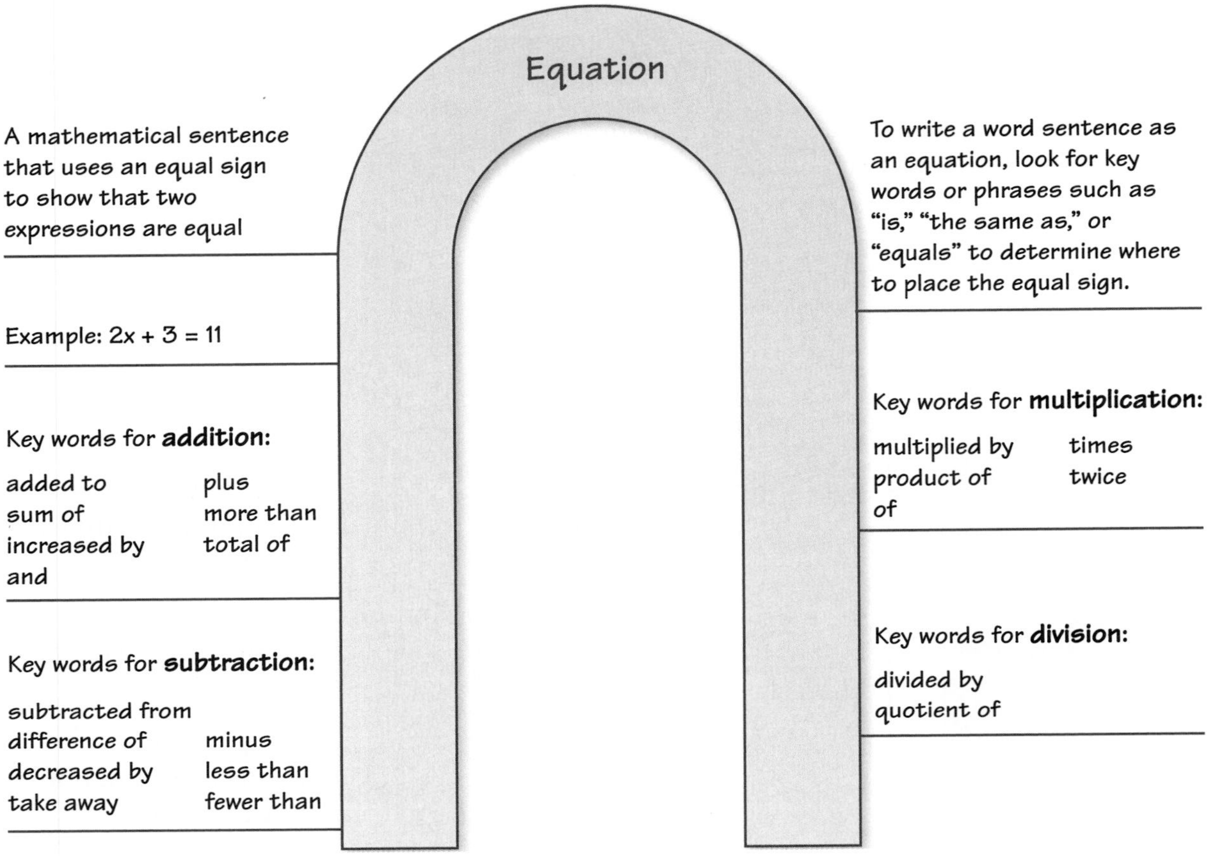

Equation
A mathematical sentence that uses an equal sign to show that two expressions are equal
Example: 2x + 3 = 11
Key words for **addition:**
added to
sum of
increased by
and
plus
more than
total of
Key words for **subtraction:**
subtracted from
difference of
decreased by
take away
minus
less than
fewer than
To write a word sentence as an equation, look for key words or phrases such as "is," "the same as," or "equals" to determine where to place the equal sign.
Key words for **multiplication:**
multiplied by
product of
of
times
twice
Key words for **division:**
divided by
quotient of

On Your Own

Make a word magnet to help you study these topics.

1. numerical expression
2. algebraic expression

After you complete this chapter, make word magnets for the following topics.

3. formula
4. term
5. like terms
6. simplest form

"I'm trying to make a word magnet for happiness, but I can only think of two words."

2.1–2.2 Quiz

Check It Out
Progress Check
BigIdeasMath.com

Evaluate the expression when $q = -36$, $r = -9$, and $s = 4$. *(Section 2.1)*

1. $11s$

2. $q \div (-3)$

3. $5(r + 15)$

4. $r - q$

5. $s^2 + 144 \div r$

6. $r^2 - 8s$

Write the phrase as an expression. *(Section 2.2)*

7. 150 plus a number t

8. a number x divided by 18

9. 7 increased by the product of 5 and a number k

10. the difference of twice a number d and negative 6

Write the word sentence as an equation. *(Section 2.2)*

11. Two-thirds multiplied by a number g equals 42.

12. Negative 4 is 9 subtracted from a number m.

13. The sum of 7.1 and the quotient of a number y and 5.3 is the same as 19.8.

14. 60 equals the sum of 3 and the product of a number b and negative 6.

15. **MOVIES** The expression $7t + 3d + 4p$ gives the cost for a family to buy t movie tickets, d drinks, and p bags of popcorn. What is the cost of the items shown? *(Section 2.1)*

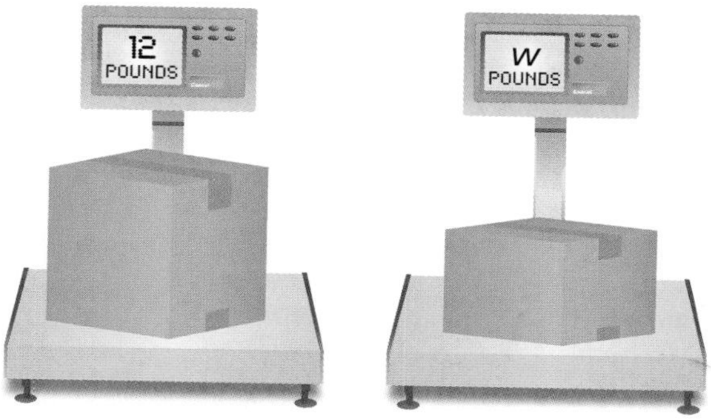

16. **PACKAGES** The weight of the larger package is 15 pounds less than 9 times the weight w of the smaller package. Write an equation that can be used to find the weight of the smaller package. *(Section 2.2)*

17. **COINS** You have q quarters and n nickels in your pocket. *(Section 2.1 and Section 2.2)*

a. Write an expression to represent the amount of money (in dollars) in your pocket.

b. You have 7 quarters and 12 nickels in your pocket. Use the expression from part (a) to find the amount of money in your pocket.

2.3 Using Formulas to Solve Problems

STANDARDS OF LEARNING
7.13

Essential Question How can you use formulas to find the area of an object with an unusual shape?

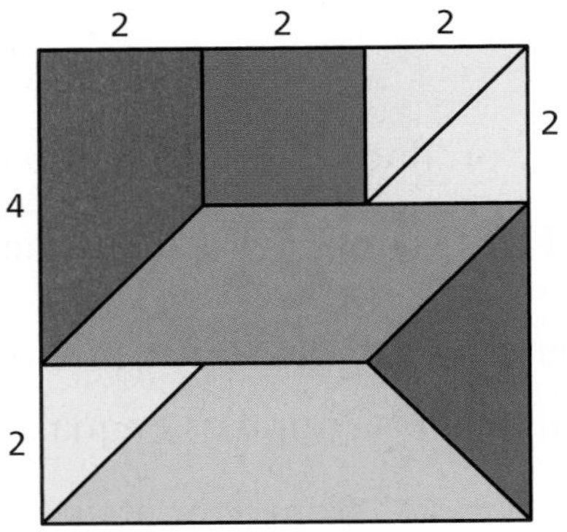

1 ACTIVITY: Using an Area Formula

Work with a partner. Copy and complete the table.

Polygon	Name	Area Formula	Area
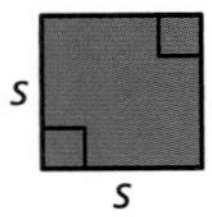	Square	$A = s^2$	$s = 2$ $A = 2^2$ $= 4$ square units
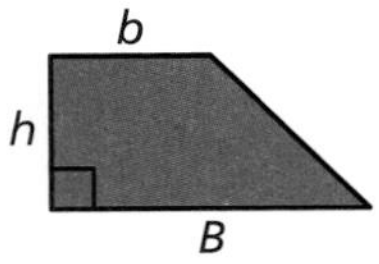	Trapezoid	$A = \frac{1}{2}h(b + B)$	
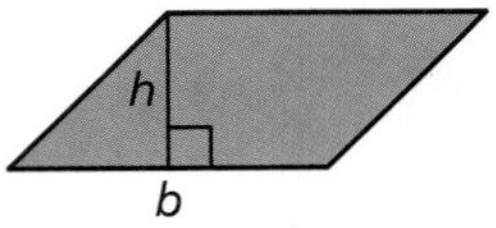			
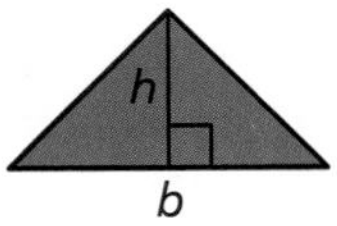			
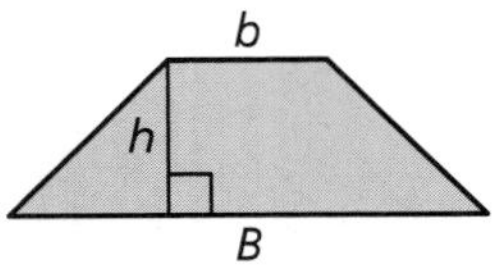			
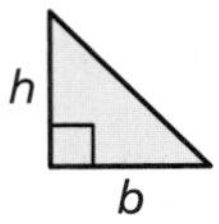			

2 ACTIVITY: Finding an Area

Work with a partner. Use the shapes from Activity 1 to find the area of the sailboat. Explain your reasoning.

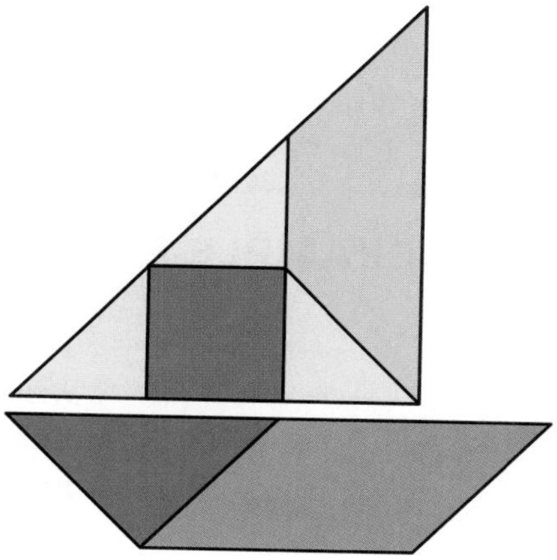

3 ACTIVITY: Finding an Area

Work with a partner. Use the shapes from Activity 1 to create the picture.

a. house

20 square units

b. rabbit

36 square units

c. bird

32 square units

What Is Your Answer?

4. **IN YOUR OWN WORDS** How can you use formulas to find the area of an object with an unusual shape?

5. Show how you can use the formula $A = bh$ for the area of a rectangle to write the formula for the area of a parallelogram.

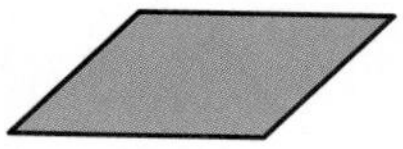

6. Show how you can use the formula $A = bh$ for the area of a rectangle to write the formula for the area of a triangle.

Practice Use what you learned about using formulas to complete Exercises 3–5 on page 62.

2.3 Lesson

Check It Out
Lesson Tutorials
BigIdeasMath.com

Key Vocabulary
formula, *p. 60*
solve a formula, *p. 60*

A **formula** is an equation that tells you how one variable is related to one or more other variables. To **solve a formula**, find the value of one variable by substituting numbers for the other variables.

EXAMPLE 1 Using a Simple Formula

The formula $M = 220 - a$ gives a person's maximum heart rate M, where a is the person's age in years. Malcolm is 12 years old. His uncle is 40 years old. What is the difference between their maximum heart rates?

Malcolm	*His Uncle*	
$M = 220 - a$	$M = 220 - a$	Write the formula.
$= 220 - 12$	$= 220 - 40$	Substitute their ages for *a*.
$= 208$	$= 180$	Subtract.

The difference between their maximum heart rates is $208 - 180$, or 28 beats per minute.

On Your Own

1. What is the difference between the maximum heart rates of Malcolm and his grandmother, who is 85 years old?

EXAMPLE 2 Using an Area Formula

Find the area of the rectangular jumping surface of the trampoline.

Use the formula for the area of a rectangle.

$A = bh$	Write the formula.
$= 14 \times 7$	Substitute 14 for *b* and 7 for *h*.
$= 98$	Multiply.

The area of the jumping surface is 98 square feet.

On Your Own

2. Find the area of a rectangular trampoline that measures 12 feet by 6 feet.

EXAMPLE 3 Using an Area Formula

A trapezoid can be used to approximate the shape of Arkansas, as shown on the map.

a. **Use the formula $A = \frac{1}{2}h(b + B)$ to find the area.**

b. **Mississippi has an area of about 46,907 square miles. Is the area of Arkansas greater than or less than the area of Mississippi?**

Remember

The corner mark ┐ in a figure means that the angle formed by the sides is a right angle.

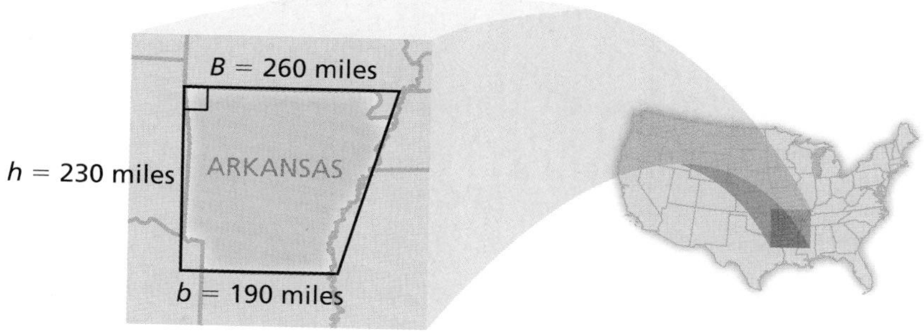

a. $A = \frac{1}{2}h(b + B)$ — Write the formula.

$= \frac{1}{2}(230)(190 + 260)$ — Substitute 230 for *h*, 190 for *b*, and 260 for *B*.

$= \frac{1}{2}(230)(450)$ — Add inside parentheses.

$= 115(450)$ — Multiply $\frac{1}{2}$ and 230.

$= 51{,}750$ — Multiply.

∴ The area of Arkansas is about 51,750 square miles.

b. Because 51,750 is greater than 46,907, the area of Arkansas is greater than the area of Mississippi.

On Your Own

Now You're Ready
Exercises 3–8

3. A trapezoid can be used to approximate the shape of Nevada, as shown on the map. How much larger is the area of Nevada than the area of Arkansas?

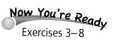

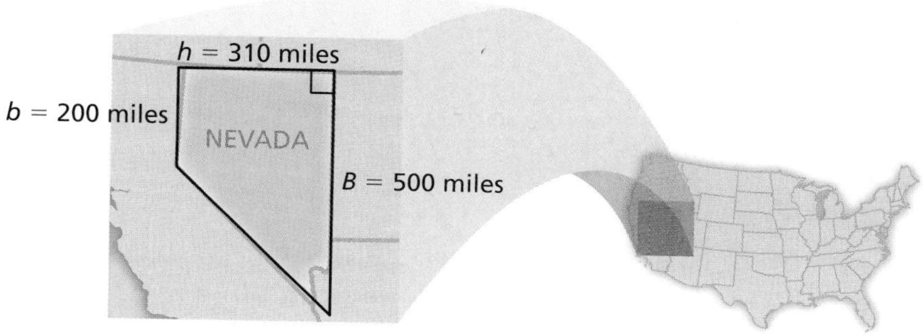

2.3 Exercises

Check It Out
Help with Homework
BigIdeasMath.com

Vocabulary and Concept Check

1. **WRITING** How is using a formula similar to evaluating an expression?
2. **REASONING** The cost C (in dollars) to make x dartboards is $C = 50 + 10x$. What do you need to know to solve this formula? Explain.

Practice and Problem Solving

Use a formula to find the area of the figure.

(2) (3) **3.**

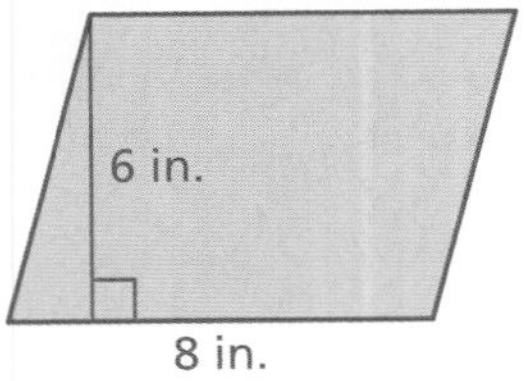

4.

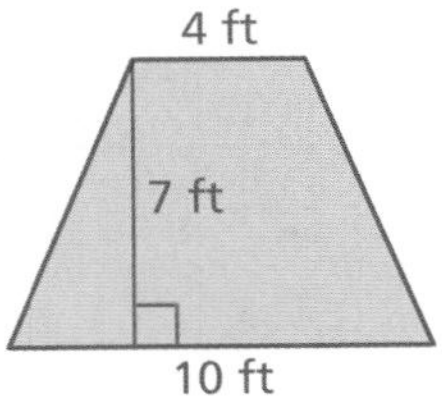

5.

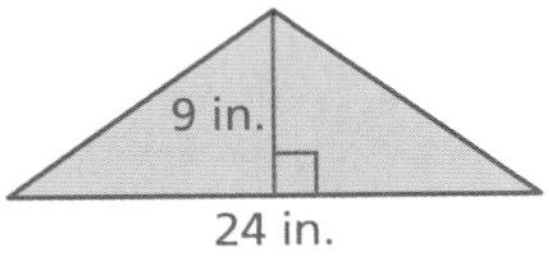

6.

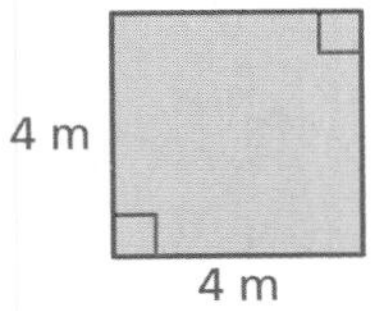

7.

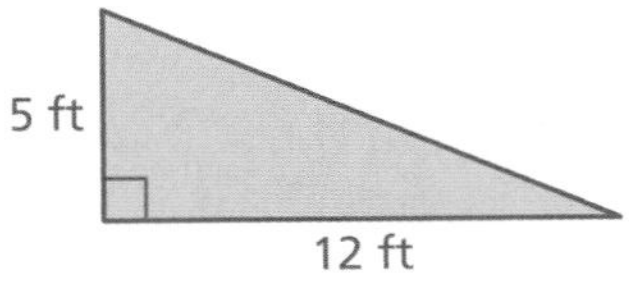

8.

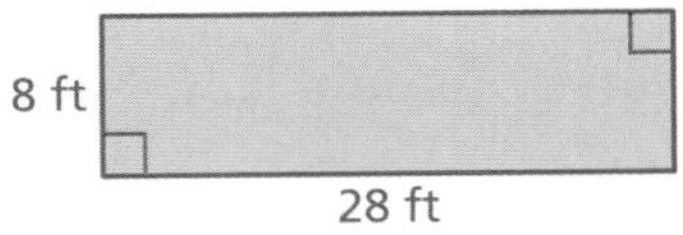

9. **PARKING SPACE** A parking space is shaped like a parallelogram with a base of 26 feet and a height of 9 feet.
   - **a.** What is the area of the parking space?
   - **b.** Draw a diagram of what the parking space might look like.
   - **c.** Use your diagram to estimate the length of the longest car that will fit in the space. Explain your reasoning.

10. **LIGHTNING** You can estimate how far you are from lightning.
    - When you see the lightning, count the number of seconds ("One one-thousand, two one-thousand, . . .") until you hear the thunder.
    - Divide the number of seconds by 5.
    - This is how many miles you are from lightning.

    You see lightning. After about a second, you hear a crack of thunder and your friend says "Wow, that was close!" Was your friend correct? How close was the lightning?

11. **VOLUME** The formula $V = \ell wh$ represents the volume of a rectangular prism with length ℓ, width w, and height h.
    - **a.** What is the volume of the cereal box?
    - **b.** The volume of a bowl is about 15 cubic inches. How many bowls of cereal does the box hold?

12. BASEBALL A pitcher's earned run average is the average number of earned runs given up per nine innings. What is the earned run average of a pitcher who gave up 75 earned runs in 225 innings?

$$\text{Earned Run Average} = \frac{9R}{I}$$

(R: Number of earned runs; I: Number of innings)

Write a formula for the area of the shaded region in terms of *x*.

13.

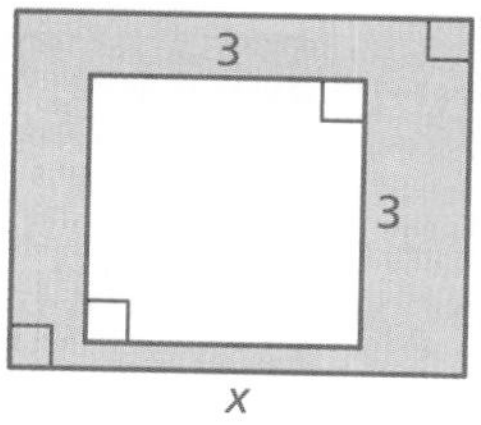

14.

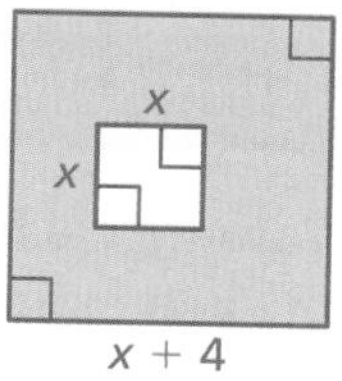

15.

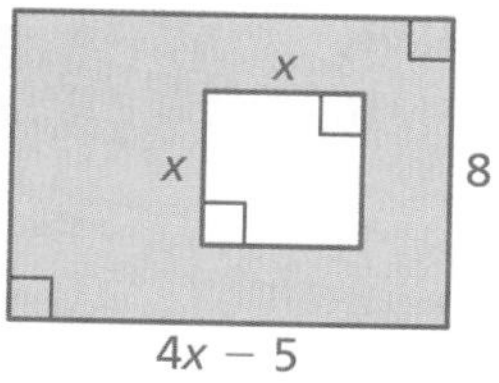

16. REASONING You know a parallelogram's area and base. Explain how you can find its height.

17. GOLD The purity of gold is measured in karats or in percent. What number of karats represents 100% pure gold? Explain your reasoning.

$$P = (25 \cdot k) \div 6$$

(P: Percent; k: Karats)

18. SNOWY TREE CRICKET To find the temperature T in degrees Fahrenheit, take the number c of chirps per minute of a snowy tree cricket and subtract 40. Then, divide by 4. Then, add 50.

a. Write a formula for the verbal description.

b. In the morning, a cricket chirps 56 times in one minute. What is the temperature?

c. Later in the afternoon, a cricket chirps 168 times in one minute. What is the temperature now?

19. Geometry Find the area of each region in the flag of the Bahamas.

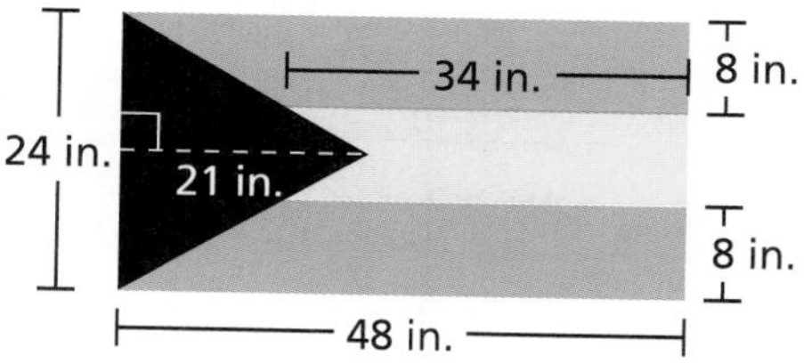

Fair Game Review What you learned in previous grades & lessons

Estimate the sum or difference. *(Skills Review Handbook)*

20. $\frac{7}{8} + \frac{9}{10}$ **21.** $\frac{1}{6} + \frac{2}{5}$ **22.** $\frac{4}{7} - \frac{7}{12}$ **23.** $\frac{4}{5} - \frac{1}{9}$

24. MULTIPLE CHOICE Which expression represents "8 more than x"? *(Section 2.2)*

Ⓐ $8 - x$ Ⓑ $8x$ Ⓒ $x + 8$ Ⓓ $\frac{8}{x}$

2.4 Simplifying Algebraic Expressions

STANDARDS OF LEARNING
7.16

Essential Question How can you simplify an algebraic expression?

1 ACTIVITY: Simplifying Algebraic Expressions

Work with a partner. Evaluate each expression when $x = 0$ and when $x = 1$. Use the results to match each expression in the left column with its equivalent expression in the right column.

	Expression	Value When	
		$x = 0$	$x = 1$
A.	$3x + 2 - x + 4$		
B.	$5(x - 3) + 2$		
C.	$x + 3 - (2x + 1)$		
D.	$-4x + 2 - x + 3x$		
E.	$-(1 - x) + 3$		
F.	$2x + x - 3x + 4$		
G.	$4 - 3 + 2(x - 1)$		
H.	$2(1 - x + 4)$		
I.	$5 - (4 - x + 2x)$		
J.	$5x - (2x + 4 - x)$		

	Expression	Value When	
		$x = 0$	$x = 1$
a.	4		
b.	$-x + 1$		
c.	$4x - 4$		
d.	$2x + 6$		
e.	$5x - 13$		
f.	$-2x + 10$		
g.	$x + 2$		
h.	$2x - 1$		
i.	$-2x + 2$		
j.	$-x + 2$		

2 ACTIVITY: Writing a Math Lesson

Work with a partner. Use your results from Activity 1 to write a lesson on simplifying an expression.

Describe steps you can use to simplify an expression.

Write 3 examples. Use expressions from Activity 1.

Write 3 exercises. Use expressions different from the ones in Activity 1.

Simplifying Expressions

Key Idea Use the following steps to simplify an algebraic expression.

1.
2.
3.

Example 1

a.

b.

c.

Exercises

Simplify the expression.

1.
2.
3.

What Is Your Answer?

3. **IN YOUR OWN WORDS** How can you simplify an algebraic expression? Give an example that demonstrates your procedure.

4. **RESEARCH** Use the Internet or some other source to find an example of an algebraic expression in real life. Is the expression simplified? Explain why or why not.

Practice Use what you learned about simplifying algebraic expressions to complete Exercises 12–14 on page 68.

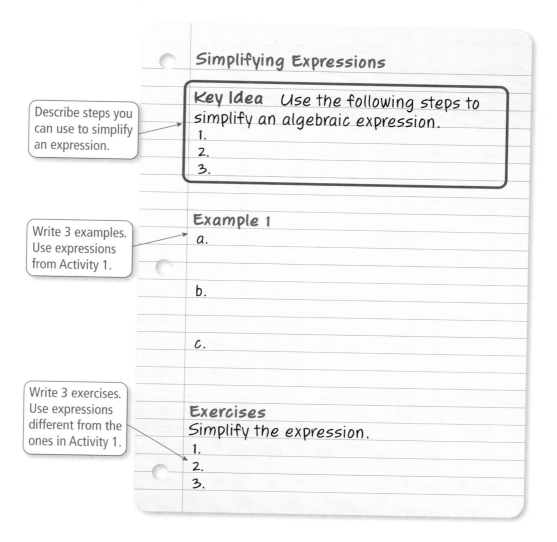

2.4 Lesson

Check It Out
Lesson Tutorials
BigIdeasMath.com

Key Vocabulary
like terms, *p. 66*
simplest form, *p. 66*

Parts of an algebraic expression are called terms. **Like terms** are terms that have the same variables raised to the same exponents. A term without a variable, such as 4, is called a *constant*. Constant terms are also like terms.

Like Terms	Unlike Terms
3 and -4	x and 5
$-2x$ and $7x$	$2x$ and $-6y$
$8x^2$ and x^2	$8x^2$ and $3x$

EXAMPLE 1 Identifying Terms and Like Terms

Identify the terms and like terms in each expression.

a. $9x - 2 + 7 - x$

$$\underbrace{9x}\ \underbrace{-\ 2}\ \underbrace{+\ 7}\ \underbrace{-\ x}$$

Terms: $9x,\ -2,\ 7,\ -x$

Like terms: $9x$ and $-x$, -2 and 7

Same variable raised to same exponent

b. $z^2 + 5z - 3z^2 + z$

$$\underbrace{z^2}\ \underbrace{+\ 5z}\ \underbrace{-\ 3z^2}\ \underbrace{+\ z}$$

Terms: $z^2,\ 5z,\ -3z^2,\ z$

Like terms: z^2 and $-3z^2$, $5z$ and z

On Your Own

Now You're Ready
Exercises 5–10

Identify the terms and like terms in the expression.

1. $y + 10 - 3y$
2. $2r^2 + 7r - r^2 - 9$
3. $7 + 4p - 5 + p + 2q$

Remember
The numerical factor of a term that contains a variable is a *coefficient*.

An algebraic expression is in **simplest form** if it has no like terms and no parentheses. To *combine* like terms that have variables, use the Distributive Property to add or subtract the coefficients.

EXAMPLE 2 Simplifying Algebraic Expressions

a. Simplify $6w + w + 5w$.

$6w$, w, and $5w$ are like terms.

$6w + w + 5w$	$= 6w + 1w + 5w$	Multiplication Property of One
	$= (6 + 1 + 5)w$	Distributive Property
	$= 12w$	Add coefficients.

b. Simplify $3y + 15 + 7y - 4$.

$3y$ and $7y$ are like terms. 15 and -4 are also like terms.

$3y + 15 + 7y - 4$	$= 3y + 7y + 15 - 4$	Commutative Property of Addition
	$= (3 + 7)y + 15 - 4$	Distributive Property
	$= 10y + 11$	Simplify.

Multi-Language Glossary at BigIdeasMath.com.

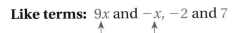

EXAMPLE 3 Standardized Test Practice

Which expression is equivalent to $5(n - 8) + 4n$?

Ⓐ $49n$ Ⓑ $9n + 40$ Ⓒ $9n - 40$ Ⓓ $5n - 40$

$5(n - 8) + 4n$	$= 5(n) - 5(8) + 4n$	Distributive Property
	$= 5n - 40 + 4n$	Multiply.
	$= 5n + 4n - 40$	Commutative Property of Addition
	$= (5 + 4)n - 40$	Distributive Property
	$= 9n - 40$	Add coefficients.

∴ The correct answer is Ⓒ.

On Your Own

Now You're Ready
Exercises 12–20

Simplify the expression.

4. $2x + 4x - 5$ **5.** $9b - 8b$ **6.** $7z^2 - 3z + 8 + z^2$

7. $3(q + 1) - 1$ **8.** $7x + 4(x - 6)$ **9.** $2(g + 4) + 5(g - 1)$

EXAMPLE 4 Real-Life Application

Each person in a group buys a ticket, a medium drink, and a large popcorn. Write an expression in simplest form that represents the amount of money the group spends at the movies.

Words Each ticket is $7.50, each medium drink is $2.75, and each large popcorn is $4.

Variable The same number of each item is purchased. So, x can represent the number of tickets, the number of medium drinks, and the number of large popcorns.

Expression $7.50x \quad + \quad 2.75x \quad + \quad 4x$

$7.50x + 2.75x + 4x$	$= (7.50 + 2.75 + 4)x$	Distributive Property
	$= 14.25x$	Add coefficients.

∴ The expression $14.25x$ represents the amount of money the group spends at the movies.

On Your Own

10. WHAT IF? Each person buys a ticket, a large drink, and a small popcorn. How does the expression change? Explain.

2.4 Exercises

Check It Out
Help with Homework
BigIdeasMath.com

Vocabulary and Concept Check

1. **WRITING** Explain how to identify the terms of $3y - 4 - 5y$.
2. **WRITING** Describe how to combine like terms in the expression $3n + 4n - 2$.
3. **VOCABULARY** Is the expression $3x + 2x - 4$ in *simplest form*? Explain.
4. **REASONING** Which algebraic expression is in simplest form? Explain.

$5x - 4 + 6y$	$4x + 8 - x$
$3(7 + y)$	$12n - n$

Practice and Problem Solving

Identify the terms and like terms in the expression.

(1) 5. $t + 8 + 3t$

6. $3z + 4 + 2 + 4z$

7. $2n - n - 4 + 7n$

8. $-x - 9x^2 + 12x^2 + 7$

9. $1.4y + 5 - 4.2 - 5y^2 + z$

10. $\frac{1}{2}s - 4 + \frac{3}{4}s + \frac{1}{8} - s^3$

11. **ERROR ANALYSIS** Describe and correct the error in identifying the like terms in the expression.

✗ $3x - 5 + 2x^2 + 9x = 3x + 2x^2 + 9x - 5$

Like Terms: $3x$, $2x^2$, and $9x$

Simplify the expression.

(2) 12. $12g + 9g$

13. $11x + 9 - 7$

14. $9s - 3s + 6$

15. $4.2v - 5 + 6.5v$

16. $8 + 9a + 6.2 - a$

17. $\frac{5}{6}y + 7 - 1\frac{1}{4} + \frac{1}{6}y$

(3) 18. $4(b - 6) + 19$

19. $6n + 5(4p - 2n)$

20. $\frac{1}{4}(8c + 16) + \frac{1}{3}(6 + 3c)$

21. **HIKING** On a hike, each hiker carries the items shown. Write an expression in simplest form that represents the weight carried by x hikers.

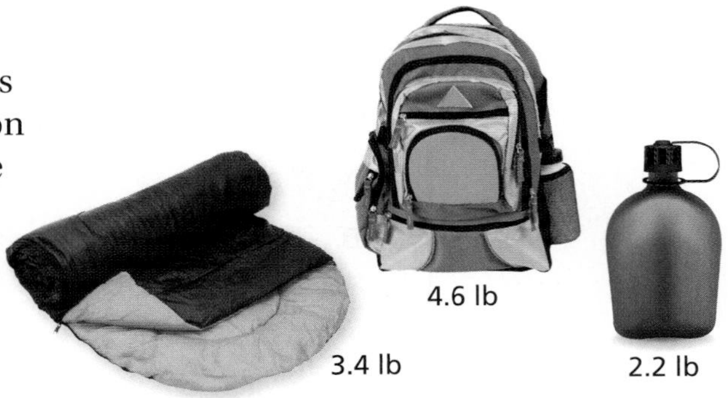

Write an expression in simplest form that represents the perimeter of the polygon.

22.

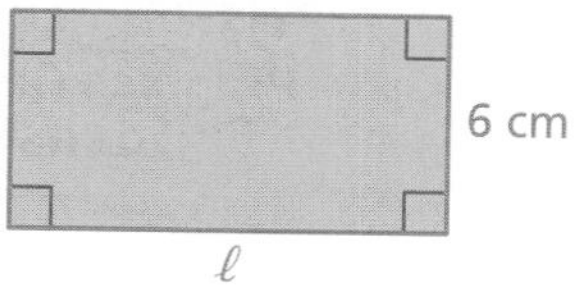

23.

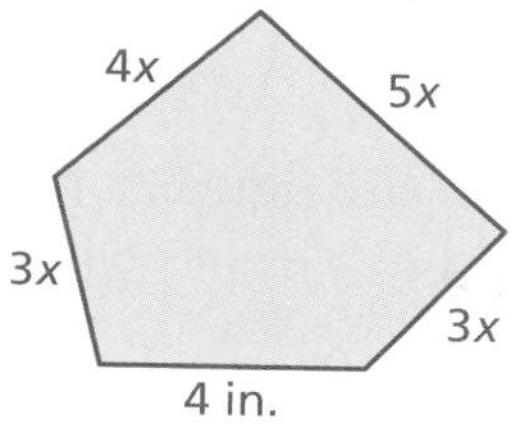

24.

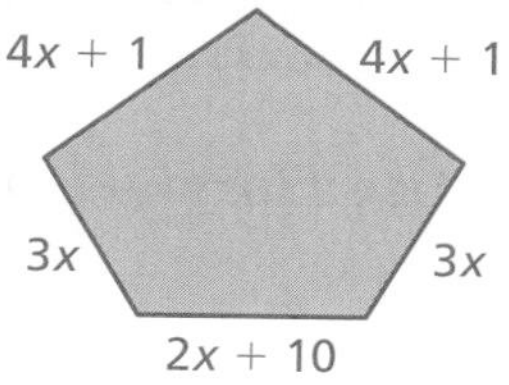

Simplify the expression.

25. $7u - 3t^2 + 4t^2$

26. $5.6y + 8z^2 + 3.4y$

27. $\frac{3}{4}(c^2 + c) + \frac{1}{4}(c - 24)$

28. REASONING Are the expressions $8x^2 + 3(x^2 + y)$ and $7x^2 + 7y + 4x^2 - 4y$ equivalent? Explain your reasoning.

29. BANNER Write an expression in simplest form that represents the area of the banner.

30. CAR WASH Write an expression in simplest form that represents the earnings for washing and waxing x cars and y trucks.

	Car	Truck
Wash	$8	$10
Wax	$12	$15

WRITING Draw a diagram that shows how the expression can represent the area of a figure. Then simplify the expression.

31. $5(2 + x + 3)$

32. $(4 + 1)(x + 2x)$

33. Critical Thinking You apply gold foil to a piece of red poster board to make the design shown.

a. Write an expression in simplest form that represents the area of the gold foil.

b. Find the area of the gold foil when $x = 3$.

c. The pattern at the right is called "St. George's Cross." Find a country that used this pattern as its flag.

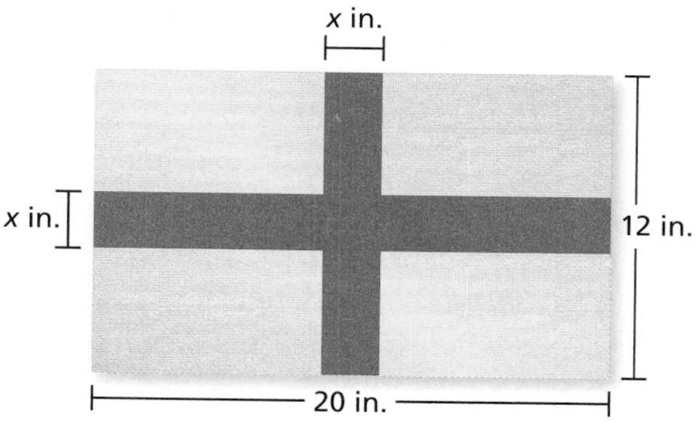

Fair Game Review What you learned in previous grades & lessons

Order the lengths from least to greatest. *(Skills Review Handbook)*

34. 15 in., 14.8 in., 15.8 in., 14.5 in., 15.3 in.

35. 0.65 m, 0.6 m, 0.52 m, 0.55 m, 0.545 m

36. MULTIPLE CHOICE A bird's nest is 12 feet above the ground. A mole's den is 12 inches below the ground. What is the difference in height of these two positions? *(Section 1.3)*

Ⓐ 24 in. Ⓑ 11 ft Ⓒ 13 ft Ⓓ 24 ft

2.3–2.4 Quiz

Check It Out
Progress Check
BigIdeasMath.com

Use a formula to find the area of the figure. *(Section 2.3)*

1.

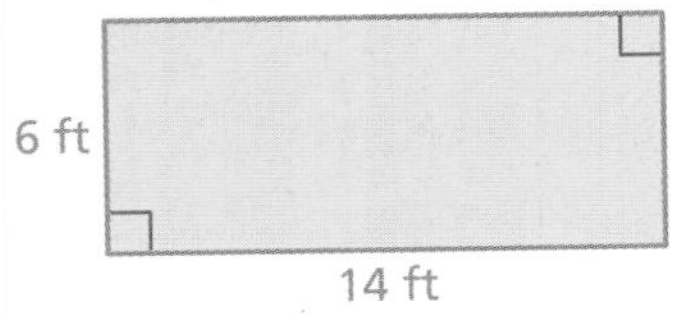

2.

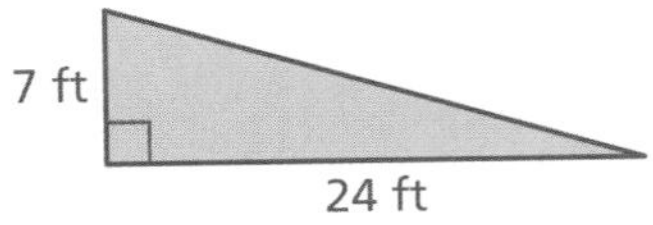

3.

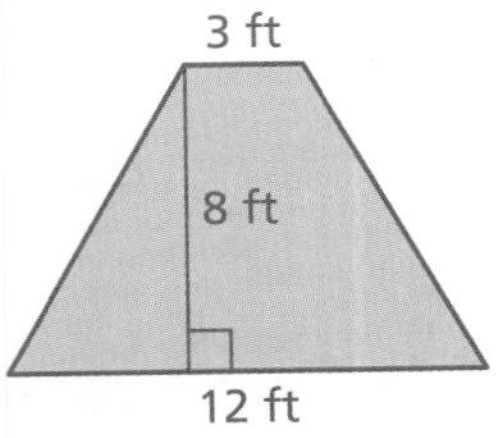

4.

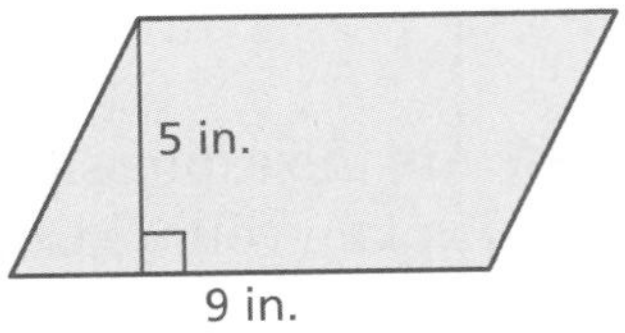

Identify the terms and like terms in the expression. *(Section 2.4)*

5. $11x + 2x$

6. $9x - 5x$

7. $21x + 6 - x - 5$

8. $8x + 14 - 3x + 1$

Simplify the expression. *(Section 2.4)*

9. $2(3x + x)$

10. $-7 + 3x + 4x$

11. $2x + 4 - 3x + 2 + 3x$

12. $7x + 6 + 3x - 2 - 5x$

13. **CALORIES** A formula for the number of calories c in s servings of a breakfast cereal is $c = 160s$. How many calories are there in 5 servings? *(Section 2.3)*

14. **PAINTING** You buy the same number of brushes, rollers, and paint cans. Write an expression in simplest form that represents the total amount of money you spend for painting supplies. *(Section 2.4)*

15. **FISH** A formula for the number of angelfish a that you keep in an aquarium is $a = g \div 3$, where the aquarium holds g gallons of water. How many angelfish can you keep in a 75-gallon aquarium? *(Section 2.3)*

16. **EXERCISE** Write an expression in simplest form for the perimeter of the exercise mat. *(Section 2.4)*

Check It Out
Vocabulary Help
BigIdeasMath.com

Review Key Vocabulary

numerical expression, *p. 46*
algebraic expression, *p. 46*
equation, *p. 52*
formula, *p. 60*
solve a formula, *p. 60*
like terms, *p. 66*
simplest form, *p. 66*

Review Examples and Exercises

2.1 Evaluating Algebraic Expressions *(pp. 44–49)*

a. Evaluate $c - d$ when $c = -22$ and $d = 5$.

Substitute −22 for *c*.

$c - d = -22 - 5$ — Substitute 5 for *d*.

$= -27$ — Subtract.

b. Evaluate $3y + 12$ when $y = -8$.

$3y + 12 = 3(-8) + 12$ — Substitute −8 for *y*.

$= -24 + 12$ — Multiply.

$= -12$ — Add.

c. Evaluate $48 \div w^2$ when $w = -4$.

$48 \div w^2 = 48 \div (-4)^2$ — Substitute −4 for *w*.

$= 48 \div 16$ — Evaluate the power.

$= 3$ — Divide.

Exercises

Evaluate the expression when $a = -12$ and $b = 3$.

1. $8 - b$
2. $-4a$
3. $-32 + a$
4. $a \div b$
5. $b + a$
6. ab
7. $b^2 + 5$
8. $2a - 75$
9. $\frac{72}{6b}$
10. **JEWELRY** The expression $5b + 8n$ represents how much you spend when you buy b bracelets and n necklaces. How much do you spend when you buy 2 bracelets and 3 necklaces?

2.2 Writing Expressions and Equations (pp. 50–55)

Write the phrase as an expression.

a. 6 less than a number x

$x - 6$ — The phrase "less than" means subtraction.

b. twice a number n increased by 12

$2n + 12$ — The phrase "twice" means multiplication. The phrase "increased by" means addition.

You are in a hot air balloon. You descend 350 feet to an altitude of 850 feet. Write an equation that can be used to find your original altitude.

Words	From your original altitude	you descend	350 feet	to	your current altitude.
Variable	Let a be your original altitude.				
Equation	a	$-$	350	$=$	850

∴ An equation is $a - 350 = 850$.

Exercises

Write the phrase as an expression.

11. 9 more than a number n

12. a number w subtracted from 25

13. 2 times the quotient of a number z and 6

14. **QUIZ** Your quiz score is 3 fewer than twice as many points as your friend scored. Write an expression for the number of points you scored on the quiz.

Write the word sentence as an equation.

15. 18 subtracted from twice a number c is 32.

16. 11 plus the product of a number g and 5 is 48.

17. One-half of the difference of a number m and 4 equals 26.

18. **RENTAL FEE** The rental fee for a kayak is \$10 less than twice the rental fee for a canoe. Write an equation that can be used to find the rental fee for a canoe.

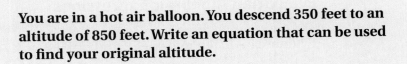

2.3 Using Formulas to Solve Problems *(pp. 58–63)*

Find the area of the triangular sail.

The sail is a triangle, so use the formula for the area of a triangle.

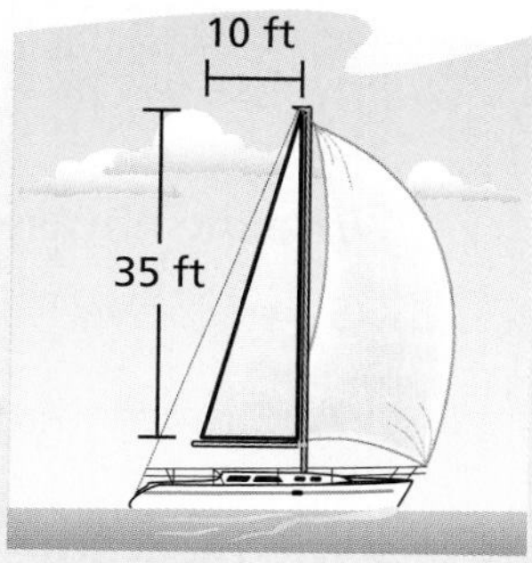

$A = \frac{1}{2}bh$	Write the formula.
$= \frac{1}{2}(10)(35)$	Substitute 10 for *b* and 35 for *h*.
$= 175$	Simplify.

∴ The area of the sail is 175 square feet.

Exercises

Use a formula to find the area of the figure.

19.

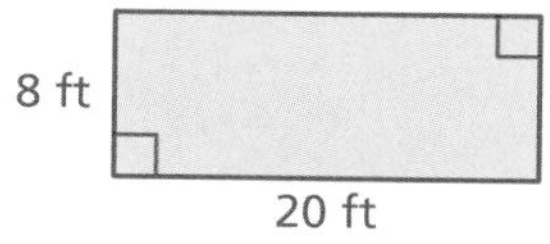

20.

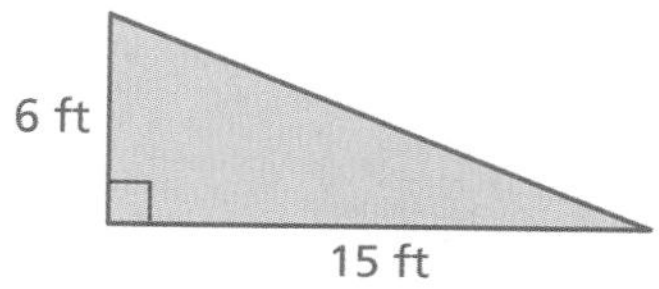

21.

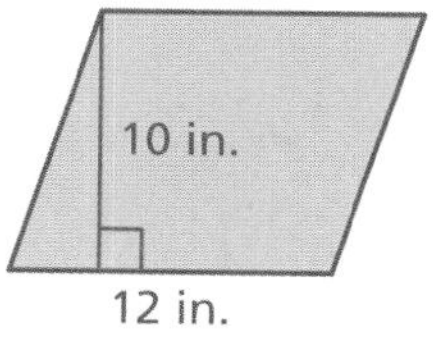

22.

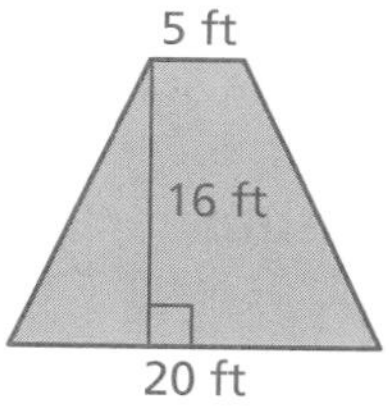

2.4 Simplifying Algebraic Expressions *(pp. 64–69)*

Identify the terms and like terms in $6y + 9 + 3y - 7$. Then simplify the expression.

$$6y + 9 + 3y - 7$$

Terms: $6y$, 9, $3y$, -7 **Like terms:** $6y$ and $3y$, 9 and -7

$6y + 9 + 3y - 7 = 6y + 3y + 9 - 7$	Commutative Property of Addition
$= (6 + 3)y + 9 - 7$	Distributive Property
$= 9y + 2$	Simplify.

Exercises

Identify the terms and like terms in the expression. Then simplify the expression.

23. $z + 8 - 4z$ **24.** $3n + 7 - n - 3$ **25.** $10x^2 - y + 12 - 3x^2$

2 Chapter Test

Check It Out
Test Practice
BigIdeasMath.com

Evaluate the expression when $x = -48$, $y = -6$, and $z = 8$.

1. $z - y$ **2.** $x + y$ **3.** $x \div z$ **4.** $\frac{y - x}{z}$

Write the phrase as an expression.

5. negative 12 divided by a number n

6. 4 less than the quotient of a number x and 3

7. 5 more than twice a number m

8. a number k plus negative 9

Write the word sentence as an equation.

9. The difference of negative 2 and a number y is 15.

10. The sum of a number w and 10 equals 22.

11. Negative 20 is the sum of 14 and one-fourth of a number c.

12. 40 decreased by the quotient of a number r and 5 is 32.

Use a formula to find the area of the figure.

13.

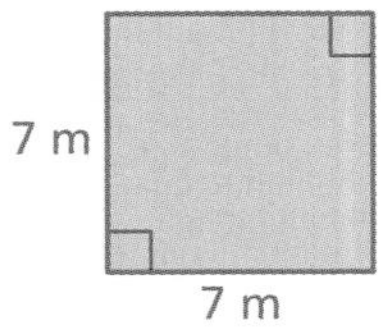

14.

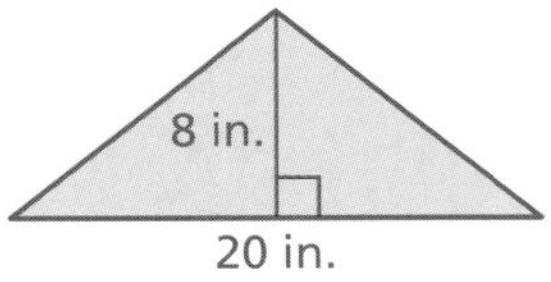

Simplify the expression.

15. $8x - 5 + 2x$

16. $2.5w - 3y + 4w$

17. $3(5 - 2n) + 9n$

18. $-4b^2 + 3c + 8b^2 - 12c$

19. TRACK AND FIELD The inside perimeter of the track is 400 meters.

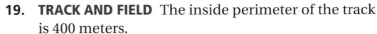

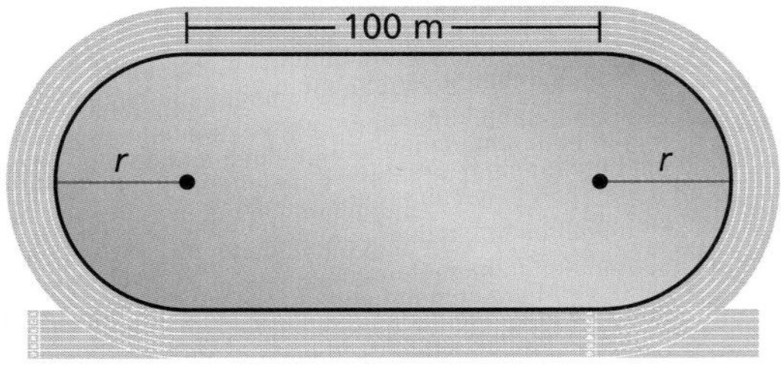

a. Write an equation that can be used to find the radius r.

b. The track coach says the radius is about 32 meters. Is the coach correct? Explain.

20. HAIR SALON Write an expression in simplest form that represents the income from w women and m men getting a haircut and a shampoo.

	Women	Men
Haircut	\$45	\$15
Shampoo	\$12	\$7

2 Standardized Test Practice

1. Which equation represents the word sentence shown below?

> The quotient of a number b and 0.3 equals negative 10.

A. $0.3b = 10$

B. $\frac{b}{0.3} = -10$

C. $\frac{0.3}{b} = -10$

D. $\frac{b}{0.3} = 10$

2. A rectangular prism and its dimensions are shown below.

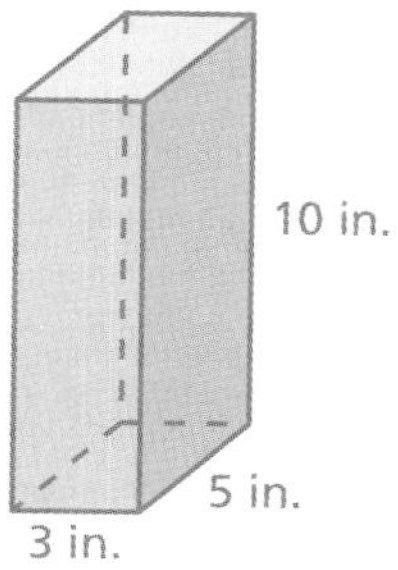

What is the volume, in cubic inches, of the rectangular prism?

3. The temperatures last week are shown below.

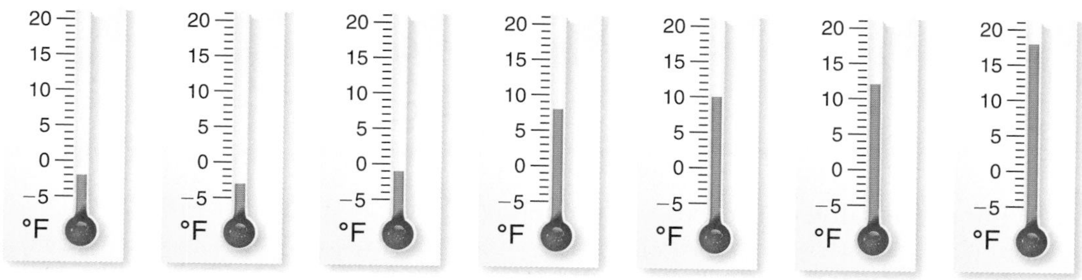

What is the mean temperature of last week?

F. $-2°\text{F}$

G. $6°\text{F}$

H. $8°\text{F}$

I. $10°\text{F}$

Test-Taking Strategy
Use Intelligent Guessing

What is the mean length of these hyena fangs: 4 in., 3 in., 3 in., 4 in., 5 in., 5 in.?
(A) 6 in. (B) 1/3 ft (C) 2 in. (D) 5 in.
MEOW!

"The mean can't be 6 or 2 or 5 inches. So, you can use intelligent guessing to find that the answer is $\frac{1}{3}$ ft, or 4 in."

4. What is the value of the expression below?

$$-28 \div (-14)$$

A. -14

B. -2

C. 2

D. 14

5. Betty was simplifying the expression in the box below.

$$\begin{aligned} -|8 + (-13)| &= -(|8| + |-13|) \\ &= -|8 + 13| \\ &= -|21| \\ &= -21 \end{aligned}$$

What should Betty do to correct the error that she made?

F. Simplify $-|8 + (-13)|$ to get $-|-5|$.

G. Find the opposite of $|21|$, which is 21.

H. Find the absolute value of 8, which is -8.

I. Distribute the negative sign to get $|-8 + (-13)|$.

6. Which integer is closest to the value of the expression below?

$$-7\frac{1}{4} \cdot \left(-9\frac{7}{8}\right)$$

A. -70

B. -63

C. 63

D. 70

7. What is the value of the expression below when $c = 0$ and $d = -6$?

$$\frac{cd - d^2}{4}$$

8. What is the value of the expression below?

$$-38 - (-14)$$

F. -52

G. -24

H. 24

I. 52

9. Jacob was evaluating the expression below when $x = -2$ and $y = 4$.

$$2.7 + x^2 \div y$$

His work is in the box below.

$$\begin{aligned} 2.7 + x^2 \div y &= 2.7 + (-2^2) \div 4 \\ &= 2.7 - 4 \div 4 \\ &= 2.7 - 1 \\ &= 1.7 \end{aligned}$$

What should Jacob do to correct the error that he made?

A. Divide 2.7 by 4 before subtracting.

B. Square -2, not 2.

C. Square 2 then subtract.

D. Subtract 4 from 2.7 before dividing.

10. Which formula should you use to find the area of the figure below?

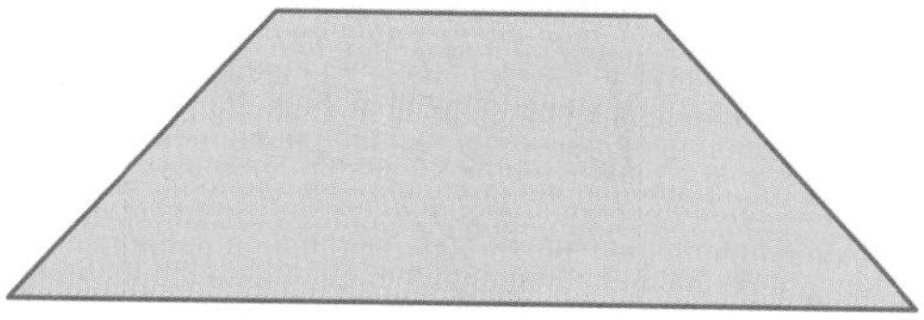

F. $A = bh$

G. $A = \frac{1}{2}bh$

H. $A = \frac{1}{2}h(b + B)$

I. $A = \pi r^2$

11. You have the following scores on math tests.

Think Solve Explain

88, 98, 94, 89, 96

Part A Find the mean, median, mode, and range of the test scores.

Part B What percentage do you need to receive on your next test to have a mean score of 93%?

Part C What effect does the test score from Part B have on the mean, median, mode, and range?

Part D Which measure best represents the six test scores? Explain your reasoning.

3 Rational Numbers, Equations, and Inequalities

3.1 Rational Numbers

3.2 Adding and Subtracting Rational Numbers

3.3 Multiplying and Dividing Rational Numbers

3.4 Solving Equations Using Addition or Subtraction

3.5 Solving Equations Using Multiplication or Division

3.6 Solving Two-Step Equations

3.7 Solving Inequalities Using Addition or Subtraction

3.8 Solving Inequalities Using Multiplication or Division

"I can't find my algebra tiles, so I am painting some of my dog biscuits."
I hope the paint is edible.
"Now I will be able to solve the equation 2x + (−2) = 2."

Nice!
"On the count of 5, I'm going to give you half of my dog biscuits."
I knew this was too good to be true.
"1, 2, 3, 4, 4½, 4¾, 4⅞,..."

What You Learned Before

"Let's play a game. The goal is to say a positive rational number that is less than the other pet's number... You go first."

Writing Decimals and Fractions

Example 1 Write 0.37 as a fraction.

$$0.37 = \frac{37}{100}$$

Example 2 Write $\frac{2}{5}$ as a decimal.

$$\frac{2}{5} = \frac{2 \cdot 2}{5 \cdot 2} = \frac{4}{10} = 0.4$$

Try It Yourself

Write the decimal as a fraction or the fraction as a decimal.

1. 0.51 **2.** 0.731 **3.** $\frac{3}{5}$ **4.** $\frac{7}{8}$

Adding and Subtracting Fractions

Example 3 Find $\frac{1}{3} + \frac{1}{5}$.

$$\frac{1}{3} + \frac{1}{5} = \frac{1 \cdot 5}{3 \cdot 5} + \frac{1 \cdot 3}{5 \cdot 3}$$

$$= \frac{5}{15} + \frac{3}{15}$$

$$= \frac{8}{15}$$

Example 4 Find $\frac{1}{4} - \frac{2}{9}$.

$$\frac{1}{4} - \frac{2}{9} = \frac{1 \cdot 9}{4 \cdot 9} - \frac{2 \cdot 4}{9 \cdot 4}$$

$$= \frac{9}{36} - \frac{8}{36}$$

$$= \frac{1}{36}$$

Multiplying and Dividing Fractions

Example 5 Find $\frac{5}{6} \cdot \frac{3}{4}$.

$$\frac{5}{6} \cdot \frac{3}{4} = \frac{5 \cdot \cancel{3}^{1}}{{}_{2}\cancel{6} \cdot 4}$$

$$= \frac{5}{8}$$

Example 6 Find $\frac{2}{3} \div \frac{9}{10}$.

$$\frac{2}{3} \div \frac{9}{10} = \frac{2}{3} \cdot \frac{10}{9}$$

Multiply by the reciprocal of the divisor.

$$= \frac{2 \cdot 10}{3 \cdot 9}$$

$$= \frac{20}{27}$$

Try It Yourself

Evaluate the expression.

5. $\frac{1}{4} + \frac{13}{20}$ **6.** $\frac{14}{15} - \frac{1}{3}$ **7.** $\frac{3}{7} \cdot \frac{9}{10}$ **8.** $\frac{4}{5} \div \frac{16}{17}$

3.1 Rational Numbers

STANDARDS OF LEARNING
7.3

Essential Question How can you use a number line to order rational numbers?

The Meaning of a Word ● Rational

The word **rational** comes from the word *ratio*.

If you sleep for 8 hours in a day, then the *ratio* of your sleeping time to the total hours in a day can be written as $\dfrac{8\text{ h}}{24\text{ h}}$.

A **rational number** is a number that can be written as the ratio of two integers.

$2 = \dfrac{2}{1}$ $\quad -3 = \dfrac{-3}{1}$ $\quad -\dfrac{1}{2} = \dfrac{-1}{2}$ $\quad 0.25 = \dfrac{1}{4}$

1 ACTIVITY: Ordering Rational Numbers

Work in groups of five. Order the numbers from least to greatest.

a. Sample: $-0.5, 1.25, -\dfrac{1}{3}, 0.5, -\dfrac{5}{3}$

- Make a number line on the floor using masking tape and a marker.

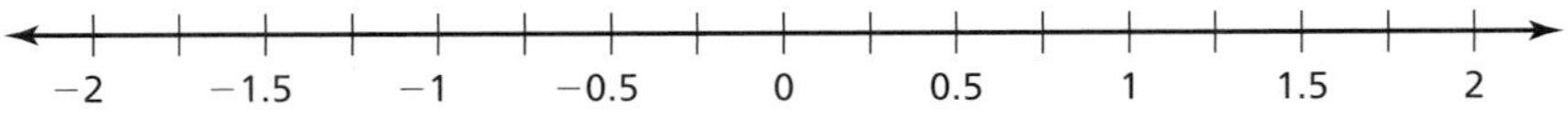

- Write the numbers on pieces of paper. Then each person should choose one.
- Stand on the location of your number on the number line.

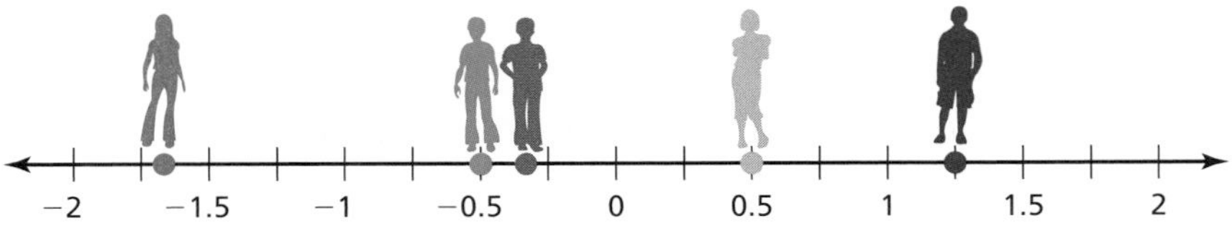

- Use your positions to order the numbers from least to greatest.

∴ So, the numbers from least to greatest are $-\dfrac{5}{3}, -0.5, -\dfrac{1}{3}, 0.5$, and 1.25.

b. $-\dfrac{7}{4}, 1.1, \dfrac{1}{2}, -\dfrac{1}{10}, -1.3$

c. $-\dfrac{1}{4}, 2.5, \dfrac{3}{4}, -1.7, -0.3$

d. $-1.4, -\dfrac{3}{5}, \dfrac{9}{2}, \dfrac{1}{4}, 0.9$

e. $\dfrac{9}{4}, 0.75, -\dfrac{5}{4}, -0.8, -1.1$

2 ACTIVITY: The Game of Math Card War

Preparation:

- Cut index cards to make 40 playing cards.
- Write each number in the table on a card.

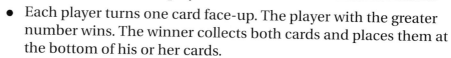

To Play:

- Play with a partner.
- Deal 20 cards to each player face-down.
- Each player turns one card face-up. The player with the greater number wins. The winner collects both cards and places them at the bottom of his or her cards.
- Suppose there is a tie. Each player lays three cards face-down, then a new card face-up. The player with the greater of these new cards wins. The winner collects all ten cards and places them at the bottom of his or her cards.
- Continue playing until one player has all the cards. This player wins the game.

$-\frac{3}{2}$	$\frac{3}{10}$	$-\frac{3}{4}$	-0.6	1.25	-0.15	$\frac{5}{4}$	$\frac{3}{5}$	-1.6	-0.3
$\frac{3}{20}$	$\frac{8}{5}$	-1.2	$\frac{19}{10}$	0.75	-1.5	$-\frac{6}{5}$	$-\frac{3}{5}$	1.2	0.3
1.5	1.9	-0.75	-0.4	$\frac{3}{4}$	$-\frac{5}{4}$	-1.9	$\frac{2}{5}$	$-\frac{3}{20}$	$-\frac{19}{10}$
$\frac{6}{5}$	$-\frac{3}{10}$	1.6	$-\frac{2}{5}$	0.6	0.15	$\frac{3}{2}$	-1.25	0.4	$-\frac{8}{5}$

What Is Your Answer?

3. **IN YOUR OWN WORDS** How can you use a number line to order rational numbers? Give an example.

The numbers are in order from least to greatest. Fill in the blank spaces with rational numbers.

4. $-\frac{1}{2}$, ___, $\frac{1}{3}$, ___, $\frac{7}{5}$, ___

5. $-\frac{5}{2}$, ___, -1.9, ___, $-\frac{2}{3}$, ___

6. $-\frac{1}{3}$, ___, -0.1, ___, $\frac{4}{5}$, ___

7. -3.4, ___, -1.5, ___, 2.2, ___

Practice Use what you learned about ordering rational numbers to complete Exercises 28–30 on page 84.

3.1 Lesson

Check It Out
Lesson Tutorials
BigIdeasMath.com

Key Vocabulary
terminating decimal, *p. 82*
repeating decimal, *p. 82*
rational number, *p. 82*

A **terminating decimal** is a decimal that ends.

1.5, –0.25, 10.625

A **repeating decimal** is a decimal that has a pattern that repeats.

$-1.333\ldots = -1.\overline{3}$

$0.151515\ldots = 0.\overline{15}$

Use *bar notation* to show which of the digits repeat.

Terminating and repeating decimals are examples of *rational numbers.*

Key Idea

Rational Numbers

A **rational number** is a number that can be written as $\frac{a}{b}$ where a and b are integers and $b \neq 0$.

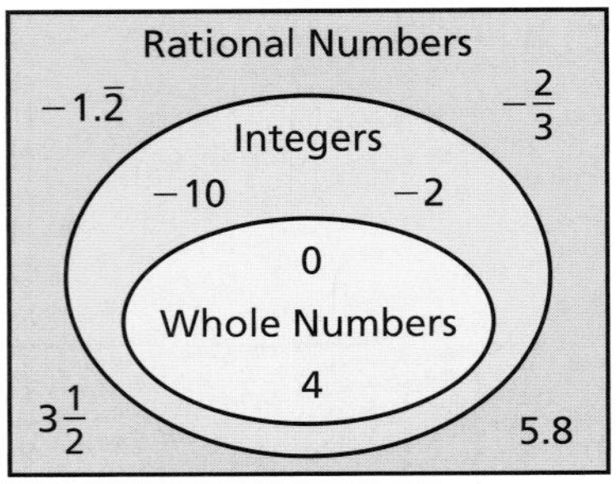

EXAMPLE 1 Writing Rational Numbers as Decimals

a. Write $-2\frac{1}{4}$ as a decimal.

Notice that $-2\frac{1}{4} = -\frac{9}{4}$.

Divide 9 by 4.

$$\begin{array}{r} 2.25 \\ 4\overline{)9.00} \\ -\underline{8} \\ 1\,0 \\ -\underline{8} \\ 20 \\ -\underline{20} \\ 0 \end{array}$$

The remainder is 0. So, it is a terminating decimal.

So, $-2\frac{1}{4} = -2.25$.

b. Write $\frac{5}{11}$ as a decimal.

Divide 5 by 11.

$$\begin{array}{r} 0.4545 \\ 11\overline{)5.0000} \\ -\underline{4\,4} \\ 60 \\ -\underline{55} \\ 50 \\ -\underline{44} \\ 60 \\ -\underline{55} \\ 5 \end{array}$$

The remainder repeats. So, it is a repeating decimal.

So, $\frac{5}{11} = 0.\overline{45}$.

On Your Own

Now You're Ready
Exercises 11–18

Write the rational number as a decimal.

1. $-\frac{6}{5}$
2. $-7\frac{3}{8}$
3. $-\frac{3}{11}$
4. $1\frac{5}{27}$

EXAMPLE 2 Writing a Decimal as a Fraction

Write -0.26 as a fraction in simplest form.

$$-0.26 = -\frac{26}{100}$$

Write the digits after the decimal point in the numerator.

The last digit is in the hundredths place. So, use 100 in the denominator.

$$= -\frac{13}{50}$$ Simplify.

On Your Own

Now You're Ready
Exercises 20–27

Write the decimal as a fraction or mixed number in simplest form.

5. -0.7 **6.** 0.125 **7.** -3.1 **8.** -10.25

EXAMPLE 3 Ordering Rational Numbers

Creature	Elevation (km)
Anglerfish	$-\frac{13}{10}$
Squid	$-2\frac{1}{5}$
Shark	$-\frac{2}{11}$
Whale	-0.8

The table shows the elevations of four sea creatures relative to sea level. Which of the sea creatures are deeper than the whale? Explain.

Write each rational number as a decimal.

$$-\frac{13}{10} = -1.3$$

$$-2\frac{1}{5} = -2.2$$

$$-\frac{2}{11} = -0.\overline{18}$$

Then graph each decimal on a number line.

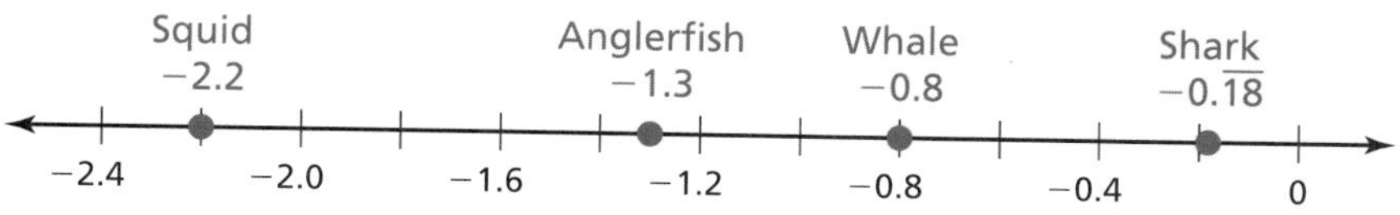

Both -2.2 and -1.3 are less than -0.8. So, the squid and the anglerfish are deeper than the whale.

On Your Own

Now You're Ready
Exercises 28–33

9. WHAT IF? The elevation of a dolphin is $-\frac{1}{10}$ kilometer. Which of the sea creatures in Example 3 are deeper than the dolphin? Explain.

3.1 Exercises

Check It Out
Help with Homework
BigIdeasMath.com

Vocabulary and Concept Check

1. **VOCABULARY** How can you tell that a number is rational?

2. **WRITING** You have to write 0.63 as a fraction. How do you choose the denominator?

Tell whether the number belongs to each of the following number sets: *rational numbers, integers, whole numbers.*

3. -5 4. $-2.1\overline{6}$ 5. 12 6. 0

Tell whether the decimal is *terminating* or *repeating*.

7. $-0.4848\ldots$ 8. -0.151 9. 72.72 10. $-5.2\overline{36}$

Practice and Problem Solving

Write the rational number as a decimal.

1 11. $\frac{7}{8}$ 12. $\frac{5}{11}$ 13. $-\frac{7}{9}$ 14. $-\frac{17}{40}$

15. $1\frac{5}{6}$ 16. $-2\frac{17}{18}$ 17. $-5\frac{7}{12}$ 18. $8\frac{15}{22}$

19. **ERROR ANALYSIS** Describe and correct the error in writing the rational number as a decimal.

✗ $-\frac{7}{11} = -0.6\overline{3}$

Write the decimal as a fraction or mixed number in simplest form.

2 20. -0.9 21. 0.45 22. -0.258 23. -0.312

24. -2.32 25. -1.64 26. 6.012 27. -12.405

Order the numbers from least to greatest.

3 28. $-\frac{3}{4}, 0.5, \frac{2}{3}, -\frac{7}{3}, 1.2$ 29. $\frac{9}{5}, -2.5, -1.1, -\frac{4}{5}, 0.8$ 30. $-1.4, -\frac{8}{5}, 0.6, -0.9, \frac{1}{4}$

31. $2.1, -\frac{6}{10}, -\frac{9}{4}, -0.75, \frac{5}{3}$ 32. $-\frac{7}{2}, -2.8, -\frac{5}{4}, \frac{4}{3}, 1.3$ 33. $-\frac{11}{5}, -2.4, 1.6, \frac{15}{10}, -2.25$

34. **COINS** You lose one quarter, two dimes, and two nickels.
 a. Write the amount as a decimal.
 b. Write the amount as a fraction in simplest form.

35. **HIBERNATION** A box turtle hibernates in sand at $-1\frac{5}{8}$ feet. A spotted turtle hibernates at $-1\frac{16}{25}$ feet. Which turtle is deeper?

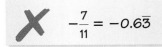

Copy and complete the statement using <, >, or =.

36. -2.2 ▢ -2.42 **37.** -1.82 ▢ -1.81 **38.** $\frac{15}{8}$ ▢ $1\frac{7}{8}$

39. $-4\frac{6}{10}$ ▢ -4.65 **40.** $-5\frac{3}{11}$ ▢ $-5.\overline{2}$ **41.** $-2\frac{13}{16}$ ▢ $-2\frac{11}{14}$

42. OPEN-ENDED Find one terminating decimal and one repeating decimal between $-\frac{1}{2}$ and $-\frac{1}{3}$.

Player	Hits	At Bats
Eva	42	90
Michelle	38	80

43. SOFTBALL In softball, a batting average is the number of hits divided by the number of times at bat. Does Eva or Michelle have the higher batting average?

44. QUIZ You miss 3 out of 10 questions on a science quiz and 4 out of 15 questions on a math quiz. Which quiz has a higher percent of correct answers?

45. SKATING Is the half pipe deeper than the skating pool? Explain.

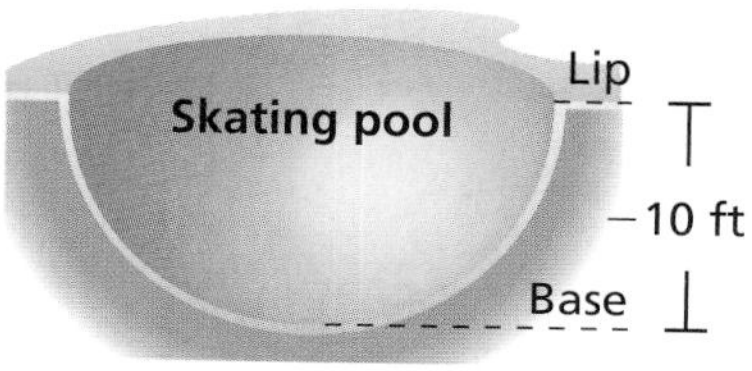

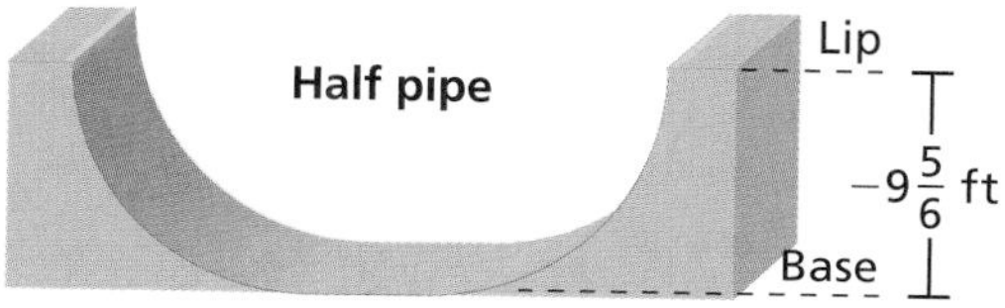

46. ENVIRONMENT The table shows the changes from the average water level of a pond over several weeks. Order the numbers from least to greatest.

Week	1	2	3	4
Change (inches)	$-\frac{7}{5}$	$-1\frac{5}{11}$	-1.45	$-1\frac{91}{200}$

47. Critical Thinking Given: a and b are integers.

a. When is $-\frac{1}{a}$ positive? **b.** When is $\frac{1}{ab}$ positive?

Fair Game Review What you learned in previous grades & lessons

Add or subtract. *(Skills Review Handbook)*

48. $\frac{3}{5} + \frac{2}{7}$ **49.** $\frac{9}{10} - \frac{2}{3}$ **50.** $8.79 - 4.07$ **51.** $11.81 + 9.34$

52. MULTIPLE CHOICE In one year, a company has a profit of −\$2 million. In the next year, the company has a profit of \$7 million. How much more money did the company make the second year? *(Section 1.3)*

Ⓐ \$2 million Ⓑ \$5 million Ⓒ \$7 million Ⓓ \$9 million

3.2 Adding and Subtracting Rational Numbers

STANDARDS OF LEARNING
7.3

Essential Question How does adding and subtracting rational numbers compare with adding and subtracting integers?

1 ACTIVITY: Adding and Subtracting Rational Numbers

Work with a partner. Use a number line to find the sum or difference.

a. Sample: $2.7 + (-3.4)$

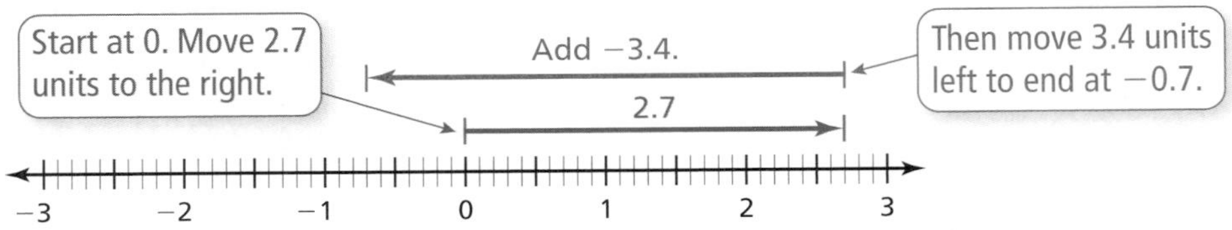

Start at 0. Move 2.7 units to the right.
Add −3.4.
2.7
Then move 3.4 units left to end at −0.7.
−3
−2
−1
0
1
2
3

So, $2.7 + (-3.4) = -0.7$.

b. $\frac{3}{10} + \left(-\frac{9}{10}\right)$

c. $-\frac{6}{10} - 1\frac{3}{10}$

d. $1.3 + (-3.4)$

e. $-1.9 - 0.8$

2 ACTIVITY: Adding and Subtracting Rational Numbers

Work with a partner. Write the numerical expression shown on the number line. Then find the sum or difference.

a.

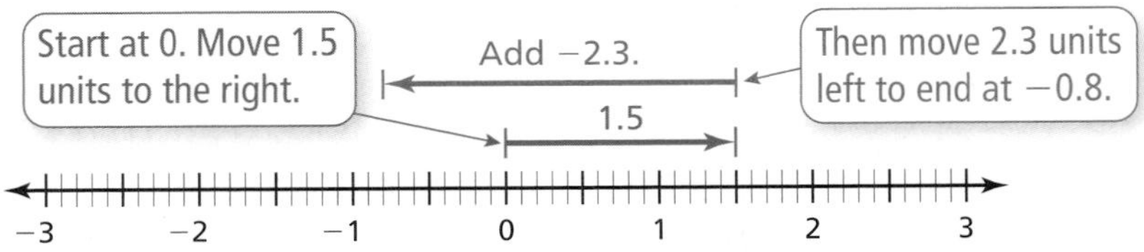

Start at 0. Move 1.5 units to the right.
Add −2.3.
1.5
Then move 2.3 units left to end at −0.8.
−3
−2
−1
0
1
2
3

b.

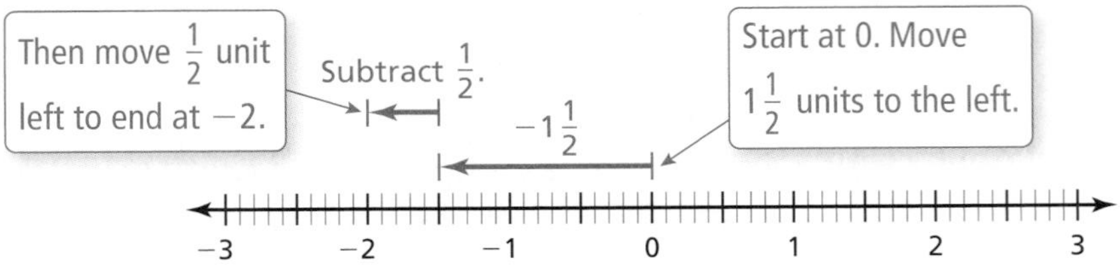

Then move $\frac{1}{2}$ unit left to end at −2.
Subtract $\frac{1}{2}$.
$-1\frac{1}{2}$
Start at 0. Move $1\frac{1}{2}$ units to the left.
−3
−2
−1
0
1
2
3

3 ACTIVITY: Financial Literacy

Work with a partner. The table shows the balance in a checkbook.

- Black numbers are amounts added to the account.
- Red numbers are amounts taken from the account.

	Date	Check #	Transaction	Amount	Balance
	--	--	Previous balance	--	100.00
➡	1/02/2009	124	Groceries	34.57	
	1/06/2009		Check deposit	875.50	
	1/11/2009		ATM withdrawal	40.00	
	1/14/2009	125	Electric company	78.43	
	1/17/2009		Music store	10.55	
	1/18/2009	126	Shoes	47.21	
	1/20/2009		Check deposit	125.00	
	1/21/2009		Interest	2.12	
➡	1/22/2009	127	Cell phone	59.99	

You can find the balance in the **second row** two different ways.

$100.00 - 34.57 = 65.43$ Subtract 34.57 from 100.00.

$100.00 + (-34.57) = 65.43$ Add −34.57 to 100.00.

a. Copy the table. Then complete the balance column.

b. How did you find the balance in the **tenth row**?

c. Use a different way to find the balance in part (b).

What Is Your Answer?

4. IN YOUR OWN WORDS How does adding and subtracting rational numbers compare with adding and subtracting integers? Give an example.

PUZZLE Find a path through the table so that the numbers add up to the sum. You can move horizontally or vertically.

5. Sum: $\frac{3}{4}$

Start → (top left); End ← (bottom right)

$\frac{1}{2}$	$\frac{2}{3}$	$-\frac{5}{7}$
$-\frac{1}{8}$	$-\frac{3}{4}$	$\frac{1}{3}$

6. Sum: -0.07

Start → (top left); End ← (bottom right)

2.43	1.75	−0.98
−1.09	3.47	−4.88

Practice Use what you learned about adding and subtracting rational numbers to complete Exercises 7–9 and 16–18 on page 90.

3.2 Lesson

Check It Out
Lesson Tutorials
BigIdeasMath.com

Key Idea

Adding and Subtracting Rational Numbers

Words To add or subtract rational numbers, use the same rules for signs as you used for integers.

Numbers $\frac{4}{5} - \frac{1}{5} = \frac{4-1}{5} = \frac{3}{5}$

$-\frac{1}{3} + \frac{1}{6} = \frac{-2}{6} + \frac{1}{6} = \frac{-2+1}{6} = \frac{-1}{6} = -\frac{1}{6}$

EXAMPLE 1 Adding Rational Numbers

Find $-\frac{8}{3} + \frac{5}{6}$. **Estimate $-3 + 1 = -2$**

$-\frac{8}{3} + \frac{5}{6} = \frac{-16}{6} + \frac{5}{6}$ Rewrite using the LCD (least common denominator).

$= \frac{-16+5}{6}$ Write the sum of the numerators over the like denominator.

$= \frac{-11}{6}$, or $-1\frac{5}{6}$ Simplify.

The sum is $-1\frac{5}{6}$. **Reasonable?** $-1\frac{5}{6} \approx -2$ ✓

Study Tip

In Example 1, notice how $-\frac{8}{3}$ is written as

$-\frac{8}{3} = \frac{-8}{3} = \frac{-16}{6}$.

EXAMPLE 2 Adding Rational Numbers

Find $-4.05 + 7.62$.

$-4.05 + 7.62 = 3.57$ $|7.62| > |-4.05|$. So, subtract $|-4.05|$ from $|7.62|$.

Use the sign of 7.62.

The sum is 3.57.

On Your Own

Now You're Ready
Exercises 4–12

Add.

1. $-\frac{7}{8} + \frac{1}{4}$
2. $-6\frac{1}{3} + \frac{20}{3}$
3. $2 + \left(-\frac{7}{2}\right)$
4. $-12.5 + 15.3$
5. $-8.15 + (-4.3)$
6. $0.65 + (-2.75)$

EXAMPLE 3 Subtracting Rational Numbers

Find $-4\frac{1}{7} - \left(-\frac{6}{7}\right)$. **Estimate $-4 - (-1) = -3$**

$$-4\frac{1}{7} - \left(-\frac{6}{7}\right) = -4\frac{1}{7} + \frac{6}{7}$$ Add the opposite of $-\frac{6}{7}$.

$$= -\frac{29}{7} + \frac{6}{7}$$ Write the mixed number as an improper fraction.

$$= \frac{-23}{7}, \text{ or } -3\frac{2}{7}$$ Simplify.

∴ The difference is $-3\frac{2}{7}$. **Reasonable? $-3\frac{2}{7} \approx -3$ ✓**

On Your Own

Subtract.

7. $\frac{1}{3} - \left(-\frac{1}{3}\right)$

8. $-3\frac{1}{3} - \frac{5}{6}$

9. $4\frac{1}{2} - 5\frac{1}{4}$

EXAMPLE 4 Real-Life Application

Clearance: 11 ft 8 in.

In the water, the bottom of a boat is 2.1 feet below the surface and the top of the boat is 8.7 feet above it. Towed on a trailer, the bottom of the boat is 1.3 feet above the ground. Can the boat and trailer pass under the bridge?

Step 1: Find the height h of the boat.

$h = 8.7 - (-2.1)$ Subtract the lowest point from the highest point.

$= 8.7 + 2.1$ Add the opposite of -2.1.

$= 10.8$ Add.

Step 2: Find the height t of the boat and trailer.

$t = 10.8 + 1.3$ Add the trailer height to the boat height.

$= 12.1$ Add.

∴ Because 12.1 feet is greater than 11 feet 8 inches, the boat and trailer cannot pass under the bridge.

On Your Own

Now You're Ready Exercises 13–21

10. WHAT IF? In Example 4, the clearance is 12 feet 1 inch. Can the boat and trailer pass under the bridge?

3.2 Exercises

Check It Out
Help with Homework
BigIdeasMath.com

Vocabulary and Concept Check

1. **WRITING** Explain how to find the sum $-8.46 + 5.31$.

2. **OPEN-ENDED** Write an addition expression using fractions that equals $-\frac{1}{2}$.

3. **DIFFERENT WORDS, SAME QUESTION** Which is different? Find "both" answers.

Add -4.8 and 3.9.	What is 3.9 less than -4.8?
What is -4.8 increased by 3.9?	Find the sum of -4.8 and 3.9.

Practice and Problem Solving

Add. Write fractions in simplest form.

1 2

4. $\frac{11}{12} + \left(-\frac{7}{12}\right)$
5. $-\frac{9}{14} + \frac{2}{7}$
6. $\frac{15}{4} + \left(-4\frac{1}{3}\right)$
7. $2\frac{5}{6} + \left(-\frac{8}{15}\right)$
8. $4 + \left(-1\frac{2}{3}\right)$
9. $-4.2 + 3.3$
10. $-3.1 + (-0.35)$
11. $12.48 + (-10.636)$
12. $20.25 + (-15.711)$

Subtract. Write fractions in simplest form.

3 4

13. $\frac{5}{8} - \left(-\frac{7}{8}\right)$
14. $\frac{1}{4} - \frac{11}{16}$
15. $-\frac{1}{2} - \left(-\frac{5}{9}\right)$
16. $-5 - \frac{5}{3}$
17. $-8\frac{3}{8} - 10\frac{1}{6}$
18. $-1 - 2.5$
19. $5.5 - 8.1$
20. $-7.34 - (-5.51)$
21. $6.673 - (-8.29)$

22. **ERROR ANALYSIS** Describe and correct the error in finding the difference.

✗ $\frac{3}{4} - \frac{9}{2} = \frac{3-9}{4-2} = \frac{-6}{2} = -3$

23. **SPORTS DRINK** Your sports drink bottle is $\frac{5}{6}$ full. After practice the bottle is $\frac{3}{8}$ full. Write the difference of the amounts after practice and before practice.

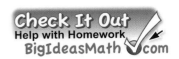

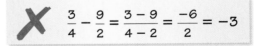

24. **BANKING** Your bank account balance is –\$20.85. You deposit \$15.50. What is your new balance?

Evaluate.

25. $2\frac{1}{6} - \left(-\frac{8}{3}\right) + \left(-4\frac{7}{9}\right)$
26. $6.3 + (-7.8) - (-2.41)$
27. $-\frac{12}{5} + \left|-\frac{13}{6}\right| + \left(-3\frac{2}{3}\right)$

28. REASONING When is the difference of two decimals an integer? Explain.

29. RECIPE A cook has $2\frac{2}{3}$ cups of flour. A recipe calls for $2\frac{3}{4}$ cups of flour. Does the cook have enough flour? If not, how much more flour is needed?

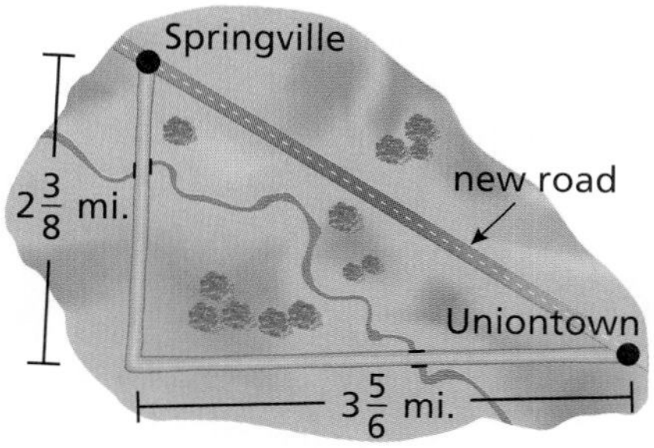

30. ROADWAY A new road that connects Uniontown to Springville is $4\frac{1}{3}$ miles long. What is the change in distance when using the new road instead of the dirt roads?

RAINFALL In Exercises 31–33, the bar graph shows the differences in a city's rainfall from the historical average.

31. What is the difference in rainfall between the wettest and driest months?

32. Find the sum of the differences for the year.

33. What does the sum in Exercise 32 tell you about the rainfall for the year?

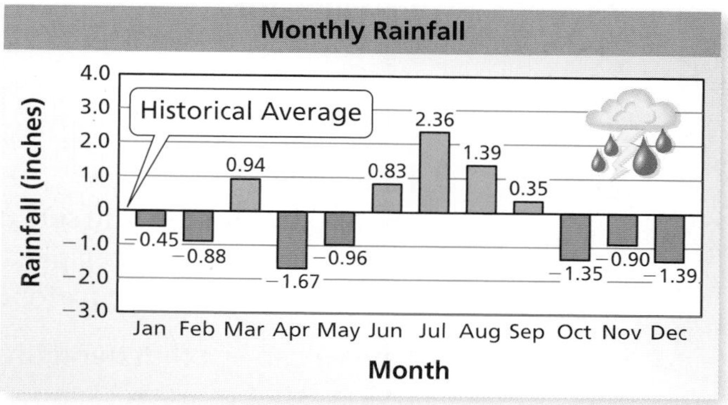

ALGEBRA Add or subtract. Write the answer in simplest form.

34. $-4x + 8x - 6x$

35. $-\frac{3n}{8} + \frac{2n}{8} - \frac{n}{8}$

36. $-4a - \frac{a}{3}$

37. $\frac{5b}{8} + \left(-\frac{2b}{3}\right)$

38. Puzzle Fill in the blanks to make the solution correct.

$$5.\square 4 - (\square.8\square) = -3.61$$

Fair Game Review What you learned in previous grades & lessons

Evaluate. *(Skills Review Handbook)*

39. 5.2×6.9

40. $7.2 \div 2.4$

41. $2\frac{2}{3} \times 3\frac{1}{4}$

42. $9\frac{4}{5} \div 3\frac{1}{2}$

43. MULTIPLE CHOICE A sports store has 116 soccer balls. Over 6 months, it sells eight soccer balls per month. How many soccer balls are in inventory at the end of the 6 months? *(Section 1.3 and Section 1.4)*

Ⓐ −48 Ⓑ 48 Ⓒ 68 Ⓓ 108

3.3 Multiplying and Dividing Rational Numbers

STANDARDS OF LEARNING
7.3

Essential Question How can you use operations with rational numbers in a story?

Share Your Work at... My.BigIdeasMath.com

LEMONADE
All You Can Drink For 50¢!

1 EXAMPLE: Writing a Story

Write a story that uses addition, subtraction, multiplication, or division of rational numbers. Draw pictures for your story.

There are many possible stories. Here is an example.

24 Lemons	-$11.75
5 cups sugar	-$1.50
30 plastic glasses	-$1.50
18 sales ($0.50 each)	$9.00
PROFIT	-$5.75

Lauryn decides to earn some extra money. She sets up a lemonade stand. To get customers, she uses big plastic glasses and makes a sign saying "All you can drink for 50¢!"

Lauryn can see that her daily profit is negative. But, she decides to keep trying. After one week, she has the same profit each day.

Sunday	Monday	Tuesday	Wednesday	Thursday	Friday	Saturday
-$5.75	-$5.75	-$5.75	-$5.75	-$5.75	-$5.75	-$5.75

Lauryn is frustrated. Her profit for the first week is

$$7(-5.75) = (-5.75) + (-5.75) + (-5.75) + (-5.75) + (-5.75) + (-5.75) + (-5.75)$$
$$= -40.25.$$

She realizes that she has too many customers who are drinking a second and even a third glass of lemonade. So, she decides to try a new strategy. Soon, she has a customer. He buys a glass of lemonade and drinks it.

He hands the empty glass to Lauryn and says "*That was great. I'll have another glass.*" Today, Lauryn says "*That will be 50¢ more, please.*" The man says "*But, you only gave me one glass and the sign says 'All you can drink for 50¢!'*" Lauryn replies, "*One glass IS all you can drink for 50¢.*"

With her new sales strategy, Lauryn starts making a profit of $8.25 per day. Her profit for the second week is

$$7(8.25) = (8.25) + (8.25) + (8.25) + (8.25) + (8.25) + (8.25) + (8.25) = 57.75.$$

Her profit for the two weeks is $-40.25 + 57.75 =$ $17.50. So, Lauryn has made some money. She decides that she is on the right track.

2 ACTIVITY: Writing a Story

Work with a partner. Write a story that uses addition, subtraction, multiplication, or division of rational numbers.

- At least one of the numbers in the story has to be negative and *not* an integer.
- Draw pictures to help illustrate what is happening in the story.
- Include the solution of the problem in the story.

If you are having trouble thinking of a story, here are some common uses of negative numbers.

- A profit of −$15 is a loss of $15.
- An elevation of −100 feet is a depth of 100 feet below sea level.
- A gain of −5 yards in football is a loss of 5 yards.
- A score of −4 in golf is 4 strokes under par.
- A balance of −$25 in your checking account means the account is overdrawn by $25.

What Is Your Answer?

3. **IN YOUR OWN WORDS** How can you use operations with rational numbers in a story? You already used rational numbers in your story. Describe another use of a negative rational number in a story.

PUZZLE Read the cartoon. Fill in the blanks using 4's or 8's to make the equation true.

"Dear Mom, I'm in a hurry. To save time I won't be typing any 4's or 8's."

4. $\left(-\frac{1}{\square}\right) + \left(-\frac{1}{\square}\right) = -\frac{1}{\square}$

5. $\left(-\frac{1}{\square}\right) \times \left(-\frac{1}{\square}\right) = \frac{1}{6\square}$

6. $1.\square \times \left(-0.\square\right) = -1.\square\square$

7. $\left(-\frac{3}{\square}\right) \div \left(\frac{3}{\square}\right) = -\frac{1}{2}$

8. $-4.\square \div 2 = -2.\square$

3.3 Lesson

Check It Out
Lesson Tutorials
BigIdeasMath.com

Key Idea

Remember

The *reciprocal* of $\frac{a}{b}$ is $\frac{b}{a}$.

Multiplying and Dividing Rational Numbers

Words To multiply or divide rational numbers, use the same rules for signs as you used for integers.

Numbers $-\frac{2}{7} \cdot \frac{1}{3} = \frac{-2 \cdot 1}{7 \cdot 3} = \frac{-2}{21} = -\frac{2}{21}$

$$-\frac{1}{2} \div \frac{4}{9} = \frac{-1}{2} \cdot \frac{9}{4} = \frac{-1 \cdot 9}{2 \cdot 4} = \frac{-9}{8} = -\frac{9}{8}$$

EXAMPLE 1 Dividing Rational Numbers

Find $-5\frac{1}{5} \div 2\frac{1}{3}$. **Estimate** $-5 \div 2 = -2\frac{1}{2}$

$-5\frac{1}{5} \div 2\frac{1}{3} = -\frac{26}{5} \div \frac{7}{3}$ Write mixed numbers as improper fractions.

$= \frac{-26}{5} \cdot \frac{3}{7}$ Multiply by the reciprocal of $\frac{7}{3}$.

$= \frac{-26 \cdot 3}{5 \cdot 7}$ Multiply the numerators and the denominators.

$= \frac{-78}{35}$, or $-2\frac{8}{35}$ Simplify.

∴ The quotient is $-2\frac{8}{35}$. **Reasonable?** $-2\frac{8}{35} \approx -2\frac{1}{2}$ ✓

EXAMPLE 2 Multiplying Rational Numbers

Find $-2.5 \cdot 3.6$.

$$\begin{array}{r} -2.5 \\ \times\ 3.6 \\ \hline 150 \\ 750 \\ \hline -9.00 \end{array}$$

The decimals have different signs.

The product is negative.

∴ The product is -9.

EXAMPLE 3 Standardized Test Practice

Which number, when multiplied by $-\frac{5}{3}$, gives a product between 5 and 6?

Ⓐ -6 Ⓑ $-3\frac{1}{4}$ Ⓒ $-\frac{1}{4}$ Ⓓ 3

Use the guess, check, and revise method.

Guess 1: Because the product is positive and the known factor is negative, choose a number that is negative. Try Choice Ⓒ.

$$-\frac{1}{4}\left(-\frac{5}{3}\right) = \frac{-1 \cdot (-5)}{4 \cdot 3} = \frac{5}{12}$$

Guess 2: The result of Choice Ⓒ is not between 5 and 6. So, choose another number that is negative. Try Choice Ⓑ.

$$-3\frac{1}{4}\left(-\frac{5}{3}\right) = -\frac{13}{4}\left(-\frac{5}{3}\right) = \frac{-13 \cdot (-5)}{4 \cdot 3} = \frac{65}{12} = 5\frac{5}{12}$$

∴ $5\frac{5}{12}$ is between 5 and 6. So, the correct answer is Ⓑ.

On Your Own

Now You're Ready
Exercises 10–33

Multiply or divide.

1. $-\frac{6}{5} \div \left(-\frac{1}{2}\right)$
2. $\frac{1}{3} \div \left(-2\frac{2}{3}\right)$
3. $\left(-\frac{1}{2}\right)^3$
4. $1.8(-5.1)$
5. $-6.3(-0.6)$
6. $(-1.3)^2$

EXAMPLE 4 Real-Life Application

Account Positions

Stock	Original Value	Current Value	Change
A	600.54	420.15	−180.39
B	391.10	518.38	127.28
C	380.22	99.70	−280.52

An investor owns stocks A, B, and C. What is the mean change in value of the stocks?

$$\text{mean} = \frac{-180.39 + 127.28 + (-280.52)}{3} = \frac{-333.63}{3} = -111.21$$

∴ The mean change in value of the stocks is −\$111.21.

On Your Own

7. In Example 4, the change in value of stock D is \$568.23. What is the mean change in value of the four stocks?

3.3 Exercises

Check It Out
Help with Homework
BigIdeasMath.com

Vocabulary and Concept Check

1. **WRITING** How is multiplying and dividing rational numbers similar to multiplying and dividing integers?

Find the reciprocal.

2. $-\frac{2}{5}$ 3. -3 4. $\frac{16}{9}$ 5. $-2\frac{1}{3}$

Tell whether the expression is *positive* or *negative* without evaluating.

6. $-\frac{3}{10} \times \left(-\frac{8}{15}\right)$ 7. $1\frac{1}{2} \div \left(-\frac{1}{4}\right)$ 8. -6.2×8.18 9. $\frac{-8.16}{-2.72}$

Practice and Problem Solving

Divide. Write fractions in simplest form.

(1) 10. $-\frac{7}{10} \div \frac{2}{5}$ 11. $\frac{1}{4} \div \left(-\frac{3}{8}\right)$ 12. $-\frac{8}{9} \div \left(-\frac{8}{9}\right)$ 13. $-\frac{1}{5} \div 20$

14. $-2\frac{4}{5} \div (-7)$ 15. $-10\frac{2}{7} \div \left(-4\frac{4}{11}\right)$ 16. $-9 \div 7.2$ 17. $8 \div 2.2$

18. $-3.45 \div (-15)$ 19. $-0.18 \div 0.03$ 20. $8.722 \div (-3.56)$ 21. $12.42 \div (-4.8)$

Multiply. Write fractions in simplest form.

(2) (3) 22. $-\frac{2}{3} \times \frac{2}{9}$ 23. $-\frac{1}{4} \times \left(-\frac{4}{3}\right)$ 24. $\frac{5}{6}\left(-\frac{8}{15}\right)$ 25. $-2\left(-1\frac{1}{4}\right)$

26. $-3\frac{1}{3} \cdot \left(-2\frac{7}{10}\right)$ 27. $\left(-1\frac{2}{3}\right)^3$ 28. $0.4 \times (-0.03)$ 29. $-0.05 \times (-0.5)$

30. $-8(0.09)$ 31. $-9.3 \cdot (-5.1)$ 32. $-95.2 \cdot (-0.12)$ 33. $(-0.4)^3$

ERROR ANALYSIS Describe and correct the error.

34. ✗ $-2.2 \times 3.7 = 8.14$

35. ✗ $-\frac{1}{4} \div \frac{3}{2} = -\frac{4}{1} \times \frac{3}{2} = -\frac{12}{2} = -6$

36. **HOUR HAND** The hour hand of a clock moves $-30°$ every hour. How many degrees does it move in $2\frac{1}{5}$ hours?

37. **SUNFLOWER SEEDS** How many 0.75-pound packages can be made with 6 pounds of sunflower seeds?

Evaluate.

38. $-4.2 + 8.1 \times (-1.9)$

39. $2.85 - 6.2 \div 2^2$

40. $-3.64 \cdot |-5.3| - 1.5^3$

41. $1\frac{5}{9} \div \left(-\frac{2}{3}\right) + \left(-2\frac{3}{5}\right)$

42. $-3\frac{3}{4} \times \frac{5}{6} - 2\frac{1}{3}$

43. $\left(-\frac{2}{3}\right)^2 - \frac{3}{4}\left(2\frac{1}{3}\right)$

44. OPEN-ENDED Write two fractions whose product is $-\frac{3}{5}$.

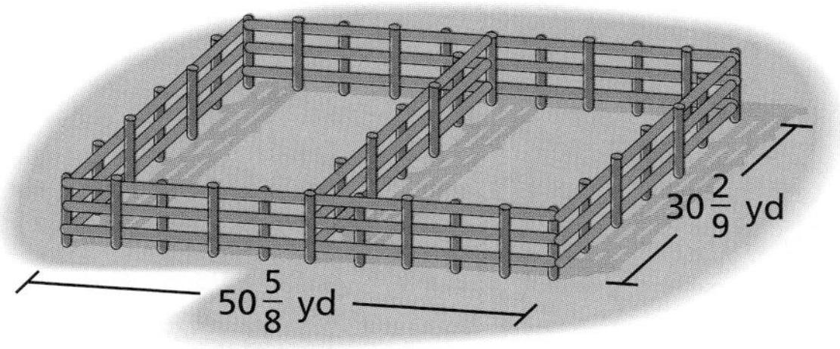

45. FENCING A farmer needs to enclose two adjacent rectangular pastures. How much fencing does the farmer need?

46. GASOLINE A 14.5-gallon gasoline tank is $\frac{3}{4}$ full. How many gallons will it take to fill the tank?

47. BOARDWALK A section of a boardwalk is made using 15 boards. Each board is $9\frac{1}{4}$ inches wide. The total width of the section is 144 inches. The spacing between each board is equal. What is the width of the spacing between each board?

48. RUNNING The table shows the changes in the times (in seconds) of four teammates. What is the mean change?

Teammate	Change
1	−2.43
2	−1.85
3	0.61
4	−1.45

49. Critical Thinking The daily changes in the barometric pressure for four days are −0.05, 0.09, −0.04, and −0.08 inches.

a. What is the mean change?

b. The mean change after five days is −0.01 inch. What is the change on the fifth day? Explain.

Fair Game Review What you learned in previous grades & lessons

Add or subtract. *(Section 3.2)*

50. $-6.2 + 4.7$

51. $-8.1 - (-2.7)$

52. $\frac{9}{5} - \left(-2\frac{7}{10}\right)$

53. $-4\frac{5}{6} + \left(-3\frac{4}{9}\right)$

54. MULTIPLE CHOICE What are the coordinates of the point in Quadrant IV? *(Skills Review Handbook)*

Ⓐ (−4, 1) Ⓑ (−3, −3)

Ⓒ (0, −2) Ⓓ (3, −3)

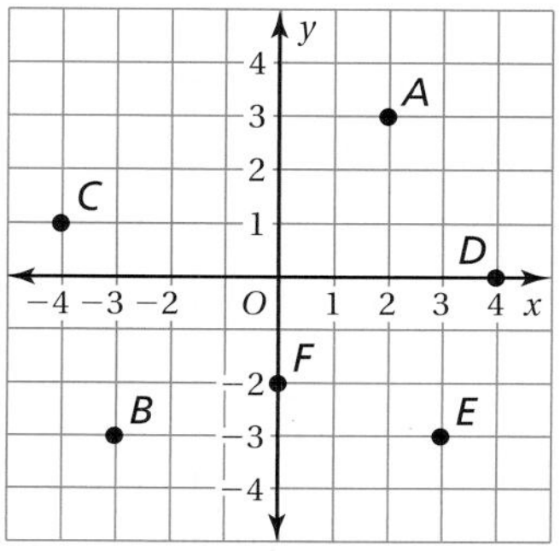

3 Study Help

Check It Out
Graphic Organizer
BigIdeasMath.com

You can use a **process diagram** to show the steps involved in a procedure. Here is an example of a process diagram for adding rational numbers.

Adding rational numbers

with the same sign

Add the absolute values of the rational numbers.

Write the sum using the common sign.

Example

$-5.5 + (-6.9)$

Because the numbers have the same sign, add $|-5.5|$ and $|-6.9|$.

$-5.5 + (-6.9) = -12.4$

Use the common sign.

with different signs

Subtract the lesser absolute value from the greater absolute value.

Write the sum using the sign of the rational number with the greater absolute value.

Example

$-5.5 + 6.9$

Because the numbers have different signs, subtract $|-5.5|$ from $|6.9|$.

$-5.5 + 6.9 = 1.4$

Use the sign of 6.9.

On Your Own

Make a process diagram with examples to help you study these topics. Your process diagram can have one or more branches.

1. writing rational numbers as decimals
2. subtracting rational numbers
3. dividing rational numbers

After you complete this chapter, make process diagrams with examples for the following topics.

4. solving equations using addition, subtraction, multiplication, or division
5. solving two-step equations
6. solving inequalities using addition, subtraction, multiplication, or division

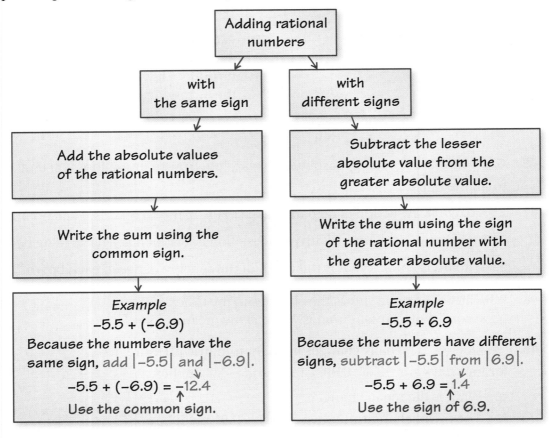

"Does this process diagram accurately show how a cat claws furniture?"

3.1–3.3 Quiz

Check It Out
Progress Check
BigIdeasMath.com

Write the rational number as a decimal. *(Section 3.1)*

1. $-\frac{3}{20}$

2. $-\frac{11}{6}$

Write the decimal as a fraction or mixed number in simplest form. *(Section 3.1)*

3. -0.325

4. -1.28

Add or subtract. Write fractions in simplest form. *(Section 3.2)*

5. $-\frac{4}{5} + \left(-\frac{3}{8}\right)$

6. $-5.8 + 2.6$

7. $\frac{12}{7} - \left(-\frac{2}{9}\right)$

8. $9.1 - 12.9$

Multiply or divide. Write fractions in simplest form. *(Section 3.3)*

9. $-2\frac{3}{8} \times \frac{8}{5}$

10. $-9.4 \times (-4.7)$

11. $-8\frac{5}{9} \div \left(-1\frac{4}{7}\right)$

12. $-8.4 \div 2.1$

13. STOCK The value of stock A changes $-\$3.68$ and the value of stock B changes $-\$3.72$. Which stock has the greater loss? Explain. *(Section 3.1)*

14. PARASAILING A parasail is at 200.6 feet above the water. After five minutes, the parasail is at 120.8 feet above the water. What is the change in height of the parasail? *(Section 3.2)*

15. FOOTBALL The table shows the statistics of a running back in a football game. How many total yards did he gain? *(Section 3.2)*

Quarter	1	2	3	4	Total
Yards	$-8\frac{1}{2}$	23	$42\frac{1}{2}$	$-2\frac{1}{4}$	?

16. LATE FEES You were overcharged \$4.52 on your cell phone bill three months in a row. The cell phone company will add $-\$4.52$ to your next bill for each month you were overcharged. How much will be added to your next bill? *(Section 3.3)*

3.4 Solving Equations Using Addition or Subtraction

STANDARDS OF LEARNING
7.14

Essential Question How can you use inverse operations to solve an equation?

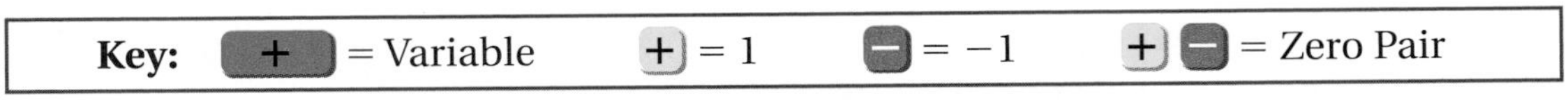

1 EXAMPLE: Using Addition to Solve an Equation

Use algebra tiles to model and solve $x - 3 = -4$.

Model the equation $x - 3 = -4$.

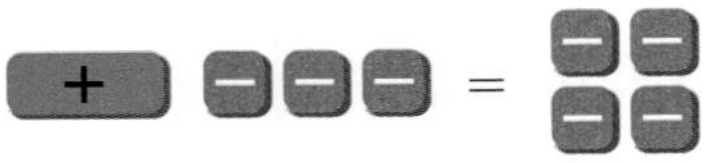

To get the green tile by itself, remove the red tiles on the left side by adding three yellow tiles to each side.

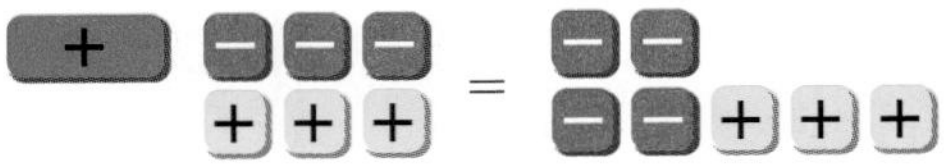

Remove the three "zero pairs" from each side.

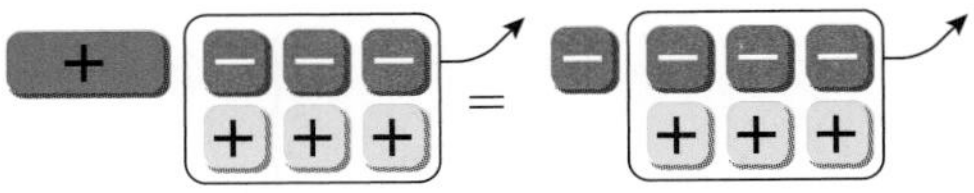

The remaining tile shows the value of x.

So, $x = -1$.

2 EXAMPLE: Using Addition to Solve an Equation

Use algebra tiles to model and solve $-5 = n + 2$.

Model the equation $-5 = n + 2$.

Remove the yellow tiles on the right side by adding two red tiles to each side.

Remove the two "zero pairs" from the right side.

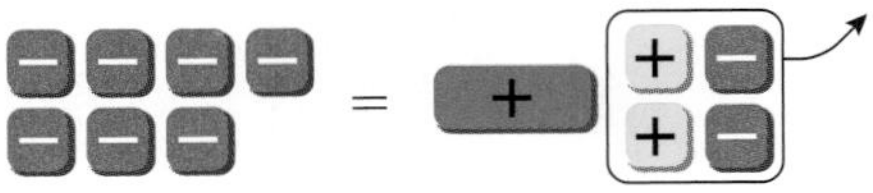

The remaining tiles show the value of n.

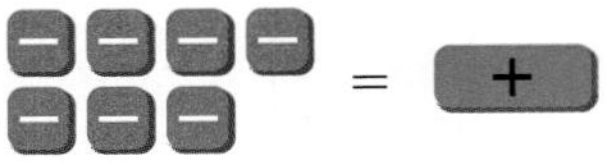

So, $-7 = n$ or $n = -7$.

3 ACTIVITY: Solving Equations Using Algebra Tiles

Work with a partner. Use algebra tiles to model and solve the equation.

a. $y + 10 = -5$

b. $p - 7 = -3$

c. $-15 = t - 5$

d. $8 = 12 + z$

4 ACTIVITY: Writing and Solving Equations

Work with a partner. Write an equation shown by the algebra tiles. Then solve.

a.

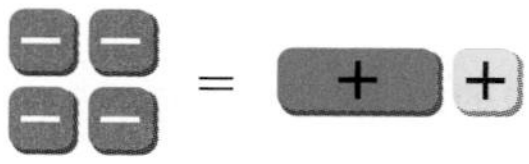

b.

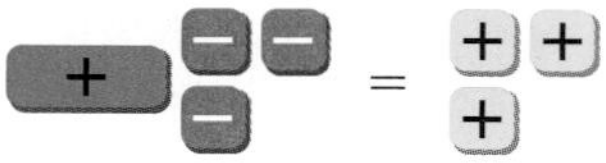

c.

d.

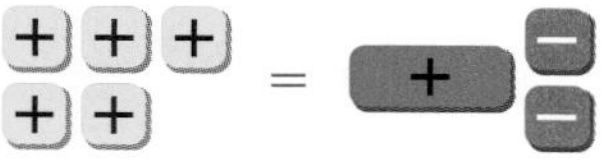

What Is Your Answer?

5. Decide whether the statement is *true* or *false*. Explain your reasoning.
   - **a.** In an equation, any letter can be used as a variable.
   - **b.** The goal in solving an equation is to get the variable by itself.
   - **c.** In the solution, the variable always has to be on the left side of the equal sign.
   - **d.** If you add a number to one side, you should add it to the other side.

6. **IN YOUR OWN WORDS** How can you use inverse operations to solve an equation without algebra tiles? Give two examples.

7. What makes the cartoon funny?

8. The word *variable* comes from the word *vary*. For example, the temperature in Maine varies a lot from winter to summer.

   Write two other English sentences that use the word *vary*.

"To vary or not to vary." That is the question.

"Dear Sir: Yesterday you said $x = 2$. Today you are saying $x = 3$. Please make up your mind."

Practice Use what you learned about solving equations using inverse operations to complete Exercises 5–8 on page 104.

3.4 Lesson

Check It Out
Lesson Tutorials
BigIdeasMath.com

Key Vocabulary
equivalent equations, p. 102

Key Ideas

Addition Property of Equality

Words Two equations are **equivalent equations** if they have the same solutions. Adding the same number to each side of an equation produces an equivalent equation.

Algebra If $a = b$, then $a + c = b + c$.

Subtraction Property of Equality

Words Subtracting the same number from each side of an equation produces an equivalent equation.

Algebra If $a = b$, then $a - c = b - c$.

EXAMPLE 1 Solving Equations

a. Solve $x - 5 = -1$.

$x - 5 = -1$ Write the equation.

Undo the subtraction. → $\underline{+5} \quad \underline{+5}$ Add 5 to each side.

$x = 4$ Simplify.

So, the solution is $x = 4$.

Check

$x - 5 = -1$

$4 - 5 \stackrel{?}{=} -1$

$-1 = -1$ ✓

b. Solve $z + \frac{3}{2} = \frac{1}{2}$.

$z + \frac{3}{2} = \frac{1}{2}$ Write the equation.

Undo the addition. → $\underline{-\frac{3}{2}} \quad \underline{-\frac{3}{2}}$ Subtract $\frac{3}{2}$ from each side.

$z = -1$ Simplify.

So, the solution is $z = -1$.

Check

$z + \frac{3}{2} = \frac{1}{2}$

$-1 + \frac{3}{2} \stackrel{?}{=} \frac{1}{2}$

$\frac{1}{2} = \frac{1}{2}$ ✓

On Your Own

Now You're Ready
Exercises 5–20

Solve the equation. Check your solution.

1. $p - 5 = -2$
2. $w + 13.2 = 10.4$
3. $x - \frac{5}{6} = -\frac{1}{6}$

EXAMPLE 2 Standardized Test Practice

A company has a profit of \$750 this week. This profit is \$900 more than the profit P last week. Which equation can be used to find P?

Ⓐ $750 = 900 - P$ Ⓑ $750 = P + 900$

Ⓒ $900 = P - 750$ Ⓓ $900 = P + 750$

Words The profit this week is \$900 more than the profit last week.

Equation $750 \quad = \quad P \quad + \quad 900$

∴ The equation is $750 = P + 900$. The correct answer is Ⓑ.

On Your Own

Now You're Ready
Exercises 22–25

4. A company has a profit of \$120.50 today. This profit is \$145.25 less than the profit P yesterday. Write an equation that can be used to find P.

EXAMPLE 3 Real-Life Application

The line graph shows the scoring while you and your friend played a video game. Write and solve an equation to find your score after Level 4.

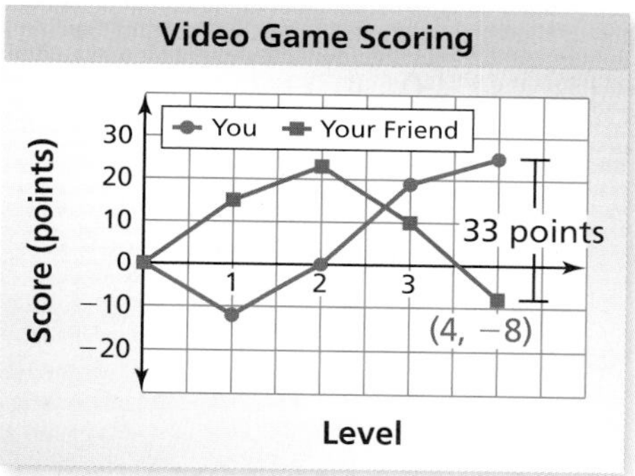

You can determine the following from the graph.

Words Your friend's score is 33 points less than your score.

Variable Let s be your score after Level 4.

Equation $-8 \quad = \quad s \quad - \quad 33$

$-8 = s - 33$ Write equation.

$\underline{+33} \quad \underline{+33}$ Add 33 to each side.

$25 = s$ Simplify.

∴ Your score after Level 4 is 25 points.

Reasonable? From the graph, your score after Level 4 is between 20 points and 30 points. So, 25 points is a reasonable answer.

On Your Own

5. **WHAT IF?** In Example 3, you have -12 points after Level 1. Your score is 27 points less than your friend's score. What is your friend's score?

3.4 Exercises

Check It Out
Help with Homework
BigIdeasMath.com

Vocabulary and Concept Check

1. **VOCABULARY** What property would you use to solve $m + 6 = -4$?
2. **VOCABULARY** Name two inverse operations.
3. **WRITING** Are the equations $m + 3 = -5$ and $m = -2$ equivalent? Explain.
4. **WHICH ONE DOESN'T BELONG?** Which equation does *not* belong with the other three? Explain your reasoning.

$x + 3 = -1$	$x + 1 = -5$	$x - 2 = -6$	$x - 9 = -13$

Practice and Problem Solving

Solve the equation. Check your solution.

1 5. $a - 6 = 13$
6. $-3 = z - 8$
7. $-14 = k + 6$
8. $x + 4 = -14$
9. $c - 7.6 = -4$
10. $-10.1 = w + 5.3$
11. $\frac{1}{2} = q + \frac{2}{3}$
12. $p - 3\frac{1}{6} = -2\frac{1}{2}$
13. $g - 9 = -19$
14. $-9.3 = d - 3.4$
15. $4.58 + y = 2.5$
16. $x - 5.2 = -18.73$
17. $q + \frac{5}{9} = \frac{1}{6}$
18. $-2\frac{1}{4} = r - \frac{4}{5}$
19. $w + 3\frac{3}{8} = 1\frac{5}{6}$
20. $4\frac{2}{5} + k = -3\frac{2}{11}$

21. **ERROR ANALYSIS** Describe and correct the error in finding the solution.

$$\begin{aligned} x + 8 &= 10 \\ +8 &\quad +8 \\ \hline x &= 18 \end{aligned}$$

Write the word sentence as an equation. Then solve.

2 22. 4 less than a number n is -15.
23. 10 more than a number c is 3.
24. The sum of a number y and -3 is -8.
25. The difference between a number p and 6 is -14.

In Exercises 26–28, write an equation. Then solve.

26. **DRY ICE** The temperature of dry ice is -109.3°F. This is 184.9°F less than the outside temperature. What is the outside temperature?
27. **PROFIT** A company makes a profit of \$1.38 million. This is \$2.54 million more than last year. What was the profit last year?
28. **HELICOPTER** The difference in elevation of a helicopter and a submarine is $18\frac{1}{2}$ meters. The elevation of the submarine is $-7\frac{3}{4}$ meters. What is the elevation of the helicopter?

GEOMETRY Write and solve an equation to find the unknown side length.

29. Perimeter = 12 cm

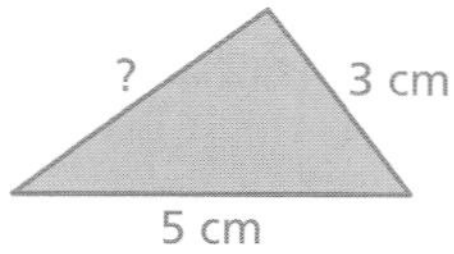

30. Perimeter = 24.2 in.

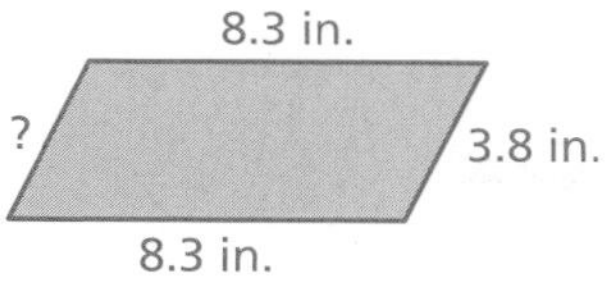

31. Perimeter = 34.6 ft

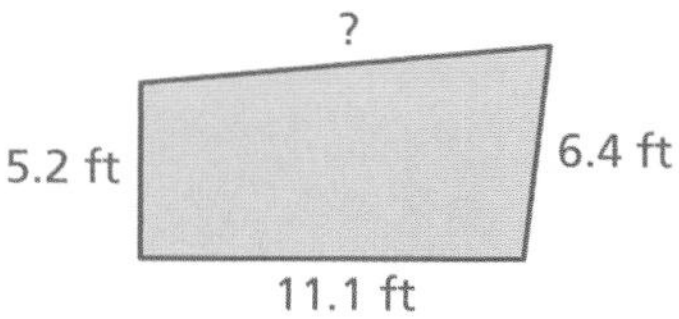

In Exercises 32–36, write an equation. Then solve.

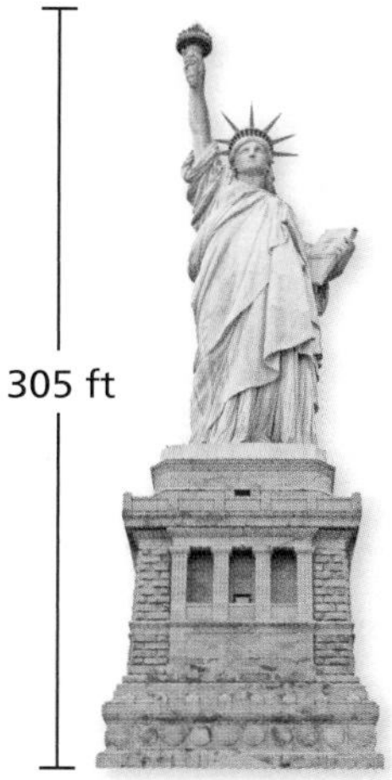

32. STATUE OF LIBERTY The total height of the Statue of Liberty and its pedestal is 153 feet more than the height of the statue. What is the height of the statue?

33. BUNGEE JUMPING Your first jump is $50\frac{1}{6}$ feet higher than your second jump. Your first jump reaches $-200\frac{2}{5}$ feet. What is the height of your second jump?

34. TRAVEL Boatesville is $65\frac{3}{5}$ kilometers from Stanton. A bus traveling from Stanton is $24\frac{1}{3}$ kilometers from Boatesville. How far has the bus traveled?

35. GEOMETRY The sum of the measures of the angles of a triangle equals 180°. What is the measure of the missing angle?

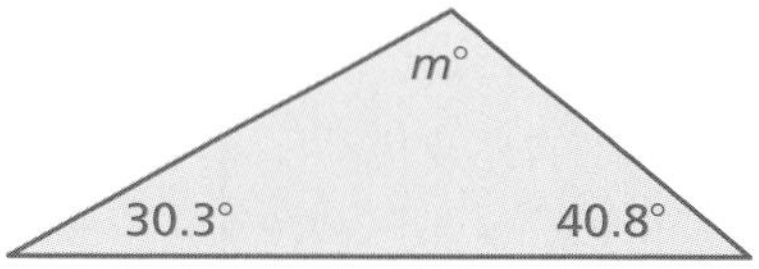

36. SKATEBOARDING The table shows your scores in a skateboarding competition. The leader has 311.62 points. What score do you need in the fourth round to win?

Round	1	2	3	4
Points	63.43	87.15	81.96	?

37. CRITICAL THINKING Find the value of $2x - 1$ when $x + 6 = 2$.

Critical Thinking **Find the values of x.**

38. $|x| = 2$

39. $|x| - 2 = 4$

40. $|x| + 5 = 18$

Fair Game Review What you learned in previous grades & lessons

Multiply or divide. *(Section 1.4 and Section 1.5)*

41. -7×8

42. $6 \times (-12)$

43. $18 \div (-2)$

44. $-26 \div 4$

45. MULTIPLE CHOICE A class of 144 students voted for a class president. Three-fourths of the students voted for you. Of the students who voted for you, $\frac{5}{9}$ are female. How many female students voted for you? *(Section 3.3)*

Ⓐ 50 Ⓑ 60 Ⓒ 80 Ⓓ 108

3.5 Solving Equations Using Multiplication or Division

Essential Question How can you use multiplication or division to solve an equation?

STANDARDS OF LEARNING
7.14
7.16

1 ACTIVITY: Using Division to Solve an Equation

Work with a partner. Use algebra tiles to model and solve the equation.

a. Sample: $3x = -12$

Model the equation $3x = -12$.

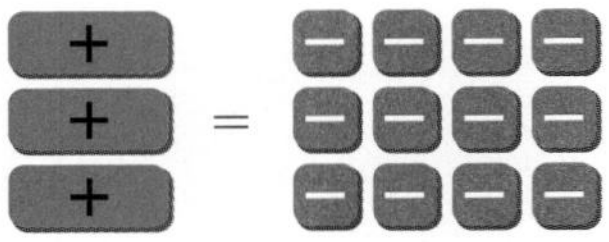

Your goal is to get one green tile by itself. Because there are three green tiles, divide the red tiles into three equal groups.

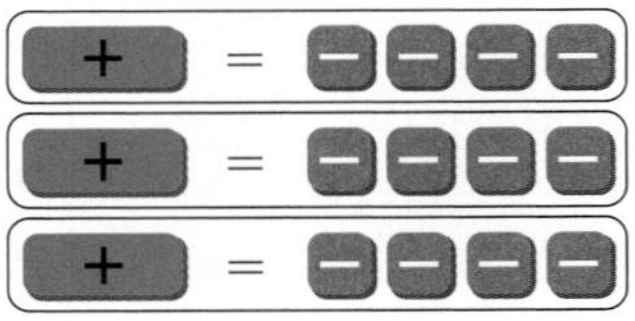

Keep one of the groups. This shows the value of x.

So, $x = -4$.

b. $2k = -8$

c. $-15 = 3t$

d. $-20 = 5m$

e. $4h = -16$

2 ACTIVITY: Writing and Solving Equations

Work with a partner. Write an equation shown by the algebra tiles. Then solve.

a.

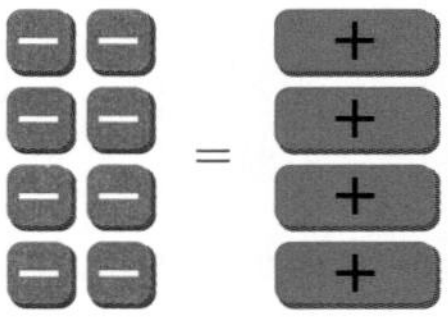

b.

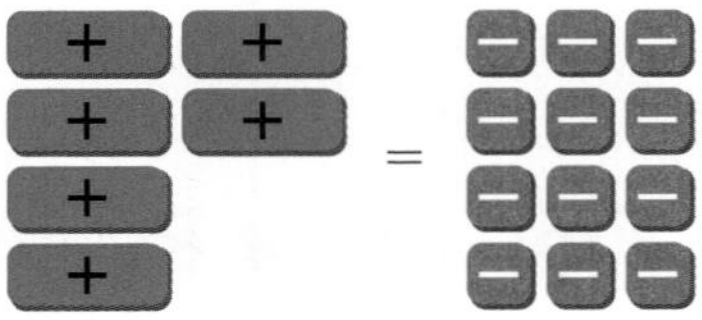

c.

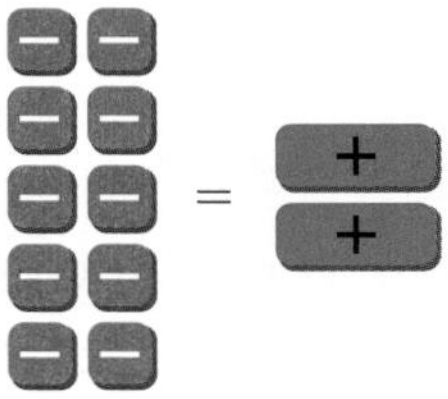

d.

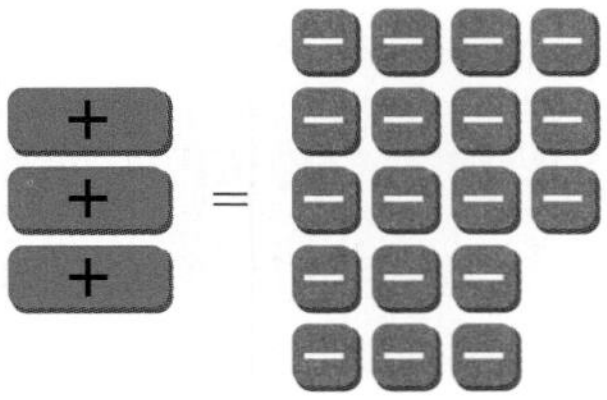

3 ACTIVITY: The Game of Math Card War

Preparation:

- Cut index cards to make 40 playing cards.
- Write each equation in the table on a card.

To Play:

- Play with a partner. Deal 20 cards to each player face-down.
- Each player turns one card face-up. The player with the greater solution wins. The winner collects both cards and places them at the bottom of his or her cards.
- Suppose there is a tie. Each player lays three cards face-down, then a new card face-up. The player with the greater solution of these new cards wins. The winner collects all ten cards, and places them at the bottom of his or her cards.
- Continue playing until one player has all the cards. This player wins the game.

$-4x = -12$	$x - 1 = 1$	$x - 3 = 1$	$2x = -10$	$-9 = 9x$
$3 + x = -2$	$x = -2$	$-3x = -3$	$\frac{x}{-2} = -2$	$x = -6$
$6x = -36$	$-3x = -9$	$-7x = -14$	$x - 2 = 1$	$-1 = x + 5$
$x = -1$	$9x = -27$	$\frac{x}{3} = -1$	$-8 = -2x$	$x = 3$
$-7 = -1 + x$	$x = -5$	$-10 = 10x$	$x = -4$	$-2 = -3 + x$
$-20 = 10x$	$x + 9 = 8$	$-16 = 8x$	$x = 2$	$x + 13 = 11$
$x = -3$	$-8 = 2x$	$x = 1$	$\frac{x}{2} = -2$	$-4 + x = -2$
$\frac{x}{5} = -1$	$-6 = x - 3$	$x = 4$	$x + 6 = 2$	$x - 5 = -4$

What Is Your Answer?

4. **IN YOUR OWN WORDS** How can you use multiplication or division to solve an equation without using algebra tiles? Give two examples.

Practice Use what you learned about solving equations to complete Exercises 7–10 on page 110.

Key Ideas

Multiplication Property of Equality

Words Multiplying each side of an equation by the same number produces an equivalent equation.

Algebra If $a = b$, then $a \cdot c = b \cdot c$.

Division Property of Equality

Words Dividing each side of an equation by the same number produces an equivalent equation.

Algebra If $a = b$, then $a \div c = b \div c$, $c \neq 0$.

EXAMPLE 1 Solving Equations

a. Solve $\frac{x}{3} = -6$.

$\frac{x}{3} = -6$ Write the equation.

Undo the division. → $3 \cdot \frac{x}{3} = 3 \cdot (-6)$ Multiply each side by 3.

$x = -18$ Simplify.

So, the solution is $x = -18$.

Check

$\frac{x}{3} = -6$

$\frac{-18}{3} \stackrel{?}{=} -6$

$-6 = -6$ ✓

b. Solve $18 = -4y$.

$18 = -4y$ Write the equation.

Undo the multiplication. → $\frac{18}{-4} = \frac{-4y}{-4}$ Divide each side by –4.

$-4.5 = y$ Simplify.

So, the solution is $y = -4.5$.

Check

$18 = -4y$

$18 \stackrel{?}{=} -4(-4.5)$

$18 = 18$ ✓

On Your Own

Now You're Ready
Exercises 7–18

Solve the equation. Check your solution.

1. $\frac{x}{5} = -2$
2. $-a = -24$
3. $3 = -1.5n$

EXAMPLE 2 Solving an Equation Using a Reciprocal

Study Tip

When you use a reciprocal to solve an equation, you are using the Multiplicative Inverse Property to make the coefficient of the variable 1.

Solve $-\frac{4}{5}x = -8$.

$-\frac{4}{5}x = -8$	Write the equation.
$-\frac{5}{4} \cdot \left(-\frac{4}{5}x\right) = -\frac{5}{4} \cdot (-8)$	Multiply each side by $-\frac{5}{4}$, the reciprocal of $-\frac{4}{5}$.
$x = 10$	Simplify.

So, the solution is $x = 10$.

On Your Own

Now You're Ready Exercises 19–22

Solve the equation. Check your solution.

4. $-14 = \frac{2}{3}x$ **5.** $-\frac{8}{5}b = 5$ **6.** $\frac{3}{8}h = -9$

EXAMPLE 3 Real-Life Application

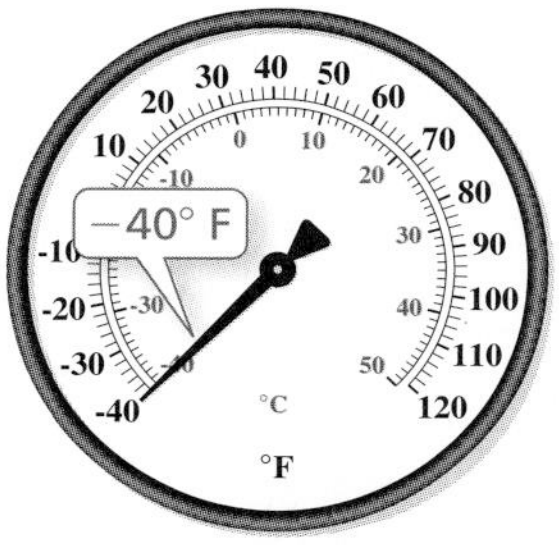

Record low temperature in Arizona

The record low temperature in Arizona is 1.6 times the record low temperature in Rhode Island. What is the record low temperature in Rhode Island?

Words The record low in Arizona is 1.6 times the record low in Rhode Island.

Variable Let t be the record low in Rhode Island.

Equation $-40 \quad = 1.6 \quad \times \quad t$

$-40 = 1.6t$	Write equation.
$\frac{-40}{1.6} = \frac{1.6t}{1.6}$	Divide each side by 1.6.
$-25 = t$	Simplify.

The record low temperature in Rhode Island is −25°F.

On Your Own

7. The record low temperature in Hawaii is –0.15 times the record low temperature in Alaska. The record low temperature in Hawaii is 12°F. What is the record low temperature in Alaska?

3.5 Exercises

Check It Out
Help with Homework
BigIdeasMath.com

Vocabulary and Concept Check

1. **WRITING** Explain why multiplication can be used to solve equations involving division.
2. **OPEN-ENDED** Turning a light on and then turning the light off are considered to be inverse operations. Describe two other real-life situations that can be thought of as inverse operations.

Describe the inverse operation that will undo the given operation.

3. Multiplying by 5
4. Subtracting 12
5. Dividing by -8
6. Adding -6

Practice and Problem Solving

Solve the equation. Check your solution.

1 7. $3h = 15$
8. $-5t = -45$
9. $\frac{n}{2} = -7$
10. $\frac{k}{-3} = 9$
11. $5m = -10$
12. $8t = -32$
13. $-0.2x = 1.6$
14. $-10 = -\frac{b}{4}$
15. $-6p = 48$
16. $-72 = 8d$
17. $\frac{n}{1.6} = 5$
18. $-14.4 = -0.6p$

2 19. $\frac{3}{4}g = -12$
20. $8 = -\frac{2}{5}c$
21. $-\frac{4}{9}f = -3$
22. $26 = -\frac{8}{5}y$

23. **ERROR ANALYSIS** Describe and correct the error in finding the solution.

✗

$$-4.2x = 21$$

$$\frac{-4.2x}{4.2} = \frac{21}{4.2}$$

$$x = 5$$

Write the word sentence as an equation. Then solve.

24. A number divided by -9 is -16.
25. A number multiplied by $\frac{2}{5}$ is $\frac{3}{20}$.
26. The product of 15 and a number is -75.
27. The quotient of a number and -1.5 is 21.

In Exercises 28 and 29, write an equation. Then solve.

28. **NEWSPAPERS** You make \$0.75 for every newspaper you sell. How many newspapers do you have to sell to buy the soccer cleats?

29. **ROCK CLIMBING** A rock climber averages $12\frac{3}{5}$ feet per minute. How many feet does the rock climber climb in 30 minutes?

OPEN-ENDED **(a) Write a multiplication equation that has the given solution. (b) Write a division equation that has the same solution.**

30. -3 **31.** -2.2 **32.** $-\frac{1}{2}$ **33.** $-1\frac{1}{4}$

34. **REASONING** Which of the methods can you use to solve $-\frac{2}{3}c = 16$?

Multiply each side by $-\frac{2}{3}$.	Multiply each side by $-\frac{3}{2}$.
Divide each side by $-\frac{2}{3}$.	Multiply each side by 3, then divide each side by -2.

35. **STOCK** A stock has a return of $-\$1.26$ per day. Write and solve an equation to find the number of days until the total return is $-\$10.08$.

36. **ELECTION** In a school election, $\frac{3}{4}$ of the students vote. There are 1464 ballots. Write and solve an equation to find the number of students.

37. **OCEANOGRAPHY** Aquarius is an underwater ocean laboratory located in the Florida Keys National Marine Sanctuary. Solve the equation $\frac{31}{25}x = -62$ to find the value of x.

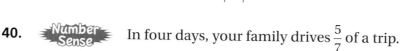

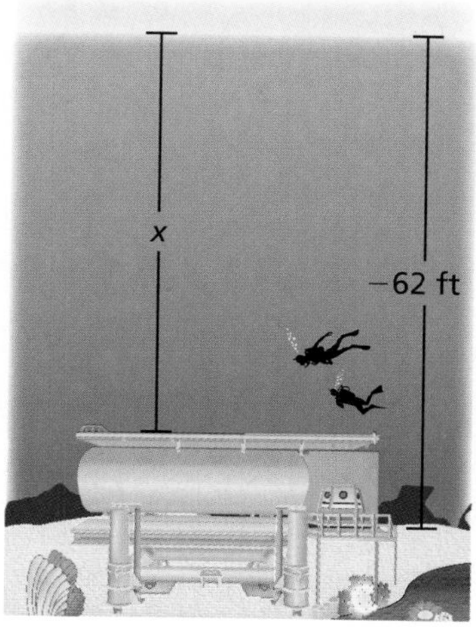

38. **SHOPPING** The price of a bike at store A is $\frac{5}{6}$ the price at store B. The price at store A is \$150.60. Write and solve an equation to find how much you save by buying the bike at store A.

39. **CRITICAL THINKING** Solve $-2|m| = -10$.

40. **Number Sense** In four days, your family drives $\frac{5}{7}$ of a trip. Your rate of travel is the same throughout the trip. The total trip is 1250 miles. In how many more days will you reach your destination?

Fair Game Review What you learned in previous grades & lessons

Subtract. *(Section 1.3)*

41. $5 - 12$ **42.** $-7 - 2$ **43.** $4 - (-8)$ **44.** $-14 - (-5)$

45. **MULTIPLE CHOICE** Of the 120 apartments in a building, 75 have been scheduled to receive new carpet. What fraction of the apartments have not been scheduled to receive new carpet? *(Section 3.1)*

Ⓐ $\frac{1}{4}$ Ⓑ $\frac{3}{8}$ Ⓒ $\frac{5}{8}$ Ⓓ $\frac{3}{4}$

3.6 Solving Two-Step Equations

STANDARDS OF LEARNING
7.14

Essential Question In a two-step equation, which step should you do first?

1 EXAMPLE: Solving a Two-Step Equation

Use algebra tiles to model and solve $2x - 3 = -5$.

Model the equation $2x - 3 = -5$.

Remove the three red tiles on the left side by adding three yellow tiles to each side.

Remove the three "zero pairs" from each side.

Because there are two green tiles, divide the red tiles into two equal groups.

Keep one of the groups. This shows the value of x.

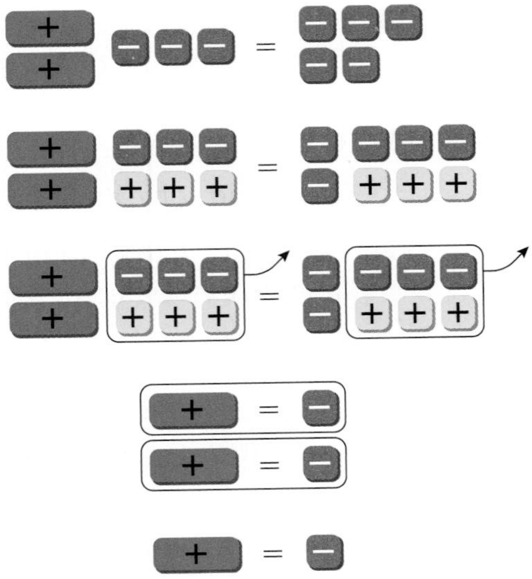

So, $x = -1$.

2 EXAMPLE: The Math Behind the Tiles

Solve $2x - 3 = -5$ without using algebra tiles. Describe each step. Which step is first, adding 3 to each side or dividing each side by 2?

Use the steps in Example 1 as a guide.

$2x - 3 = -5$	Write the equation.
$2x - 3 + 3 = -5 + 3$	Add 3 to each side.
$2x = -2$	Simplify.
$\frac{2x}{2} = \frac{-2}{2}$	Divide each side by 2.
$x = -1$	Simplify.

So, $x = -1$. Adding 3 to each side is the first step.

3 ACTIVITY: Solving Equations Using Algebra Tiles

Work with a partner.

- **Write an equation shown by the algebra tiles.**
- **Use algebra tiles to model and solve the equation.**
- **Check your answer by solving the equation without using algebra tiles.**

a.

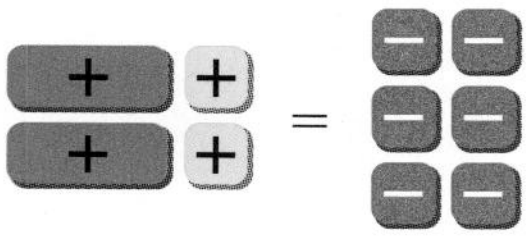

b.

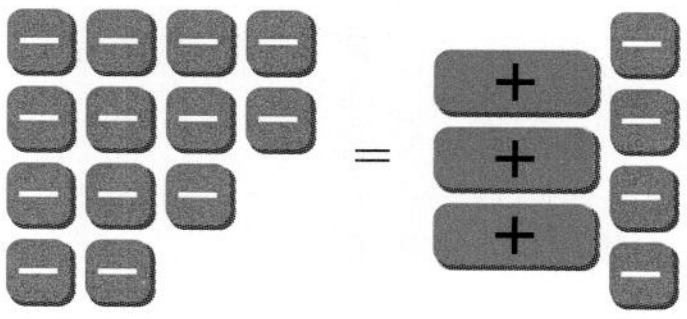

4 ACTIVITY: Working Backwards

Work with a partner.

a. **Sample:** Your friend pauses a video game to get a drink. You continue the game. You double the score by saving a princess. Then you lose 75 points because you do not collect the treasure. You finish the game with -25 points. How many points did you start with?

One way to solve the problem is to work backwards. To do this, start with the end result and retrace the events.

You have -25 points at the end of the game.	$\mathbf{-25}$
You lost 75 points for not collecting the treasure, so add 75 to -25.	$\mathbf{-25 + 75 = 50}$
You doubled your score for saving the princess, so find half of 50.	$\mathbf{50 \div 2 = 25}$

So, you started the game with 25 points.

b. You triple your account balance by making a deposit. Then you withdraw \$127.32 to buy groceries. Your account is now overdrawn by \$10.56. By working backwards, find your account balance before you made the deposit.

What Is Your Answer?

5. **IN YOUR OWN WORDS** In a two-step equation, which step should you do first? Give four examples.

6. Solve the equation $2x - 75 = -25$. How do your steps compare with the strategy of working backwards in Activity 4?

Practice Use what you learned about solving two-step equations to complete Exercises 6–11 on page 116.

3.6 Lesson

Check It Out
Lesson Tutorials
BigIdeasMath.com

EXAMPLE 1 Solving a Two-Step Equation

Solve $-3x + 5 = 2$. Check your solution.

	$-3x + 5 = 2$	Write the equation.
Undo the addition.	$-5 \quad -5$	Subtract 5 from each side.
	$-3x = -3$	Simplify.
Undo the multiplication.	$\frac{-3x}{-3} = \frac{-3}{-3}$	Divide each side by -3.
	$x = 1$	Simplify.

Check

$-3x + 5 = 2$

$-3(1) + 5 \stackrel{?}{=} 2$

$-3 + 5 \stackrel{?}{=} 2$

$2 = 2$ ✓

So, the solution is $x = 1$.

On Your Own

Now You're Ready Exercises 6–17

Solve the equation. Check your solution.

1. $2x + 12 = 4$
2. $-5c + 9 = -16$
3. $3(x - 4) = 9$

EXAMPLE 2 Solving a Two-Step Equation

Solve $\frac{x}{8} - \frac{1}{2} = -\frac{7}{2}$.

$\frac{x}{8} - \frac{1}{2} = -\frac{7}{2}$	Write the equation.
$+\frac{1}{2} \quad +\frac{1}{2}$	Add $\frac{1}{2}$ to each side.
$\frac{x}{8} = -3$	Simplify.
$8 \cdot \frac{x}{8} = 8 \cdot (-3)$	Multiply each side by 8.
$x = -24$	Simplify.

Check

$\frac{x}{8} - \frac{1}{2} = -\frac{7}{2}$

$\frac{-24}{8} - \frac{1}{2} \stackrel{?}{=} -\frac{7}{2}$

$-3 - \frac{1}{2} \stackrel{?}{=} -\frac{7}{2}$

$-\frac{7}{2} = -\frac{7}{2}$ ✓

So, the solution is $x = -24$.

Study Tip

You can simplify the equation in Example 2 before solving. Multiply each side by the LCD of the fractions, 8.

$\frac{x}{8} - \frac{1}{2} = -\frac{7}{2}$

$x - 4 = -28$

$x = -24$

On Your Own

Now You're Ready Exercises 20–25

Solve the equation. Check your solution.

4. $\frac{m}{2} + 6 = 10$
5. $-\frac{z}{3} + 5 = 9$
6. $\frac{2}{5} + 4a = -\frac{6}{5}$

EXAMPLE 3 Combining Like Terms Before Solving

Solve $3y - 8y = 25$.

$3y - 8y = 25$	Write the equation.
$-5y = 25$	Combine like terms.
$y = -5$	Divide each side by -5.

So, the solution is $y = -5$.

EXAMPLE 4 Real-Life Application

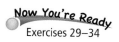

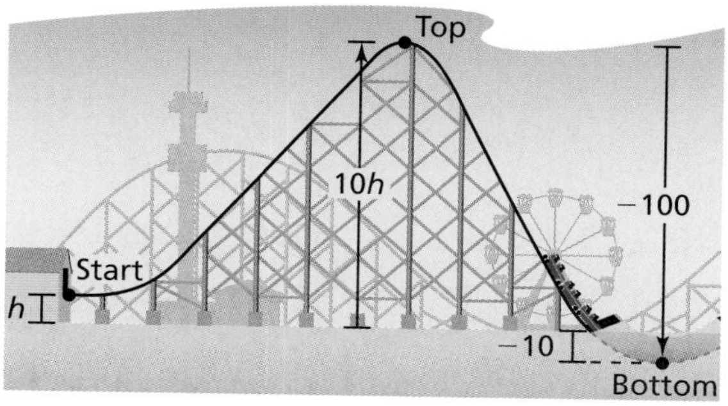

The height at the top of a roller coaster hill is 10 times the height h of the starting point. The height decreases 100 feet from the top to the bottom of the hill. The height at the bottom of the hill is -10 feet. Find h.

Location	Verbal Description	Expression
Start	The height at the start is h.	h
Top of hill	The height at the top of the hill is 10 times the starting height h.	$10h$
Bottom of hill	Height decreases by 100 feet. So, subtract 100.	$10h - 100$

The height at the bottom of the hill is -10 feet. Solve $10h - 100 = -10$ to find h.

$10h - 100 = -10$	Write equation.
$10h = 90$	Add 100 to each side.
$h = 9$	Divide each side by 10.

The height at the start is 9 feet.

On Your Own

Now You're Ready
Exercises 29–34

Solve the equation. Check your solution.

7. $4 - 2y + 3 = -9$ **8.** $7x - 10x = 15$ **9.** $-8 = 1.3m - 2.1m$

10. WHAT IF? In Example 4, the height at the bottom of the hill is -5 feet. Find the height h.

3.6 Exercises

Vocabulary and Concept Check

1. **WRITING** How do you solve two-step equations?

Match the equation with the first step to solve it.

2. $4 + 4n = -12$ 3. $4n = -12$ 4. $\frac{n}{4} = -12$ 5. $\frac{n}{4} - 4 = -12$

A. Add 4. B. Subtract 4. C. Multiply by 4. D. Divide by 4.

Practice and Problem Solving

Solve the equation. Check your solution.

① 6. $2v + 7 = 3$ 7. $4b + 3 = -9$ 8. $17 = 5k - 2$

9. $-6t - 7 = 17$ 10. $8n + 16.2 = 1.6$ 11. $-5g + 2.3 = -18.8$

12. $2t - 5 = -10$ 13. $-4p + 9 = -5$ 14. $11 = -5x - 2$

15. $4 + 2.2h = -3.7$ 16. $-4.8f + 6.4 = -8.48$ 17. $7.3y - 5.18 = -51.9$

ERROR ANALYSIS **Describe and correct the error in finding the solution.**

18.

✗

$$-6 + 2x = -10$$
$$-6 + \frac{2x}{2} = -\frac{10}{2}$$
$$-6 + x = -5$$
$$x = 1$$

19.

✗

$$-3x + 2 = -7$$
$$-3x = -9$$
$$-\frac{3x}{3} = \frac{-9}{3}$$
$$x = -3$$

Solve the equation. Check your solution.

② 20. $\frac{3}{5}g - \frac{1}{3} = -\frac{10}{3}$ 21. $\frac{a}{4} - \frac{5}{6} = -\frac{1}{2}$ 22. $-\frac{1}{3} + 2z = -\frac{5}{6}$

23. $2 - \frac{b}{3} = -\frac{5}{2}$ 24. $-\frac{2}{3}x + \frac{3}{7} = \frac{1}{2}$ 25. $-\frac{9}{4}v + \frac{4}{5} = \frac{7}{8}$

In Exercises 26–28, write an equation. Then solve.

26. **WEATHER** Starting at 1:00 P.M., the temperature changes −4 degrees per hour. How long will it take to reach −1°?

27. **BOWLING** It costs \$2.50 to rent bowling shoes. Each game costs \$2.25. You have \$9.25. How many games can you bowl?

28. **CELL PHONES** A cell phone company charges a monthly fee plus \$0.25 for each text message. The monthly fee is \$30.00 and you owe \$59.50. How many text messages did you have?

Temperature at 1:00 P.M.

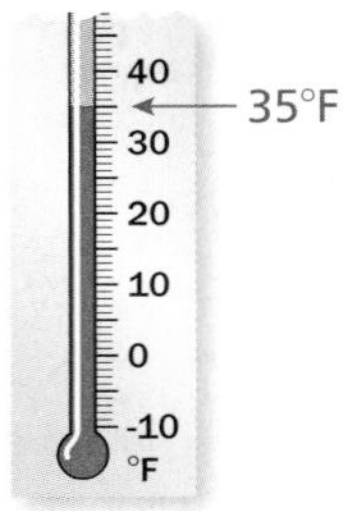

Solve the equation. Check your solution.

3 **29.** $3v - 9v = 30$ **30.** $12t - 8t = -52$ **31.** $-8d - 5d + 7d = 72$

32. $6(x - 2) = -18$ **33.** $-4(m + 3) = 24$ **34.** $-8(y + 9) = -40$

35. **WRITING** Write a real-world problem that can be modeled by $\frac{1}{2}x - 2 = 8$. Then solve the equation.

36. **GEOMETRY** The perimeter of the parallelogram is 102 feet. Find m.

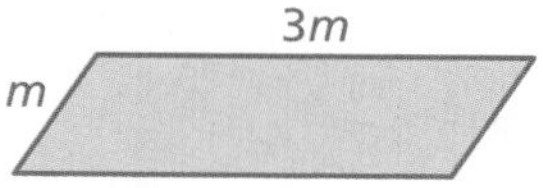

REASONING **Exercises 37 and 38 are missing information. Tell what information is needed to solve the problem.**

37. **TAXI** A taxi service charges an initial fee plus $1.80 per mile. How far can you travel for $12?

38. **EARTH** The coldest surface temperature on the moon is 57 degrees colder than twice the coldest surface temperature on Earth. What is the coldest surface temperature on Earth?

39. **SCIENCE** On Saturday, you catch insects for your science class. Five of the insects escape. The remaining insects are divided into three groups to share in class. Each group has nine insects. How many insects did you catch on Saturday?

a. Solve the problem by working backwards.

b. Solve the equation $\frac{x - 5}{3} = 9$. How does the answer compare with the answer to part (a)?

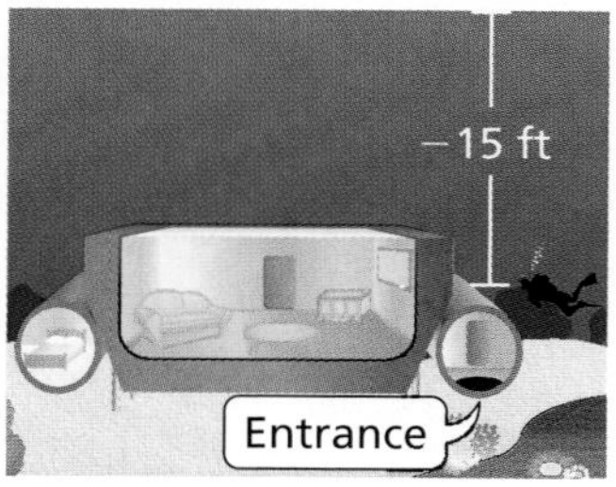

40. **UNDERWATER HOTEL** You must scuba dive to the entrance of your room at Jule's Undersea Lodge in Key Largo, Florida. The diver is 1 foot deeper than $\frac{2}{3}$ of the elevation of the entrance. What is the elevation of the entrance?

41. Geometry How much should you change the length of the rectangle so that the perimeter is 54 centimeters? Write an equation that shows how you found your answer.

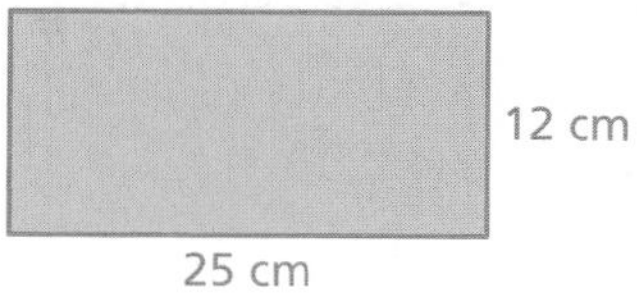

Fair Game Review What you learned in previous grades & lessons

Multiply or divide. *(Section 3.3)*

42. -6.2×5.6 **43.** $\frac{8}{3} \times \left(-2\frac{1}{2}\right)$ **44.** $\frac{5}{2} \div \left(-\frac{4}{5}\right)$ **45.** $-18.6 \div (-3)$

46. **MULTIPLE CHOICE** Which fraction is *not* equivalent to 0.75? *(Skills Review Handbook)*

Ⓐ $\frac{15}{20}$ Ⓑ $\frac{9}{12}$ Ⓒ $\frac{6}{9}$ Ⓓ $\frac{3}{4}$

3.7 Solving Inequalities Using Addition or Subtraction

Essential Question How can you use an inequality to describe a real-life situation?

STANDARDS OF LEARNING
7.15

1 ACTIVITY: Writing an Inequality

Work with a partner. In 3 years, your friend will still not be old enough to apply for a driver's license.

a. Which of the following represents your friend's situation? What does x represent? Explain your reasoning.

$x + 3 < 16$ $\quad$ $x + 3 \leq 16$ $\quad$ $x + 3 > 16$ $\quad$ $x + 3 \geq 16$

b. Graph the possible ages of your friend on a number line.

2 ACTIVITY: The Triangle Inequality

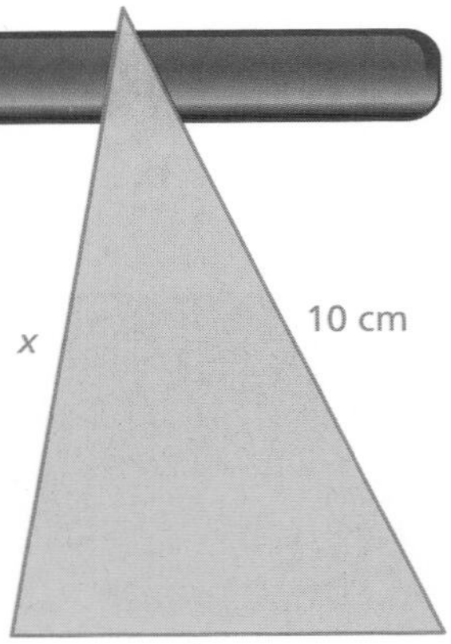

Work with a partner. Draw different triangles whose sides have lengths 10 cm, 6 cm, and x cm.

a. Which of the following describes how *small* x can be?

$6 + x < 10$ $\quad$ $6 + x \leq 10$

$6 + x > 10$ $\quad$ $6 + x \geq 10$

b. Which of the following describes how *large* x can be?

$x - 6 < 10$ $\quad$ $x - 6 \leq 10$ $\quad$ $x - 6 > 10$ $\quad$ $x - 6 \geq 10$

3 ACTIVITY: Writing an Inequality

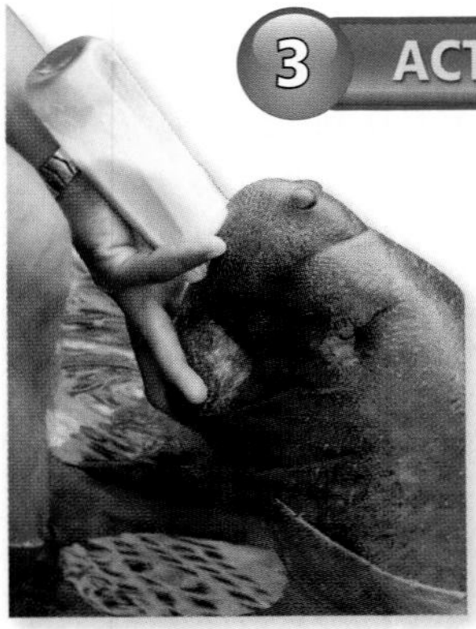

Work with a partner. Baby manatees are about 4 feet long at birth. They grow to a maximum length of 13 feet.

a. Which of the following can represent a baby manatee's growth? What does x represent? Explain your reasoning.

$x + 4 < 13$ $\quad$ $x + 4 \leq 13$ $\quad$ $x - 4 > 13$ $\quad$ $x - 4 \geq 13$

b. Graph the solution on a number line.

4 ACTIVITY: Puzzles

Work with a partner.

a. Use the clues to find the word that is spelled by ●●●●. Assume A $= 1$, B $= 2$, and so on.

CLUES	CLUES	CLUES	CLUES
$4 + h \le 7$	$9 \le h - 6$	$h - 5 \le 10$	$7 > h - 6$
$h + 1 > 3$	$3 + h < 19$	$h + 12 > 26$	$11 + h \ge 23$

b. Trace the pieces and cut them out. Rearrange them to make a square without overlapping the pieces.

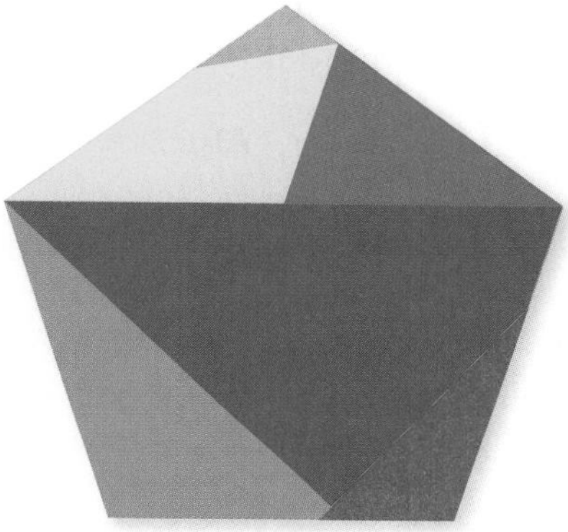

Pentagon to Square

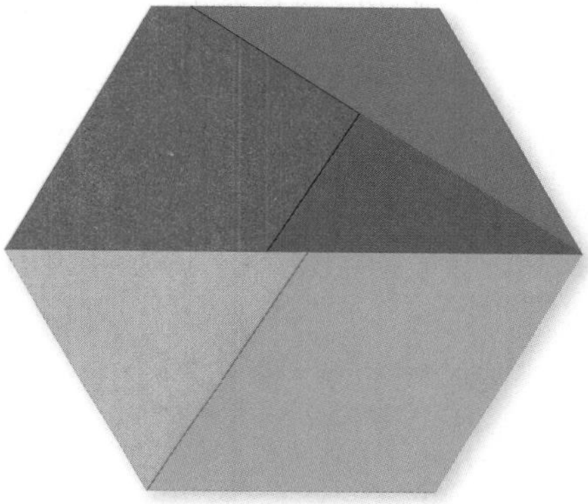

Hexagon to Square

c. Use exactly four 4's and the operations $+$, $-$, $\times$, and $\div$ to write expressions that have values of 0, 1, 2, 3, 4, 5, 6, 7, 8, 9, and 10. For instance, $44 - 44 = 0$.

What Is Your Answer?

5. **IN YOUR OWN WORDS** How can you use an inequality to describe a real-life situation?

6. Write a real-life situation that you can represent with an inequality. Write the inequality. Graph the solution on a number line.

Practice Use what you learned about solving inequalities using addition or subtraction to complete Exercises 3–5 on page 122.

3.7 Lesson

Check It Out
Lesson Tutorials
BigIdeasMath.com

Key Ideas

Addition Property of Inequality

Words If you add the same number to each side of an inequality, the inequality remains true.

Numbers

$$\begin{array}{rcr} -3 & < & 2 \\ \underline{+4} & & \underline{+4} \\ 1 & < & 6 \end{array}$$

Algebra

$$\begin{array}{rcr} x-3 & > & -10 \\ \underline{+3} & & \underline{+3} \\ x & > & -7 \end{array}$$

Subtraction Property of Inequality

Words If you subtract the same number from each side of an inequality, the inequality remains true.

Numbers

$$\begin{array}{rcr} -3 & < & 1 \\ \underline{-5} & & \underline{-5} \\ -8 & < & -4 \end{array}$$

Algebra

$$\begin{array}{rcr} x+7 & > & -20 \\ \underline{-7} & & \underline{-7} \\ x & > & -27 \end{array}$$

These properties are also true for $\leq$ and $\geq$.

Study Tip

You can solve inequalities in much the same way you solve equations. Use inverse operations to get the variable by itself.

EXAMPLE 1 Solving an Inequality Using Addition

Solve $x - 6 \geq -10$. Graph the solution.

$$\begin{array}{rcrl} x-6 & \geq & -10 & \text{Write the inequality.} \\ \underline{+6} & & \underline{+6} & \text{Add 6 to each side.} \\ x & \geq & -4 & \text{Simplify.} \end{array}$$

Undo the subtraction. (points to $+6$)

The solution is $x \geq -4$.

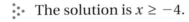

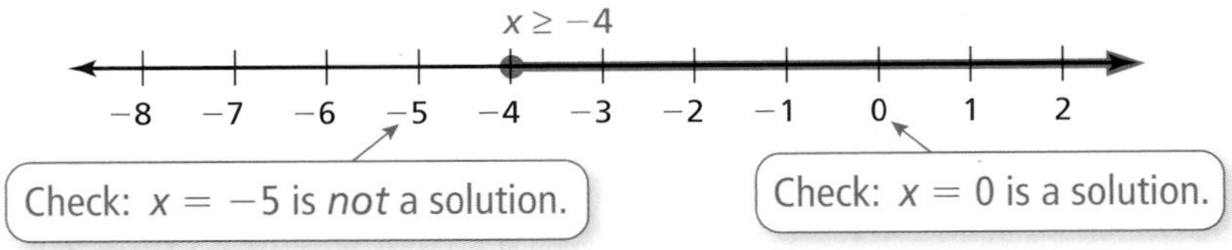

x ≥ −4
−8
−7
−6
−5
−4
−3
−2
−1
0
1
2
Check: x = −5 is not a solution.
Check: x = 0 is a solution.

Study Tip

To check the solution in Example 1, choose convenient numbers to the left and right of -4.

On Your Own

Solve the inequality. Graph the solution.

1. $b - 2 > -9$
2. $m - 3.8 \leq 5$
3. $\frac{1}{4} > y - \frac{1}{4}$

EXAMPLE 2 Solving an Inequality Using Subtraction

Solve $-8 > 1.4 + x$. Graph the solution.

$-8 > 1.4 + x$		Write the inequality.
Undo the addition. →	$\underline{-1.4} \quad \underline{-1.4}$	Subtract 1.4 from each side.
$-9.4 > x$		Simplify.

∴ The solution is $x < -9.4$.

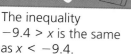

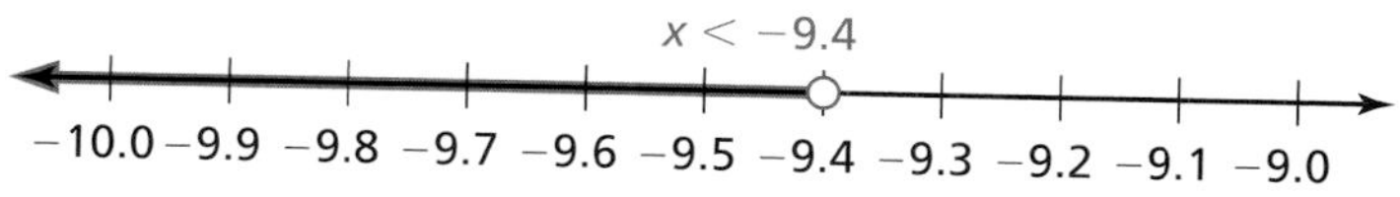

Reading

The inequality $-9.4 > x$ is the same as $x < -9.4$.

On Your Own

Now You're Ready
Exercises 3–17

Solve the inequality. Graph the solution.

4. $k + 5 \leq -3$

5. $\frac{5}{6} \leq z + \frac{2}{3}$

6. $p + 0.7 > -2.3$

EXAMPLE 3 Real-Life Application

On a train, carry-on bags can weigh no more than 50 pounds. Your bag weighs 24.8 pounds. Write and solve an inequality that represents the amount of weight you can add to your bag.

Words	Weight of your bag	plus	amount of weight you can add	is no more than	the weight limit.
Variable	Let w be the possible weight you can add.				
Inequality	24.8	$+$	w	$\leq$	50

$24.8 + w \leq 50$	Write the inequality.
$\underline{-24.8} \qquad \underline{-24.8}$	Subtract 24.8 from each side.
$w \leq 25.2$	Simplify.

∴ You can add no more than 25.2 pounds to your bag.

On Your Own

7. **WHAT IF?** Your carry-on bag weighs 32.5 pounds. Write and solve an inequality that represents the possible weight you can add to your bag.

3.7 Exercises

Check It Out
Help with Homework
BigIdeasMath.com

Vocabulary and Concept Check

1. **REASONING** Is the inequality $r - 5 \le 8$ the same as $8 \le r - 5$? Explain.

2. **WHICH ONE DOESN'T BELONG?** Which inequality does *not* belong with the other three? Explain your reasoning.

$c + \frac{7}{2} \le \frac{3}{2}$ $\qquad$ $c + \frac{7}{2} \ge \frac{3}{2}$ $\qquad$ $\frac{3}{2} \ge c + \frac{7}{2}$ $\qquad$ $c - \frac{3}{2} \le -\frac{7}{2}$

Practice and Problem Solving

Solve the inequality. Graph the solution.

(1) (2)

3. $x - 4 < 5$
4. $5 + h > 7$
5. $3 \ge y - 2$
6. $y - 3 \ge 7$
7. $t - 8 > -4$
8. $n + 11 \le 20$
9. $a + 7 > -1$
10. $5 < v - \frac{1}{2}$
11. $\frac{1}{5} > d + \frac{4}{5}$
12. $-\frac{2}{3} \le g - \frac{1}{3}$
13. $m + \frac{7}{4} \le \frac{11}{4}$
14. $11.2 \le k + 9.8$
15. $h - 1.7 < -3.2$
16. $0 > s + \pi$
17. $5 \ge u - 4.5$

18. **ERROR ANALYSIS** Describe and correct the error in graphing the solution of the inequality.

✗

$5 \ge x - 5$

$10 \ge x$

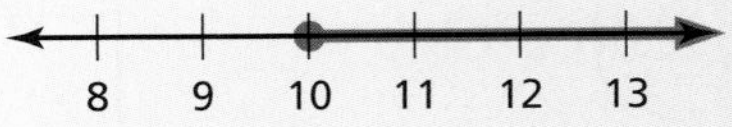

19. **PELICAN** The maximum volume of a great white pelican's bill is about 700 cubic inches.

    a. A pelican scoops up 100 cubic inches of water. Write and solve an inequality that represents the additional volume the bill can contain.

    b. A pelican's stomach can contain about one-third the maximum amount that its bill can contain. Write an inequality that represents the volume of the pelican's stomach.

Write and solve an inequality that represents the value of x.

20. The perimeter is less than 16 feet.

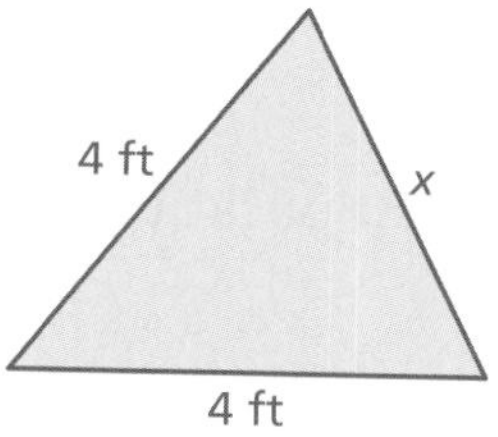

21. The base is greater than the height.

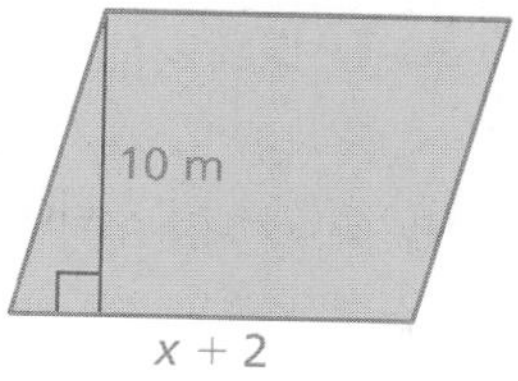

22. The perimeter is less than or equal to 5 feet.

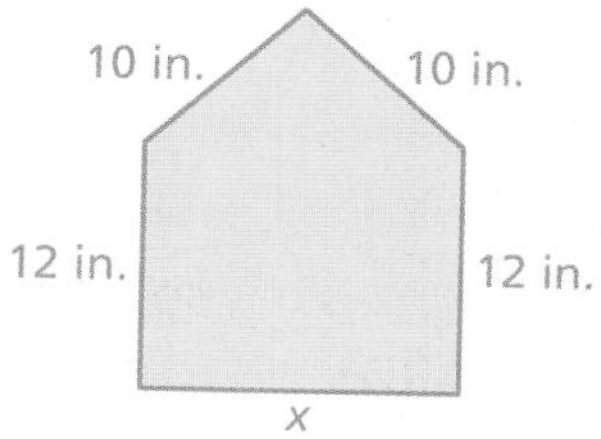

23. REASONING The solution of $w + c \leq 8$ is $w \leq 3$. What is the value of c?

24. FENCE The hole for a fence post is 2 feet deep. The top of the fence post needs to be at least 4 feet above the ground. Write and solve an inequality that represents the required length of the fence post.

25. VIDEO GAME You need at least 12,000 points to advance to the next level of a video game.

a. Write and solve an inequality that represents the number of points you need to advance.

b. You find a treasure chest that increases your score by 60%. How does this change the inequality?

26. POWER A circuit overloads at 1800 watts of electricity. A microwave that uses 1100 watts of electricity is plugged into the circuit.

a. Write and solve an inequality that represents the additional number of watts you can plug in without overloading the circuit.

b. In addition to the microwave, what two appliances in the table can you plug in without overloading the circuit?

Appliance	Watts
Clock radio	50
Blender	300
Hot plate	1200
Toaster	800

27. Number Sense The possible values of x are given by $x - 3 \geq 2$. What is the least possible value of $5x$?

Fair Game Review What you learned in previous grades & lessons

Solve the equation. Check your solution. *(Section 3.5)*

28. $6 = 3x$

29. $\frac{r}{5} = 2$

30. $4c = 15$

31. $8 = \frac{2}{3}b$

32. MULTIPLE CHOICE Which fraction is equivalent to 3.8? *(Skills Review Handbook)*

Ⓐ $\frac{5}{19}$ Ⓑ $\frac{19}{5}$ Ⓒ $\frac{12}{15}$ Ⓓ $\frac{12}{5}$

3.8 Solving Inequalities Using Multiplication or Division

STANDARDS OF LEARNING
7.15

Essential Question How can you use multiplication or division to solve an inequality?

1 ACTIVITY: Using a Table to Solve an Inequality

Work with a partner.

- **Copy and complete the table.**
- **Decide which graph represents the solution of the inequality.**
- **Write the solution of the inequality.**

a. $3x \le 6$

x	-1	0	1	2	3	4	5
$3x$							
$3x \overset{?}{\le} 6$							

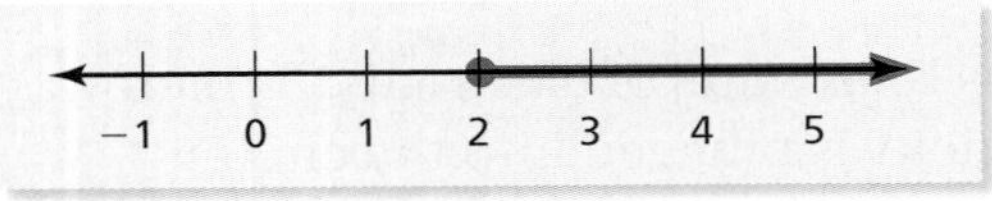

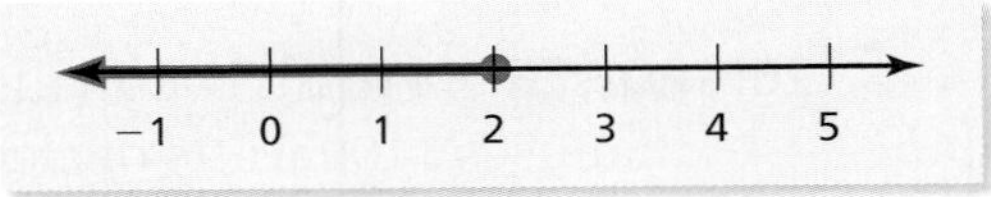

b. $-2x > 4$

x	-5	-4	-3	-2	-1	0	1
$-2x$							
$-2x \overset{?}{>} 4$							

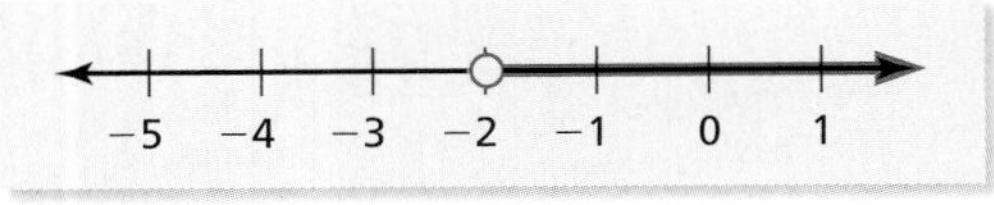

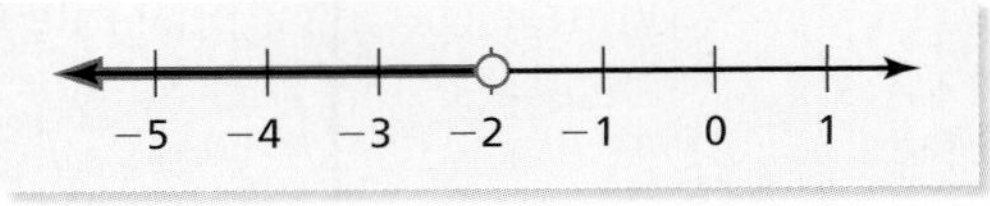

2 ACTIVITY: Writing a Rule

Work with a partner. Use a table to solve each inequality.

a. $3x > 3$ **b.** $4x \le 4$ **c.** $-2x \ge 6$ **d.** $-5x < 10$

Write a rule that describes how to solve inequalities like those in Activity 1. Then use your rule to solve each of the four inequalities above.

3 ACTIVITY: Using a Table to Solve an Inequality

Work with a partner.

- **Copy and complete the table.**
- **Decide which graph represents the solution of the inequality.**
- **Write the solution of the inequality.**

a. $\frac{x}{2} \geq 1$

x	-1	0	1	2	3	4	5
$\frac{x}{2}$							
$\frac{x}{2} \overset{?}{\geq} 1$							

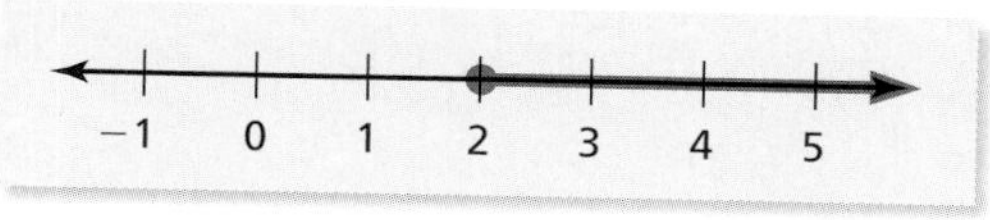

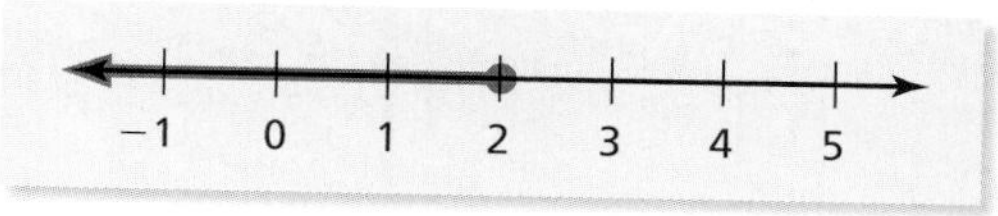

b. $\frac{x}{-3} < \frac{2}{3}$

x	-5	-4	-3	-2	-1	0	1
$\frac{x}{-3}$							
$\frac{x}{-3} \overset{?}{<} \frac{2}{3}$							

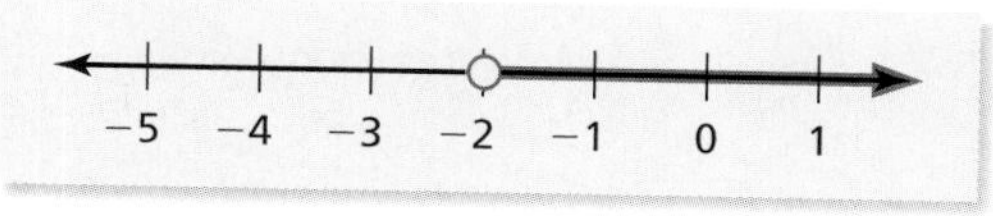

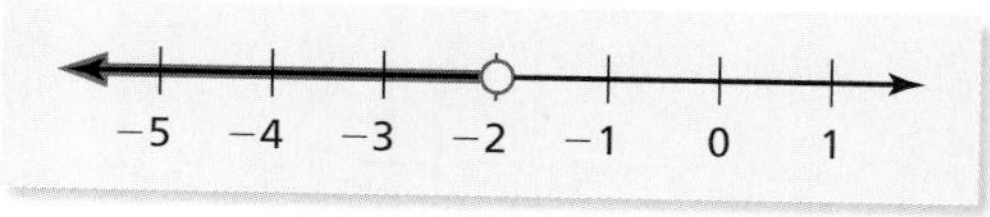

4 ACTIVITY: Writing a Rule

Work with a partner. Use a table to solve each inequality.

a. $\frac{x}{4} \geq 1$ **b.** $\frac{x}{2} < \frac{3}{2}$ **c.** $\frac{x}{-2} > 2$ **d.** $\frac{x}{-5} \leq \frac{1}{5}$

Write a rule that describes how to solve inequalities like those in Activity 3. Then use your rule to solve each of the four inequalities above.

What Is Your Answer?

5. IN YOUR OWN WORDS How can you use multiplication or division to solve an inequality?

Practice Use what you learned about solving inequalities using multiplication or division to complete Exercises 4–9 on page 129.

3.8 Lesson

Check It Out
Lesson Tutorials
BigIdeasMath.com

Key Idea

Multiplication and Division Properties of Inequality (Case 1)

Words If you multiply or divide each side of an inequality by the same *positive* number, the inequality remains true.

Numbers

$$-6 < 8 \qquad\qquad 6 > -8$$

$$2 \cdot (-6) < 2 \cdot 8 \qquad\qquad \frac{6}{2} > \frac{-8}{2}$$

$$-12 < 16 \qquad\qquad 3 > -4$$

Algebra

$$\frac{x}{2} < -9 \qquad\qquad 4x > -12$$

$$2 \cdot \frac{x}{2} < 2 \cdot (-9) \qquad\qquad \frac{4x}{4} > \frac{-12}{4}$$

$$x < -18 \qquad\qquad x > -3$$

These properties are also true for ≤ and ≥.

EXAMPLE 1 Solving an Inequality Using Multiplication

Solve $\frac{x}{8} > 5$. Graph the solution.

$\frac{x}{8} > 5$	Write the inequality.
Undo the division. → $8 \cdot \frac{x}{8} > 8 \cdot 5$	Multiply each side by 8.
$x > 40$	Simplify.

∴ The solution is $x > 40$.

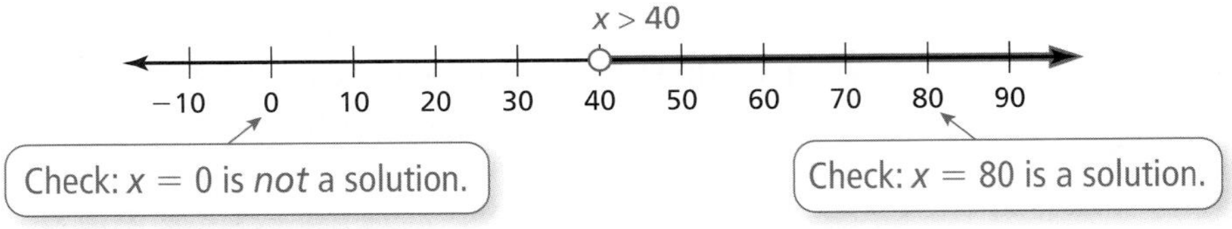

x > 40
−10
0
10
20
30
40
50
60
70
80
90
Check: x = 0 is *not* a solution.
Check: x = 80 is a solution.

On Your Own

Solve the inequality. Graph the solution.

1. $a \div 2 < 4$ **2.** $\frac{n}{7} \geq -1$ **3.** $-6.4 \geq \frac{w}{5}$

EXAMPLE 2 Solving an Inequality Using Division

Solve $3x \le -24$. Graph the solution.

$3x \le -24$ Write the inequality.

Undo the multiplication. → $\dfrac{3x}{3} \le \dfrac{-24}{3}$ Divide each side by 3.

$x \le -8$ Simplify.

∴ The solution is $x \le -8$.

$x \le -8$

```
◄——————+——+——+——+——+——+——+——+——+——+——►
      -16  -14  -12  -10  -8   -6   -4   -2   0    2    4
```

Check: $x = -10$ is a solution.

Check: $x = 0$ is *not* a solution.

On Your Own

Exercises 10–18

Solve the inequality. Graph the solution.

4. $4b \ge 36$

5. $2k > -10$

6. $-18 > 1.5q$

🔑 Key Idea

Multiplication and Division Properties of Inequality (Case 2)

Words If you multiply or divide each side of an inequality by the same *negative* number, the direction of the inequality symbol must be reversed for the inequality to remain true.

Common Error ⚠

A negative sign in an inequality does not necessarily mean you must reverse the inequality symbol.

Only reverse the inequality symbol when you multiply or divide both sides by a negative number.

Numbers $-6 < 8$ $6 > -8$

$(-2) \cdot (-6) \;>\; (-2) \cdot 8$ $\dfrac{6}{-2} \;<\; \dfrac{-8}{-2}$

$12 > -16$ $-3 < 4$

Algebra $\dfrac{x}{-6} < 3$ $-5x > 30$

$-6 \cdot \dfrac{x}{-6} \;>\; -6 \cdot 3$ $\dfrac{-5x}{-5} \;<\; \dfrac{30}{-5}$

$x > -18$ $x < -6$

These properties are also true for \le and \ge.

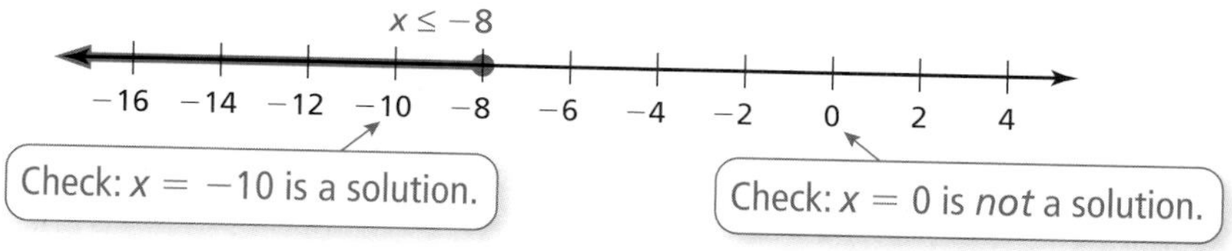

EXAMPLE 3 Solving an Inequality Using Multiplication

Solve $\frac{y}{-3} > 2$. Graph the solution.

$$\frac{y}{-3} > 2$$ Write the inequality.

Undo the division. → $$-3 \cdot \frac{y}{-3} < -3 \cdot 2$$ Multiply each side by −3. Reverse the inequality symbol.

$$y < -6$$ Simplify.

∴ The solution is $y < -6$.

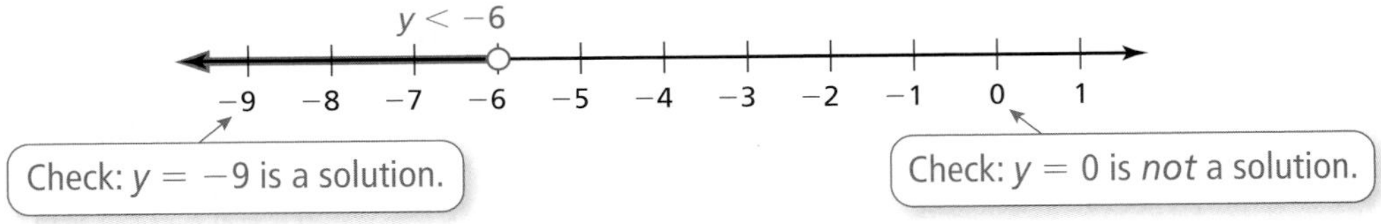

y < −6
−9 −8 −7 −6 −5 −4 −3 −2 −1 0 1
Check: y = −9 is a solution.
Check: y = 0 is *not* a solution.

EXAMPLE 4 Solving an Inequality Using Division

Solve $-7y \leq -35$. Graph the solution.

$$-7y \leq -35$$ Write the inequality.

Undo the multiplication. → $$\frac{-7y}{-7} \geq \frac{-35}{-7}$$ Divide each side by −7. Reverse the inequality symbol.

$$y \geq 5$$ Simplify.

∴ The solution is $y \geq 5$.

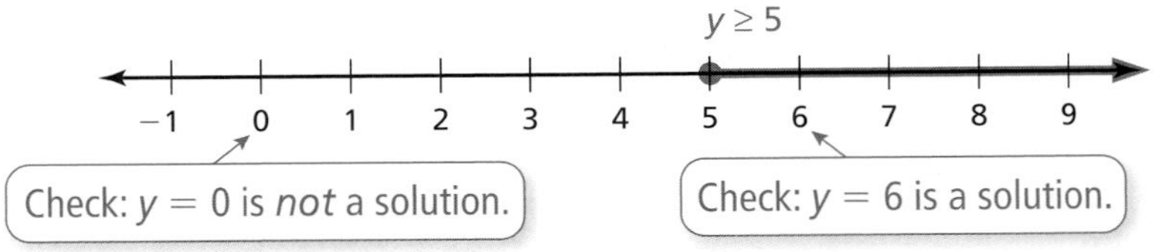

y ≥ 5
−1 0 1 2 3 4 5 6 7 8 9
Check: y = 0 is *not* a solution.
Check: y = 6 is a solution.

On Your Own

Now You're Ready
Exercises 27–35

Solve the inequality. Graph the solution.

7. $\frac{p}{-4} < 7$

8. $\frac{x}{-5} \leq -5$

9. $1 \geq -\frac{1}{10}z$

10. $-9m > 63$

11. $-2r \geq -22$

12. $-0.4y \geq -12$

3.8 Exercises

Check It Out
Help with Homework
BigIdeasMath.com

Vocabulary and Concept Check

1. **VOCABULARY** Explain how to solve $\frac{x}{6} < -5$.

2. **WRITING** Explain how solving $2x < -8$ is different from solving $-2x < 8$.

3. **OPEN-ENDED** Write two inequalities that have the same solution set: one that can be solved using division and one that can be solved using multiplication.

Practice and Problem Solving

Use a table to solve the inequality.

4. $4x < 4$

5. $-2x \le 2$

6. $-5x > 15$

7. $\frac{x}{-3} \ge 1$

8. $\frac{x}{-2} > \frac{5}{2}$

9. $\frac{x}{4} \le \frac{3}{8}$

Solve the inequality. Graph the solution.

(1) (2) 10. $3n > 18$

11. $\frac{c}{4} \le -9$

12. $1.2m < 12$

13. $-14 > x \div 2$

14. $\frac{w}{5} \ge -2.6$

15. $5 < 2.5k$

16. $4x \le -\frac{3}{2}$

17. $2.6y \le -10.4$

18. $10.2 > \frac{b}{3.4}$

19. **ERROR ANALYSIS** Describe and correct the error in solving the inequality.

✗

$$\frac{x}{2} < -5$$

$$2 \cdot \frac{x}{2} > 2 \cdot (-5)$$

$$x > -10$$

Write the word sentence as an inequality. Then solve the inequality.

20. The quotient of a number and 3 is at most 4.

21. A number divided by 8 is less than -2.

22. Four times a number is at least -12.

23. The product of 5 and a number is greater than 20.

24. **CAMERA** You earn \$9.50 per hour at your summer job. Write and solve an inequality that represents the number of hours you need to work in order to buy a digital camera that costs \$247.

25. **COPIES** You have \$3.65 to make copies. Write and solve an inequality that represents the number of copies you can make.

26. **PLAYGROUND** Students at a playground are divided into five equal groups with at least six students in each group. Write and solve an inequality to represent the number of students at the playground.

Solve the inequality. Graph the solution.

③ ④ 27. $-2n \le 10$

28. $-5w > 30$

29. $\dfrac{h}{-6} \ge 7$

30. $-8 < -\dfrac{1}{3}x$

31. $-2y < -11$

32. $-7d \ge 56$

33. $2.4 > -\dfrac{m}{5}$

34. $\dfrac{k}{-0.5} \le 18$

35. $-2.5 > \dfrac{b}{-1.6}$

36. **ERROR ANALYSIS** Describe and correct the error in solving the inequality.

✗

$-4m \ge 16$

$\dfrac{-4m}{-4} \ge \dfrac{16}{-4}$

$m \ge -4$

37. **CRITICAL THINKING** Are all numbers greater than zero solutions of $-x > 0$? Explain.

38. **TRUCKING** In many states, the maximum height (including freight) of a vehicle is 13.5 feet.

a. Write and solve an inequality that represents the number of crates that can be stacked vertically on the bed of the truck.

b. Five crates are stacked vertically on the bed of the truck. Is this legal? Explain.

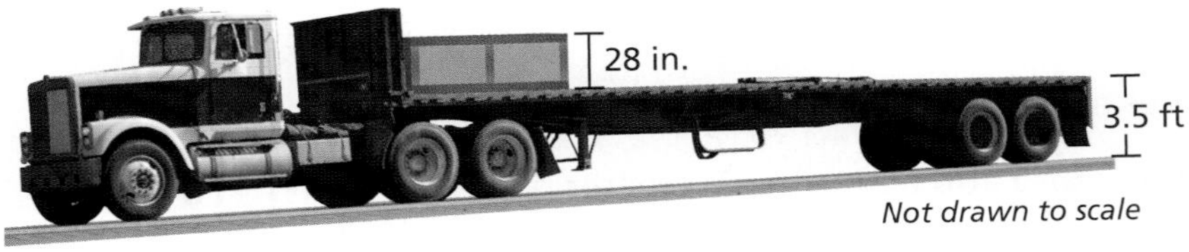

Write and solve an inequality that represents the value of *x*.

39. Area $\ge 102\text{ cm}^2$

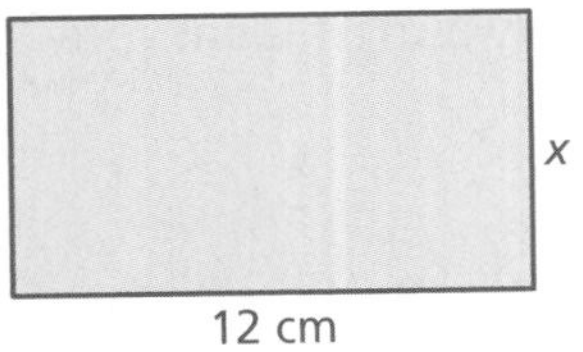

40. Area $< 30\text{ ft}^2$

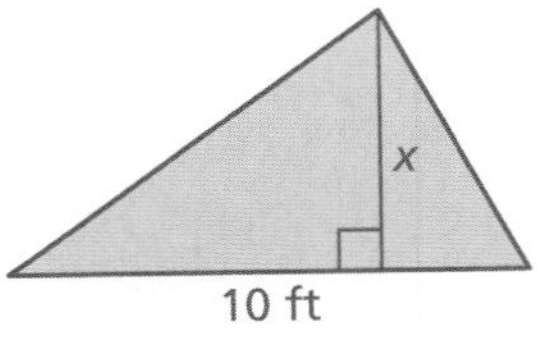

41. **TRIP** You and three friends are planning a trip. You want to keep the cost below $80 per person. Write and solve an inequality that represents the total cost of the trip.

42. **REASONING** Explain why the direction of the inequality symbol must be reversed when multiplying or dividing by the same negative number.

43. **PROJECT** Choose two musical artists to research.

a. Use the Internet or a magazine to complete the table.

b. Find the average number of copies sold per month for each CD.

c. Use the release date to write and solve an inequality that represents the minimum average number of copies sold per month for each CD.

d. In how many months do you expect the number of copies of the second top selling CD to surpass the current number of copies of the top selling CD?

Artist	Name of CD	Release Date	Current Number of Copies Sold
1.			
2.			

Number Sense **Describe all numbers that satisfy *both* inequalities. Include a graph with your description.**

44. $3m > -12$ and $2m < 12$

45. $\frac{n}{2} \geq -3$ and $\frac{n}{-4} \geq 1$

46. $2x \geq -4$ and $2x \geq 4$

47. $\frac{m}{-4} > -5$ and $\frac{m}{4} < 10$

Fair Game Review What you learned in previous grades & lessons

Solve the equation. *(Section 3.6)*

48. $-4w + 5 = -11$

49. $4(x - 3) = 21$

50. $\frac{v}{6} - 7 = 4$

51. $\frac{m + 300}{4} = 96$

52. **MULTIPLE CHOICE** Which measure can have more than one value for a given data set? *(Skills Review Handbook)*

Ⓐ mean Ⓑ median Ⓒ mode Ⓓ range

3.4–3.8 Quiz

Check It Out
Progress Check
BigIdeasMath.com

Solve the equation. Check your solution. *(Section 3.4, Section 3.5, and Section 3.6)*

1. $-6.5 + x = -4.12$

2. $4\frac{1}{2} + p = -5\frac{3}{4}$

3. $-\frac{b}{7} = 4$

4. $-2w + 3.7 = -0.5$

Solve the inequality. Graph the solution. *(Section 3.7 and Section 3.8)*

5. $h - 1 \le -9$

6. $\frac{3}{2} < p + \frac{1}{2}$

7. $-2.3 \ge \frac{x}{5}$

8. $-4y \ge 60$

9. **VACATION** During a family vacation, you spend \$29.79 on souvenirs. You have \$20.21 left. Write and solve an equation to find the amount of money you had at the beginning of your vacation. *(Section 3.4)*

10. **WATER LEVEL** During a drought, the water level of a lake changes $-\frac{1}{8}$ inch per day. Write and solve an equation to find how long it takes for the water level to change -3 inches. *(Section 3.5)*

11. **SCRAPBOOKING** The mat needs to be cut to have a 0.5-inch border on all four sides. *(Section 3.6)*

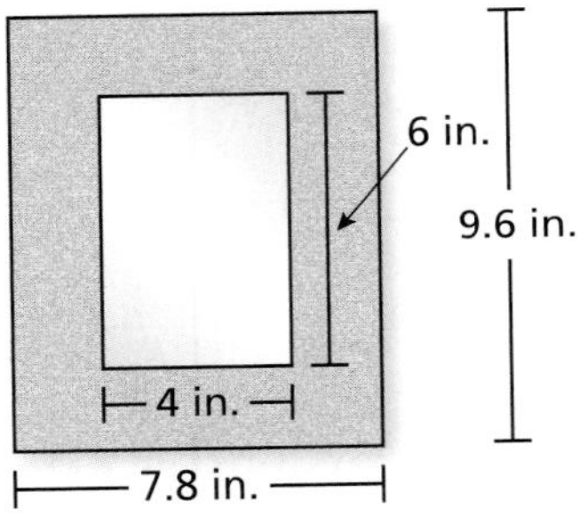

a. How much should you cut from the left and right sides?

b. How much should you cut from the top and bottom?

12. **REASONING** The solution of $x - a > 4$ is $x > 11$. What is the value of a? *(Section 3.7)*

13. **FLOWERS** A soccer team needs to raise \$200 for new uniforms. The team earns \$0.50 for each flower sold. Write and solve an inequality to find the number of flowers it must sell to meet or exceed its fundraising goal. *(Section 3.8)*

3 Chapter Review

Check It Out
Vocabulary Help
BigIdeasMath.com

Review Key Vocabulary

terminating decimal, *p. 82*
repeating decimal, *p. 82*
rational number, *p. 82*
equivalent equations, *p. 102*

Review Examples and Exercises

3.1 Rational Numbers *(pp. 80–85)*

a. Write $4\frac{3}{5}$ as a decimal.

Notice that $4\frac{3}{5} = \frac{23}{5}$.

Divide 23 by 5.

$$\begin{array}{r} 4.6 \\ 5\overline{)23.0} \\ -\,20 \\ \hline 3\,0 \\ -\,3\,0 \\ \hline 0 \end{array}$$

The remainder is 0. So, it is a terminating decimal.

∴ So, $4\frac{3}{5} = 4.6$.

b. Write −0.14 as a fraction in simplest form.

$-0.14 = -\frac{14}{100}$

Write the digits after the decimal point in the numerator.

The last digit is in the hundredths place. So, use 100 in the denominator.

$= -\frac{7}{50}$ Simplify.

Exercises

Write the rational number as a decimal.

1. $-\frac{8}{15}$ **2.** $\frac{5}{8}$ **3.** $-\frac{13}{6}$ **4.** $1\frac{7}{16}$

Write the decimal as a fraction or mixed number in simplest form.

5. −0.6 **6.** −0.35 **7.** −5.8 **8.** 24.23

3.2 Adding and Subtracting Rational Numbers (pp. 86–91)

Find −8.18 + 3.64.

$-8.18 + 3.64 = -4.54$ $\quad |-8.18| > |3.64|$. So, subtract $|3.64|$ from $|-8.18|$.

Use the sign of −8.18.

Exercises

Add or subtract. Write fractions in simplest form.

9. $-4\frac{5}{9} + \frac{8}{9}$ **10.** $-\frac{5}{12} - \frac{3}{10}$ **11.** $-2.53 + 4.75$ **12.** $3.8 - (-7.45)$

13. TURTLE A turtle is $20\frac{5}{6}$ inches below the surface of a pond. It dives to a depth of $32\frac{1}{4}$ inches. What is the change in the turtle's position?

3.3 Multiplying and Dividing Rational Numbers (pp. 92–97)

Find $-4\frac{1}{6} \div 1\frac{1}{3}$.

$-4\frac{1}{6} \div 1\frac{1}{3} = -\frac{25}{6} \div \frac{4}{3}$ Write mixed numbers as improper fractions.

$= \frac{-25}{6} \bullet \frac{3}{4}$ Multiply by the reciprocal of $\frac{4}{3}$.

$= \frac{-25 \bullet 3}{6 \bullet 4}$ Multiply the numerators and the denominators.

$= \frac{-25}{8}$, or $-3\frac{1}{8}$ Simplify.

Exercises

Multiply or divide. Write fractions in simplest form.

14. $-\frac{4}{9}\left(-\frac{7}{9}\right)$ **15.** $\frac{9}{10} \div \left(-\frac{6}{5}\right)$ **16.** $\frac{8}{15}\left(-\frac{2}{3}\right)$ **17.** $-\frac{4}{11} \div \frac{2}{7}$

18. $-5.9(-9.7)$ **19.** $6.4 \div (-3.2)$ **20.** $4.5(-5.26)$ **21.** $-15.4 \div (-2.5)$

22. SUNKEN SHIP The elevation of a sunken ship is −120 feet. Your elevation is $\frac{5}{8}$ of the ship's elevation. What is your elevation?

3.4 Solving Equations Using Addition or Subtraction (pp. 100–105)

Solve $x - 9 = -6$.

	$x - 9 = -6$	Write the equation.
Undo the subtraction. →	$+9 \quad +9$	Add 9 to each side.
	$x = 3$	Simplify.

Check

$x - 9 = -6$

$3 - 9 \stackrel{?}{=} -6$

$-6 = -6$ ✓

Exercises

Solve the equation. Check your solution.

23. $p - 3 = -4$ **24.** $6 + q = 1$ **25.** $-2 + j = -22$ **26.** $b - 19 = -11$

27. $n + \frac{3}{4} = \frac{1}{4}$ **28.** $v - \frac{5}{6} = -\frac{7}{8}$ **29.** $t - 3.7 = 1.2$ **30.** $\ell + 15.2 = -4.5$

31. GIFT CARD After using a gift card as a partial payment to buy a shirt, you still owe \$9.99. What is the value of the gift card?

3.5 Solving Equations Using Multiplication or Division (pp. 106–111)

Solve $\frac{x}{5} = -7$.

	$\frac{x}{5} = -7$	Write the equation.
Undo the division. →	$5 \cdot \frac{x}{5} = 5 \cdot (-7)$	Multiply each side by 5.
	$x = -35$	Simplify.

Check

$\frac{x}{5} = -7$

$\frac{-35}{5} \stackrel{?}{=} -7$

$-7 = -7$ ✓

Exercises

Solve the equation. Check your solution.

32. $\frac{x}{3} = -8$ **33.** $-7 = \frac{y}{7}$ **34.** $-\frac{z}{4} = -\frac{3}{4}$ **35.** $-\frac{w}{20} = -2.5$

36. $4x = -8$ **37.** $-10 = 2y$ **38.** $-5.4z = -32.4$ **39.** $-6.8w = 3.4$

40. TEMPERATURE The mean temperature change is -3.2°F per day for five days. What is the total change over the five-day period?

3.6 Solving Two-Step Equations *(pp. 112–117)*

Solve $\frac{x}{5} + \frac{7}{10} = -\frac{3}{10}$.

$\frac{x}{5} + \frac{7}{10} = -\frac{3}{10}$ Write the equation.

$\frac{x}{5} = -1$ Subtract $\frac{7}{10}$ from each side.

$x = -5$ Multiply each side by 5.

Check

$\frac{x}{5} + \frac{7}{10} = -\frac{3}{10}$

$\frac{-5}{5} + \frac{7}{10} \stackrel{?}{=} -\frac{3}{10}$

$-\frac{3}{10} = -\frac{3}{10}$ ✓

Exercises

Solve the equation. Check your solution.

41. $-2c + 6 = -8$

42. $3(3w - 4) = -20$

43. $\frac{w}{6} + \frac{5}{8} = -1\frac{3}{8}$

44. $-3x - 4.6 = 5.9$

45. **EROSION** The floor of a canyon has an elevation of -14.5 feet. Erosion causes the elevation to change by -1.5 feet per year. How many years will it take for the canyon floor to have an elevation of -31 feet?

3.7 Solving Inequalities Using Addition or Subtraction *(pp. 118–123)*

Solve $-4 < n - 3$. Graph the solution.

$-4 < n - 3$ Write the inequality.

Undo the subtraction. → $+3 \quad +3$ Add 3 to each side.

$-1 < n$ Simplify.

The solution is $n > -1$.

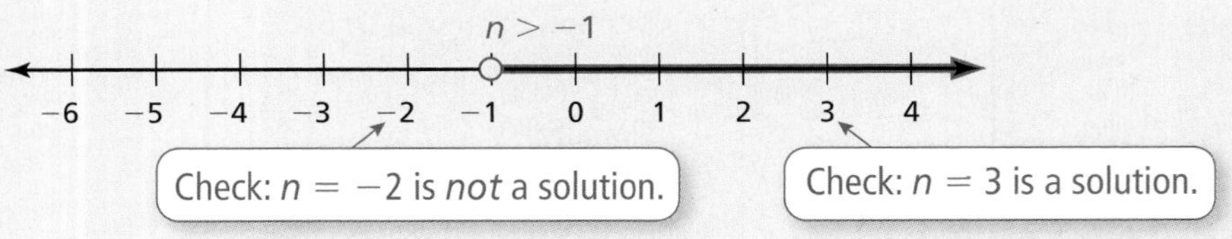

n > −1
−6 −5 −4 −3 −2 −1 0 1 2 3 4
Check: n = −2 is not a solution.
Check: n = 3 is a solution.

Exercises

Solve the inequality. Graph the solution.

46. $b + 13 < 18$

47. $x - 3 \leq 10$

48. $y + 1 \geq -2$

49. $s - 1.5 > -2.5$

50. $k - 7 \leq 0$

51. $\frac{1}{4} + m \leq \frac{1}{2}$

3.8 Solving Inequalities Using Multiplication or Division *(pp. 124–131)*

a. Solve $\frac{d}{4} > -3$. Graph the solution.

$\frac{d}{4} > -3$ — Write the inequality.

Undo the division. → $4 \cdot \frac{d}{4} > 4 \cdot (-3)$ — Multiply each side by 4.

$d > -12$ — Simplify.

∴ The solution is $d > -12$.

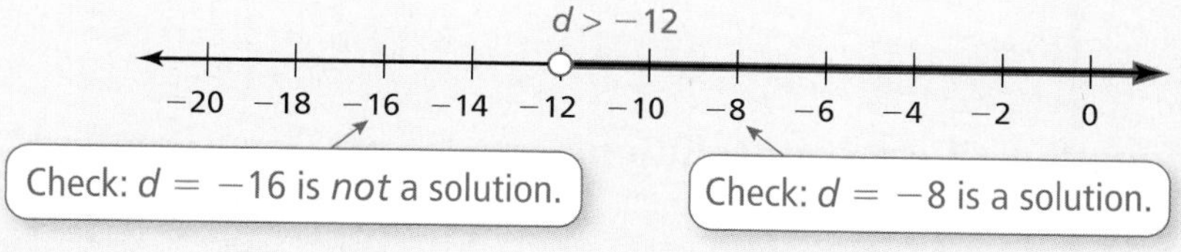
d > −12
−20 −18 −16 −14 −12 −10 −8 −6 −4 −2 0
Check: d = −16 is not a solution.
Check: d = −8 is a solution.

b. Solve $-8a \geq -48$. Graph the solution.

$-8a \geq -48$ — Write the inequality.

Undo the multiplication. → $\frac{-8a}{-8} \leq \frac{-48}{-8}$ — Divide each side by −8. Reverse the inequality symbol.

$a \leq 6$ — Simplify.

∴ The solution is $a \leq 6$.

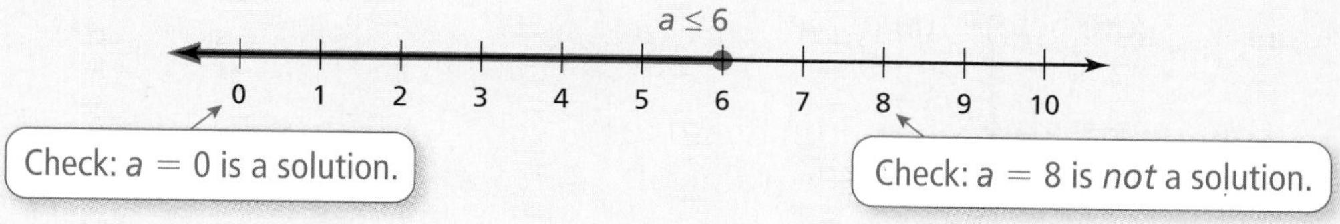
a ≤ 6
0 1 2 3 4 5 6 7 8 9 10
Check: a = 0 is a solution.
Check: a = 8 is not a solution.

Exercises

Solve the inequality. Graph the solution.

52. $\frac{x}{2} \geq 4$

53. $4z < -44$

54. $-2q \geq -18$

55. $\frac{m}{-6} < 3$

56. $8 \leq \frac{1}{4}a$

57. $-7 > -0.5y$

58. GUMBALLS You have \$2.15. Each gumball in a gumball machine costs \$0.25. Write and solve an inequality that represents the number of gumballs you can buy.

3 Chapter Test

Check It Out
Test Practice
BigIdeasMath.com

Write the rational number as a decimal.

1. $-\frac{1}{9}$

2. $\frac{21}{16}$

Write the decimal as a fraction or mixed number in simplest form.

3. -0.122

4. -7.09

Add, subtract, multiply, or divide. Write fractions in simplest form.

5. $-\frac{4}{9} + \frac{23}{18}$

6. $2.86 - 12.1$

7. $-4.4 \times (-6.02)$

8. $-1\frac{5}{6} \div 4\frac{1}{6}$

Solve the equation. Check your solution.

9. $\frac{2}{9}g = -8$

10. $-14 = 6c$

11. $2(x + 1) = -2$

12. $\frac{2}{7}k - \frac{3}{8} = -\frac{19}{8}$

Solve the inequality. Graph the solution.

13. $-60 > -5t$

14. $x - \frac{7}{8} \le \frac{9}{8}$

15. GYMNASTICS You lose 0.3 point for stepping out of bounds during a floor routine. Your final score is 9.124. Write and solve an equation to find your score before the penalty.

16. PERIMETER The perimeter of the triangle is 45. Find the value of x.

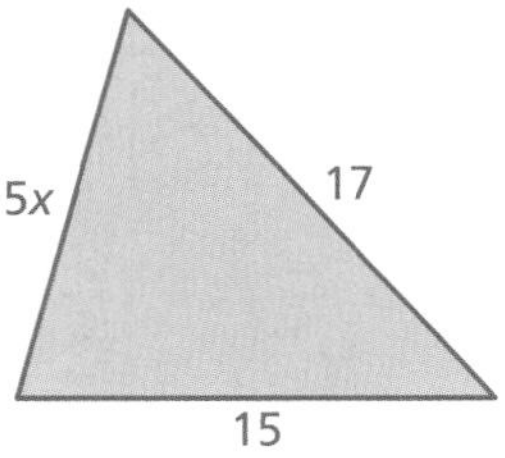

17. GARAGE The vertical clearance for a hotel parking garage is 10 feet. Write and solve an inequality that represents the possible height (in inches) of a roof cargo box.

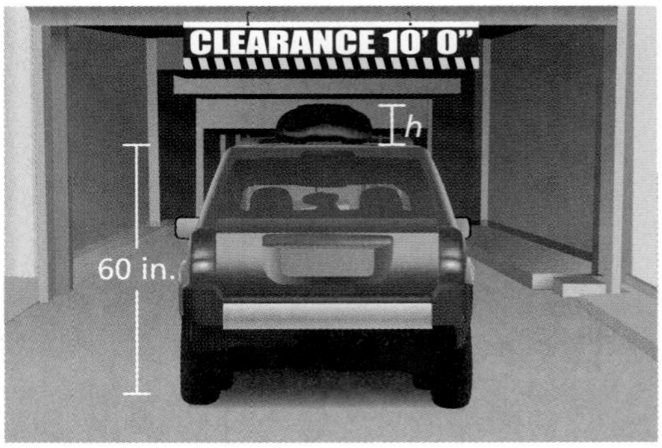

18. LUNCH BILL A lunch bill, including tax, is divided equally among you and five friends. Everyone pays less than $8.75. Write and solve an inequality that represents the total amount of the bill.

Test-Taking Strategy
Estimate the Answer

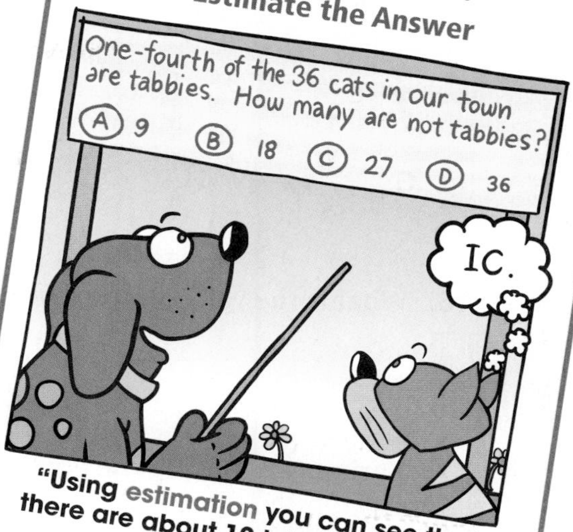

One-fourth of the 36 cats in our town are tabbies. How many are not tabbies?
(A) 9 (B) 18 (C) 27 (D) 36
IC.

"Using estimation you can see that there are about 10 tabbies. So about 30 are not tabbies."

1. When José and Sean were each 5 years old, José was $1\frac{1}{2}$ inches taller than Sean. José grew at an average rate of $2\frac{3}{4}$ inches per year from the time that he was 5 years old until the time he was 13 years old. José was 63 inches tall when he was 13 years old. How tall was Sean when he was 5 years old?

A. $39\frac{1}{2}$ in.

B. $42\frac{1}{2}$ in.

C. $44\frac{3}{4}$ in.

D. $47\frac{3}{4}$ in.

2. Which graph represents the inequality below?

$$\frac{x}{-4} \geq -1$$

F.

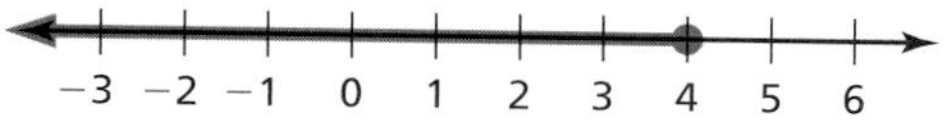

G.

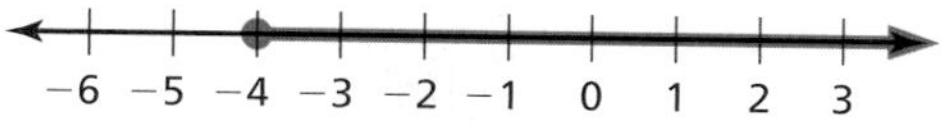

H.

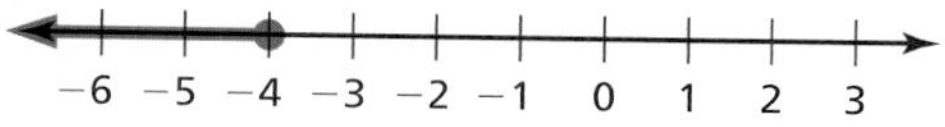

I.

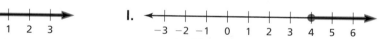

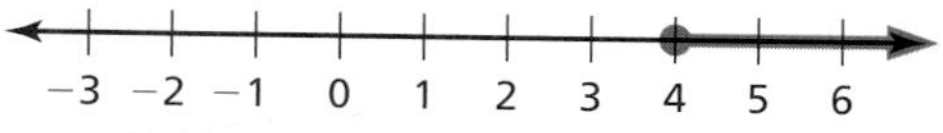

3. What is the missing number in the sequence below?

$$\frac{1}{16}, \frac{1}{8}, \frac{1}{4}, \frac{1}{2}, ____$$

4. What is the value of the expression below?

$$|-2 - (-2.5)|$$

A. -4.5

B. -0.5

C. 0.5

D. 4.5

5. Which equation is equivalent to the equation shown below?

$$-\frac{3}{4}x + \frac{1}{8} = -\frac{3}{8}$$

F. $-\frac{3}{4}x = -\frac{3}{8} - \frac{1}{8}$

G. $-\frac{3}{4}x = -\frac{3}{8} + \frac{1}{8}$

H. $x + \frac{1}{8} = -\frac{3}{8} \bullet \left(-\frac{4}{3}\right)$

I. $x + \frac{1}{8} = -\frac{3}{8} \bullet \left(-\frac{3}{4}\right)$

6. What is the value of the expression below?

$$-5 \div 20$$

7. Karina was solving the equation in the box below.

$$-96 = -6(15 - 2x)$$
$$-96 = -90 - 12x$$
$$-96 + 90 = -90 + 90 - 12x$$
$$-6 = -12x$$
$$\frac{-6}{-12} = \frac{-12x}{-12}$$
$$\frac{1}{2} = x$$

What should Karina do to correct the error that she made?

A. First add 6 to both sides of the equation.

B. First add $2x$ to both sides of the equation.

C. Distribute the -6 to get $90 - 12x$.

D. Distribute the -6 to get $-90 + 12x$.

8. Current, voltage, and resistance are related according to the formula below, where I represents the current, in amperes, V represents the voltage, in volts, and R represents the resistance, in ohms.

$$I = \frac{V}{R}$$

What is the voltage when the current is 0.5 ampere and the resistance is 0.8 ohm?

F. 4.0 volts

G. 1.3 volts

H. 0.4 volt

I. 0.3 volt

9. What is the area of a triangle with a base length of $2\frac{1}{2}$ inches and a height of 3 inches?

A. $2\frac{3}{4}$ in.^{2}

B. $3\frac{3}{4}$ in.^{2}

C. $5\frac{1}{2}$ in.^{2}

D. $7\frac{1}{2}$ in.^{2}

10. What is the circumference of the circle below? (Use 3.14 for π.)

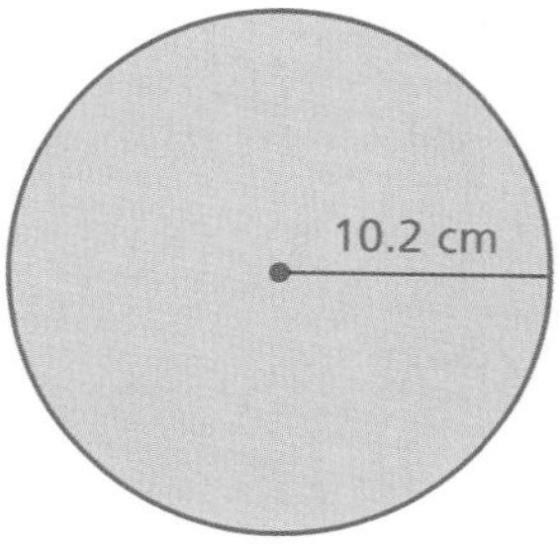

F. 64.056 cm

G. 60.028 cm

H. 32.028 cm

I. 30.028 cm

11. Four points are graphed on the number line below.

Think Solve Explain

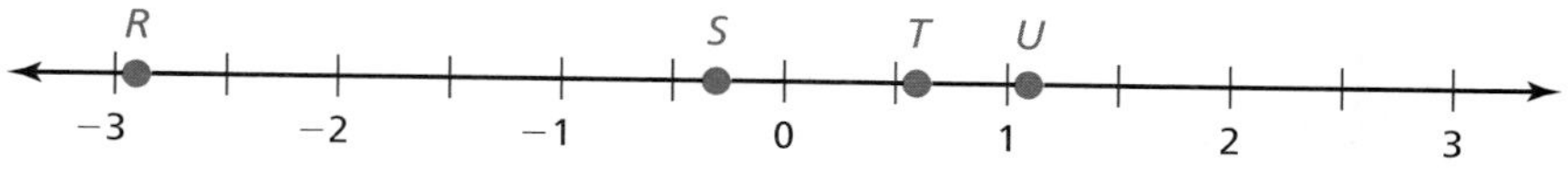

Part A Choose the two points whose values have the greatest sum. Approximate this sum. Explain your reasoning.

Part B Choose the two points whose values have the greatest difference. Approximate this difference. Explain your reasoning.

Part C Choose the two points whose values have the greatest product. Approximate this product. Explain your reasoning.

Part D Choose the two points whose values have the greatest quotient. Approximate this quotient. Explain your reasoning.

12. What number belongs in the box to make the equation true?

$$\frac{-0.4}{\square} + 0.8 = -1.2$$

A. 1

B. 0.2

C. −0.2

D. −1

4 Tables, Graphs, and Functions

What You Learned Before

"Here's an interesting survey about favorite dog toys."

Identifying Patterns

Example 1 **Using the numbers from the In and Out Table, find and state the rule in words.**

In	Out
30	0
40	10
50	20
60	30

Each Out value is 30 less than the In value.

∴ The In value minus 30 equals the Out value.

Try It Yourself

Using the numbers from the In and Out Table, find and state the rule in words.

1.

In	Out
5	10
7	14
10	20
40	80

2.

In	Out
0.5	1
1.5	2
3	3.5
9.5	10

3.

In	Out
13	0
15	2
30	17
45	32

Plotting Points

Example 2 **Write an ordered pair corresponding to Point S.**

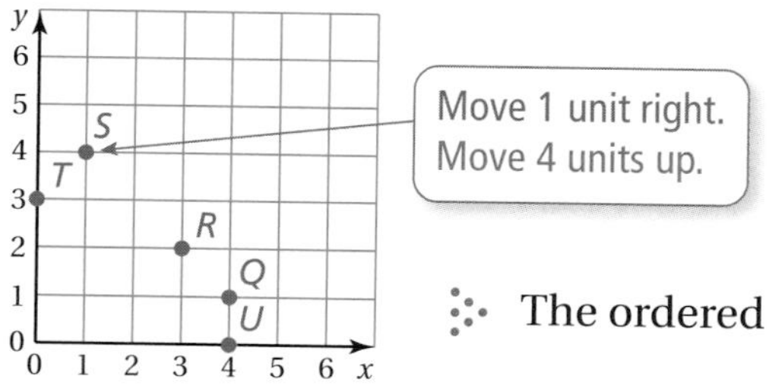

∴ The ordered pair $(1, 4)$ corresponds to Point S.

Try It Yourself

Use the graph in Example 2 to write an ordered pair corresponding to the point.

4. Point Q **5.** Point R **6.** Point U **7.** Point T

4.1 Mapping Diagrams

STANDARDS OF LEARNING
7.12

Essential Question What is a mapping diagram? How can it be used to represent a function?

1 ACTIVITY: Constructing Mapping Diagrams

Work with a partner. Copy and complete the mapping diagram.

a. Area A

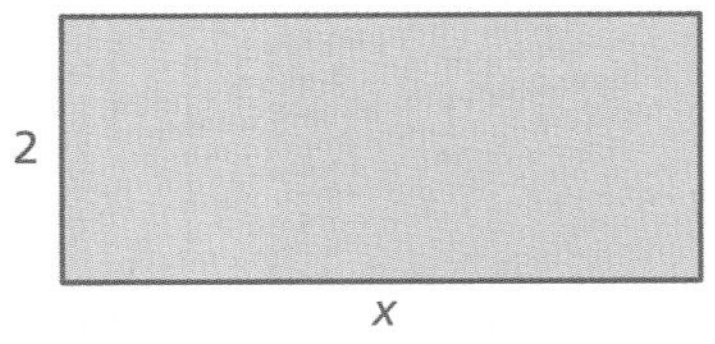

Input, x **Output, A**

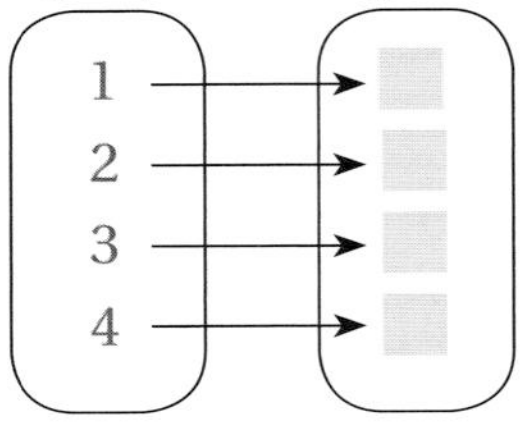

b. Perimeter P

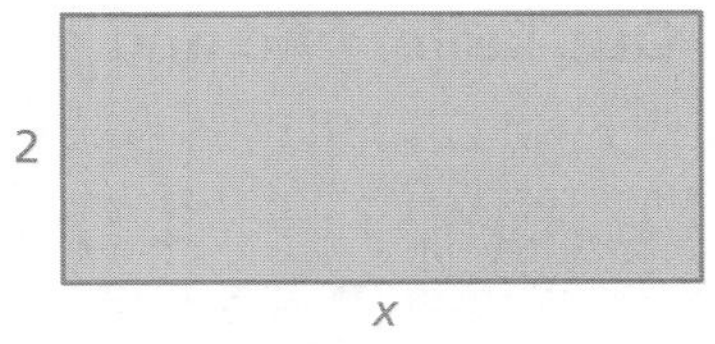

Input, x **Output, P**

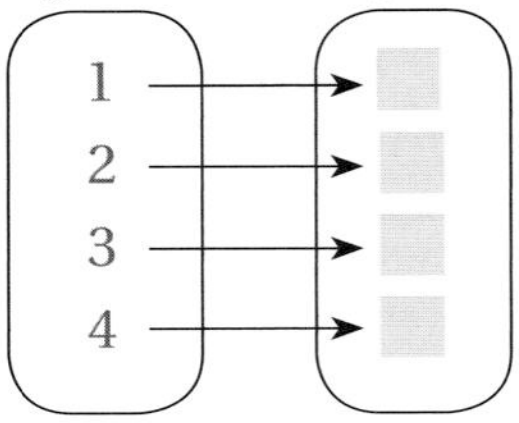

c. Circumference C

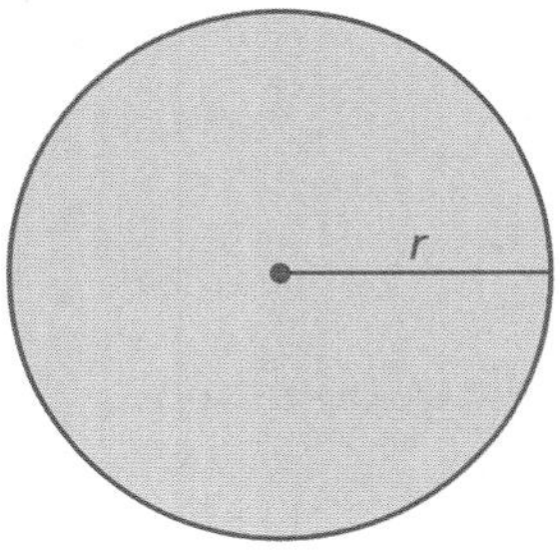

Input, r **Output, C**

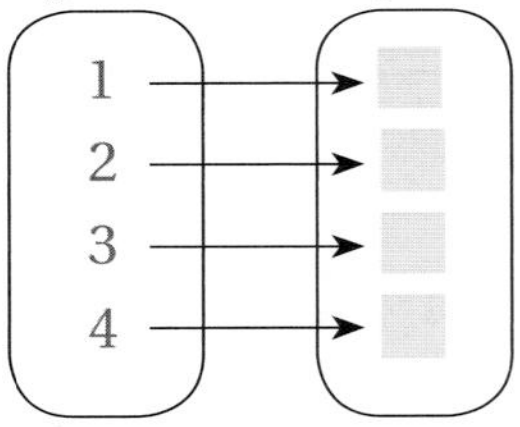

d. Volume V

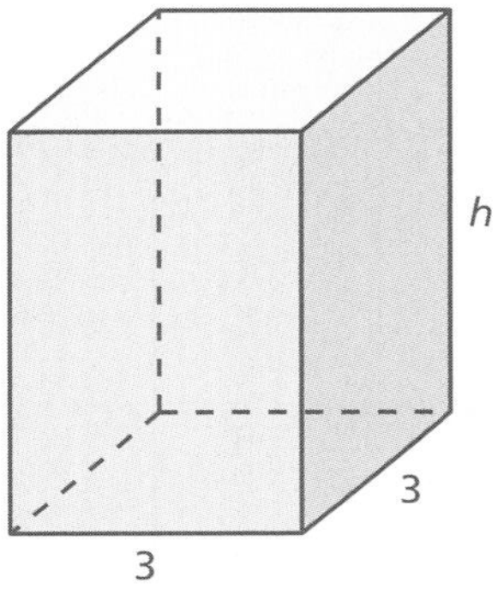

Input, h **Output, V**

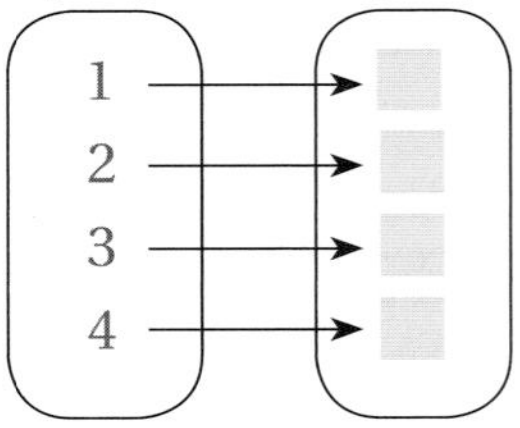

2 ACTIVITY: Interpreting Mapping Diagrams

Work with a partner. Describe the pattern in the mapping diagram. Copy and complete the diagram.

a.

Input, d	Output, P
0.1	10
0.2	20
0.3	30
0.4	
0.5	
0.6	
0.7	

b.

Input, t	Output, M
1	8
2	10
3	12
4	
5	
6	
7	

c.

Input, n	Output, S
1/2	3/2
3/2	9/2
5/2	15/2
7/2	
9/2	

d.

Input, x	Output, A
1	4/3
2	5/3
3	2
4	
5	
6	
7	

What Is Your Answer?

3. **IN YOUR OWN WORDS** What is a mapping diagram? How can it be used to represent a function?
4. Construct a mapping diagram that represents a function you have studied.

"I made a mapping diagram."

"It shows how I feel about my skateboard with each passing day."

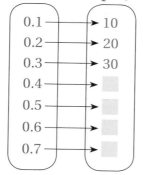

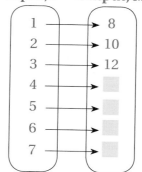

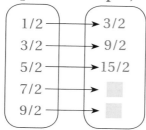

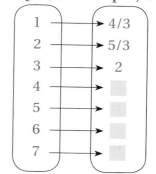

Practice Use what you learned about mapping diagrams to complete Exercises 3–5 on page 148.

4.1 Lesson

Check It Out
Lesson Tutorials
BigIdeasMath.com

Key Vocabulary
input, *p. 146*
output, *p. 146*
function, *p. 146*
mapping diagram, *p. 146*

Ordered pairs can be used to show **inputs** and **outputs**.

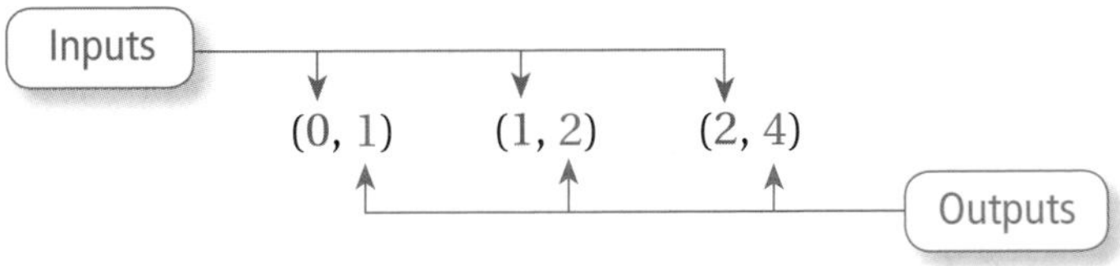

Key Idea

Functions and Mapping Diagrams

A **function** is a relationship that pairs each input with exactly one output. A function can be represented by ordered pairs or a **mapping diagram**.

Ordered Pairs

(0, 1)

(1, 2)

(2, 4)

Mapping Diagram

Input **Output**

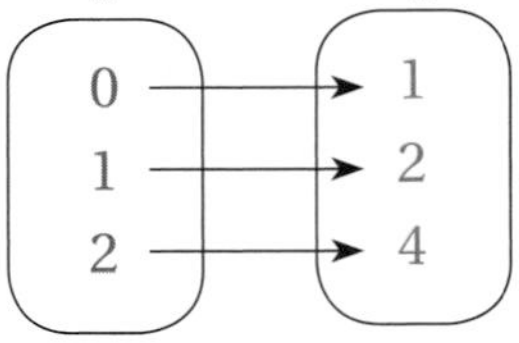

EXAMPLE 1 Listing Ordered Pairs

List the ordered pairs shown in the mapping diagram.

a. **Input** **Output**

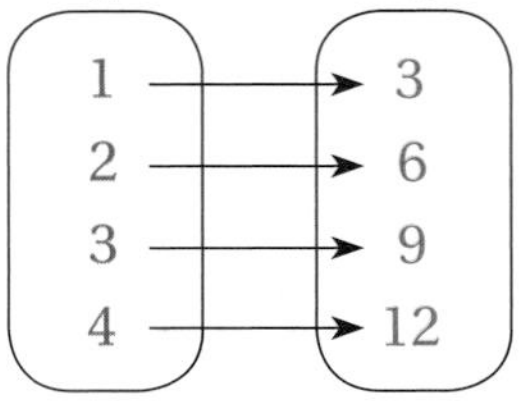

b. **Input** **Output**

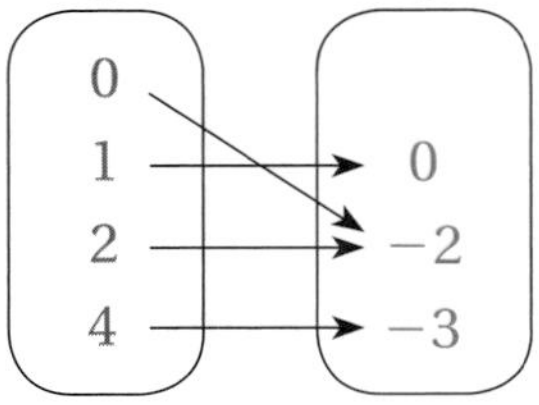

a. The ordered pairs are (1, 3), (2, 6), (3, 9), and (4, 12).

b. The ordered pairs are (0, −2), (1, 0), (2, −2), and (4, −3).

On Your Own

Now You're Ready
Exercises 6–8

List the ordered pairs shown in the mapping diagram.

1. **Input** **Output**

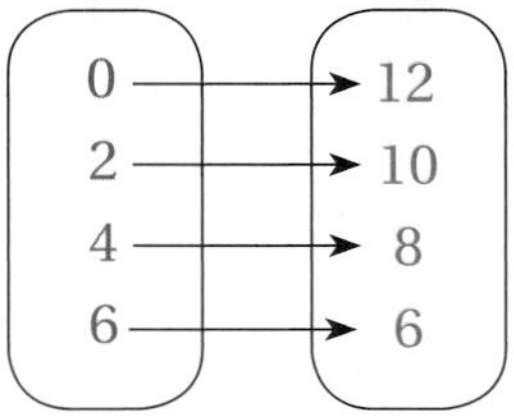

2. **Input** **Output**

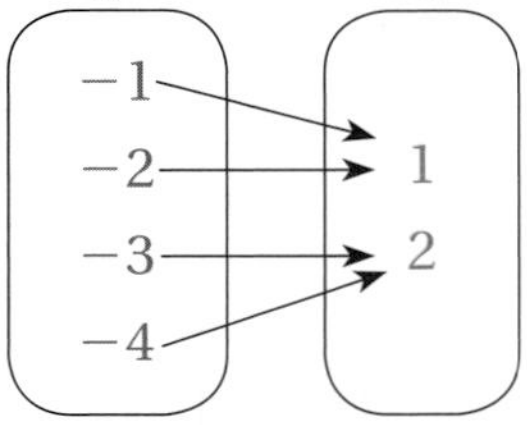

EXAMPLE 2 Drawing a Mapping Diagram

Draw a mapping diagram of $(-1, -1)$, $(2, 6)$, $(3, 5)$, and $(5, 9)$.

List the inputs and outputs in order from least to greatest.

Inputs: $-1, 2, 3, 5$

Outputs: $-1, 5, 6, 9$

Draw arrows from the inputs to their outputs.

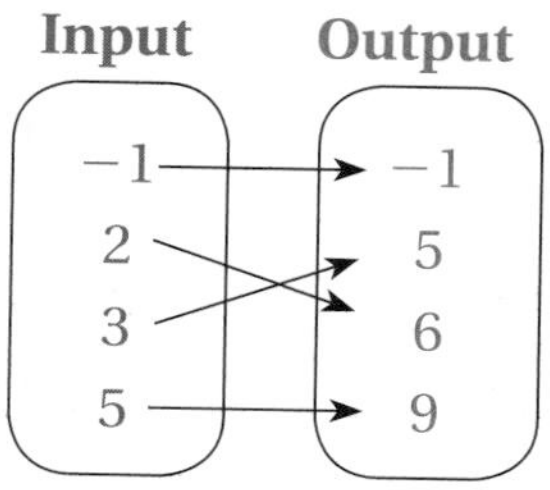

EXAMPLE 3 Describing a Mapping Diagram

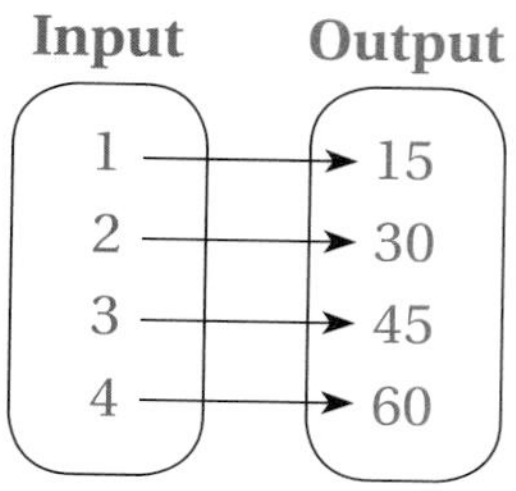

Describe the pattern of inputs and outputs in the mapping diagram on the left.

Look at the relationship between the inputs and outputs.

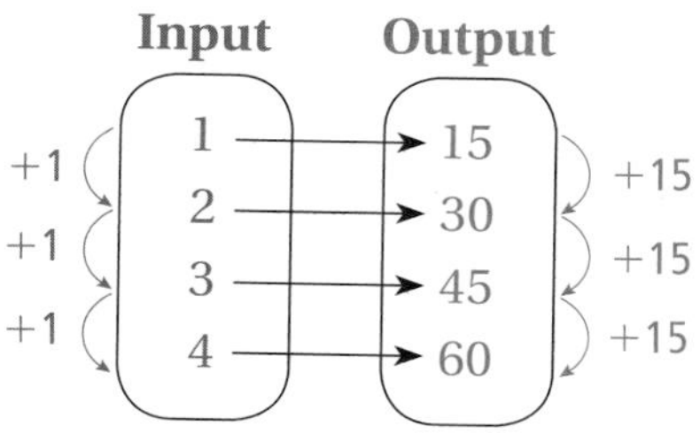

∴ As each input increases by 1, the output increases by 15.

On Your Own

Now You're Ready
Exercises 9–17

3. Draw a mapping diagram of $(0, -2)$, $(2, 4)$, $(5, 3)$, and $(8, 1)$.
4. Describe the pattern of inputs and outputs in the mapping diagram shown.

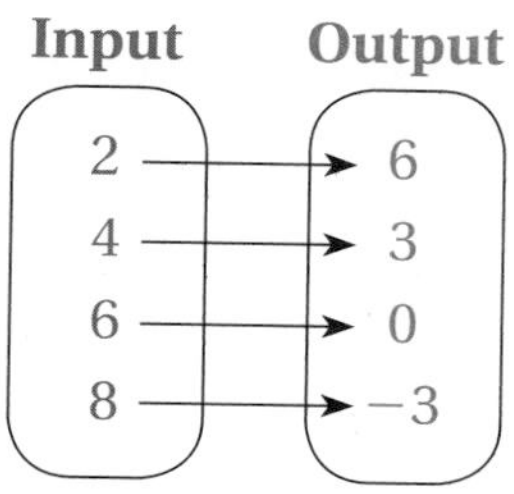

EXAMPLE 4 Real-Life Application

Number of Songs Played	Time Onstage (minutes)
8	45
10	60
7	45
14	90

The table shows the number of songs played by four bands at a festival and the amount of time each band played. Use the table to draw a mapping diagram.

Let the number of songs played be the inputs and the times onstage be the outputs.

Inputs: 7, 8, 10, 14

Outputs: 45, 60, 90

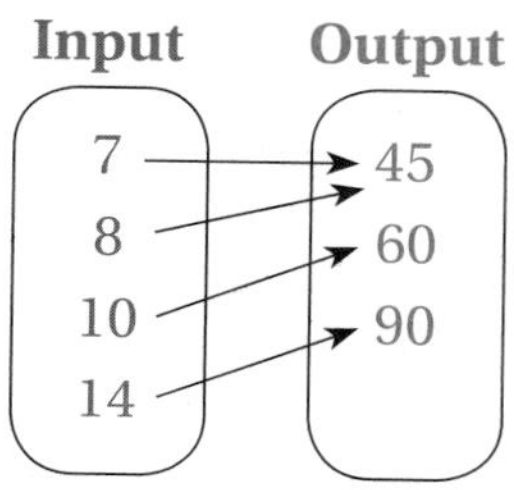

On Your Own

5. **WHAT IF?** A fifth band plays 12 songs and is onstage for 70 minutes. Draw a mapping diagram for the five bands.

4.1 Exercises

Vocabulary and Concept Check

1. **VOCABULARY** In an ordered pair, which number represents the input? the output?

2. **OPEN-ENDED** Draw a mapping diagram where the number of inputs is greater than the number of outputs.

Practice and Problem Solving

Describe the pattern in the mapping diagram. Copy and complete the diagram.

3.
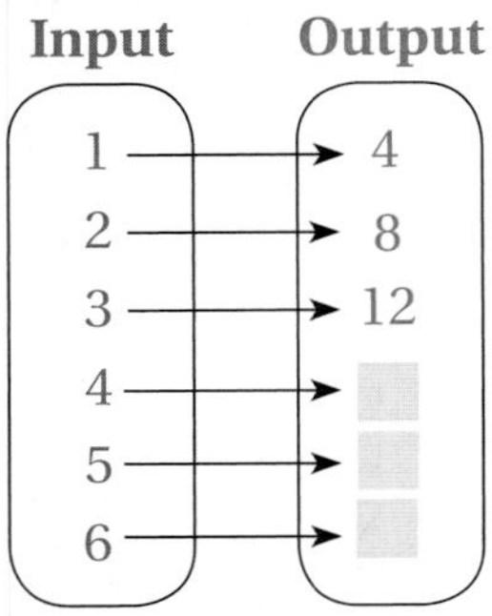

4.
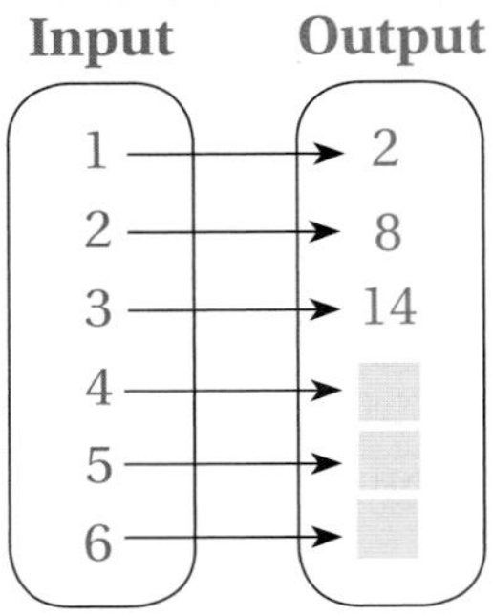

5.
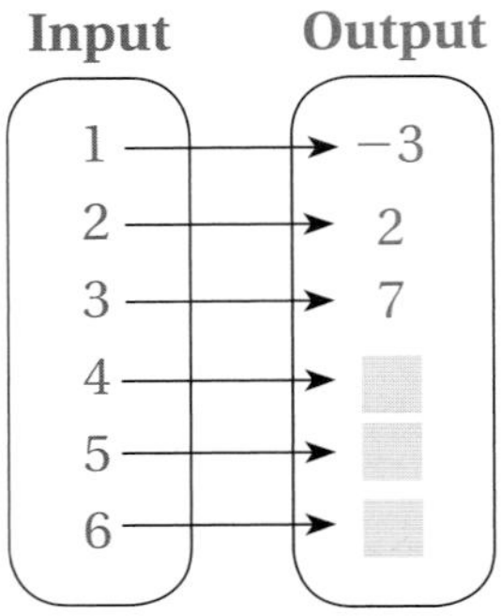

List the ordered pairs shown in the mapping diagram.

1 6.
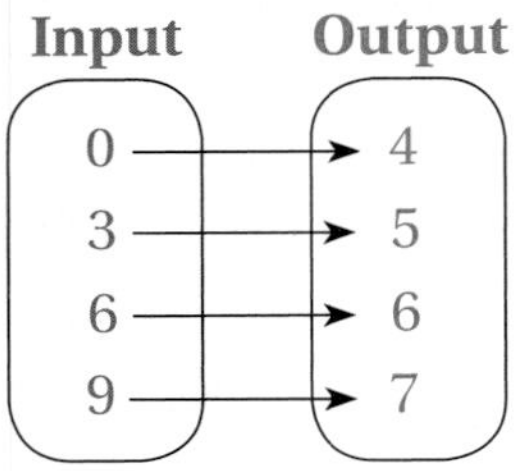

7.
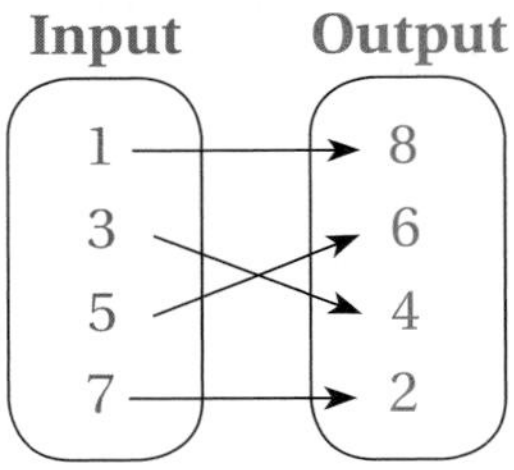

8.
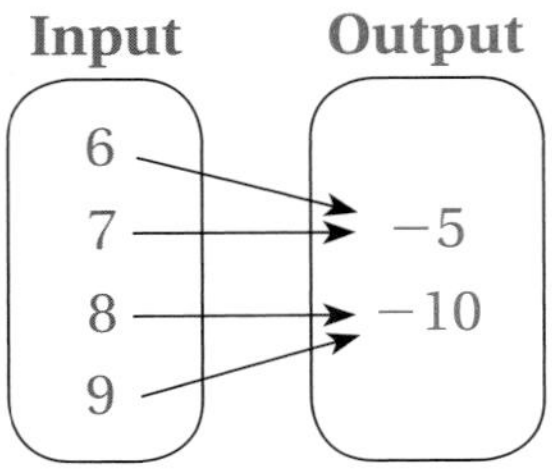

Draw a mapping diagram of the set of ordered pairs.

2 9. (1, 3), (5, 7), (8, 10), (14, 16)

10. (0, 10), (4, 6), (6, 4), (7, 3)

11. (0, −2), (1, −4), (4, 15), (6, 19)

12. (1, −2), (2, −2), (3, −2), (4, 2), (5, 2)

13. **ERROR ANALYSIS** Describe and correct the error in drawing a mapping diagram of the set of ordered pairs.

(5, 4), (6, 4), (7, 4), (8, 4)

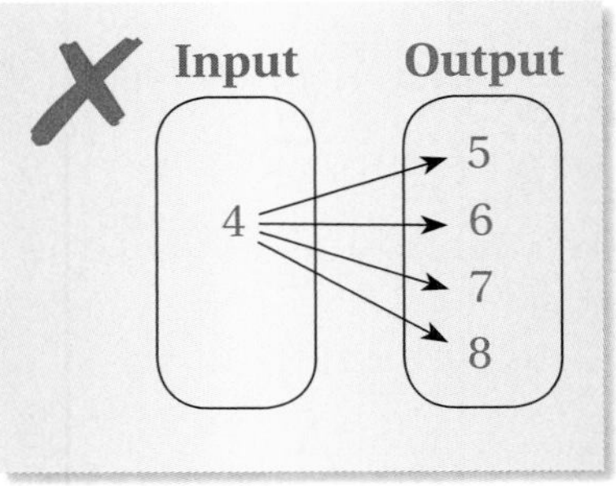

14. **AREA** The table shows the radius and approximate area for four circles. Use the table to draw a mapping diagram.

Radius (ft)	Area (ft^2)
2	12.56
3	28.26
7	153.86
9	254.34

Draw a mapping diagram for the graph. Then describe the pattern of inputs and outputs.

③ **15.**

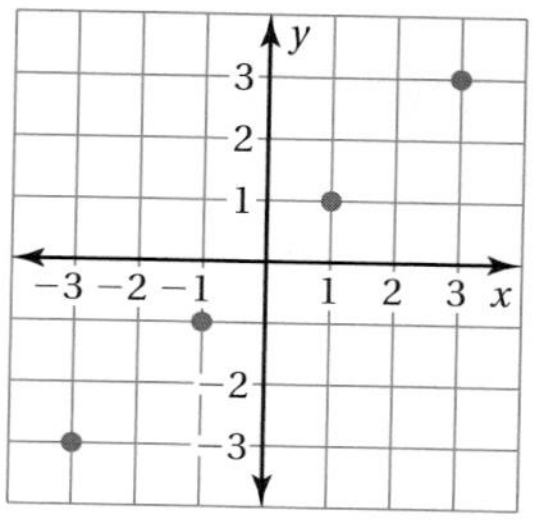

16.

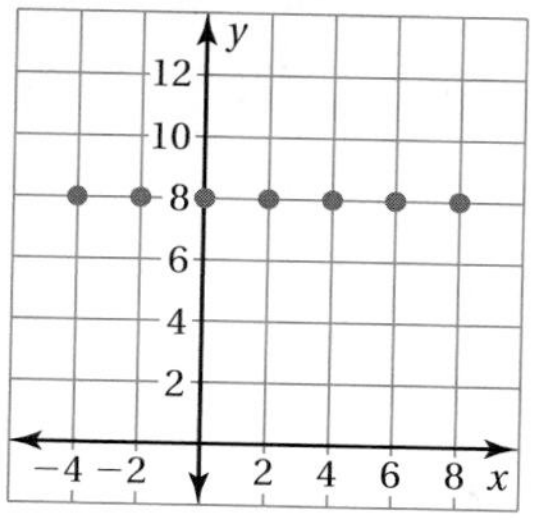

17.

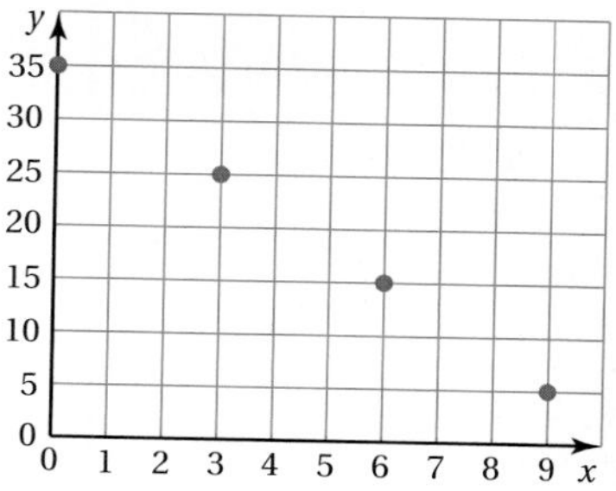

18. SCUBA DIVING The normal pressure at sea level is one atmosphere of pressure (1 ATM). As you dive below sea level, the pressure increases by 1 ATM for each 10 meters of depth.

a. Complete the mapping diagram.

b. List the ordered pairs. Then plot the ordered pairs in a coordinate plane.

c. Compare the mapping diagram and graph. Which do you prefer? Why?

d. RESEARCH What are common depths for people who are just learning to scuba dive? What are common depths for experienced scuba divers?

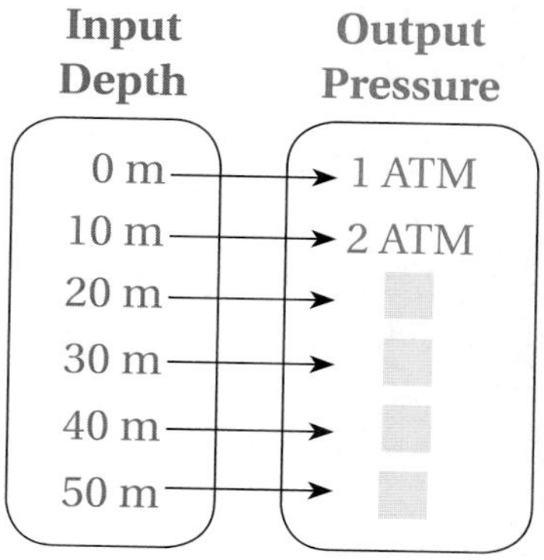

19. MOVIES A store sells previously viewed movies. The table shows the cost of buying 1, 2, 3, or 4 movies.

a. Use the table to draw a mapping diagram.

b. Describe the pattern. How does the cost per movie change as you buy more movies?

Movies	Cost
1	$10
2	$18
3	$24
4	$28

20. Critical Thinking The table shows the outputs for several inputs. What do you think the output would be for an input of 200? Explain.

Input, x	0	1	2	3	4
Output, y	25	30	35	40	45

Fair Game Review What you learned in previous grades & lessons

Write the word sentence as an equation. Then solve. *(Section 3.4 and Section 3.5)*

21. The sum of a number x and 7 is 15.

22. 3 times a number n is 24.

23. MULTIPLE CHOICE Which inequality represents the word sentence? *(Section 3.7)*

"The sum of a number x and 7 is at least 25."

Ⓐ $x + 7 > 25$ Ⓑ $x + 7 \leq 25$ Ⓒ $x + 7 \geq 25$ Ⓓ $x + 7 < 25$

4.2 Functions as Words and Equations

STANDARDS OF LEARNING
7.12

Essential Question How can you describe a function with words? How can you describe a function with an equation?

1 ACTIVITY: Describing a Function

Work with a partner. Two mapping diagrams related to the rectangle are shown. Describe each function in words. Then write an equation for each function.

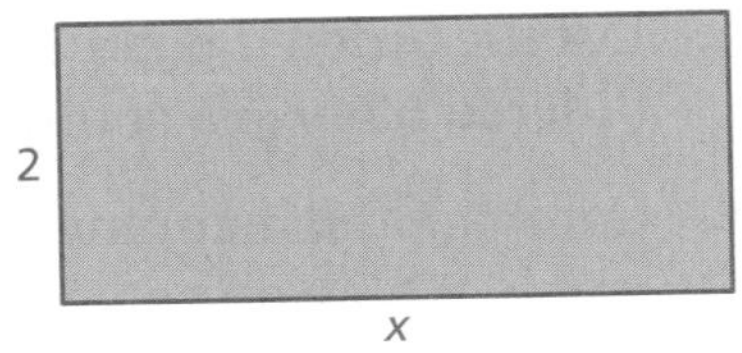

a. Area A

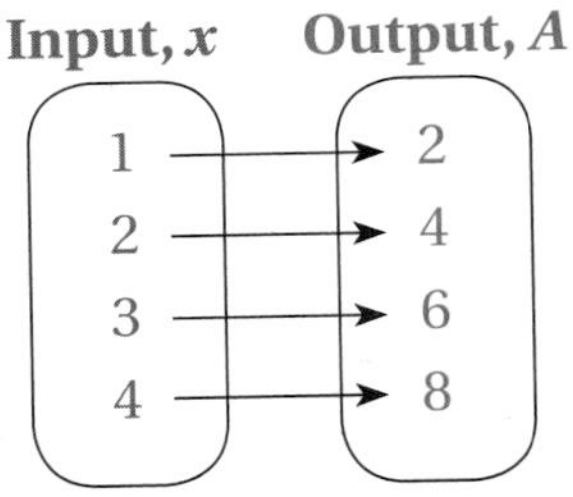

b. Perimeter P

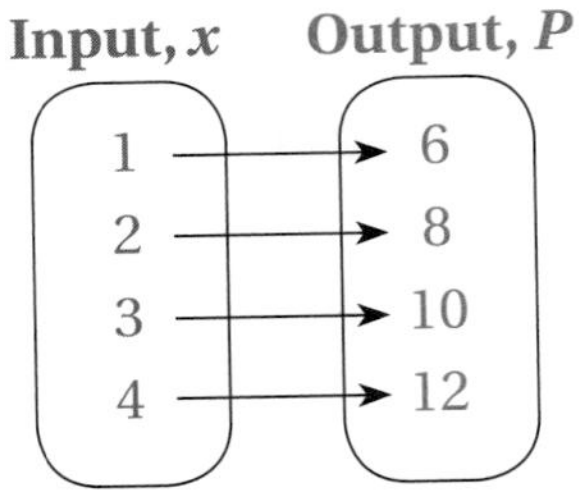

2 ACTIVITY: Describing a Function

Work with a partner. Copy and complete the mapping diagram for the area of the figure. Then write an equation that describes the function.

a.

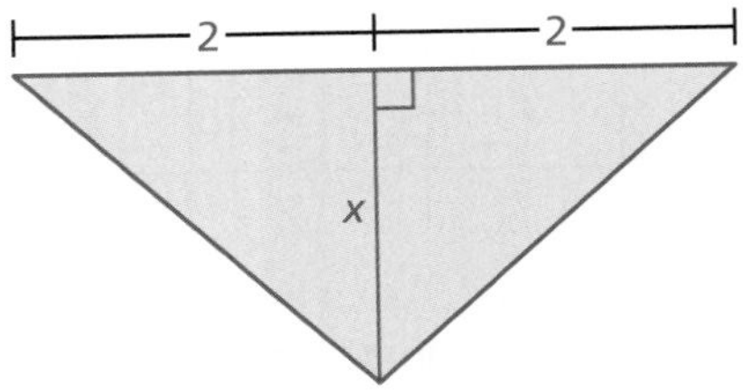

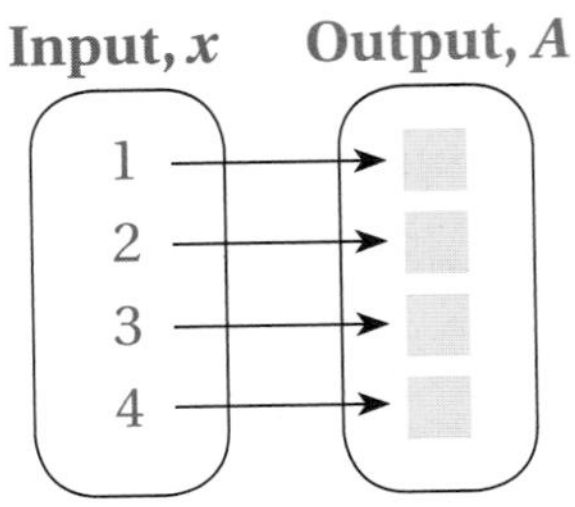

b.

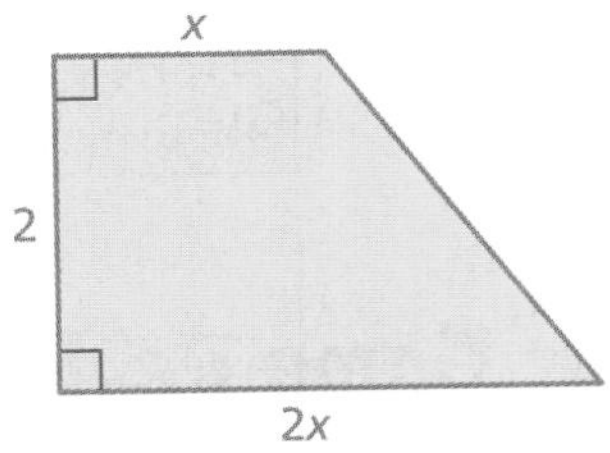

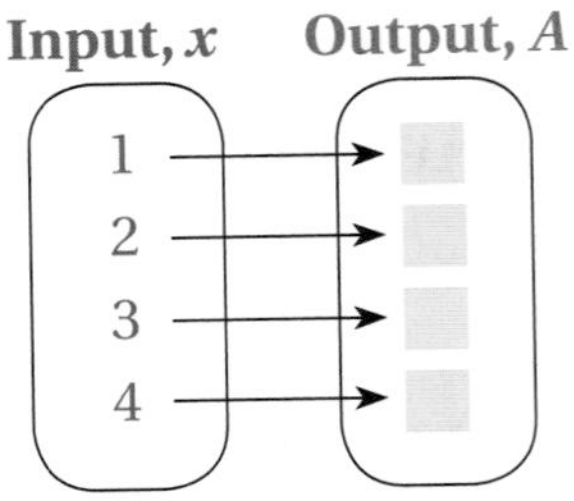

3 ACTIVITY: Describing a Function

Work with a partner. Copy and complete the mapping diagram. Then write an equation that describes the function.

a. Sales tax of 6%

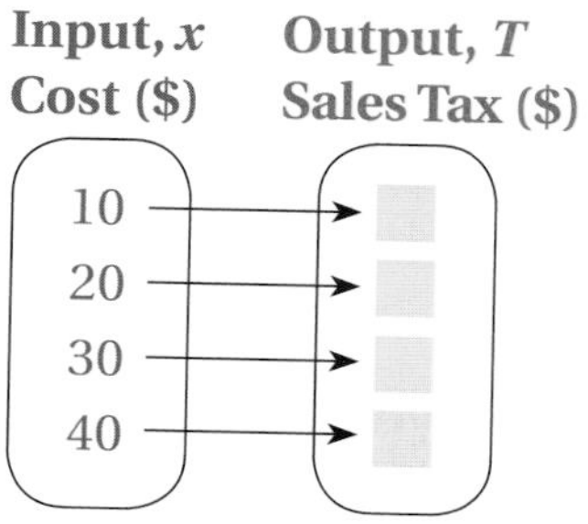

b. Phone bill of $5 per hour

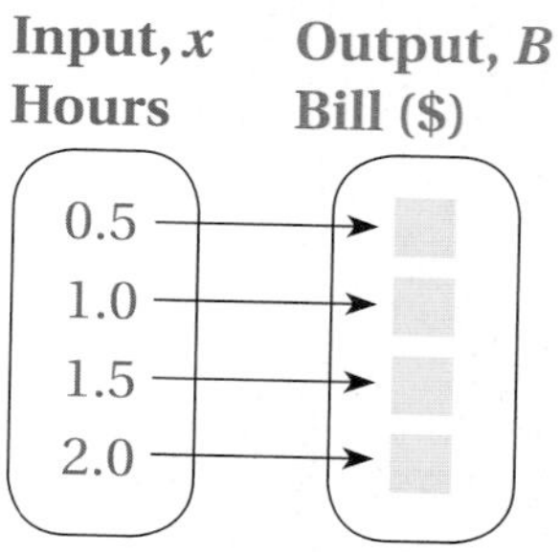

c. Perimeter of a square

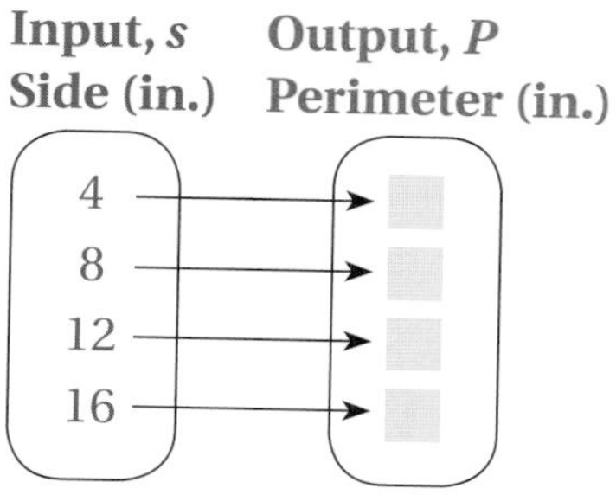

d. Surface area of a cube

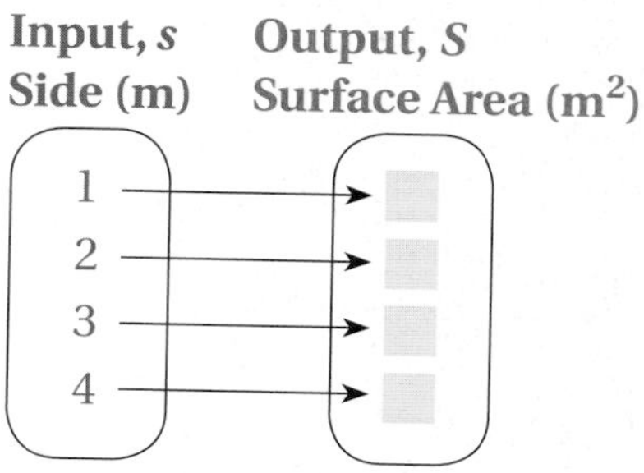

e. Circumference of a circle

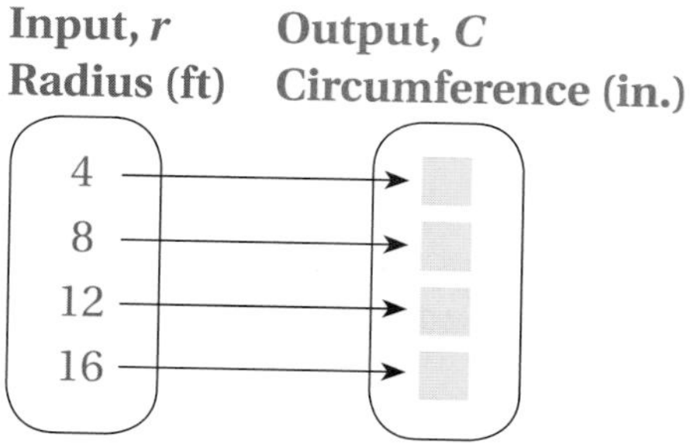

f. Area of a semicircle

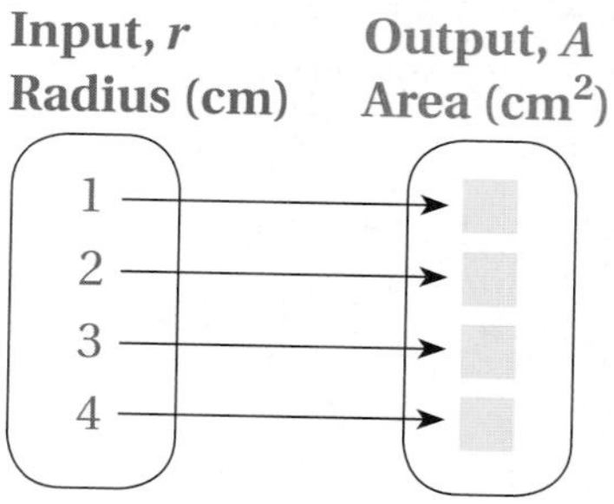

What Is Your Answer?

4. IN YOUR OWN WORDS How can you describe a function with words? How can you describe a function with an equation? Give some examples from lessons you have studied this year.

Practice Use what you learned about writing functions as words and equations to complete Exercises 3–5 on page 154.

4.2 Lesson

Check It Out
Lesson Tutorials
BigIdeasMath.com

Key Vocabulary
function rule, *p. 152*

Key Idea

Function Rule

A **function rule** describes the relationship between inputs and outputs. A function can be written as an equation in two variables.

Output → $y = 3x$ ← Input

EXAMPLE 1 Writing an Equation in Two Variables

Write an equation for "The output is five less than the input."

Words The output is five less than the input.

Equation $y \quad = \quad x - 5$

∴ An equation is $y = x - 5$.

On Your Own

Now You're Ready
Exercises 6–9

Write an equation that describes the function.

1. The output is eight more than the input.
2. The output is four times the input.
3. The output is the area of the rectangle.

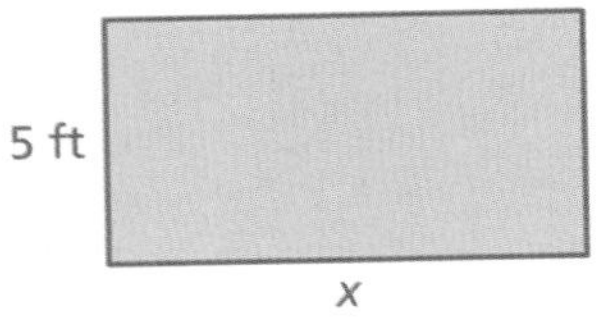

EXAMPLE 2 Evaluating a Function

What is the value of $y = 2x + 5$ when $x = 3$?

$y = 2x + 5$	Write the equation.
$= 2(3) + 5$	Substitute 3 for x.
$= 11$	Simplify.

∴ When $x = 3$, $y = 11$.

On Your Own

Now You're Ready
Exercises 10–15

Find the value of y when $x = 5$.

4. $y = 4x - 1$ 5. $y = 10x$ 6. $y = 7 - 3x$

Multi-Language Glossary at BigIdeasMath.com.

EXAMPLE 3 Checking Solutions

Study Tip

A solution of an equation with two variables is an ordered pair that makes the equation true.

Tell whether $(x, y) = (-2, 0)$ is a solution of the equation.

a. $y = x - 2$

$y = x - 2$	Write equation.
$0 \stackrel{?}{=} -2 - 2$	Substitute.
$0 \neq -4$ ✗	Simplify.

$(-2, 0)$ is *not* a solution.

b. $y = 5x + 10$

$y = 5x + 10$	Write equation.
$0 \stackrel{?}{=} 5(-2) + 10$	Substitute.
$0 = 0$ ✓	Simplify.

$(-2, 0)$ is a solution.

EXAMPLE 4 Real-Life Application

Cedar Point Park Hours
10:00 A.M.–10:00 P.M.

The maXair ride can accommodate 1000 riders in an hour.

a. Write an equation that relates riders and time.

b. How many people can ride the maXair each day?

a. **Words** Number of riders equals the number of riders per hour times the number of hours.

Variables Let r be the number of riders.
Let h be the number of hours.

Equation $r \quad = \quad 1000 \quad \cdot \quad h$

An equation is $r = 1000h$.

b. The park is open for 12 hours. Substitute 12 for h in the equation from part (a).

$r = 1000h$	Write the equation.
$= 1000(12)$	Substitute 12 for h.
$= 12{,}000$	Multiply.

Each day, 12,000 people can ride the maXair.

On Your Own

Now You're Ready
Exercises 16–21

7. Is $(2, 5)$ a solution of $y = x + 3$? Explain.
8. The Millennium Force roller coaster at Cedar Point can accommodate 1500 riders in an hour.
   a. Write an equation that relates riders and time.
   b. How many people can ride the Millennium Force each day?

4.2 Exercises

Check It Out
Help with Homework
BigIdeasMath.com

Vocabulary and Concept Check

1. **VOCABULARY** Identify the input variable and the output variable for the function rule $y = 2x + 5$.

2. **REASONING** Explain why (2, 6) is a solution of $y = x + 4$, but (6, 2) is *not* a solution.

Practice and Problem Solving

Write an equation that describes the function.

3.

Input	Output
0	0
1	4
2	8
3	12

4.

Input	Output
1	8
2	9
3	10
4	11

5.

Input	Output
10	5
20	15
30	25
40	35

Write an equation that describes the function.

1 6. The output is three less than the input.

7. The output is six times the input.

8. The output is half of the input.

9. The output is eleven more than the input.

Find the value of *y* for the given value of *x*.

2 10. $y = x + 5;\ x = 3$

11. $y = 7x;\ x = -5$

12. $y = 1 - 2x;\ x = 9$

13. $y = 3x + 2;\ x = 0.5$

14. $y = 4x + 7;\ x = \frac{5}{2}$

15. $y = \frac{x}{2} + 9;\ x = -12$

Tell whether the ordered pair is a solution of the equation.

3 16. $y = x - 6;\ (6, 0)$

17. $y = x + 7;\ (-1, 8)$

18. $y = 11x;\ (2, 22)$

19. $y = \frac{x}{5};\ (5, 25)$

20. $y = 9x + 8;\ (1, 18)$

21. $y = \frac{x}{3} - 6;\ (12, -2)$

22. **ERROR ANALYSIS** Describe and correct the error in checking whether (13, 3) is a solution of $y = 3x + 4$.

✗
$y = 3x + 4$
$13 = 3(3) + 4$
$13 = 13$ ✓
(13, 3) is a solution.

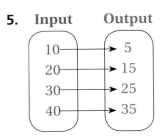

23. **SPEED** A Scrub-Jay flies at a rate of 18 feet per second.
   a. Write an equation that relates the distance *d* traveled in *s* seconds.
   b. How many feet does a Scrub-Jay fly in 30 seconds?

Find the value of *x* for the given value of *y*.

24. $y = 5x - 7$; $y = -22$ **25.** $y = 9 - 7x$; $y = 37$ **26.** $y = \dfrac{x}{4} - 7$; $y = 2$

27. BRACELETS You decide to make and sell bracelets. The cost of your materials is $84. You charge $3.50 for each bracelet.

 a. Write an equation you can use to find the profit *P* for selling *b* bracelets.

 b. You will *break even* if the cost of your materials equals your income. How many bracelets must you sell to break even?

28. AIRBOAT TOURS You want to take a two-hour airboat tour.

 a. Write an equation that represents the cost *G* of a tour at Gator Tours.

 b. Write an equation that represents the cost *S* of a tour at Snake Tours.

 c. Which is a better deal? Explain.

| Tools - Repairing |
| Smith Wood & Metal Works Inc. |
| 51 Penn Ave. Lockwood 45845 845-5845 |
| Lake Mead Power Tools |
| 2514 Sun Dr. Meadville 45895 845-3145 |
| **Tools - Sharpening** |
| See Sharpening Services |
| **Tools - Steel Distrs** |
| See Steel Distributing & Warehousing |

Gator Tours
$35 boarding fee plus $5 each 1/2 hour
all rates are per person

| Top Soil |
| CONNIE'S LANDSCAPE SUPL. |
| 8745 Wattsburg Rd. 16547 485-3254 |
| Sunnyville Peat Products |
| 512 Turnpike Rd. Wonderland 16454 .. 458-3251 |
| Lordstown Landscape Supl |
| 6589 W. Town Rd. 14585 489-1125 |

| Tours |

Snake Tours
$25 per hour

All rates are per person

| Towing - Auto |
| A & B Service Center |
| 5485 East Way Lane 54845 485-3695 |
| A-1 Towing |
| 5845 Shipping Lane 45845 548-1451 |
| Ace Towing |
| 4856 Airport Road 45854 125-1584 |
| Armstrong Auto and Repair |
| 4584 Creek Rd. 58452 584-3147 |
| Bennett's Towing Service |
| 125 Penny Lane 45848 458-2158 |
| Tom's Towing |
| 485 Grahamville Rd. 48546 589-7588 |
| Vince Auto Repair |
| 5486 Walnut Ave. 48543 |

| Tours |
| Our Town's Visual Tours |
| 484 W. County Rd. 48451 985-3231 |
| Get-A-Way Travel & Tours |
| 4845 Conway Ave. 48643 479-3641 |
| Our Town's Visual Tours |
| 484 W. County Rd. 48451 985-3231 |
| Get-A-Way Travel & Tours |
| 4845 Conway 48° |

29. CRITICAL THINKING Can you write a function for the area of any rectangle given the perimeter of the rectangle? Explain.

30. **Puzzle** The blocks that form the diagonals of each square are shaded. Each block is one square unit. Find the "green area" of Square 20. Find the "green area" of Square 21. Explain your reasoning.

Square 1

Square 2

Square 3

Square 4

Square 5

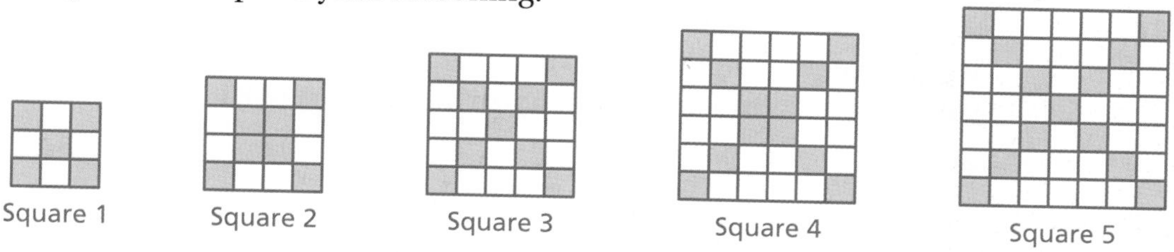

Fair Game Review What you learned in previous grades & lessons

Copy and complete the table. *(Skills Review Handbook)*

31.

x	1	2	3
x + 7			

32.

x	6	8	13
x − 3			

33. MULTIPLE CHOICE You want to volunteer for at most 20 hours each month. So far, you have volunteered for 7 hours this month. Which inequality represents the number of hours you can volunteer for the rest of this month? *(Section 3.7)*

 Ⓐ $h \geq 13$ **Ⓑ** $h \geq 27$ **Ⓒ** $h \leq 13$ **Ⓓ** $h < 27$

4.3 Input-Output Tables

STANDARDS OF LEARNING
7.12

Essential Question How can you use a table to describe a function?

1 ACTIVITY: Using a Function Table

Work with a partner.

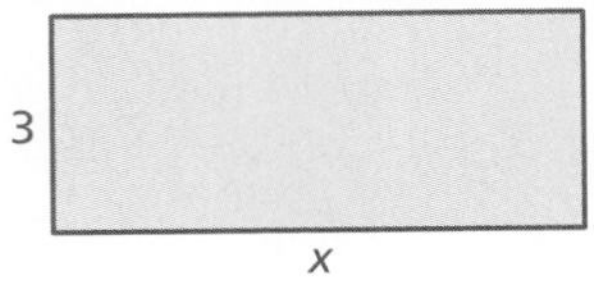

a. Copy and complete the table for the perimeter of the rectangle.

Input, x	1	2	3	4	5
Output, P					

b. Write an equation that describes the function.

c. Use your equation to find the value of x for which the perimeter is 50.

2 ACTIVITY: Using a Function Table

Work with a partner. Use the strategy shown in Activity 1 to make a table that shows the pattern for the area. Write an equation that describes the function. Then use your equation to find which figure has an area of 81.

1 square unit

a.

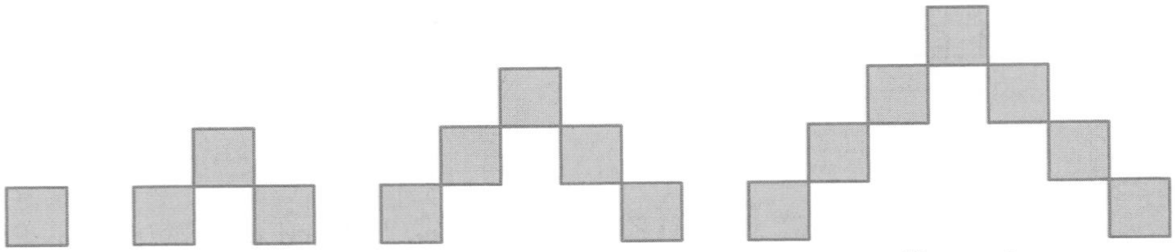

Figure 1 Figure 2 Figure 3 Figure 4

b.

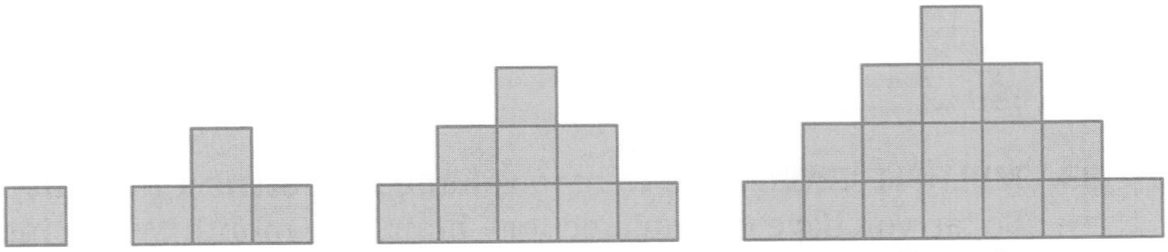

Figure 1 Figure 2 Figure 3 Figure 4

3 ACTIVITY: Making a Function Table

Work with a partner. Copy and complete a sales tax table for each of the four cities.

Madison, WI, 5.50%

Sale, *x*	\$20	\$30	\$40	\$50	\$60
Sales Tax, *T*					

Ann Arbor, MI, 6.00%

Sale, *x*	\$20	\$30	\$40	\$50	\$60
Sales Tax, *T*					

Edison, NJ, 7.00%

Sale, *x*	\$20	\$30	\$40	\$50	\$60
Sales Tax, *T*					

Norman, OK, 7.50%

Sale, *x*	\$20	\$30	\$40	\$50	\$60
Sales Tax, *T*					

What Is Your Answer?

4. IN YOUR OWN WORDS How can you use a table to describe a function? Describe an example of a function table in real life.

Amount of Sale	Tax
.10 – .29	.01
.30 – .49	.02
.50 – .69	.03
.70 – .89	.04
.90 – 1.09	.05

"Dear Sir: Yesterday, I bought a piece of 9-cent candy eight times and paid NO tax. Today, I bought eight pieces at once and you charged me \$0.04 tax. What's going on?"

Practice Use what you learned about input-output tables to complete Exercises 3 and 4 on page 160.

4.3 Lesson

Check It Out
Lesson Tutorials
BigIdeasMath.com

Key Vocabulary
input-output table, *p. 158*

Key Idea

Input-Output Tables

A function can be represented by an **input-output table**. The table below is for the function $y = x + 2$.

Input, x	Output, y	
		← $y = x + 2$
1	3	← $3 = 1 + 2$
2	4	← $4 = 2 + 2$
3	5	← $5 = 3 + 2$
4	6	← $6 = 4 + 2$

EXAMPLE 1 Completing Input-Output Tables

Write an equation for the function. Then copy and complete the table.

a. The output is 4 less than the input.

Input, x	2	3	4	5
Output, y				

b. The output is twice the input.

Input, x	−7	−5	−3	−1
Output, y				

a. An equation is $y = x - 4$.

Input, x	2	3	4	5
Output, y	−2	−1	0	1

$y = x - 4$

b. An equation is $y = 2x$.

Input, x	−7	−5	−3	−1
Output, y	−14	−10	−6	−2

$y = 2x$

On Your Own

Now You're Ready
Exercises 5 and 6

Write an equation for the function. Then copy and complete the table.

1. The output is 5 more than the input.

Input, x	1	3	5	7
Output, y				

2. The output is the product of 7 and the input.

Input, x	−4	−2	0	2
Output, y				

EXAMPLE 2 Standardized Test Practice

Which function rule is shown by the table?

Input, x	Output, y
-1	-5
2	10
4	20
8	40

Ⓐ $y = 5x$ Ⓑ $y = \frac{x}{5}$

Ⓒ $y = x + 4$ Ⓓ $y = 10x$

Look at the relationship between the inputs and outputs. Each output y is 5 times the input x. So, the function rule is $y = 5x$.

∴ The correct answer is Ⓐ.

EXAMPLE 3 Finding a Missing Input

Input, x	Output, y
1	7
5	15
10	25
20	45
?	53

Each output in the table is 5 more than twice the input. Find the missing input.

Step 1: Write an equation for the function shown by the table.

Words Output is five more than twice the input.

Variables Let y be the output value and x be the input value.

Equation $y \quad = \quad 5 + \quad 2 \cdot \quad x$

An equation is $y = 5 + 2x$.

Step 2: Substitute 53 for y. Then solve for x.

$y = 5 + 2x$	Write the equation.
$53 = 5 + 2x$	Substitute 53 for y.
$48 = 2x$	Subtract 5 from each side.
$24 = x$	Divide each side by 2.

∴ The missing input is 24.

Check

$2x + 5 = 53$

$2(24) + 5 \stackrel{?}{=} 53$

$48 + 5 \stackrel{?}{=} 53$

$53 = 53$ ✓

On Your Own

Now You're Ready
Exercises 7–13

Use the first three input values to write an equation for the function shown by the table. Then find the missing input.

3.

Input, x	Output, y
1	5
3	7
7	11
?	25

4.

Input, x	Output, y
-10	-5
-4	-2
2	1
?	4

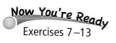

4.3 Exercises

Check It Out
Help with Homework
BigIdeasMath.com

Vocabulary and Concept Check

1. **VOCABULARY** Explain how you can use an input-output table to represent a function.
2. **DIFFERENT WORDS, SAME QUESTION** Which is different? Find "both" answers.

What output is 4 more than twice the input 3?	What output is twice the sum of the input 3 and 4?
What output is the sum of 2 times the input 3 and 4?	What output is 4 increased by twice the input 3?

Practice and Problem Solving

Copy and complete the input-output table for the function.

3. $y = x + 5$

Input, x	1	2	3	4
Output, y				

4. $y = 7x$

Input, x	-6	-4	-2	0
Output, y				

Write an equation for the function. Then copy and complete the table.

1 5. The output is 3 more than the input.

Input, x	-6	-3	0	3
Output, y				

6. The output is 5 times the input.

Input, x	1	3	5	7
Output, y				

Write an equation for the function shown by the table.

2 7.

Input, x	1	2	3	4
Output, y	9	10	11	12

8.

Input, x	-2	-4	-6	-8
Output, y	-4	-8	-12	-16

9.

Input, x	-3	0	3	6
Output, y	-1	0	1	2

10.

Input, x	3	5	7	9
Output, y	-3	-1	1	3

11. **ERROR ANALYSIS** Describe and correct the error in writing an equation for the function shown by the table.

✗

Input, x	0	4	8	12
Output, y	0	1	2	3

$y = 4x$

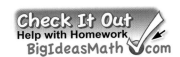

In Exercises 12 and 13, copy and complete the table.

3 **12.** For each output, multiply the input by 4, then subtract 5.

Input, x	2	3	4	7		
Output, y	3	7	11	23	35	55

13. For each output, divide the input by 2, then add 4.

Input, x		−6	−4	−2		
Output, y	0	1	2	3	8	13

14. GEOGRAPHY You travel along US Highway 1 from mile marker 0 in Key West to mile marker 100 in Key Largo.

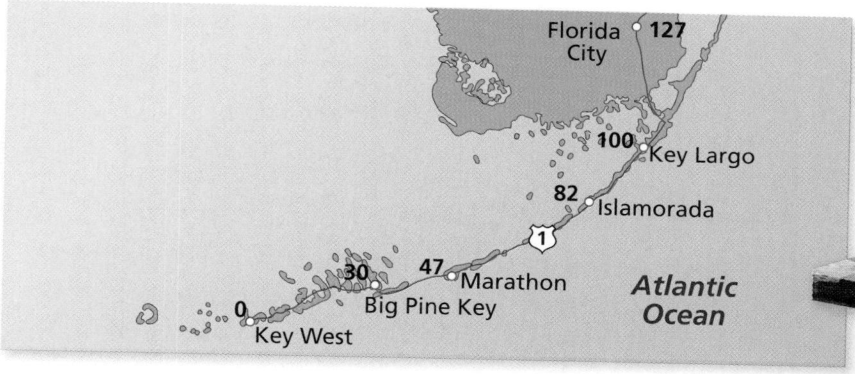

a. Copy and complete the input-output table.

Distance from Key West, x	0	30	47	82	100
Distance to Key Largo, y					

b. Write a function rule in which x is the input and y is the output.

c. Can you use your function rule to find the distance to Florida City? If not, write a function rule that you can use.

15. TIME Make an input-output table with the Greenwich Mean Time (GMT) hourly times as inputs, and times where you live as outputs. Write a function rule for the data.

16. Critical Thinking Write an equation with the same outputs as $y = 2x + 3$ for $x = 0, 1, 2, 3,$ and 4.

Fair Game Review What you learned in previous grades & lessons

Plot the ordered pairs in the same coordinate plane. *(Skills Review Handbook)*

17. (1, 2) **18.** (−1, −4) **19.** (2, −3) **20.** (−2, 4)

21. MULTIPLE CHOICE Which is the solution of the inequality $6x \leq 24$? *(Section 3.8)*

Ⓐ $x < 4$ Ⓑ $x \leq 4$ Ⓒ $x < 144$ Ⓓ $x \leq 144$

Check It Out
Graphic Organizer
BigIdeasMath.com

You can use a **four square** to organize information about a topic. Each of the four squares can be a category, such as *definition, vocabulary, example, non-example, words, algebra, table, numbers, visual, graph,* or *equation*. Here is an example of a four square for a function.

Function

Definition
A relationship that pairs each input with exactly one output

Example
in words:
The output is four times the input.
as an equation:
$y = 4x$

Visual
mapping diagram

Input		Output
0	→	0
1	→	4
2	→	8
3	→	12

Table
input-output table

Input, x	Output, y
0	0
1	4
2	8
3	12

On Your Own

Make a four square to help you study these topics.

1. function
2. mapping diagram
3. input-output table

After you complete this chapter, make four squares for the following topics.

4. graph of a function
5. linear function
6. arithmetic sequence
7. geometric sequence

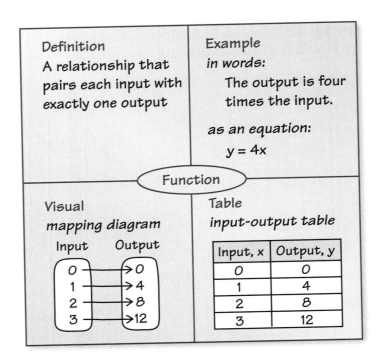

"My four square shows that my new red skateboard is faster than my old blue skateboard."

4.1–4.3 Quiz

Check It Out
Progress Check
BigIdeasMath.com

List the ordered pairs shown in the mapping diagram. *(Section 4.1)*

1.

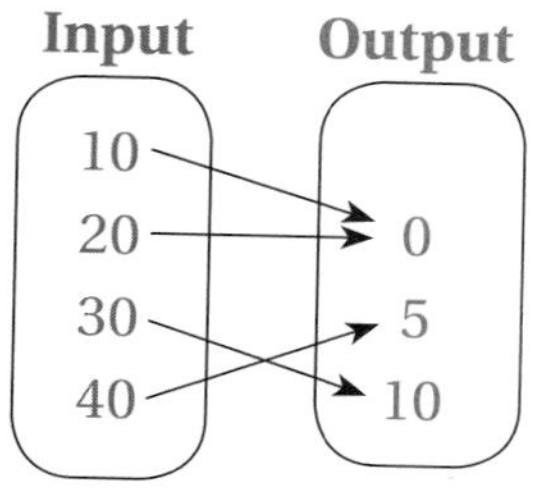

2.

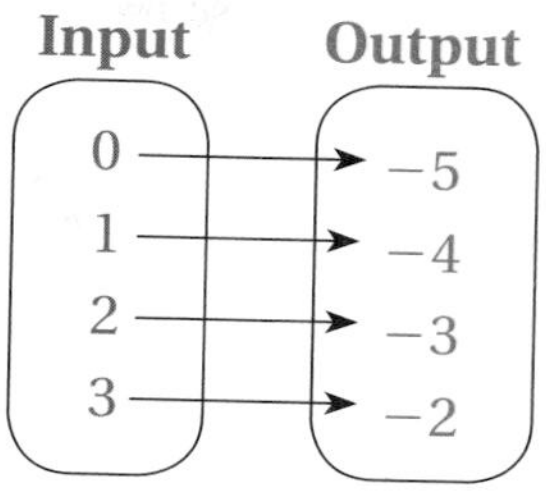

Find the value of *y* for the given value of *x*. *(Section 4.2)*

3. $y = 10x;\ x = -3$

4. $y = 6 - 2x;\ x = 11$

5. $y = 4x + 5;\ x = \frac{1}{2}$

Tell whether the ordered pair is a solution of the equation. *(Section 4.2)*

6. $y = x - 1;\ (5, 6)$

7. $y = 5x + 3;\ (-2, -7)$

8. $y = \frac{x}{7};\ (42, 6)$

9. Write an equation for the function "The output is the product of 9 and the input." Then copy and complete the table. *(Section 4.3)*

Input, *x*	1	2	3	4
Output, *y*				

10. Write an equation for the function shown by the table. *(Section 4.3)*

Input, *x*	0	8	16	24
Output, *y*	0	−1	−2	−3

11. RACE You run a 10-kilometer race at a steady pace of 1 kilometer every 6 minutes. Copy and complete the input-output table. Then write a function rule in which *x* is the input and *y* is the output. *(Section 4.3)*

Distance, *x*	1	2	6		10
Time, *y*	6	12	36	48	

12. PUPPIES The table shows the ages of four puppies and their weights. Use the table to draw a mapping diagram. *(Section 4.1)*

Age (weeks)	Weight (oz)
3	11
4	85
6	85
10	480

13. GIFT CARD You have a \$45 gift card for an online music store. Each song costs \$0.90. *(Section 4.2)*

a. Write an equation you can use to find the number of dollars *d* remaining on the card after you buy *s* songs.

b. What is the greatest number of songs you can buy with the gift card?

4.4 Graphs

STANDARDS OF LEARNING 7.12

Essential Question How can you use a graph to describe a function?

1 ACTIVITY: Interpreting a Graph

Work with a partner. Use the graph to test the truth of each statement. If the statement is true, write an equation that shows how to get one measurement from the other measurement.

a. "You can find the horsepower of a race car engine if you know its volume in cubic inches."

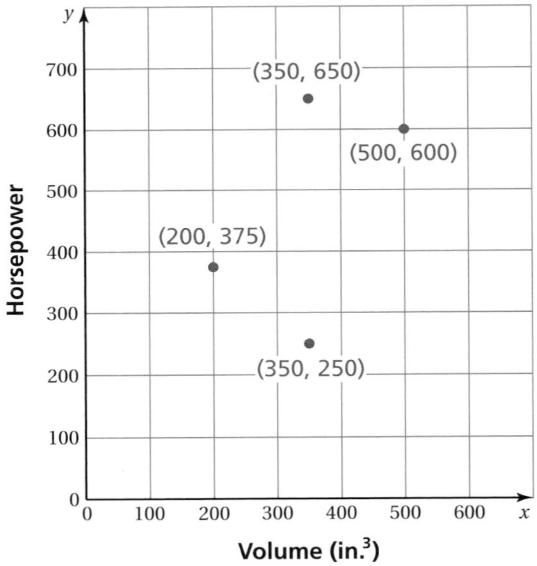

b. "You can find the volume of a race car engine in cubic centimeters if you know its volume in cubic inches."

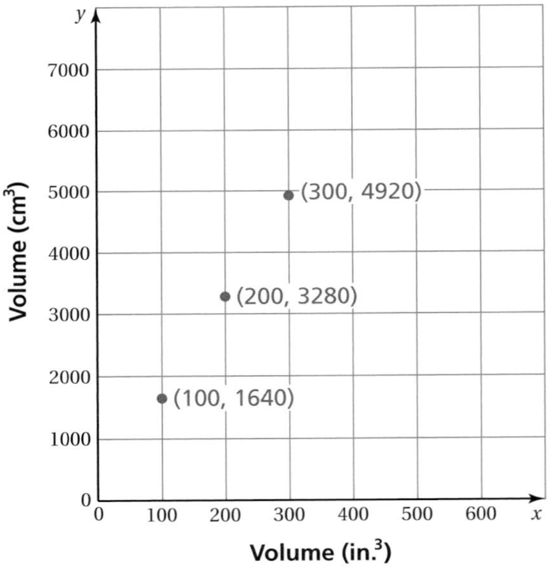

2 ACTIVITY: Interpreting a Graph

Work with a partner. The table shows the average speeds of the winners of the Daytona 500. Graph the data. Does the graph allow you to predict future winning speeds? Explain why or why not.

Year	2000	2001	2002	2003	2004	2005	2006	2007	2008
Speed (mi/h)	156	162	143	134	156	135	143	149	153

3 ACTIVITY: Conducting an Experiment

Work with a partner.

Collect Materials:

- Metal washer
- String (at least 15 in. long)
- Stopwatch

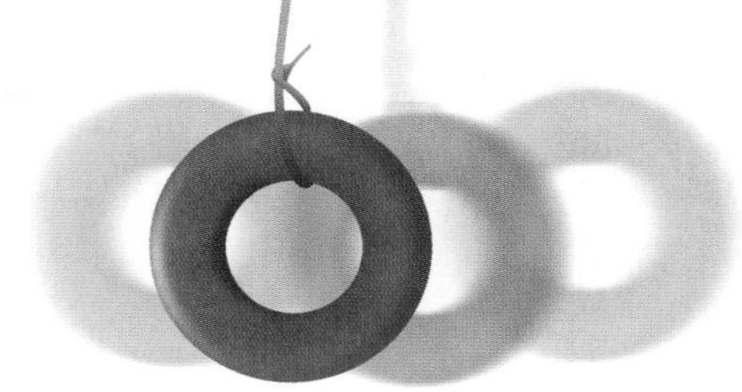

Perform the Experiment:

- Tie one end of the string securely around the washer.
- Hold the string 6 inches from the washer. Swing the washer and measure the time it takes to swing back and forth 10 times.
- Record your result in a table.
- Repeat the experiment when holding the string at lengths of 7 in., 8 in., 9 in., 10 in., 11 in., and 12 in.

Analyze the Results:

- Make a graph of your data.
- Describe the graph.

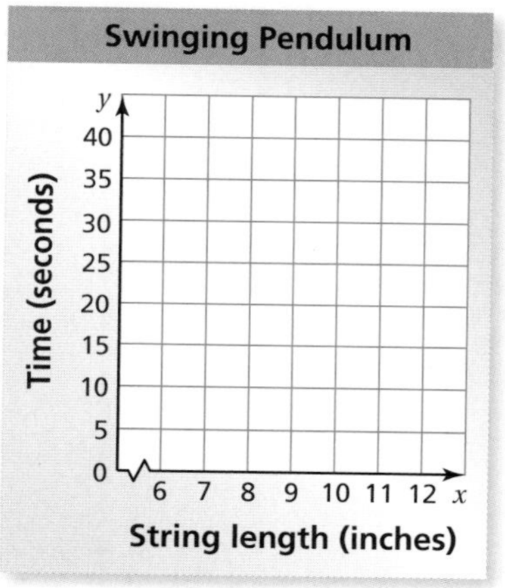

Use Your Results to Predict:

- Use your graph to predict how long it will take a 14-inch pendulum to swing 10 times.

Test Your Prediction:

- Hold the string 14 inches from the washer and repeat the experiment. How close was your prediction?

What Is Your Answer?

4. **IN YOUR OWN WORDS** How can you use a graph to describe a function? Find a graph in a magazine, in a newspaper, or on the Internet that allows you to predict the future.

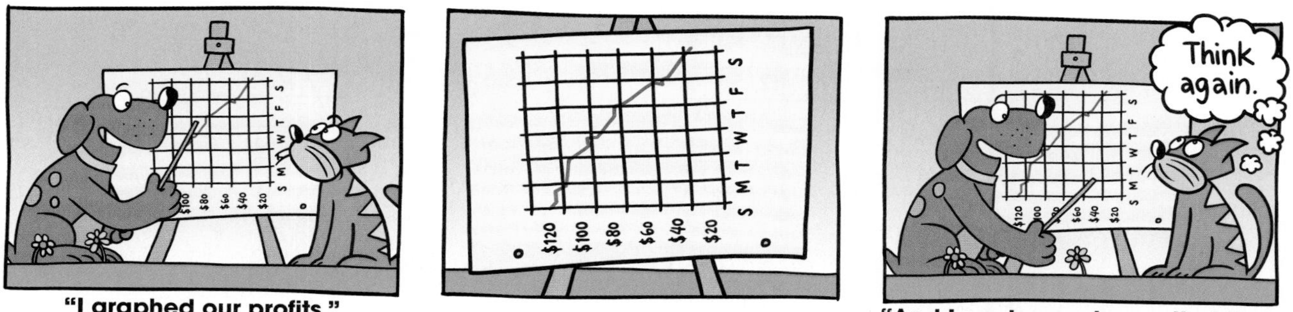

"I graphed our profits."

"And I am happy to say that they are going up every day!"

Practice Use what you learned about graphs to complete Exercises 4–7 on page 168.

4.4 Lesson

Check It Out
Lesson Tutorials
BigIdeasMath.com

Key Vocabulary
graph, *p. 166*

Key Idea

Graph of a Function

A function can be represented by a **graph**. The graph below is for the function $y = x + 2$.

Input, x	Output, y	Ordered Pair, (x, y)
1	3	(1, 3)
2	4	(2, 4)
3	5	(3, 5)

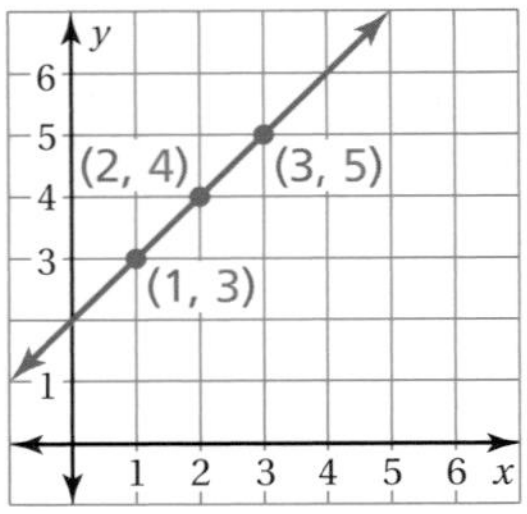

By drawing a line through the points, you graph *all* of the solutions of $y = x + 2$.

EXAMPLE 1 Graphing a Function

Graph $y = 2x - 3$.

Make an input-output table. Use the values 1, 2, and 3 for x.

x	$y = 2x - 3$	y	(x, y)
1	$y = 2(1) - 3$	-1	$(1, -1)$
2	$y = 2(2) - 3$	1	$(2, 1)$
3	$y = 2(3) - 3$	3	$(3, 3)$

Plot the ordered pairs.

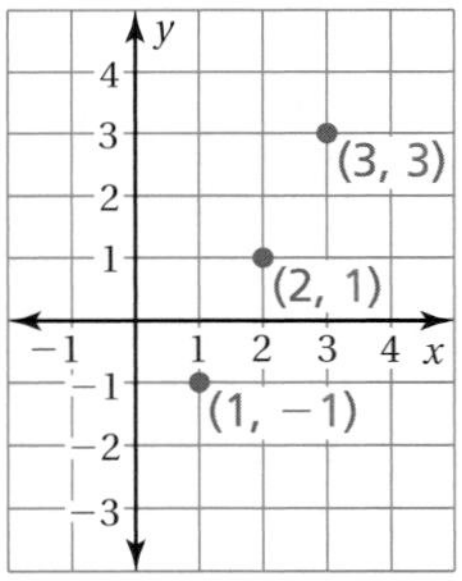

Draw a line through the points.

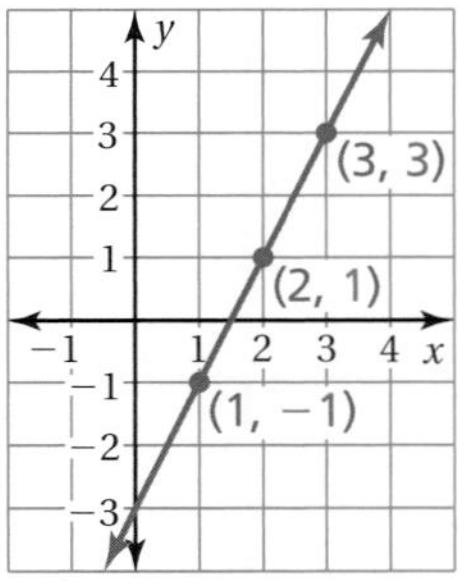

On Your Own

Now You're Ready
Exercises 8–15

Graph the function.

1. $y = x + 1$
2. $y = -3x$
3. $y = 3x + 2$

EXAMPLE 2 Graphing a Function

Use the function $p = 20g$ to find the number of pounds p of carbon dioxide produced by burning g gallons of gasoline. Graph the function.

Make an input-output table.

Input, g	$p = 20g$	Output, p	Ordered Pair, (g, p)
1	$p = 20(1)$	20	(1, 20)
2	$p = 20(2)$	40	(2, 40)
3	$p = 20(3)$	60	(3, 60)

Plot the ordered pairs.

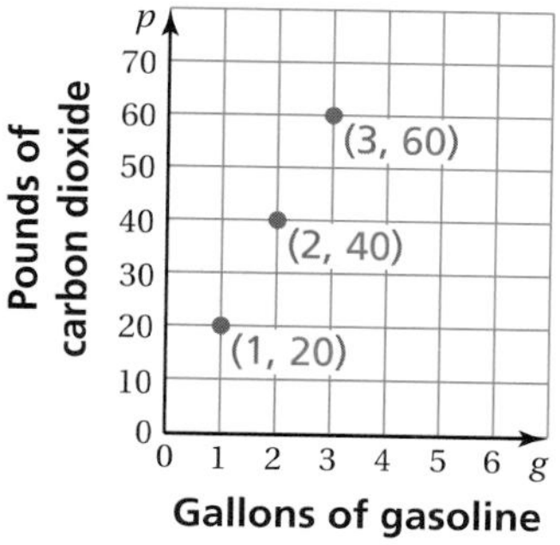

Draw a line through the points.

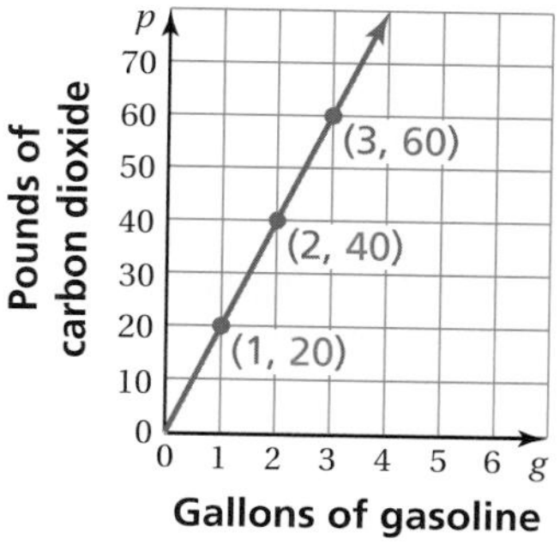

Reading

Because you cannot have a negative number of gallons, use only positive values of g.

On Your Own

Now You're Ready
Exercise 17

4. Use the function $d = 35t$ to find the distance d (in miles) that a boat travels in time t (in hours). Graph the function.

Summary

Representing a Function

There are several ways to represent a function.

Words Each output is 2 more than the input.

Equation $y = x + 2$

Input-Output Table

Input, x	Output, y
1	3
2	4
3	5
4	6

Mapping Diagram

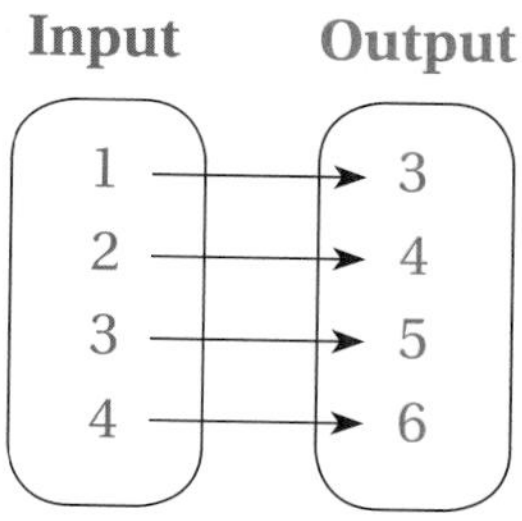

Graph

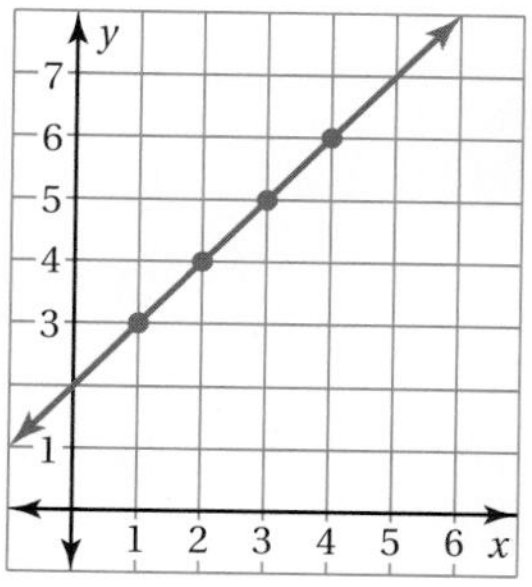

4.4 Exercises

Check It Out
Help with Homework
BigIdeasMath.com

Vocabulary and Concept Check

1. **VOCABULARY** Describe steps you can use to draw a graph of a function.
2. **OPEN-ENDED** Describe four ways to represent the function $y = 4x + 5$.
3. **WRITING** You are given the graph of a function. Explain how you could find the function rule that describes the graph.

Practice and Problem Solving

Graph the data.

4.

Input, *x*	0	2	4	6
Output, *y*	0	4	8	12

5.

Input, *x*	0	1	2	3
Output, *y*	1	2	3	4

6.

Input, *x*	Output, *y*
1	0
3	−2
5	−4
7	−6

7.

Input, *x*	Output, *y*
2	0.5
4	1
6	1.5
8	2

Graph the function.

1 8. $y = x + 4$ 9. $y = 2x$ 10. $y = -5x + 3$ 11. $y = \frac{x}{4}$

12. $y = 2x + 5$ 13. $y = x - 8$ 14. $y = \frac{3}{2}x + 1$ 15. $y = 1 + 0.5x$

16. **ERROR ANALYSIS** Describe and correct the error in graphing the line from the input-output table.

Input, *x*	−4	−2	0	2
Output, *y*	−1	1	3	5

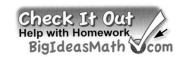

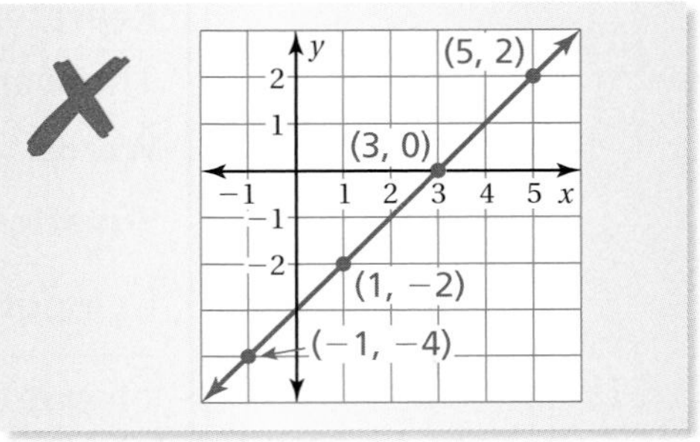

2 17. **DOLPHIN** Use the function $p = 30d$ to find the number of pounds p of fish that a dolphin eats in d days. Graph the function.

Match each graph with a function.

18.

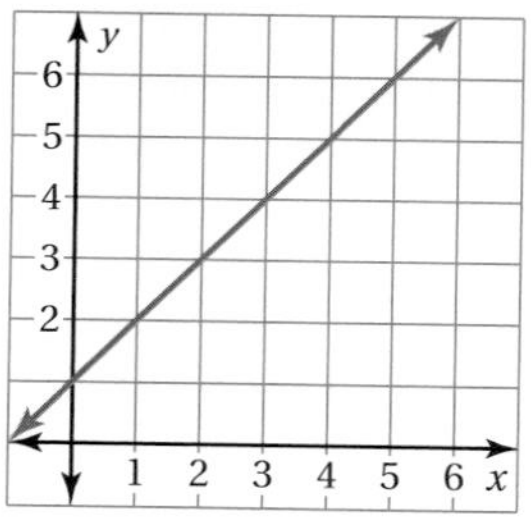

19.

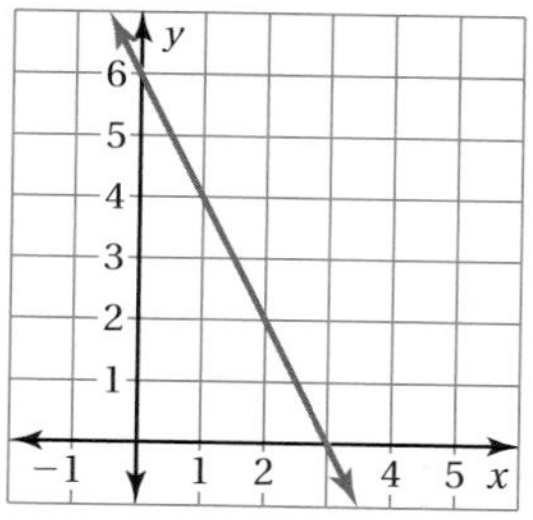

20.

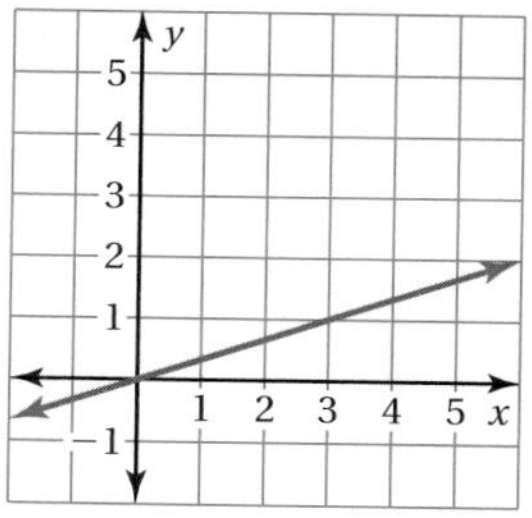

A. $y = \frac{x}{3}$　　B. $y = x + 1$　　C. $y = -2x + 6$

21. SALE A furniture store is having a sale where everything is 40% off.

a. Write a function that you can use to find the amount of discount d on an item with a regular price p.

b. Graph the function using the inputs 100, 200, 300, 400, and 500 for p.

c. You buy a bookshelf that has a regular price of $85. What is the sale price of the bookshelf?

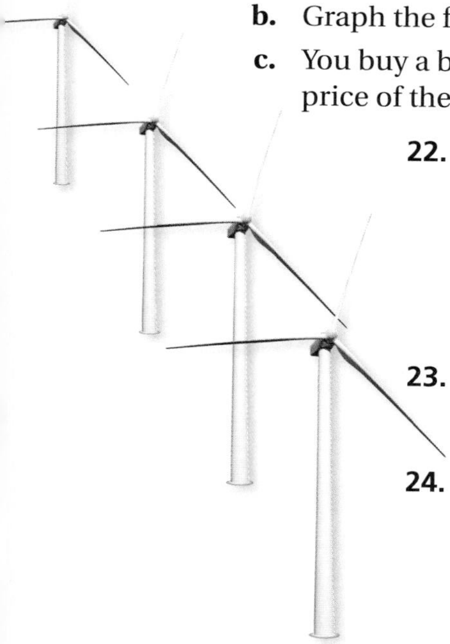

22. WIND TURBINE The table shows the number of rotations a wind turbine makes in x minutes.

Minutes, x	1	2	3	4	5
Rotations, y	25	50	75	100	125

a. Graph the data.

b. Find a function to describe the data.

23. REASONING The graph of a function is a straight line that goes through the points (3, 2), (5, 8), and (8, y). What is the value of y?

24. Critical Thinking Make an input-output table where the input is the side length of a square and the output is the *perimeter*. Make a second input-output table where the input is the side length of a square and the output is the *area*. Graph both functions in the same coordinate plane. Compare the functions and graphs.

Fair Game Review What you learned in previous grades & lessons

Find the value of *y* for the given value of *x*. *(Section 4.2)*

25. $y = x + 2;\ x = -4$

26. $y = 4x - 3;\ x = 3$

27. $y = 3x + 1;\ x = 6$

28. $y = 16x;\ x = -7$

29. MULTIPLE CHOICE The rectangular rug has a perimeter of 224 inches. What is the width of the rug? *(Skills Review Handbook)*

Ⓐ 3.5 in.　Ⓑ 36 in.

Ⓒ 48 in.　Ⓓ 96 in.

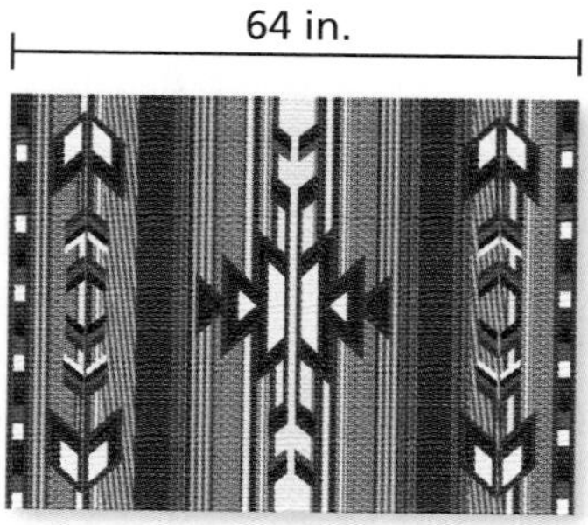

4.5 Analyzing Graphs

STANDARDS OF LEARNING
7.12

Essential Question How can you analyze a function from its graph?

1 ACTIVITY: Analyzing Graphs

Work with a partner. Copy and complete the table for the given situation. Then make a graph of the data. Write an equation for the function. Describe the characteristics of the graph.

a. Find the area of a square with side length *s*.

Side, *s*	1	2	3	4
Area, *A*				

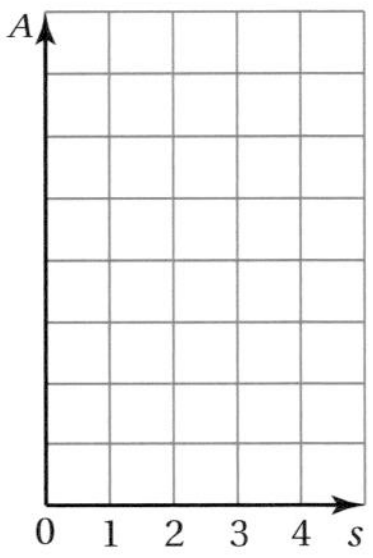

b. Find the amount earned for working *h* hours at $3 per hour.

Hour, *h*	1	2	3	4
Amount, *A*				

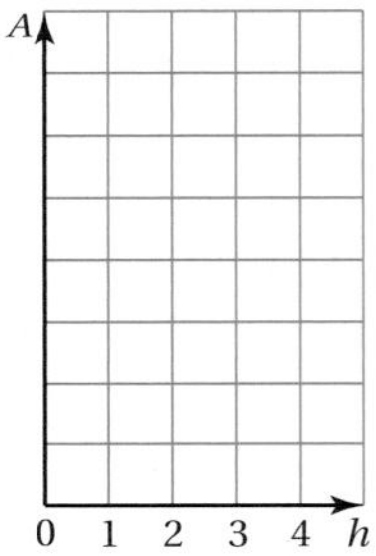

c. You start with $20 in a savings account. Find the amount left in the account when you withdraw $2 each day *d*.

Day, *d*	1	2	3	4
Amount, *A*				

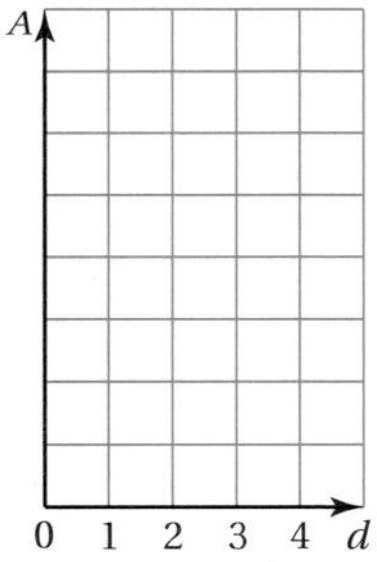

d. You start with $10 in a savings account. Find the amount in the account when you deposit $2 each day *d*.

Day, *d*	1	2	3	4
Amount, *A*				

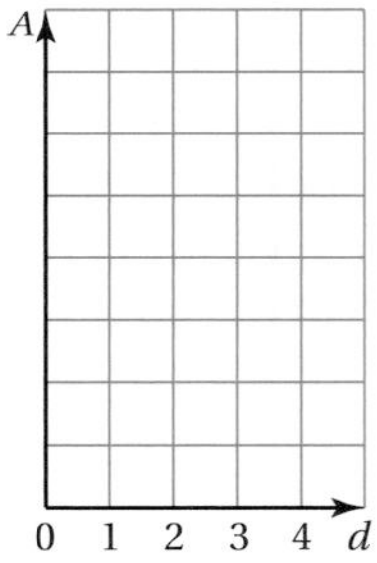

2 ACTIVITY: Conducting an Experiment

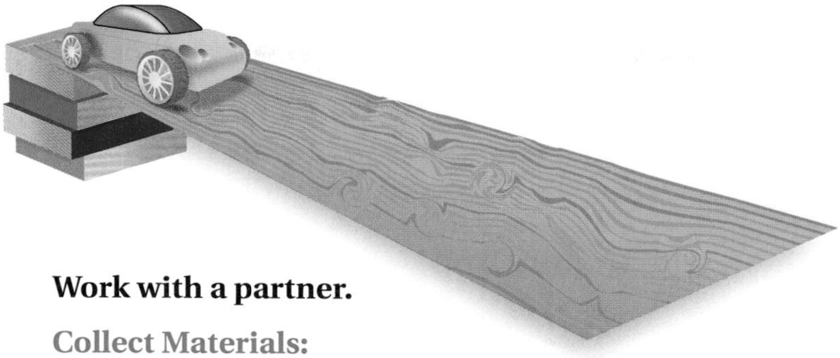

Work with a partner.

Collect Materials:

- A board at least 8 feet long
- Five books of the same thickness
- Toy car
- Stopwatch

Perform the Experiment:

- Place one book underneath one end of the board.
- Put the car at the top of the ramp. Measure the time (in seconds) it takes the car to roll down the ramp.
- Record your result in a table.
- Repeat the experiment with two, three, and four books.

Analyze the Results:

- Make a graph of your data.
- Does the graph have the characteristics of any of the graphs in Activity 1? Explain.

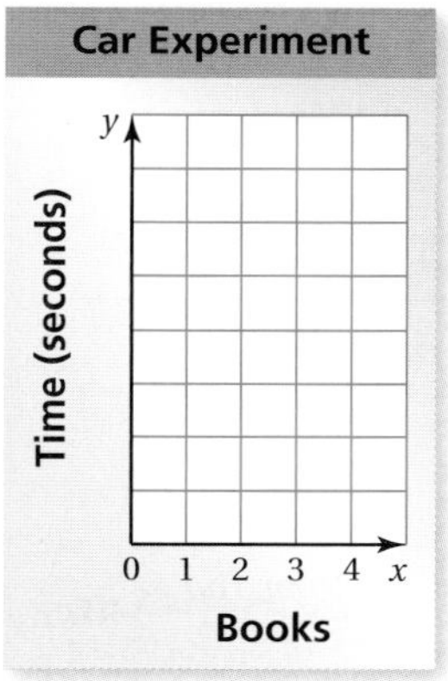

Use Your Results to Predict:

- Use your graph to predict how long it will take the car to roll down the ramp when five books are placed under the board.

Test Your Prediction:

- Repeat the experiment with five books. How close was your prediction?

What Is Your Answer?

3. **IN YOUR OWN WORDS** How can you analyze a function from its graph? Give a real-life example of how a graph can help you make a decision.

Practice Use what you learned about analyzing graphs to complete Exercises 3 and 4 on page 174.

4.5 Lesson

Check It Out
Lesson Tutorials
BigIdeasMath.com

A function whose graph is a straight line is a **linear function**.

EXAMPLE 1 Identifying Linear Functions

Key Vocabulary
linear function, *p. 172*

Does the graph represent a linear function? Explain.

a.

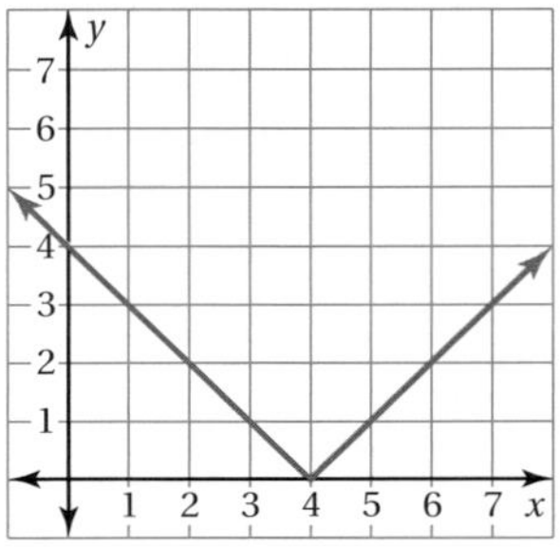

The graph is not a straight line. So, the graph does *not* represent a linear function.

b.

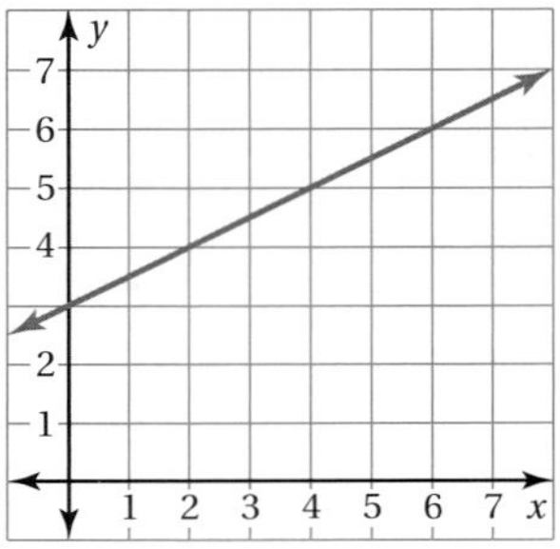

The graph is a straight line. So, the graph does represent a linear function.

EXAMPLE 2 Identifying a Linear Function

Input, *x*	Output, *y*
0	0
1	1
4	2
9	3

Does the input-output table represent a linear function? Explain.

The ordered pairs in the table are (0, 0), (1, 1), (4, 2), and (9, 3). Plot the ordered pairs and draw a graph through the points.

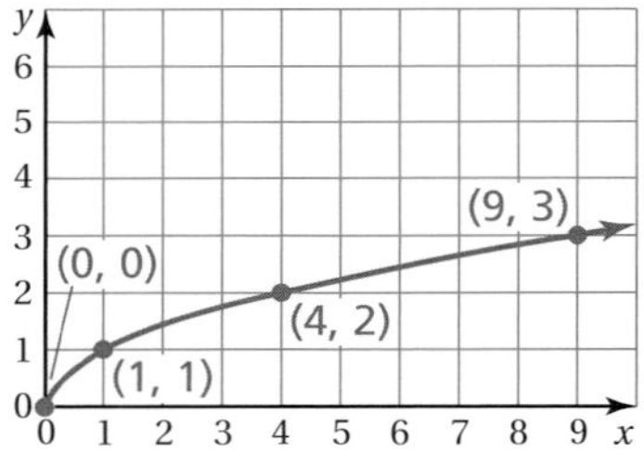

The graph is not a straight line. So, the function is *not* linear.

On Your Own

Now You're Ready
Exercises 5–14

Does the graph or table represent a linear function? Explain.

1.

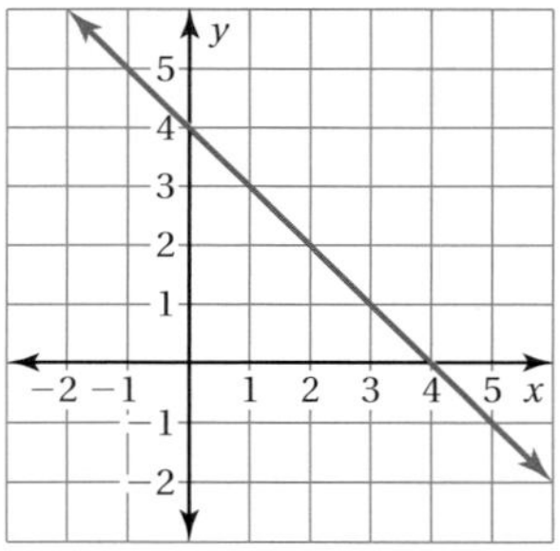

2.

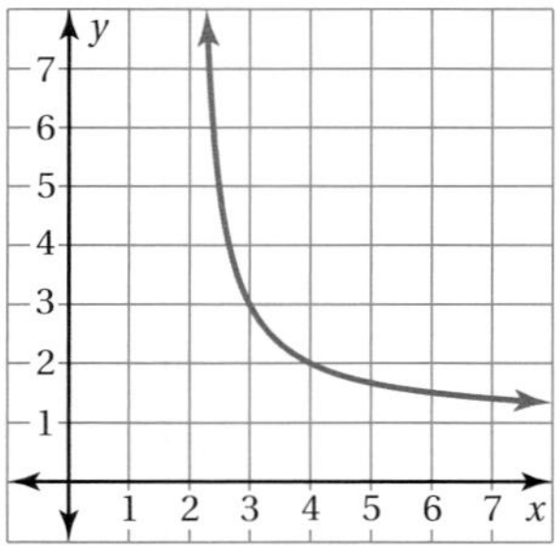

3.

Input, *x*	−3	−1	0	2
Output, *y*	−2	0	1	3

4.

Input, *x*	0	1	2	3
Output, *y*	−1	−2	−5	−10

EXAMPLE 3 Identifying a Linear Function

Is the function relating the diagram number x to the number of dots y linear?

Make an input-output table. Then plot the ordered pairs and draw the graph.

Diagram, x	Dots, y	(x, y)
1	1	(1, 1)
2	3	(2, 3)
3	6	(3, 6)
4	10	(4, 10)

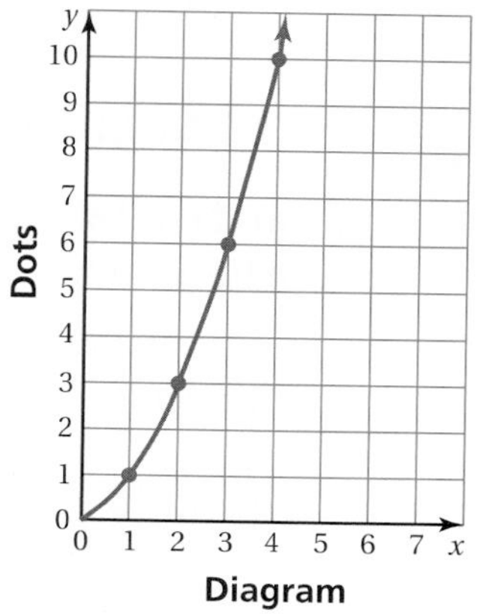

The graph is not a straight line. So, the function is *not* linear.

EXAMPLE 4 Comparing Linear Functions

Your sister earns \$10 per hour.
Your brother earns \$7 per hour.

The functions $m = 10h$ and $m = 7h$ show the relationship between the numbers of hours h they work and the money m they earn. Which graph is steeper? Explain.

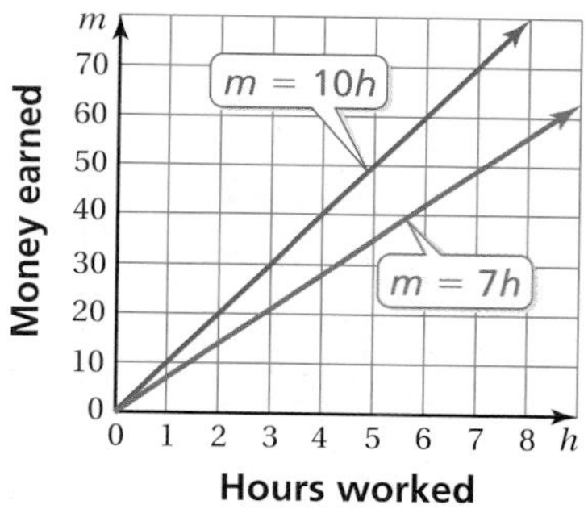

The graph of $m = 10h$ is steeper. The reason it is steeper is that your sister's hourly rate is greater than your brother's hourly rate.

On Your Own

Now You're Ready
Exercises 15–19

5. Make an input-output table for the pattern. Is the function relating the diagram number x to the number of dots y linear? Explain.

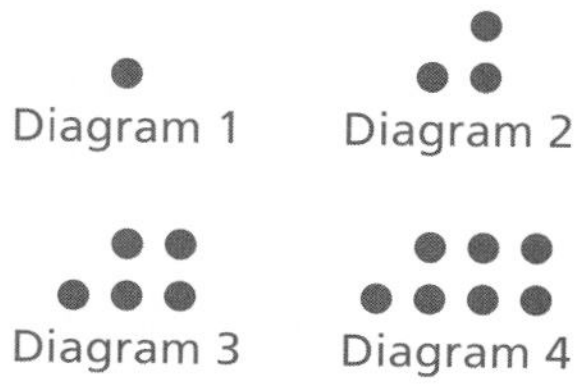

6. The functions $d = 65t$ and $d = 55t$ show the relationship between the distances d (in miles) traveled and the times t (in hours) for two cars. Graph the functions. Which graph is steeper? Explain.

4.5 Exercises

Check It Out
Help with Homework
BigIdeasMath.com

Vocabulary and Concept Check

1. **VOCABULARY** Why are some functions called *linear functions*?
2. **WRITING** How can you decide whether or not an input-output table represents a linear function?

Practice and Problem Solving

Copy and complete the table. Then make a graph of the data. Write an equation for the function.

3. Find the diameter of a circle with radius r.

Radius, *r*	1	2	3	4
Diameter, *d*				

4. Find the cost of renting roller blades for h hours at \$6 per hour.

Hours, *h*	1	2	3	4
Cost, *c*				

Does the graph represent a linear function? Explain.

1 5.

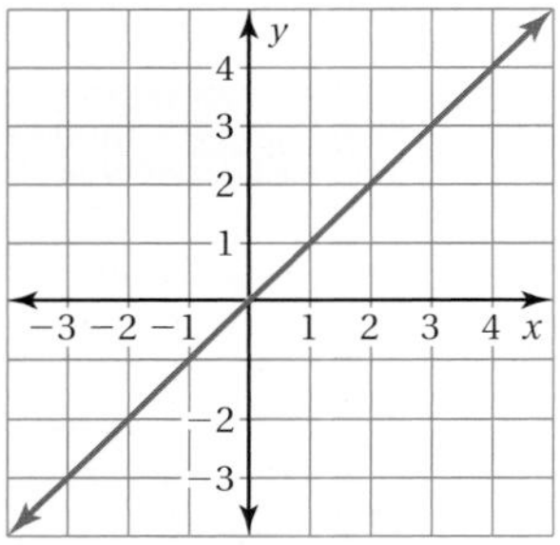

6.

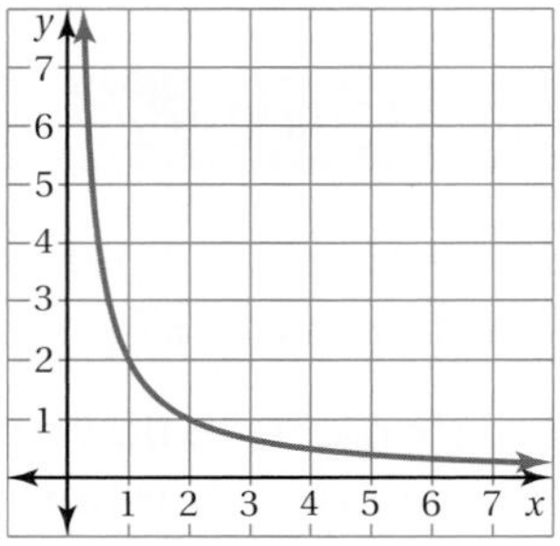

7.

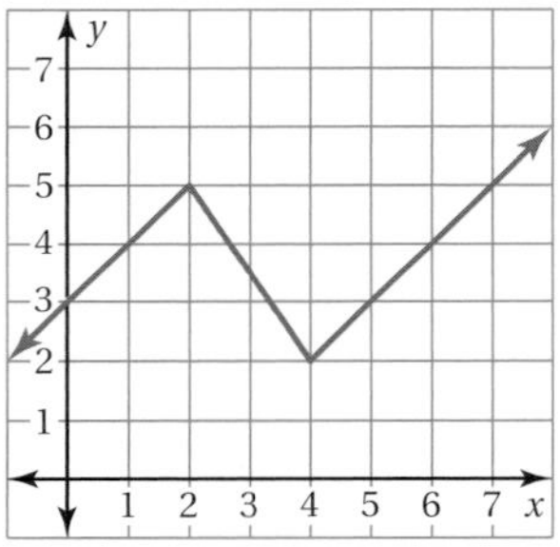

8.

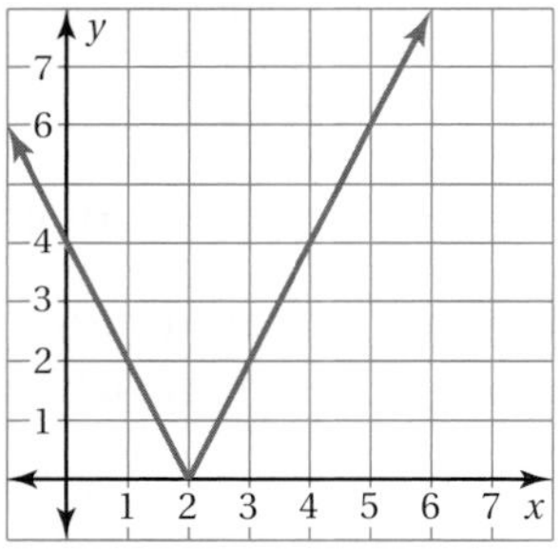

9.

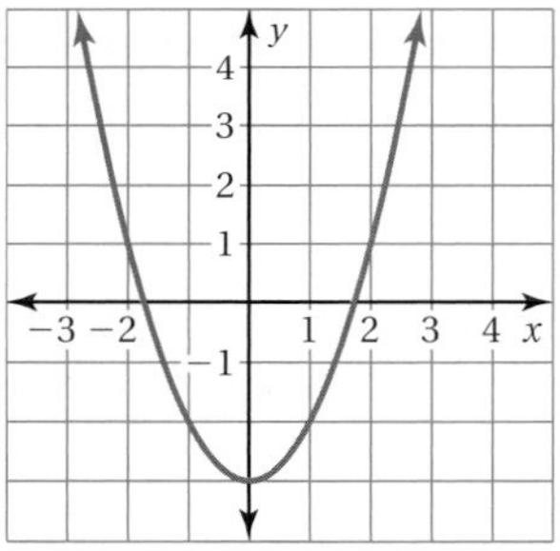

10.

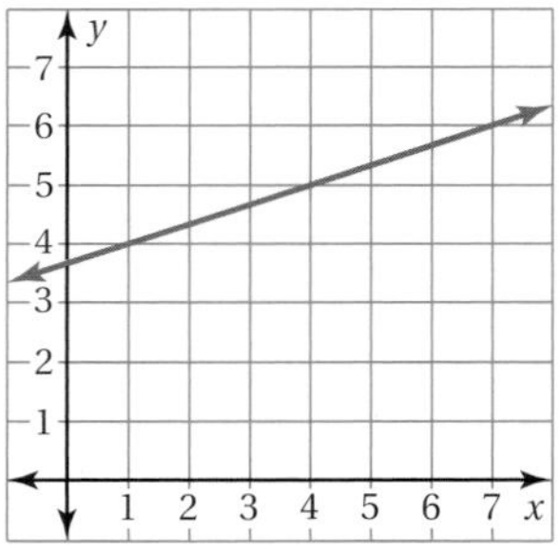

Does the input-output table represent a linear function? Explain.

2 11.

Input, *x*	1	2	3	4
Output, *y*	1	3	5	7

12.

Input, *x*	0	2	4	6
Output, *y*	−10	−9	−8	−7

13.

Input, *x*	1	4	7	10
Output, *y*	5	2	2	5

14.

Input, *x*	−3	−2	−1	0
Output, *y*	6	1	−2	−3

Graph each linear function. Which graph is steeper? Explain.

4 **15.** $y = 5x$ and $y = \frac{1}{5}x$ **16.** $y = \frac{4}{5}x$ and $y = \frac{3}{5}x$ **17.** $y = x$ and $y = 2x + 1$

Make an input-output table for the pattern. Is the function relating the figure number x to the area y linear? Explain.

3 **18.**

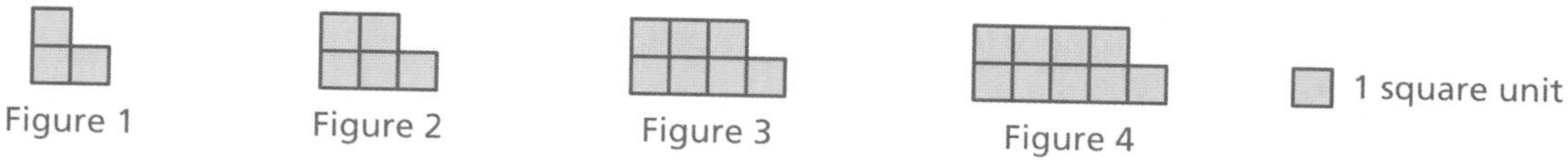

19.

20. FOOTRACE In a 100-meter race, you give your friend a head start. Your speeds are shown.

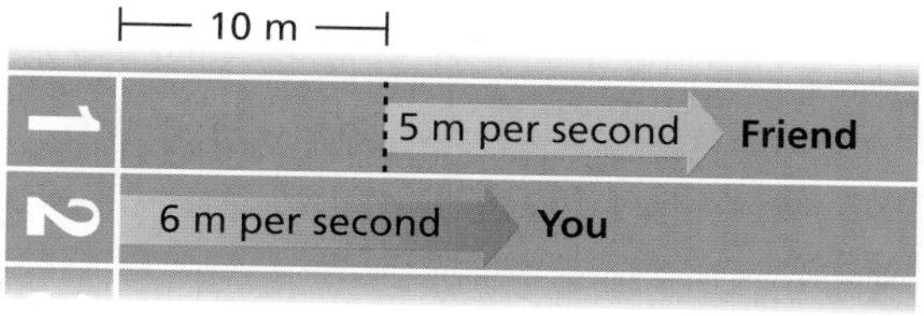

a. Write a function that represents your distance d after t seconds.

b. Write a function that represents your friend's distance d after t seconds.

c. Graph your distance and your friend's distance in the same coordinate plane.

d. What does the intersection of the two graphs represent?

21. AIRPORT USE The graph shows the numbers of flights that arrive at and depart from two regional airports. Which airport has more flights in a day? How many more? Explain.

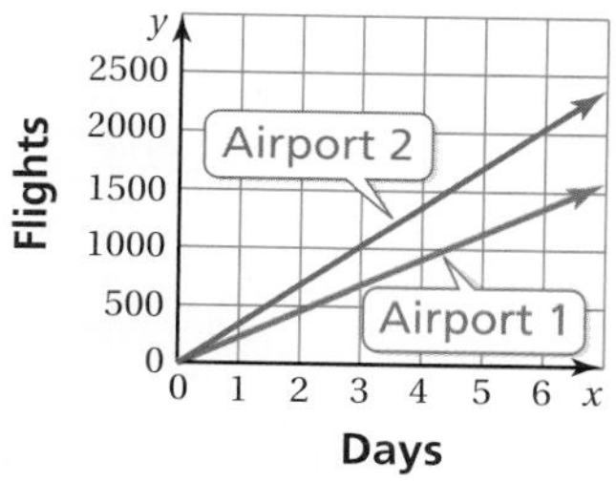

22. Reasoning Use the graph.

a. Copy and complete the table. Then find the differences.

x	0	1	2	3	4	5
y	1	1.5	2	2.5		

Differences: 0.5 0.5 ? ? ?

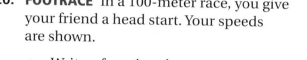

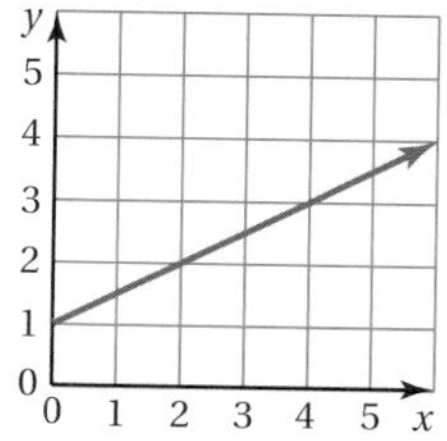

b. Graph other lines and find the "differences." Describe a property suggested by your results.

Fair Game Review What you learned in previous grades & lessons

Tell which number is greater. *(Skills Review Handbook)*

23. 40%, $\frac{11}{25}$ **24.** 0.27, 2.8% **25.** $\frac{4}{5}$, 0.802 **26.** $\frac{33}{50}$, $66\frac{2}{3}\%$

27. MULTIPLE CHOICE For which inequality is $x = 7$ a solution? *(Section 3.7 and Section 3.8)*

Ⓐ $x < 7$ Ⓑ $x + 4 \geq 12$ Ⓒ $21 \leq 3x$ Ⓓ $4x - 5 > 23$

4.6 Arithmetic Sequences

Essential Question How are arithmetic sequences used to describe patterns?

Share Your Work at... My.BigIdeasMath.com

STANDARDS OF LEARNING
7.2

1 EXAMPLE: Writing a Story

Write a story that uses the following arithmetic sequence. Draw pictures for your story.

52, 46, 40, . . .

There are many possible stories. Here is a student's story about temperature.

Natalie and her family went to a camp in the Blue Ridge Mountains. On Saturday morning she and her family decided to go for a long hike. They started at 8:00 A.M. and by noon had hiked 7 miles.

At noon the temperature was 52 degrees Fahrenheit and a storm started to roll in. Natalie and her family decided to return to their cabin. By 1 P.M. the temperature had dropped to 46 degrees. Natalie said that if the temperature continued to drop at that rate, it would be freezing before they got back to the cabin.

Her father asked how she had come to that conclusion. Here is how she explained it. "We hiked for 4 hours before turning around. If the temperature drops by 6 degrees each hour, the temperature will have dropped to 28 degrees by 4 P.M. Because 32 degrees is freezing, it will be below freezing."

Time	12:00	1:00	2:00	3:00	4:00
Temperature	52°	46°	40°	34°	28°

2 ACTIVITY: Writing a Story

Work with a partner. Think of a story that uses an arithmetic sequence.

a. Write your story. Then draw pictures for your story.

b. Include a table in your story and show how the table can be used to solve the problem that is described in your story.

You can use any topic and sequence you want. If you can't think of any, here are some possible ideas.

A parachutist falls at the rate of 15 feet per second.

A tree frog jumps 14 inches per jump.

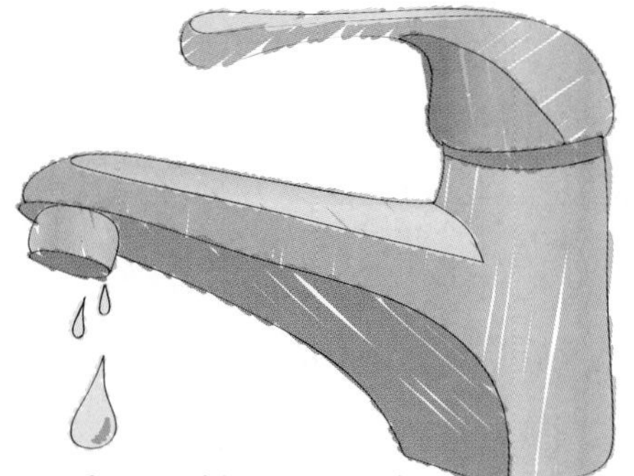

A leaky faucet drips 4 cups of water each hour.

A corn plant grows 4 inches each week.

What Is Your Answer?

3. IN YOUR OWN WORDS How are arithmetic sequences used to describe patterns? Give an example from real life.

Practice Use what you learned about arithmetic sequences to complete Exercises 3–5 on page 180.

Check It Out
Lesson Tutorials
BigIdeasMath.com

Key Vocabulary
sequence, *p. 178*
term, *p. 178*
arithmetic sequence, *p. 178*
common difference, *p. 178*

A **sequence** is an ordered list of numbers. Each number in a sequence is called a **term**. Each term has a specific position in the sequence.

5, 10, 15, 20, 25, . . .

1st position (points to 5) — 5th position (points to 25)

Key Idea

Arithmetic Sequence

In an **arithmetic sequence**, the difference between consecutive terms is the same. This difference is called the **common difference**. Each term is found by adding the common difference to the previous term.

5, 10, 15, 20, . . . — Terms of an arithmetic sequence

+5 +5 +5 — Common difference

EXAMPLE 1 Extending an Arithmetic Sequence

Write the next three terms of the arithmetic sequence −7, −14, −21, −28, ...

Use a table to organize the terms and find the pattern.

Position	1	2	3	4
Term	−7	−14	−21	−28

+(−7) +(−7) +(−7)

Each term is 7 less than the previous term. So, the common difference is −7.

Add −7 to a term to find the next term.

Position	1	2	3	4	5	6	7
Term	−7	−14	−21	−28	−35	−42	−49

+(−7) +(−7) +(−7)

∴ The next three terms are −35, −42, and −49.

On Your Own

Now You're Ready
Exercises 6–11

Write the next three terms of the arithmetic sequence.

1. −12, 0, 12, 24, . . .
2. 0.2, 0.6, 1, 1.4, . . .
3. $4, 3\frac{3}{4}, 3\frac{1}{2}, 3\frac{1}{4}, \ldots$

Multi-Language Glossary at BigIdeasMath.com.

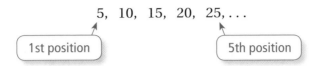

To write a function that describes an arithmetic sequence, find the relationship between each term and its position. Let n be the term's position in the sequence and y be the value of the term.

EXAMPLE 2 Writing a Function that Describes an Arithmetic Sequence

Write a function that describes the arithmetic sequence −7, −14, −21, −28, ...

Use a table to find the relationship between each term and its position.

Position, n	Term, y	Rule
1	−7	$1 \cdot (-7) = -7$
2	−14	$2 \cdot (-7) = -14$
3	−21	$3 \cdot (-7) = -21$
4	−28	$4 \cdot (-7) = -28$
n	y	$n \cdot (-7) = y$

Each term is found by multiplying its position by −7.

So, a function that describes the arithmetic sequence is $y = -7n$.

EXAMPLE 3 Graphing an Arithmetic Sequence

Graph the arithmetic sequence 4, 8, 12, 16, ... What do you notice?

Make a table. Match the position of each term with its value.

Position, n	Term, y
1	4
2	8
3	12
4	16

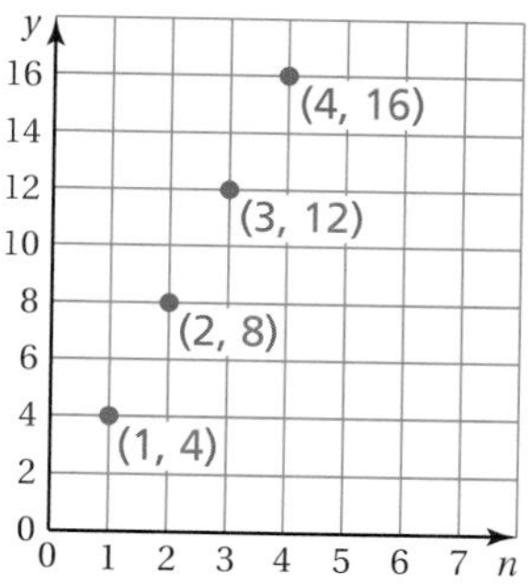

Plot the ordered pairs (1, 4), (2, 8), (3, 12), and (4, 16).

The points of the graph lie on a line.

Study Tip

The points of any arithmetic sequence lie on a line.

On Your Own

Now You're Ready Exercises 13–23

Write a function that describes the arithmetic sequence.

4. 4, 5, 6, 7, . . . **5.** 8, 16, 24, 32, . . . **6.** −2, −1, 0, 1, . . .

Write the next three terms of the arithmetic sequence. Then graph the sequence.

7. 3, 6, 9, 12, . . . **8.** 4, 2, 0, −2, . . . **9.** 1, 0.8, 0.6, 0.4, . . .

4.6 Exercises

Check It Out
Help with Homework
BigIdeasMath.com

Vocabulary and Concept Check

1. **WRITING** Describe how to find the common difference of an arithmetic sequence.
2. **WHICH ONE DOESN'T BELONG?** Which one does *not* belong with the other three? Explain your reasoning.

$3, -1, -5, -9, \ldots$	$2, 8, 32, 128, \ldots$
$-1, 5, 11, 17, \ldots$	$12, 7, 2, -3, \ldots$

Practice and Problem Solving

Make a table to represent the arithmetic sequence described.

3. An inchworm crawls 10 inches each minute.
4. A teacher assigns 8 math problems each night.
5. A collector has 75 souvenir spoons. She receives 5 new spoons each year.

Write the next three terms of the arithmetic sequence.

1 6. $-4, 3, 10, 17, \ldots$

7. $60, 30, 0, -30, \ldots$

8. $1.3, 1, 0.7, 0.4, \ldots$

9. $2, 2\frac{2}{3}, 3\frac{1}{3}, 4, \ldots$

10. $\frac{5}{6}, \frac{2}{3}, \frac{1}{2}, \frac{1}{3}, \ldots$

11. $-2.1, -0.3, 1.5, 3.3, \ldots$

12. **MOVIE REVENUE** A movie earns \$100 million the first week it is released. The movie earns \$20 million less each additional week. Write an arithmetic sequence for the movie earnings.

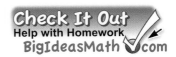

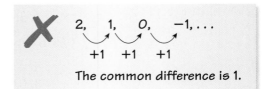

Write a function that describes the arithmetic sequence.

2 13. $-5, -4, -3, -2, \ldots$

14. $-3, -6, -9, -12, \ldots$

15. $\frac{1}{2}, 1, 1\frac{1}{2}, 2, \ldots$

16. $10, 11, 12, 13, \ldots$

17. $-10, -20, -30, -40, \ldots$

18. $\frac{1}{7}, \frac{2}{7}, \frac{3}{7}, \frac{4}{7}, \ldots$

19. **ERROR ANALYSIS** Describe and correct the error in finding the common difference of the arithmetic sequence.

✗ $2, \quad 1, \quad 0, \quad -1, \ldots$

$+1 \quad +1 \quad +1$

The common difference is 1.

Write the next three terms of the arithmetic sequence. Then graph the sequence.

3 **20.** 7, 6.4, 5.8, 5.2, . . .

21. $-15, 0, 15, 30, \ldots$

22. $\frac{1}{2}, \frac{5}{8}, \frac{3}{4}, \frac{7}{8}, \ldots$

23. $-1, -3, -5, -7, \ldots$

24. NUMBER SENSE Each term of an arithmetic sequence can be found by multiplying the term's position by -5. Write the first 7 terms of the sequence.

25. SPEED On a highway, you take 3 seconds to increase your speed from 62 to 65 miles per hour. Your speed increases the same amount each second.

a. Write the first 4 terms of the sequence that represents your speed.

b. Write a function that describes the arithmetic sequence.

26. REASONING Describe how to find the 20th term of the sequence $-4, -8, -12, -16, \ldots$. Then find the 20th term.

27. Critical Thinking The number of births in the United States each minute after midnight January 1, 2012 can be estimated by the sequence in the table.

Minutes after Midnight January 1, 2012	1	2	3	4
Babies Born	5	10	15	20

a. Write a function that describes the arithmetic sequence that can be used to estimate the number of births after a given number of minutes.

b. Explain how to use your function to estimate the number of births in a day.

c. Explain how to use your function to estimate the number of births in a year.

Fair Game Review What you learned in previous grades & lessons

List the ordered pairs shown in the mapping diagram. *(Section 4.1)*

28.

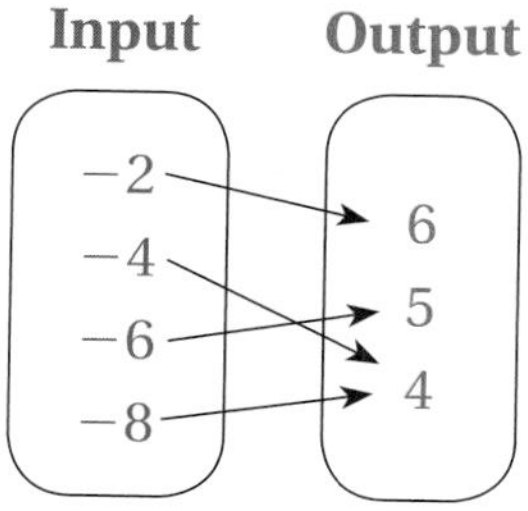

29.

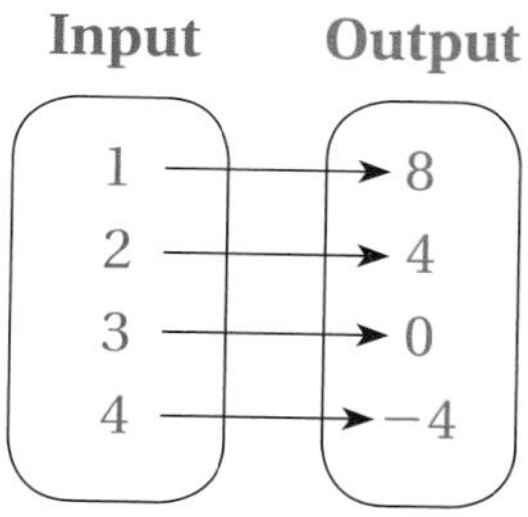

30. MULTIPLE CHOICE Use a formula to find the area of the trapezoid. *(Section 2.3)*

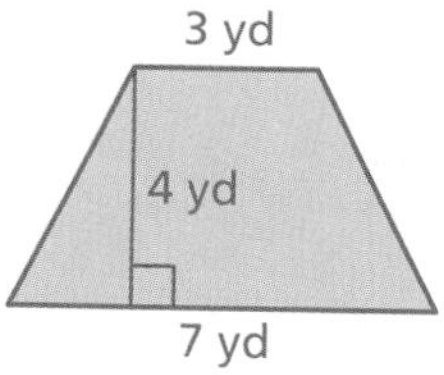

Ⓐ 14 yd^2 Ⓑ 20 yd^2 Ⓒ 22 yd^2 Ⓓ 40 yd^2

4.7 Geometric Sequences

STANDARDS OF LEARNING
7.2

Essential Question How can you find the balance in an account that earns simple interest or compound interest?

1 ACTIVITY: Comparing Simple and Compound Interest

Work with a partner. Interest that is calculated only on principal is *simple interest*. Interest that is calculated on principal *and* previously earned interest is *compound interest*.

You deposit $1000 in a savings account that earns 6% interest per year.

a. Copy and complete the first table that shows the balance after 10 years with simple interest.

b. Copy and complete the second table that shows the balance after 10 years with interest that is compounded annually.

c. Which type of interest gives the greater balance?

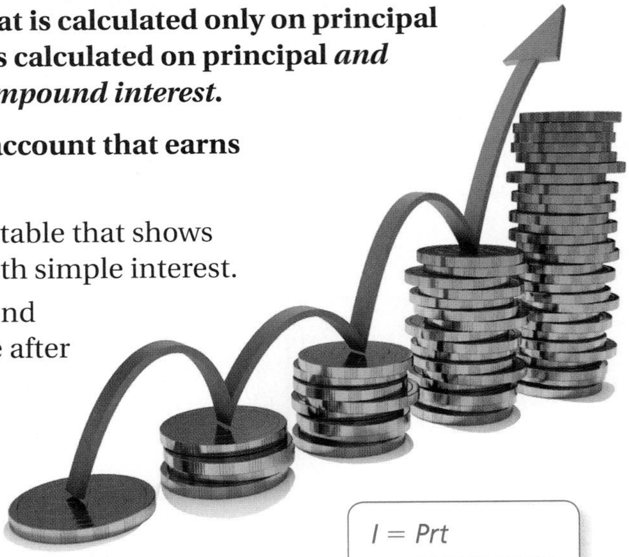

$I = Prt$
$= 1000(0.06)(1)$

Simple Interest			
t	Principal	Annual Interest	Balance at End of Year
1	$1000.00	$60.00	$1060.00
2	$1000.00	$60.00	$1120.00
3			
4			
5			
6			
7			
8			
9			
10			

$I = Prt$
$= 1060(0.06)(1)$

Compound Interest			
t	Principal and Interest	Annual Interest	Balance at End of Year
1	$1000.00	$60.00	$1060.00
2	$1060.00	$63.60	$1123.60
3			
4			
5			
6			
7			
8			
9			
10			

Principal and Interest = Balance at End of Previous Year

2 ACTIVITY: Comparing Simple and Compound Interest

Work with a partner.

a. Graph the end-of-year balances for each type of interest in Activity 1.

b. Which graph shows an arithmetic sequence? Explain your reasoning.

c. Which graph shows a geometric sequence? Explain your reasoning.

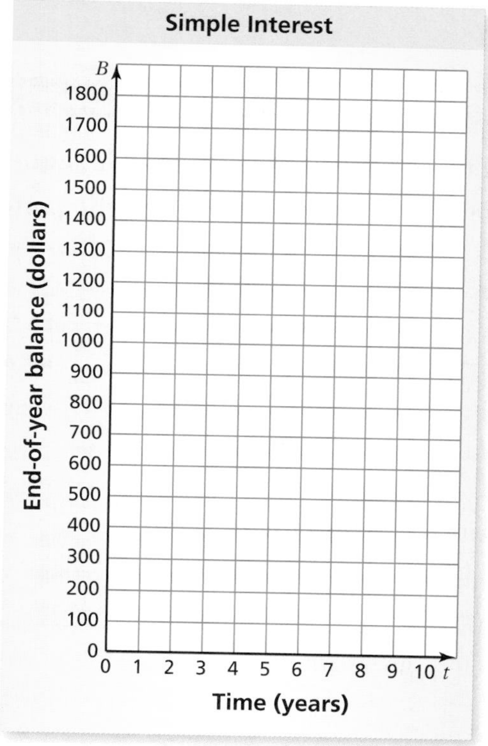

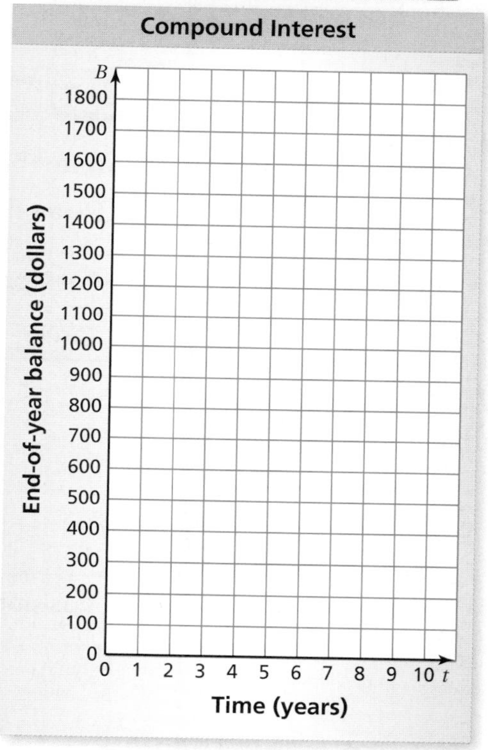

What Is Your Answer?

3. **IN YOUR OWN WORDS** How can you find the balance in an account that earns simple interest or compound interest?

4. Use what you learned in Activity 2. About how many years will it take for the balance to double with simple interest? with compound interest?

Practice

Use what you learned about simple and compound interest to complete Exercises 4 and 5 on page 186.

4.7 Lesson

Check It Out
Lesson Tutorials
BigIdeasMath.com

Key Vocabulary
geometric sequence, *p. 184*
common ratio, *p. 184*

Key Idea

Geometric Sequence

In a **geometric sequence**, the ratio between consecutive terms is the same. This ratio is called the **common ratio**. Each term is found by multiplying the previous term by the common ratio.

$$1, \quad 5, \quad 25, \quad 125, \ldots$$

$\times 5 \quad \times 5 \quad \times 5$ ← Common ratio

Terms of a geometric sequence

EXAMPLE 1 Extending a Geometric Sequence

Write the next three terms of the geometric sequence 160, 80, 40, 20, ...

Use a table to organize the terms and find the pattern.

Position	1	2	3	4
Term	160	80	40	20

$\times \frac{1}{2} \quad \times \frac{1}{2} \quad \times \frac{1}{2}$

Each term is one-half the previous term. So, the common ratio is $\frac{1}{2}$.

Multiply a term by $\frac{1}{2}$ to find the next term.

Position	1	2	3	4	5	6	7
Term	160	80	40	20	10	5	$\frac{5}{2}$

$\times \frac{1}{2} \quad \times \frac{1}{2} \quad \times \frac{1}{2}$

∴ The next three terms are 10, 5, and $\frac{5}{2}$.

On Your Own

Now You're Ready
Exercises 6–11

Write the next three terms of the geometric sequence.

1. 0.1, 1, 10, 100, . . .

2. 4, −12, 36, −108, . . .

3. 256, 64, 16, 4, . . .

4. 9, 3, 1, $\frac{1}{3}$, . . .

EXAMPLE 2 Identifying Sequences

Tell whether each sequence is *geometric, arithmetic,* or *neither.*

Study Tip

A sequence is neither arithmetic nor geometric when there is no common difference or common ratio.

a. 40, 30, 20, 10, . . .

Position	1	2	3	4
Term	40	30	20	10

$+(-10)+(-10)+(-10)$

Each term is 10 less than the previous term. The common difference is -10.

So, the sequence is arithmetic.

b. 2, -4, 8, -16, . . .

Position	1	2	3	4
Term	2	-4	8	-16

$\times(-2)\times(-2)\times(-2)$

Each term is -2 times the previous term. The common ratio is -2.

So, the sequence is geometric.

On Your Own

Now You're Ready
Exercises 13–21

Tell whether the sequence is *geometric, arithmetic,* or *neither.*

5. 2, 8, 32, 128, . . .

6. 1, 2, 6, 24, . . .

7. -6, -3, 0, 3, . . .

8. 1, -4, 16, -64, . . .

EXAMPLE 3 Graphing a Geometric Sequence

Graph the geometric sequence 3, 6, 12, 24, ... What do you notice?

Make a table. Match the position of each term with its value.

Study Tip

In Example 3, as n increases by 1, y increases by larger and larger amounts.

Position, *n*	Term, *y*
1	3
2	6
3	12
4	24

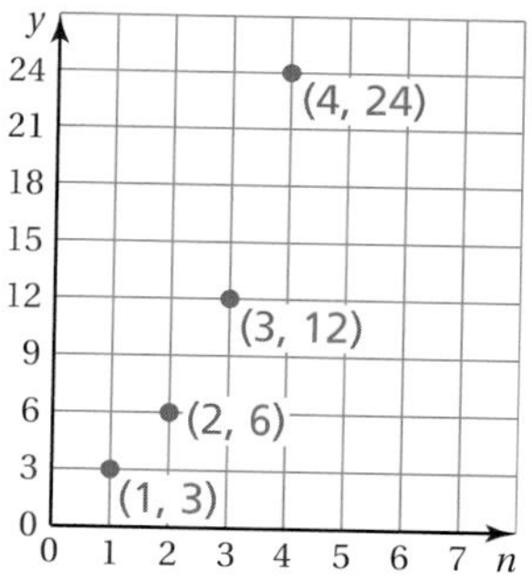

Plot the ordered pairs (1, 3), (2, 6), (3, 12), and (4, 24).

The points of the graph do *not* lie on a line.

On Your Own

Now You're Ready
Exercises 22–25

Write the next three terms of the geometric sequence. Then graph the sequence.

9. 1, 3, 9, 27, . . .

10. 128, 64, 32, 16, . . .

11. 80, -40, 20, -10, . . .

4.7 Exercises

Check It Out
Help with Homework
BigIdeasMath.com

Vocabulary and Concept Check

1. **VOCABULARY** Describe the relationship between any two consecutive terms of a *geometric sequence*.

2. **REASONING** Compare and contrast the two sequences.

   2, 4, 6, 8, 10, . . .   2, 4, 8, 16, 32, . . .

3. **WHICH ONE DOESN'T BELONG?** Which one does *not* belong with the other three? Explain your reasoning.

   1, −3, 9, −27, . . .   1, −3, 5, −7, . . . $\frac{1}{9}, \frac{1}{3}$, 1, 3, . . .   27, 9, 3, 1, . . .

Practice and Problem Solving

Copy and complete the table using an interest rate of 7%.

4.

t	Principal	Annual Interest	Balance at End of Year
1	$100.00	$7.00	$107.00
2	$100.00	$7.00	$114.00
3			
4			

5.

t	Principal and Interest	Annual Interest	Balance at End of Year
1	$100.00	$7.00	$107.00
2	$107.00	$7.49	$114.49
3			
4			

Write the next three terms of the geometric sequence.

1 6. 2, 10, 50, 250, . . .

7. −7, 14, −28, 56, . . .

8. 81, −27, 9, −3, . . .

9. −375, −75, −15, −3, . . .

10. 10,000, 100, 1, 0.01, . . .

11. 8, 12, 18, 27, . . .

12. **ERROR ANALYSIS** Describe and correct the error in writing the next three terms of the geometric sequence.

   ✗ −8, 4, −2, 1, . . .
   ×(−2) ×(−2) ×(−2)
   The next three terms are −2, 4, and −8.

Tell whether the sequence is *geometric*, *arithmetic*, or *neither*.

2 13. −8, 0, 8, 16, . . .

14. −1, 3, −5, 7, . . .

15. 1, 4, 9, 16, . . .

16. $\frac{3}{49}, \frac{3}{7}$, 3, 21, . . .

17. 192, 24, 3, $\frac{3}{8}$, . . .

18. −25, −18, −12, −7, . . .

19. −64, 16, −4, 1, . . .

20. 5, 8, 13, 21, . . .

21. 1.3, 0.6, −0.1, −0.8, . . .

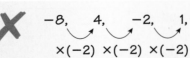

Write the next three terms of the geometric sequence. Then graph the sequence.

3 **22.** $\frac{1}{49}, \frac{1}{7}, 1, 7, \ldots$

23. $36, 6, 1, \frac{1}{6}, \ldots$

24. $\frac{1}{27}, \frac{1}{3}, 3, 27, \ldots$

25. $-2, 1, -\frac{1}{2}, \frac{1}{4}, \ldots$

26. CATCH PHRASE A sequence is described by the saying:

Each time I take two steps forward, I fall one step back.

a. The first four terms of the sequence are 0, 2, 1, and 3. Write the next three terms of the sequence.

b. Is the sequence *arithmetic*, *geometric*, or *neither*? Explain.

27. REASONING The first two terms of a sequence are 3 and 6.

a. Suppose the sequence is geometric. Write the next three terms.

b. Suppose the sequence is arithmetic. Write the next three terms.

Training On Your Own

Day 1: Run 1 km.

Each day after Day 1: Run 20% farther than the previous day.

28. TRAINING You follow the training schedule from your coach.

a. What type of sequence is formed by your daily distances?

b. On what day do you run approximately 3 kilometers?

29. Critical Thinking A soup kitchen makes 16 gallons of soup. Each day, $\frac{1}{4}$ of the soup is served and the rest is saved for the next day.

a. Write the first five terms of the sequence of the number of fluid ounces of soup left each day.

b. Write a function to represent the sequence.

c. When is all the soup gone? Explain.

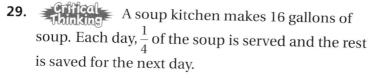

Fair Game Review What you learned in previous grades & lessons

Simplify the fraction. *(Skills Review Handbook)*

30. $\frac{10}{12}$

31. $\frac{12}{16}$

32. $\frac{20}{45}$

33. $\frac{72}{96}$

34. MULTIPLE CHOICE Which expression is equivalent to $3(x - 5) + 4x$? *(Section 2.4)*

Ⓐ $8x$

Ⓑ $-x - 15$

Ⓒ $7x - 5$

Ⓓ $7x - 15$

4.4–4.7 Quiz

Check It Out
Progress Check
BigIdeasMath.com

Graph the function. *(Section 4.4)*

1. $y = x - 10$

2. $y = 2x + 3$

3. $y = \frac{x}{2}$

Does the graph represent a linear function? Explain. *(Section 4.5)*

4.

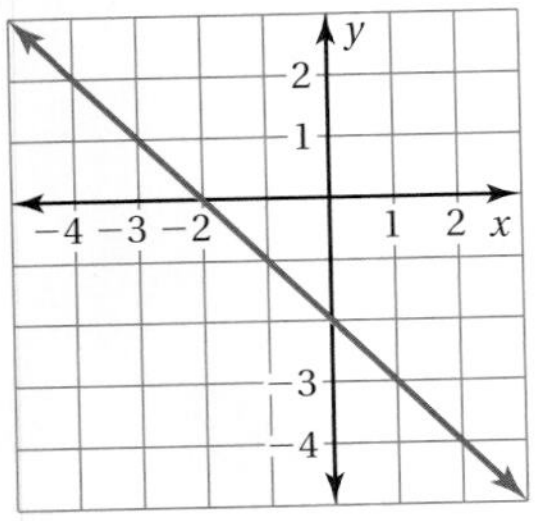

5.

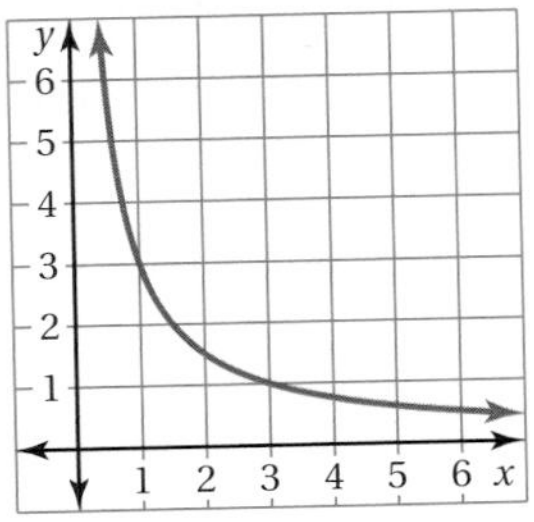

6.

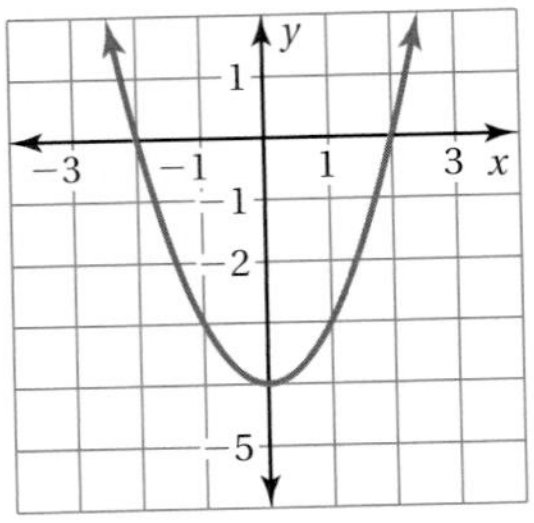

Does the input-output table represent a linear function? Explain. *(Section 4.5)*

7.

Input, *x*	−3	−2	−1	0
Output, *y*	−11	−6	−3	−2

8.

Input, *x*	2	4	6	8
Output, *y*	−7	−9	−11	−13

Write a function that describes the arithmetic sequence. *(Section 4.6)*

9. 5, 6, 7, 8, . . .

10. −3, −2, −1, 0, . . .

11. 4, 8, 12, 16, . . .

12. −1.5, −0.5, 0.5, 1.5, . . .

Write the next three terms of the geometric sequence. *(Section 4.7)*

13. −15, 1.5, −0.15, 0.015, . . .

14. 2, −6, 18, −54, . . .

15. 1, 4, 16, 64, . . .

16. 2, 5, 12.5, 31.25, . . .

17. SALES TAX The sales tax rate is 6%. *(Section 4.4)*

- **a.** Write a function you can use to find the amount of sales tax t on the price p of each item at a gift shop.
- **b.** Graph the function using the inputs 0, 10, 20, 30, and 40 for p.
- **c.** You buy a souvenir that costs \$4.00. How much do you pay for the souvenir after sales tax is added?

18. HELICOPTER The graph shows the numbers of gallons of gasoline that two helicopters use after 5 hours of flying. Which helicopter uses more fuel in an hour? Explain. *(Section 4.5)*

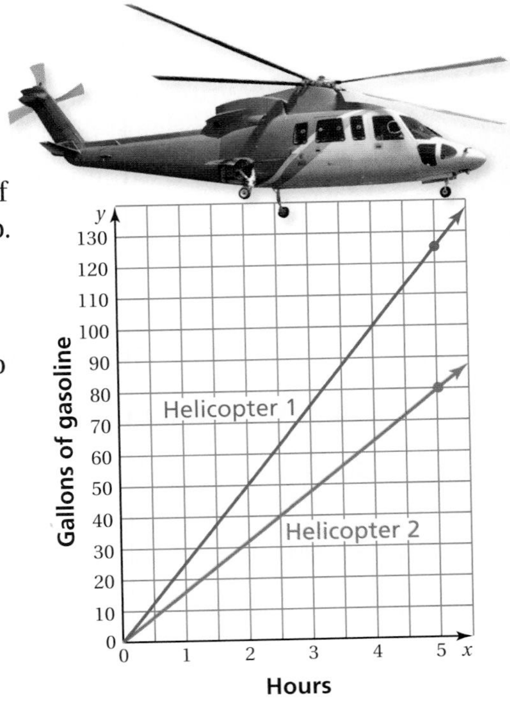

Check It Out
Vocabulary Help
BigIdeasMath.com

Review Key Vocabulary

input, *p. 146*
output, *p. 146*
function, *p. 146*
mapping diagram, *p. 146*
function rule, *p. 152*
input-output table, *p. 158*
graph, *p. 166*
linear function, *p. 172*
sequence, *p. 178*
term, *p. 178*
arithmetic sequence, *p. 178*
common difference, *p. 178*
geometric sequence, *p. 184*
common ratio, *p. 184*

Review Examples and Exercises

4.1 Mapping Diagrams *(pp. 144–149)*

Draw a mapping diagram of (−2, −1), (0, 3), (2, −1), and (4, 5).

Inputs: −2, 0, 2, 4

Outputs: −1, 3, 5

Draw arrows from the inputs to their outputs.

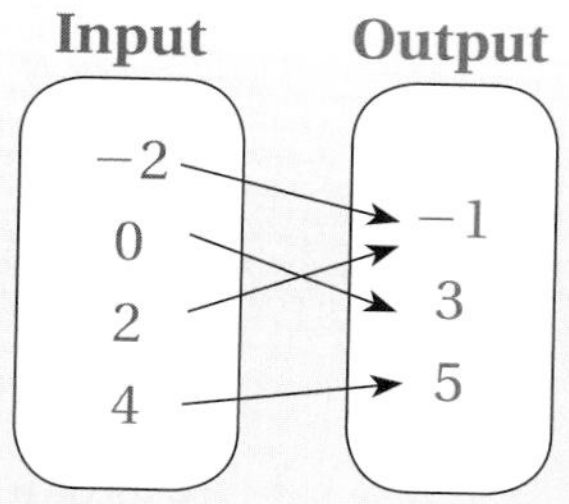

Exercises

Draw a mapping diagram of the set of ordered pairs.

1. (1, 30), (2, 60), (3, 90), (4, 120)

2. (−1, 5), (0, −5), (1, 10), (2, 5)

4.2 Functions as Words and Equations *(pp. 150–155)*

What is the value of $y = 3x + 4$ when $x = 8$?

$y = 3x + 4$ Write the equation.

$= 3(8) + 4$ Substitute 8 for *x*.

$= 28$ Simplify.

∴ When $x = 8$, $y = 28$.

Exercises

Find the value of *y* for the given value of *x*.

3. $y = 2x - 3;\ x = -4$

4. $y = 2 - 9x;\ x = \frac{2}{3}$

5. $y = \frac{x}{3} + 5;\ x = 6$

Tell whether $(x, y) = (-2, 3)$ is a solution of the equation.

6. $y = x + 4$

7. $y = 7 - 2x$

8. $y = 3x + 9$

4.3 Input-Output Tables (pp. 156–161)

Write an equation for the function shown by the table.

Input, x	1	3	5	7
Output, y	7	9	11	13

Look at the relationship between the inputs and outputs. Each output y is 6 more than the input x.

∴ So, the function rule is $y = x + 6$.

Exercises

Write an equation for the function shown by the table.

9.

Input, x	1	2	3	4
Output, y	7	14	21	28

10.

Input, x	−16	−12	−8	−4
Output, y	−4	−3	−2	−1

4.4 Graphs (pp. 164–169)

Graph $y = x - 1$.

Make an input-output table. Use the values −1, 0, 1, and 2 for x.

x	$y = x - 1$	y	(x, y)
−1	$y = (-1) - 1$	−2	$(-1, -2)$
0	$y = (0) - 1$	−1	$(0, -1)$
1	$y = (1) - 1$	0	$(1, 0)$
2	$y = (2) - 1$	1	$(2, 1)$

Plot the ordered pairs. Draw a line through the points.

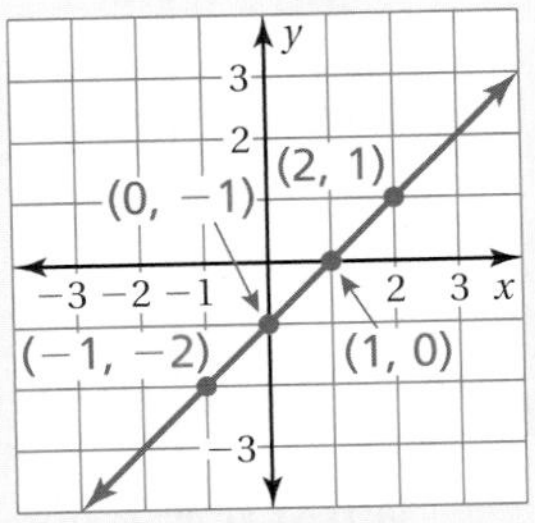

Exercises

Graph the function.

11. $y = x + 3$

12. $y = -5x$

13. $y = 3 - 3x$

4.5 Analyzing Graphs (pp. 170–175)

Does the graph represent a linear function? Explain.

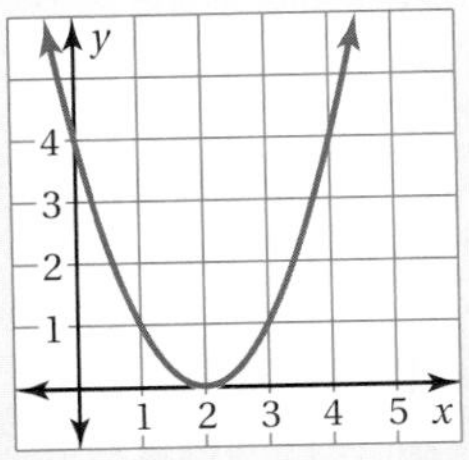

∴ The graph is not a straight line. So, the graph does *not* represent a linear function.

Exercises

Does the table represent a linear function? Explain.

14.

Input, x	-3	-2	-1	0
Output, y	0	2	4	6

15.

Input, x	3	4	5	6
Output, y	2	1	0	1

4.6 Arithmetic Sequences (pp. 176–181)

Write a function that describes the arithmetic sequence $-2, -4, -6, -8, \ldots$.

Use a table to find the relationship between each term and its position.

Position, n	Term, y	Rule
1	-2	$1 \cdot (-2) = -2$
2	-4	$2 \cdot (-2) = -4$
3	-6	$3 \cdot (-2) = -6$
4	-8	$4 \cdot (-2) = -8$
n	y	$n \cdot (-2) = y$

Each term is found by multiplying its position by -2.

∴ So, a function that describes the arithmetic sequence is $y = -2n$.

Exercises

Write a function that describes the arithmetic sequence.

16. $8, 9, 10, 11, \ldots$

17. $6, 12, 18, 24, \ldots$

18. $-9, -8, -7, -6, \ldots$

4.7 Geometric Sequences (pp. 182–187)

Tell whether each sequence is *geometric, arithmetic,* or *neither.*

a. $10, 5, 0, -5, \ldots$

Position	1	2	3	4
Term	10	5	0	-5

$+(-5)$ $+(-5)$ $+(-5)$

The common difference is -5.

∴ So, the sequence is arithmetic.

b. $1, -3, 9, -12, \ldots$

Position	1	2	3	4
Term	1	-3	9	-27

$\times(-3)$ $\times(-3)$ $\times(-3)$

The common ratio is -3.

∴ So, the sequence is geometric.

Exercises

Tell whether the sequence is *geometric, arithmetic,* or *neither.*

19. $9, 7, 5, 3, \ldots$

20. $-2, 6, -24, 120, \ldots$

21. $32, 16, 8, 4, \ldots$

4 Chapter Test

Check It Out
Test Practice
BigIdeasMath.com

Draw a mapping diagram of the set of ordered pairs.

1. (1, 0), (2, 1), (3, 0), (4, 1)

2. (−6, 5), (−4, 3), (−2, −1), (0, 1)

3. Tell whether (1, −2) is a solution of $y = 2x - 4$.

4. Write an equation for the function shown by the table.

Input, x	3	4	5	6
Output, y	9	10	11	12

Graph the function.

5. $y = x + 8$

6. $y = 1 - 3x$

7. $y = x - 4$

Does the graph or table represent a linear function? Explain.

8.

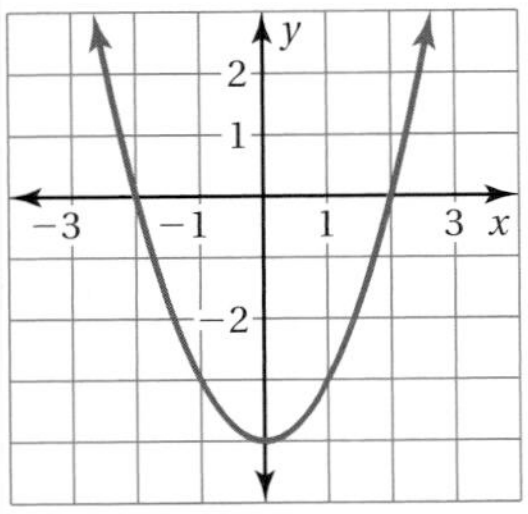

9.

Input, x	Output, y
6	4
7	1
8	0
9	1

10.

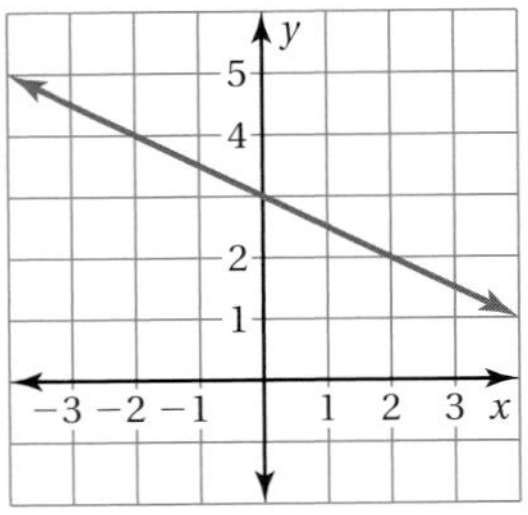

11. Write a function that describes the arithmetic sequence $-\frac{1}{4}, -\frac{1}{2}, -\frac{3}{4}, -1, \ldots$.

Tell whether the sequence is *geometric, arithmetic,* or *neither*.

12. 3, 2, 5, 4, . . .

13. 3, −3, −9, −15, . . .

14. 3, −15, 75, −375, . . .

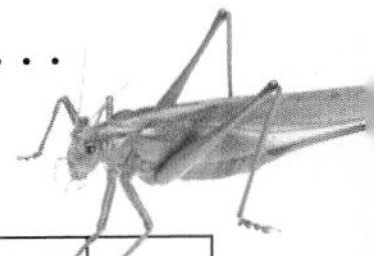

15. GRASSHOPPER The table shows the lengths of four grasshoppers and their jump heights. Use the table to draw a mapping diagram.

Body Length (mm)	5	6	7	8
Jump Height (mm)	50	60	70	80

16. WATER SKI The table shows the number of meters a water skier travels in x minutes.

Minutes, x	1	2	3	4	5
Meters, y	600	1200	1800	2400	3000

a. Graph the data.

b. Find a function to describe the data.

c. At this rate, how many *kilometers* would the water skier travel in 12 minutes?

17. PATTERN Make an input-output table for the pattern. Is the function relating the figure number x to the number of dots y linear? Explain.

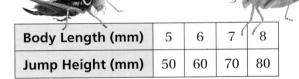

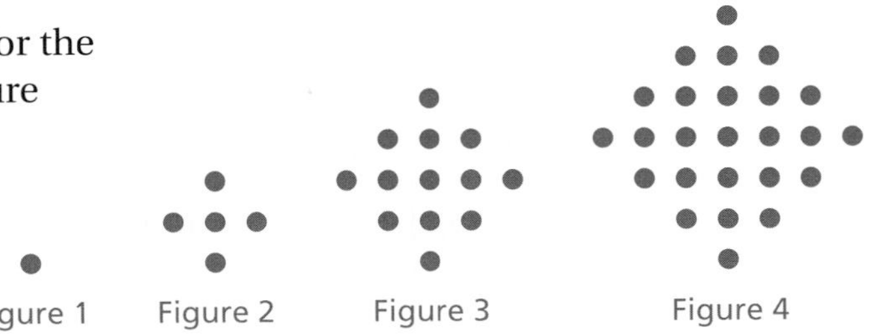

1. Which description is a correct way to solve the equation below?

$$4k - 8 = \frac{2}{3}$$

A. Add 8 to both sides and then divide both sides by 4.

B. Add 8 to both sides then multiply both sides by 4.

C. Divide both sides by 4 and then add 8 to both sides.

D. Subtract 8 from both sides and then divide both sides by 4.

Test-Taking Strategy

Work Backwards

"Work backwards by trying 1, 2, 3, and 4. You will see that $5 \cdot 3 + 3 = 18$. So, C is correct."

2. What is the volume of the rectangular prism below?

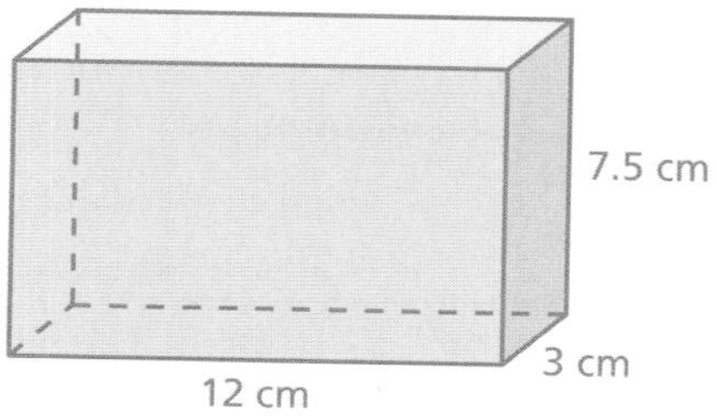

F. $22.5\ \text{cm}^3$

G. $90\ \text{cm}^3$

H. $270\ \text{cm}^3$

I. $288\ \text{cm}^3$

3. A button and its radius are shown below.

What is the circumference of the button? (Use 3.14 for π.)

A. 4.71 in.

B. 6.14 in.

C. 9.42 in.

D. 18.84 in.

4. A mapping diagram is shown.

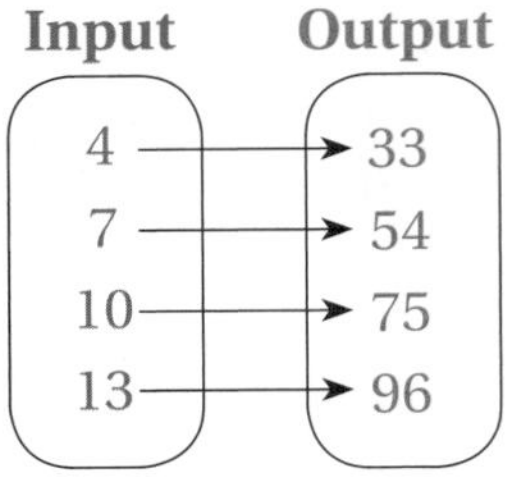

What number belongs in the box below so that the equation will correctly describe the function represented by the mapping diagram?

$$y = \square x + 5$$

5. Which number line is a graph of the solution to the inequality below?

$$x + 7 \geq 12$$

F. 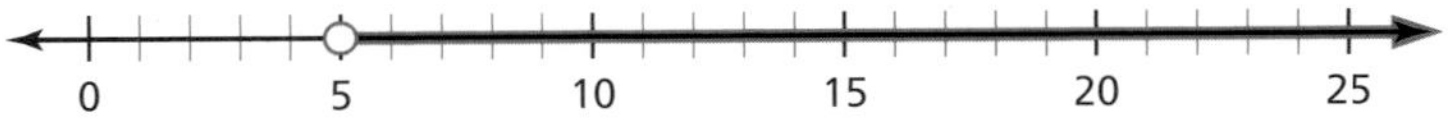

G. 

H. 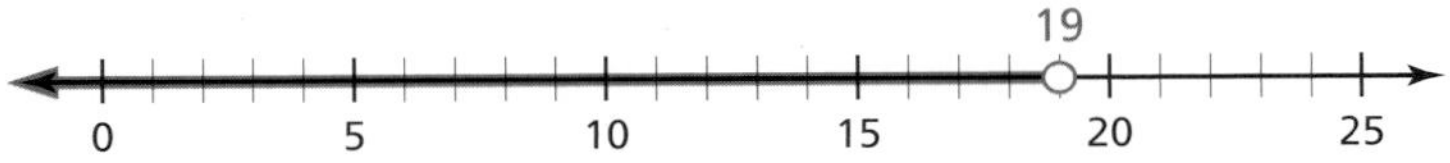

I. 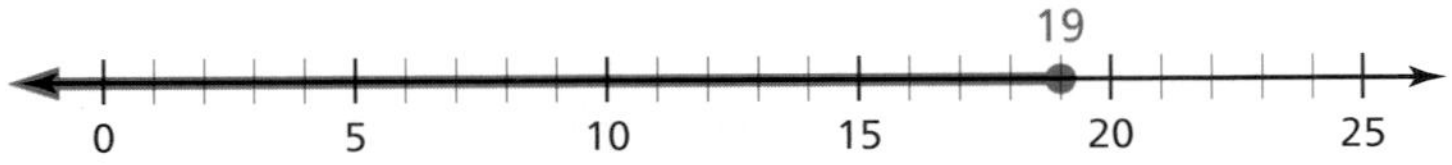

6. Your friend was solving the inequality in the box below.

$15 \leq 6n - 9$

$24 \leq 6n$

$n \leq 4$

What should your friend do to correct the error that she made?

A. Rewrite the answer as $n \geq 4$.

B. Multiply both sides of the inequality by 6.

C. Subtract 9 from both sides of the inequality.

D. Rewrite the given inequality as $6n - 9 \leq 15$.

7. Each output in the input-output table below is equal to 8 less than $\frac{1}{2}$ the input.

Input, x	20	34	50	?	106
Output, y	2	9	17	30	45

What is the value of the missing input?

8. What value of p makes the equation below true?

$$-2p - 8 = -32$$

F. -20

G. -12

H. 12

I. 20

9. *Part A* Draw a mapping diagram for (5, 14), (6, 17), (7, 20), and (8, 23).

Part B Describe any patterns that you see in the mapping diagram.

Part C Determine another input and output that follows the patterns you describe in Part B. Explain your reasoning.

10. Which input-output table represents a linear function?

A.

Input, x	1	2	3	4
Output, y	1000	200	30	4

B.

Input, x	1	2	3	4
Output, y	1	4	9	16

C.

Input, x	2	4	6	8
Output, y	7	13	19	25

D.

Input, x	3	4	5	6
Output, y	8	4	2	1

5 Ratios and Proportions

"I am doing an experiment with slope. I want you to run up and down the board 10 times."

"Now with 2 more dog biscuits, do it again and we'll compare your rates."

"Dear Sir: I counted the number of bacon, cheese, and chicken dog biscuits in the box I bought."

"There were 16 bacon, 12 cheese, and only 8 chicken. That's a ratio of 4:3:2. Please go back to the original ratio of 1:1:1."

What You Learned Before

...or possibly proportional to our stupidity!

"I wonder if our rate is proportional to the slope of the hill."

Simplifying Fractions

Example 1 **Simplify $\frac{4}{8}$.**

$$\frac{4 \div 4}{8 \div 4} = \frac{1}{2}$$

Simplify fractions by using the Greatest Common Factor.

Example 2 **Simplify $\frac{10}{15}$.**

$$\frac{10 \div 5}{15 \div 5} = \frac{2}{3}$$

Identifying Equivalent Fractions

Example 3 **Is $\frac{1}{4}$ equivalent to $\frac{13}{52}$?**

$$\frac{13 \div 13}{52 \div 13} = \frac{1}{4}$$

∴ $\frac{1}{4}$ is equivalent to $\frac{13}{52}$.

Example 4 **Is $\frac{30}{54}$ equivalent to $\frac{5}{8}$?**

$$\frac{30 \div 6}{54 \div 6} = \frac{5}{9}$$

∴ $\frac{30}{54}$ is *not* equivalent to $\frac{5}{8}$.

Solving Equations

Example 5 **Solve $12x = 168$.**

$12x = 168$ Write the equation.

$\frac{12x}{12} = \frac{168}{12}$ Divide each side by 12.

$x = 14$ Simplify.

Check

$12x = 168$

$12(14) \stackrel{?}{=} 168$

$168 = 168$ ✓

Try It Yourself

Simplify.

1. $\frac{12}{144}$ **2.** $\frac{15}{45}$ **3.** $\frac{75}{100}$ **4.** $\frac{16}{24}$

Are the fractions equivalent? Explain.

5. $\frac{15}{60} \stackrel{?}{=} \frac{3}{4}$ **6.** $\frac{2}{5} \stackrel{?}{=} \frac{24}{144}$ **7.** $\frac{15}{20} \stackrel{?}{=} \frac{3}{5}$ **8.** $\frac{2}{8} \stackrel{?}{=} \frac{16}{64}$

Solve the equation. Check your solution.

9. $\frac{y}{-5} = 3$ **10.** $0.6 = 0.2a$ **11.** $-2w = -9$ **12.** $\frac{1}{7}n = -4$

5.1 Ratios and Rates

STANDARDS OF LEARNING
7.12

Essential Question How do rates help you describe real-life problems?

Share Your Work at... My.BigIdeasMath.com

The Meaning of a Word ● Rate

When you rent snorkel gear at the beach, you should pay attention to the rental **rate**. The rental rate is in dollars per hour.

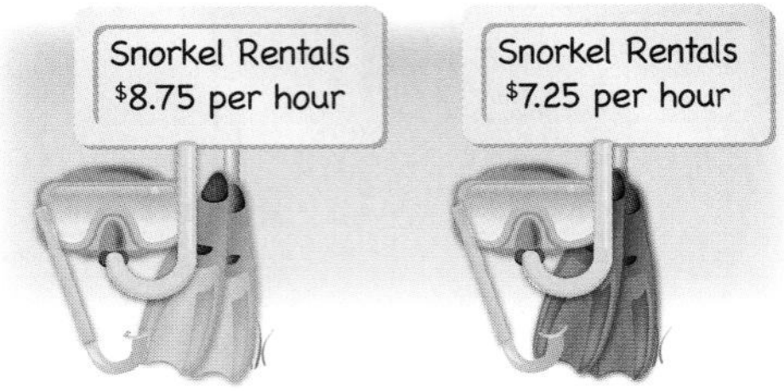

1 ACTIVITY: Finding Reasonable Rates

Work with a partner.

a. Match each description with a verbal rate.

b. Match each verbal rate with a numerical rate.

c. Give a reasonable numerical rate for each description. Then give an unreasonable rate.

Description	***Verbal Rate***	***Numerical Rate***
Your pay rate for washing cars	inches per month	$\frac{\square\ \text{m}}{\text{sec}}$
The average rainfall rate in a rain forest	pounds per acre	$\frac{\square\ \text{people}}{\text{yr}}$
Your average driving rate along an interstate	meters per second	$\frac{\square\ \text{lb}}{\text{acre}}$
The growth rate for the length of a baby alligator	people per year	$\frac{\square\ \text{mi}}{\text{h}}$
Your running rate in a 100-meter dash	dollars per hour	$\frac{\square\ \text{in.}}{\text{yr}}$
The population growth rate of a large city	dollars per year	$\frac{\square\ \text{in.}}{\text{mo}}$
The average pay rate for a professional athlete	miles per hour	$\frac{\$\ \square}{\text{h}}$
The fertilization rate for an apple orchard	inches per year	$\frac{\$\ \square}{\text{yr}}$

2 ACTIVITY: Unit Analysis

Work with a partner. Some real-life problems involve the product of an amount and a rate. Find each product. List the units.

a. **Sample:** $6\text{ h} \times \frac{\$12}{\text{h}} = 6\not{\text{h}} \times \frac{\$12}{\not{\text{h}}}$ — Divide out "hours."

$= \$72$ — Multiply. Answer is in dollars.

b. $6\text{ mo} \times \frac{\$700}{\text{mo}}$

c. $10\text{ gal} \times \frac{22\text{ mi}}{\text{gal}}$

d. $9\text{ lb} \times \frac{\$3}{\text{lb}}$

e. $13\text{ min} \times \frac{60\text{ sec}}{\text{min}}$

3 ACTIVITY: Writing a Story

Work with a partner.

- **Think of a story that compares two different rates.**
- **Write the story.**
- **Draw pictures for the story.**

What Is Your Answer?

4. **RESEARCH** Use newspapers, the Internet, or magazines to find examples of salaries. Try to find examples of each of the following ways to write salaries.

   a. dollars per hour **b.** dollars per month **c.** dollars per year

5. **IN YOUR OWN WORDS** How do rates help you describe real-life problems? Give two examples.

6. To estimate the annual salary for a given hourly pay rate, multiply by 2 and insert "000" at the end.

   Sample: \$10 per hour is about \$20,000 per year.

   a. Explain why this works. Assume the person is working 40 hours a week.

   b. Estimate the annual salary for an hourly pay rate of \$8 per hour.

   c. You earn \$1 million per month. What is your annual salary?

   d. Why is the cartoon funny?

"We had someone apply for the job. He says he would like \$1 million a month, but will settle for \$8 an hour."

Practice Use what you discovered about ratios and rates to complete Exercises 7–10 on page 202.

5.1 Lesson

Check It Out
Lesson Tutorials
BigIdeasMath.com

Key Vocabulary
ratio, *p. 200*
rate, *p. 200*
unit rate, *p. 200*

A **ratio** is a comparison of two quantities using division.

$\frac{3}{4}$, 3 to 4, 3 : 4

A **rate** is a ratio of two quantities with different units.

$\frac{60 \text{ miles}}{2 \text{ hours}}$

A rate with a denominator of 1 is called a **unit rate**.

$\frac{30 \text{ miles}}{1 \text{ hour}}$

EXAMPLE 1 Finding Ratios and Rates

There are 45 males and 60 females in a subway car. The subway car travels 2.5 miles in 5 minutes.

a. Find the ratio of males to females.

b. Find the speed of the subway car.

a. $\frac{\text{males}}{\text{females}} = \frac{45}{60} = \frac{3}{4}$

The ratio of males to females is $\frac{3}{4}$.

b. 2.5 miles in 5 minutes $= \frac{2.5 \text{ mi}}{5 \text{ min}} = \frac{2.5 \text{ mi} \div 5}{5 \text{ min} \div 5} = \frac{0.5 \text{ mi}}{1 \text{ min}}$

The speed is 0.5 mile per minute.

EXAMPLE 2 Finding a Rate from a Table

The table shows the amount of money you can raise by walking for a charity. Find your unit rate in dollars per mile.

Distance (miles)	2	4	6	8
Money (dollars)	24	48	72	96

(Distance: +2, +2, +2; Money: +24, +24, +24)

Use the table to find the unit rate.

$\frac{\text{change in money}}{\text{change in distance}} = \frac{\$24}{2 \text{ mi}}$ The money raised increases by \$24 every 2 miles.

$= \frac{\$12}{1 \text{ mi}}$ Simplify.

Your unit rate is \$12 per mile.

Multi-Language Glossary at BigIdeasMath.com.

On Your Own

Now You're Ready
Exercises 11–26

1. In Example 1, find the ratio of females to males.
2. In Example 1, find the ratio of females to total passengers.
3. The table shows the distance that the International Space Station travels while orbiting Earth. Find the speed in miles per second.

Time (seconds)	3	6	9	12
Distance (miles)	14.4	28.8	43.2	57.6

EXAMPLE 3 Finding a Rate from a Line Graph

The graph shows the distance that sound travels through water. Find the speed of sound in kilometers per second.

Sound through Water

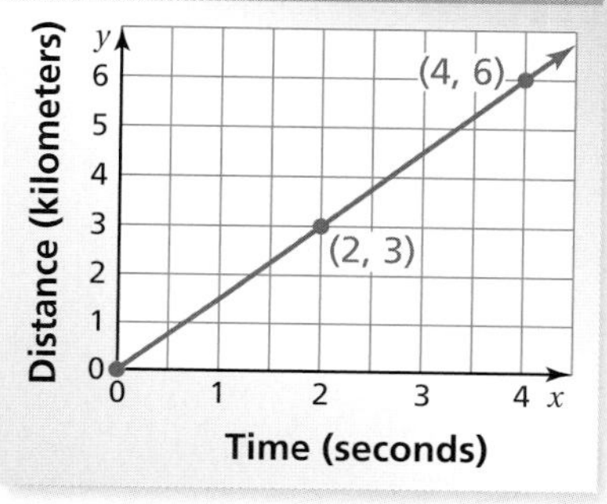

Step 1: Choose a point on the line.

The point (2, 3) shows you that sound travels 3 kilometers in 2 seconds.

Step 2: Find the speed.

$$\frac{\text{distance traveled}}{\text{elapsed time}} = \frac{3}{2} \quad \begin{matrix}\leftarrow \text{kilometers} \\ \leftarrow \text{seconds}\end{matrix}$$

$$= \frac{1.5 \text{ km}}{1 \text{ sec}} \qquad \text{Simplify.}$$

∴ The speed is 1.5 kilometers per second.

On Your Own

4. **WHAT IF?** In Example 3, you use the point (4, 6) to find the speed. Does your answer change? Why or why not?
5. The graph shows the distance that sound travels through air. Find the speed of sound in kilometers per second.
6. Does sound travel faster in water or in air? Explain.

Sound through Air

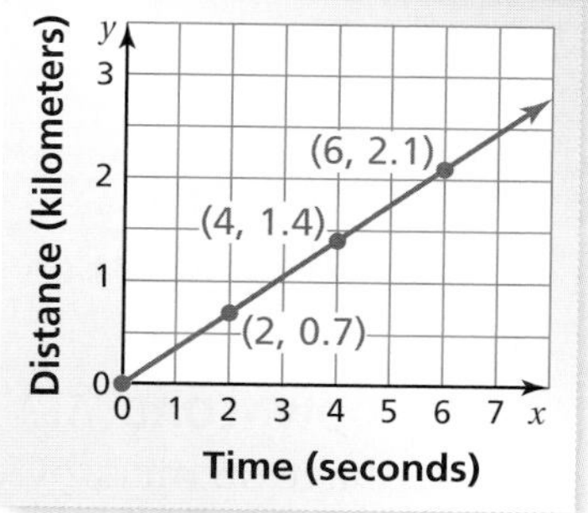

5.1 Exercises

Check It Out
Help with Homework
BigIdeasMath.com

Vocabulary and Concept Check

1. **VOCABULARY** How can you tell when a rate is a unit rate?
2. **WRITING** Why do you think rates are usually written as unit rates?
3. **OPEN-ENDED** Write a real-life rate that applies to you.

Estimate the unit rate.

4. \$74.75

5. \$1.19

6. \$2.35

Practice and Problem Solving

Find the product. List the units.

7. $8 \text{ h} \times \frac{\$9}{\text{h}}$
8. $8 \text{ lb} \times \frac{\$3.50}{\text{lb}}$
9. $14 \text{ sec} \times \frac{60 \text{ MB}}{\text{sec}}$
10. $6 \text{ h} \times \frac{19 \text{ mi}}{\text{h}}$

Write the ratio as a fraction in simplest form.

(1) 11. 25 to 45
12. 63 : 28
13. 35 girls : 15 boys
14. 2 feet : 8 feet
15. 16 dogs to 12 cats
16. 51 correct : 9 incorrect

Find the unit rate.

17. 180 miles in 3 hours
18. 256 miles per 8 gallons
19. \$9.60 for 4 pounds
20. \$4.80 for 6 cans
21. 297 words in 5.5 minutes
22. 54 meters in 2.5 hours

Use the table to find the rate.

(2) 23.

Servings	0	1	2	3
Calories	0	90	180	270

24.

Days	0	1	2	3
Liters	0	1.6	3.2	4.8

25.

Packages	3	6	9	12
Servings	13.5	27	40.5	54

26.

Years	2	6	10	14
Feet	7.2	21.6	36	50.4

27. **DOWNLOAD** At 1 P.M., you have 24 megabytes of a movie. At 1:15 P.M., you have 96 megabytes. What is the download rate in megabytes per minute?

28. **POPULATION** In 2000, the U.S. population was 281 million people. In 2008, it was 305 million. What was the rate of population change per year?

29. **TICKETS** The graph shows the cost of buying tickets to a concert.

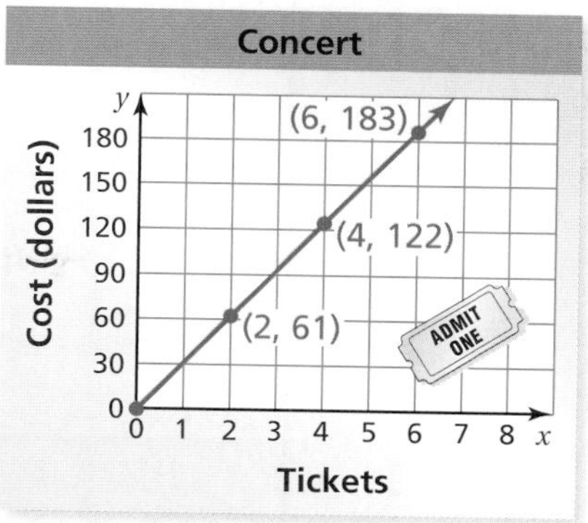

a. What does the point (4, 122) represent?

b. What is the unit rate?

c. What is the cost of buying 10 tickets?

30. **CRITICAL THINKING** Are the two statements equivalent? Explain your reasoning.

- The ratio of boys to girls is 2 to 3.
- The ratio of girls to boys is 3 to 2.

31. **TENNIS** A sports store sells three different packs of tennis balls. Which pack is the best buy? Explain.

Beverage	Serving Size	Calories	Sodium
Whole milk	1 cup	146	98 mg
Orange juice	1 pt	210	10 mg
Apple juice	24 fl oz	351	21 mg

32. **NUTRITION** The table shows nutritional information for three beverages.

a. Which has the most calories per fluid ounce?

b. Which has the least sodium per fluid ounce?

33. **Open-Ended** Fire hydrants are painted four different colors to indicate the rate at which water comes from the hydrant.

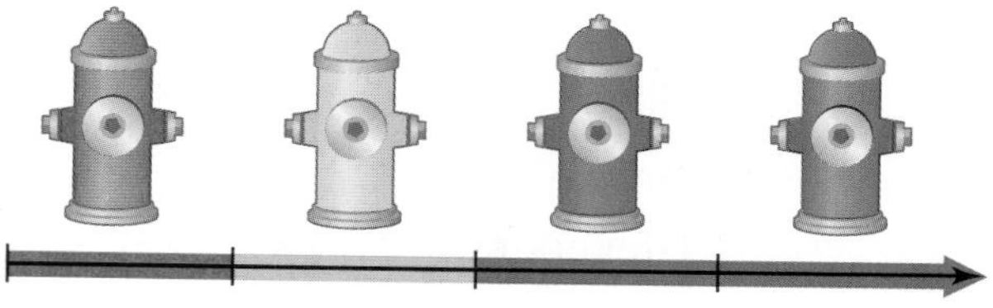

a. **RESEARCH** Use the Internet to find the ranges of the rates for each color.

b. Research why a firefighter needs to know the rate at which water comes out of the hydrant.

Fair Game Review What you learned in previous grades & lessons

Plot the ordered pair in a coordinate plane. *(Skills Review Handbook)*

34. $A(-5, -2)$ 35. $B(-3, 0)$ 36. $C(-1, 2)$ 37. $D(1, 4)$

38. **MULTIPLE CHOICE** Which fraction is greater than $-\frac{2}{3}$ and less than $-\frac{1}{2}$? *(Section 3.1)*

Ⓐ $-\frac{3}{4}$ Ⓑ $-\frac{7}{12}$ Ⓒ $-\frac{5}{12}$ Ⓓ $-\frac{3}{8}$

5.2 Slope

STANDARDS OF LEARNING
7.12

Essential Question How can you compare two rates graphically?

1 ACTIVITY: Comparing Unit Rates

Work with a partner. The table shows the maximum speeds of several animals.

a. Find the missing speeds. Round your answers to the nearest tenth.

b. Which animal is fastest? Which animal is slowest?

c. Explain how you convert between the two units of speed.

Animal	Speed (miles per hour)	Speed (feet per second)
Antelope	61.0	
Black Mamba Snake		29.3
Cheetah		102.6
Chicken		13.2
Coyote	43.0	
Domestic Pig		16.0
Elephant		36.6
Elk		66.0
Giant Tortoise	0.2	
Giraffe	32.0	
Gray Fox		61.6
Greyhound	39.4	
Grizzly Bear		44.0
Human		41.0
Hyena	40.0	
Jackal	35.0	
Lion		73.3
Peregrine Falcon	200.0	
Quarter Horse	47.5	
Spider		1.76
Squirrel	12.0	
Thomson's Gazelle	50.0	
Three-Toed Sloth		0.2
Tuna	47.0	

2 ACTIVITY: Comparing Two Rates Graphically

Work with a partner. A cheetah and a Thomson's gazelle are running at constant speeds.

a. Find the missing distances.

	Cheetah	Gazelle
Time (seconds)	Distance (feet)	Distance (feet)
0	0	0
1	102.6	
2		
3		
4		
5		
6		
7		

b. Use the table to complete the line graph for each animal.

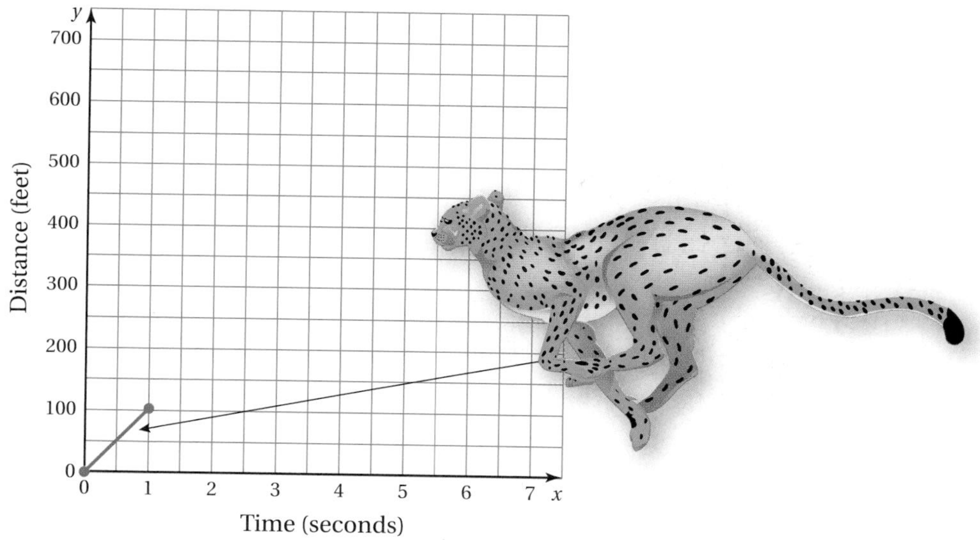

c. Which graph is steeper? The speed of which animal is greater?

What Is Your Answer?

3. IN YOUR OWN WORDS How can you compare two rates graphically? Explain your reasoning. Give some examples with your answer.

4. Choose 10 animals from Activity 1.

a. Make a table for each animal similar to the table in Activity 2.

b. Sketch a graph of the distances for each animal.

c. Compare the steepness of the 10 graphs. What can you conclude?

5.2 Lesson

Check It Out
Lesson Tutorials
BigIdeasMath.com

Key Vocabulary
slope, *p. 206*

Key Idea

Slope

Slope is the rate of change between any two points on a line. It is a measure of the *steepness* of a line.

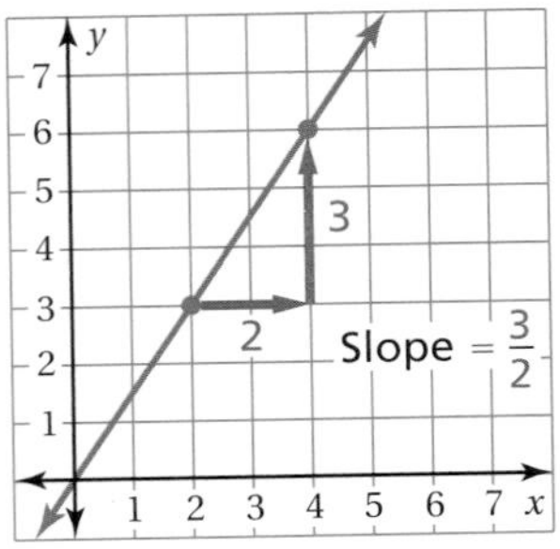

To find the slope of a line, find the ratio of the change in y (vertical change) to the change in x (horizontal change).

$$\text{slope} = \frac{\text{change in } y}{\text{change in } x}$$

EXAMPLE 1 Finding Slopes

Find the slope of each line.

a.

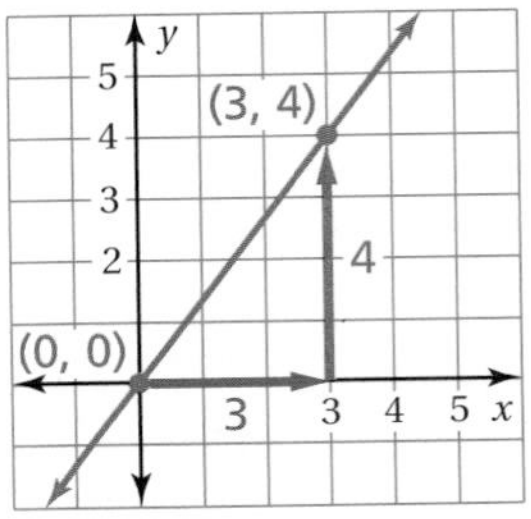

$$\text{slope} = \frac{\text{change in } y}{\text{change in } x}$$

$$= \frac{4}{3}$$

The slope of the line is $\frac{4}{3}$.

b.

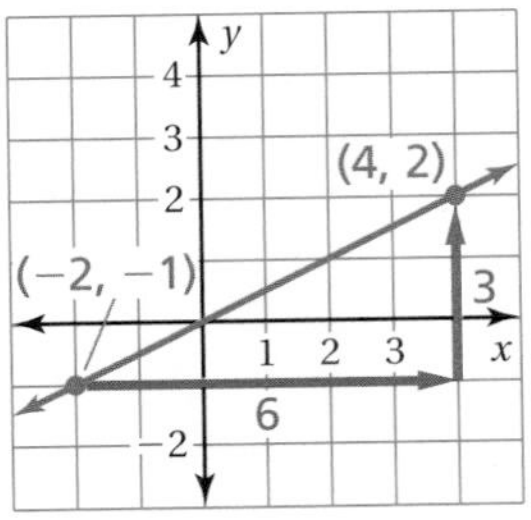

$$\text{slope} = \frac{\text{change in } y}{\text{change in } x}$$

$$= \frac{3}{6} = \frac{1}{2}$$

The slope of the line is $\frac{1}{2}$.

On Your Own

Now You're Ready
Exercises 4–9

Find the slope of the line.

1.

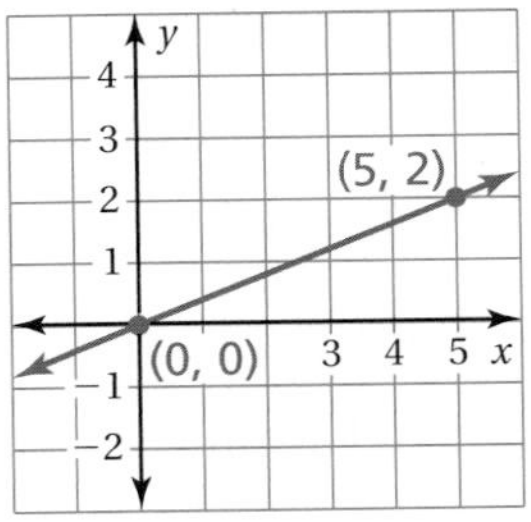

2.

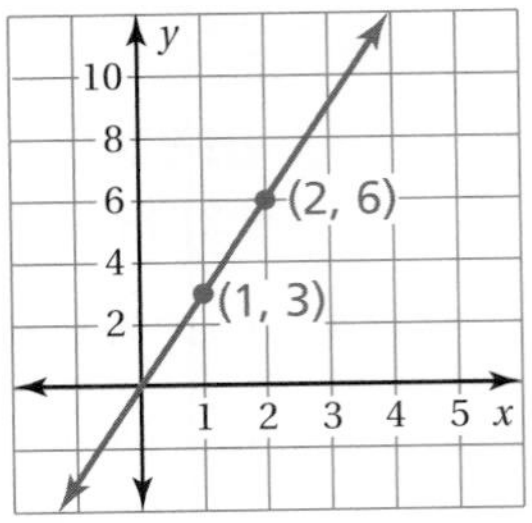

EXAMPLE 2 Finding a Slope

The table shows your earnings for babysitting.

a. Graph the data.

b. Find and interpret the slope of the line through the points.

Hours, x	0	2	4	6	8	10
Earnings, y (dollars)	0	10	20	30	40	50

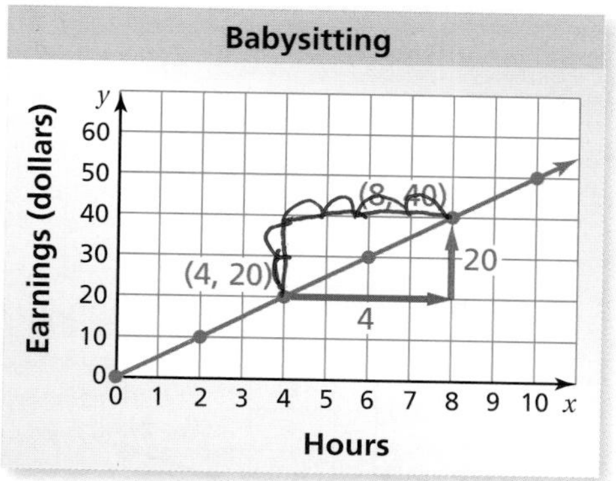

a. Graph the data. Draw a line through the points.

b. Choose any two points to find the slope of the line.

$$\text{slope} = \frac{\text{change in } y}{\text{change in } x}$$

$$= \frac{20}{4} \quad \begin{matrix} \leftarrow \text{dollars} \\ \leftarrow \text{hours} \end{matrix}$$

$$= 5$$

The slope of the line is 5. So, you earn $5 per hour babysitting.

On Your Own

Now You're Ready
Exercises 10 and 11

3. In Example 2, use two other points to find the slope. Does the slope change?

4. The graph shows the earnings of you and your friend for babysitting.

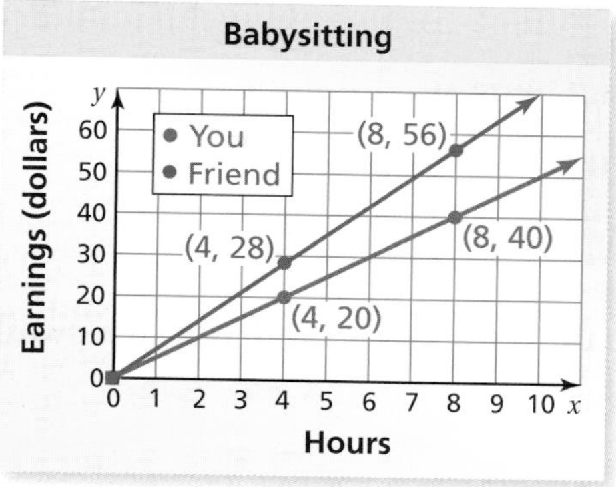

a. Compare the steepness of the lines. What does this mean in the context of the problem?

b. Find and interpret the slope of the blue line.

5.2 Exercises

Check It Out
Help with Homework
BigIdeasMath.com

Vocabulary and Concept Check

1. **VOCABULARY** Is there a connection between rate and slope? Explain.
2. **REASONING** Which line has the greatest slope?
3. **REASONING** Is it more difficult to run up a ramp with a slope of $\frac{1}{5}$ or a ramp with a slope of 5? Explain.

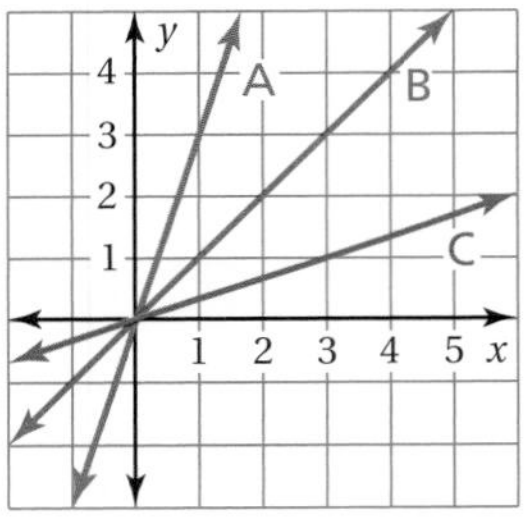

Practice and Problem Solving

Find the slope of the line.

(1) 4.

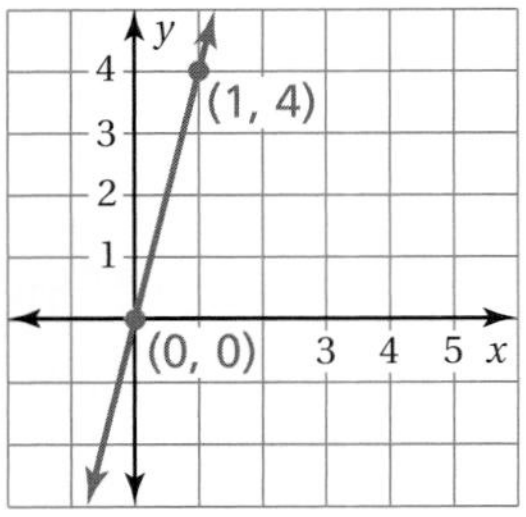

5.

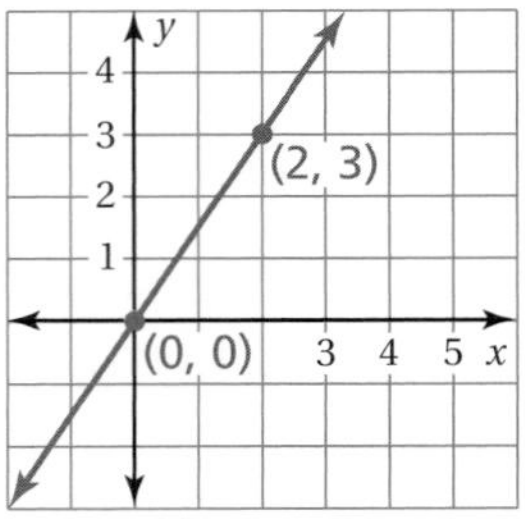

6.

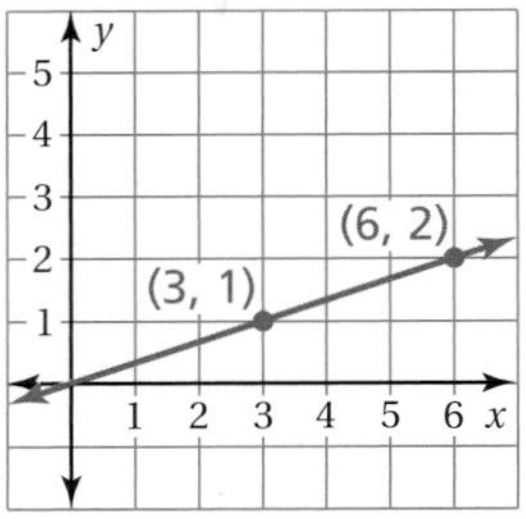

7.

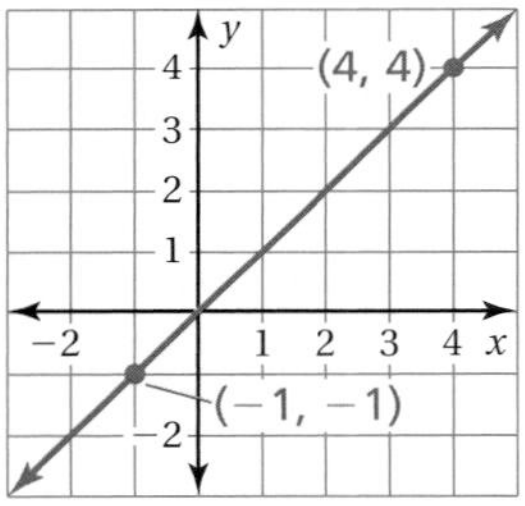

8.

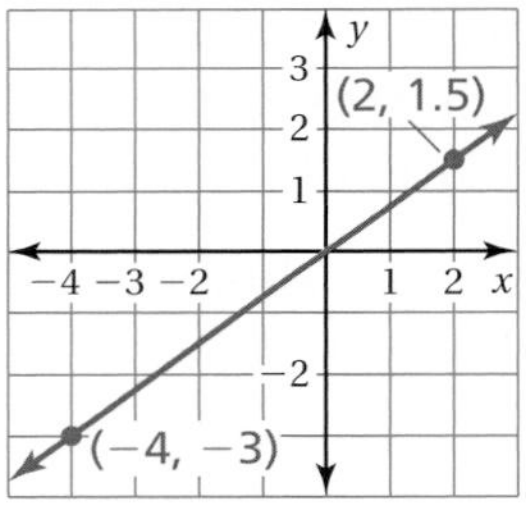

9.

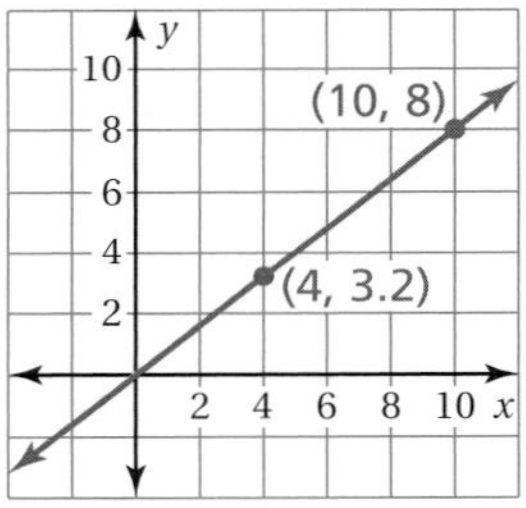

Graph the data. Then find the slope of the line through the points.

(2) 10.

Minutes, *x*	3	5	7	9
Words, *y*	135	225	315	405

11.

Gallons, *x*	5	10	15	20
Miles, *y*	162.5	325	487.5	650

Graph the line that passes through the two points. Then find the slope of the line.

12. (0, 0), (5, 8)
13. (−2, −2), (2, 2)
14. (10, 4), (−5, −2)

15. **ERROR ANALYSIS** Describe and correct the error in finding the slope of the line passing through (0, 0) and (4, 5).

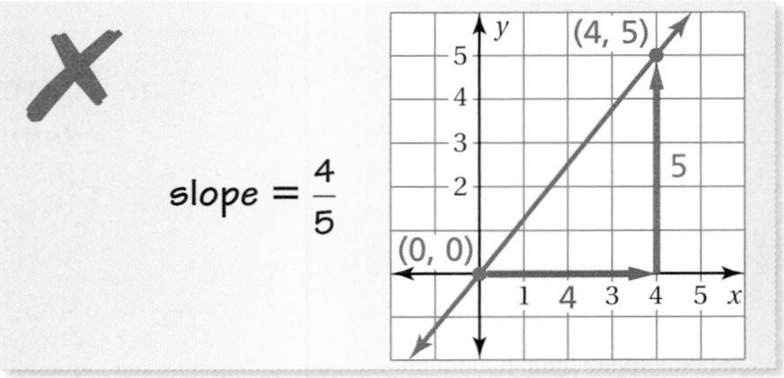

16. CAMPING The graph shows the amount of money you and a friend are saving for a camping trip.

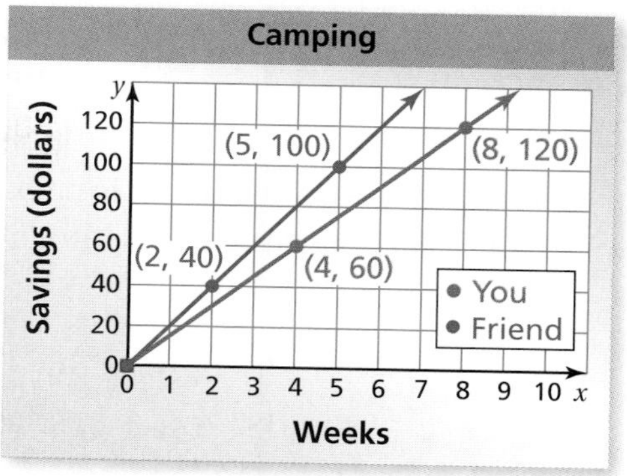

a. Compare the steepness of the lines. What does this mean in the context of the problem?

b. Find the slope of each line.

c. How much more money does your friend save each week than you?

d. The camping trip costs \$165. How long will it take you to save enough money?

17. MAPS The table shows data from a key to a map of Ohio.

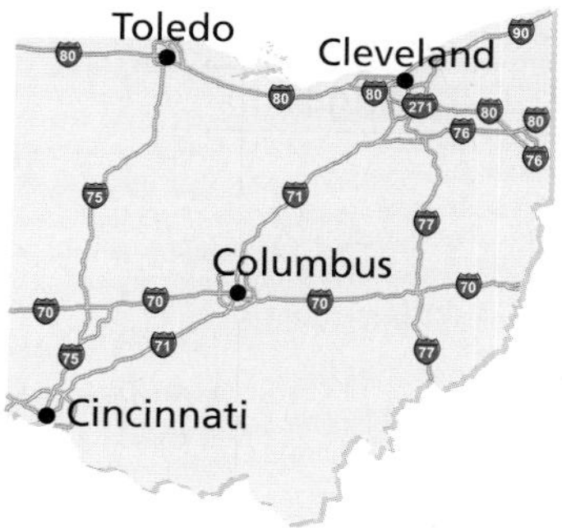

Distance on Map (mm), x	10	20	30	40
Actual Distance (mi), y	25	50	75	100

a. Graph the data.

b. Find the slope of the line. What does this mean in the context of the problem?

c. The map distance between Toledo and Columbus is 48 millimeters. What is the actual distance?

d. Cincinnati is about 225 miles from Cleveland. What is the distance between these cities on the map?

18. CRITICAL THINKING What is the slope of a line that passes through the points (2, 0) and (5, 0)? Explain.

19. Number Sense A line has a slope of 2. It passes through the points (1, 2) and (3, y). What is the value of y?

Fair Game Review What you learned in previous grades & lessons

Copy and complete the statement using <, >, or =. *(Section 3.1)*

20. $\frac{9}{2}$ ▢ $\frac{8}{3}$

21. $-\frac{8}{15}$ ▢ $\frac{10}{18}$

22. $\frac{-6}{24}$ ▢ $\frac{-2}{8}$

Multiply. *(Section 3.3)*

23. $-\frac{3}{5} \times \frac{8}{6}$

24. $1\frac{1}{2} \times \left(-\frac{6}{15}\right)$

25. $-2\frac{1}{4} \times -1\frac{1}{3}$

26. MULTIPLE CHOICE You have 18 stamps from Mexico in your stamp collection. These stamps are $\frac{3}{8}$ of your collection. The rest of the stamps are from the United States. How many stamps are from the United States? *(Section 3.5)*

Ⓐ 12 Ⓑ 24 Ⓒ 30 Ⓓ 48

Check It Out
Graphic Organizer
BigIdeasMath ✓.com

You can use an **information wheel** to organize information about a concept. Here is an example of an information wheel for slope.

Rate of change between any two points on a line

A measure of the steepness of a line

$\dfrac{\text{vertical change}}{\text{horizontal change}}$

Slope

$\dfrac{\text{change in } y}{\text{change in } x}$

Through (0, 0) and (4, 3): slope $= \dfrac{3}{4}$

Through (2, 3) and (−6, −3): slope $= \dfrac{-6}{-8} = \dfrac{3}{4}$

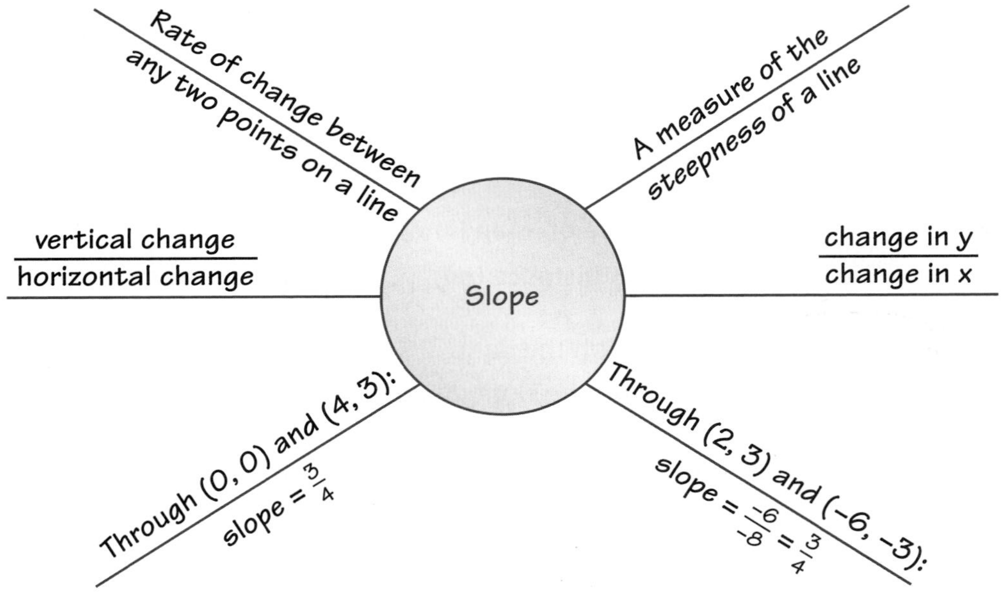

On Your Own

Make an information wheel to help you study these topics.

1. ratio

2. rate

3. unit rate

After you complete this chapter, make information wheels for the following topics.

4. proportion

5. cross products

6. writing proportions

7. solving proportions

"My **information wheel** summarizes how cats act when they get baths."

Write the ratio as a fraction in simplest form. *(Section 5.1)*

1. 18 red buttons : 12 blue buttons

2. 30 inches to 3 inches

Use the table to find the rate. *(Section 5.1)*

3.

Songs	0	2	4	6
Cost	\$0	\$1.98	\$3.96	\$5.94

4.

Hour	3	6	9	12
Gallons	10.5	21	31.5	42

Find the slope of the line. *(Section 5.2)*

5.

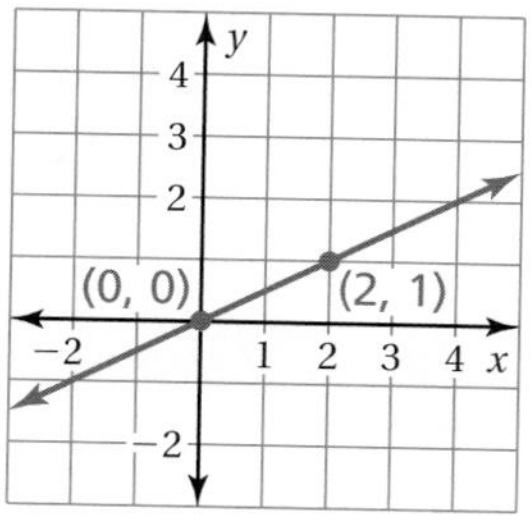

6.

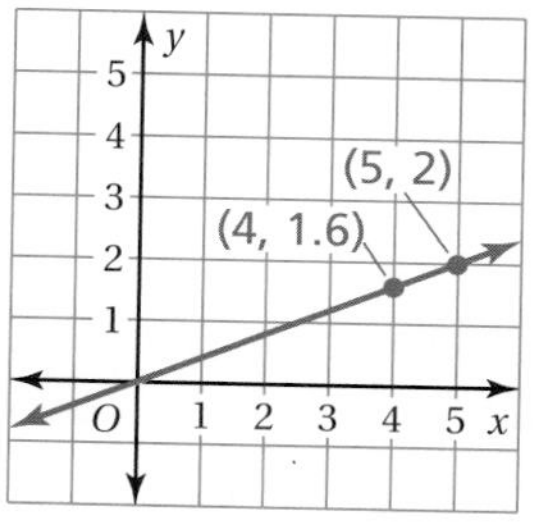

Graph the data. Then find the slope of the line through the points. *(Section 5.2)*

7.

Hours, x	2	4	6	8
Miles, y	10	20	30	40

8.

Packages, x	6	10	14	18
Servings, y	9	15	21	27

9. MUSIC DOWNLOAD The amount of time needed to download music is shown in the table. Find the rate in megabytes per second. *(Section 5.1)*

Seconds	6	12	18	24
Megabytes	2	4	6	8

10. SOUND The graph shows the distance that sound travels through steel. Find the speed of sound through steel in kilometers per second. *(Section 5.1)*

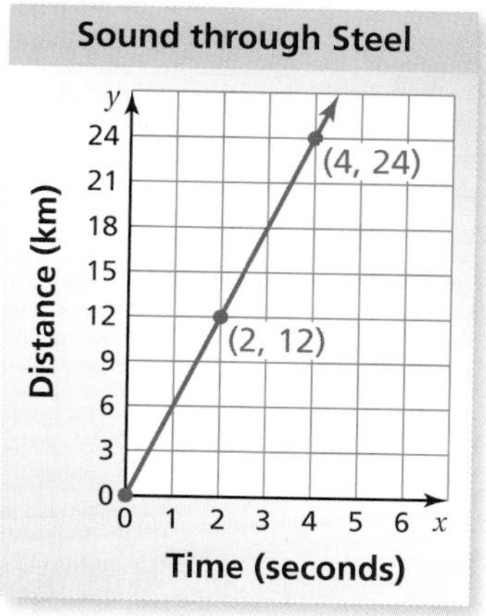

11. LAWN MOWING The graph shows how much you and your friend each earn mowing lawns. *(Section 5.2)*

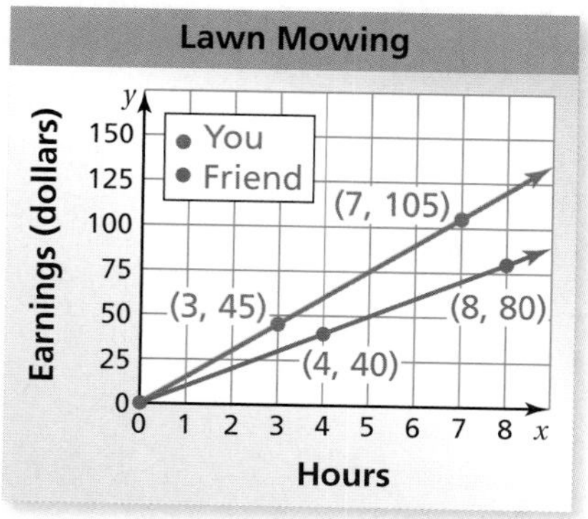

a. Compare the steepness of the lines. What does this mean in the context of the problem?

b. Find and interpret the slope of each line.

c. How much more money do you earn per hour than your friend?

5.3 Proportions

STANDARDS OF LEARNING
7.4

Essential Question How can proportions help you decide when things are "fair"?

The Meaning of a Word ● Proportional

When you work toward a goal, your success is usually **proportional** to the amount of work you put in.

An equation stating that two ratios are equal is a **proportion**.

1 ACTIVITY: Determining Proportions

Work with a partner. Tell whether the two ratios are equivalent. If they are not equivalent, change the second day to make the ratios equivalent. Explain your reasoning.

a. On the first day, you pay $5 for 2 boxes of popcorn. The next day, you pay $7.50 for 3 boxes.

First Day

$$\frac{\$5.00}{\$7.50} \stackrel{?}{=} \frac{2 \text{ boxes}}{3 \text{ boxes}}$$

Next Day

b. On the first day, it takes you 3 hours to drive 135 miles. The next day, it takes you 5 hours to drive 200 miles.

First Day

$$\frac{3 \text{ h}}{5 \text{ h}} \stackrel{?}{=} \frac{135 \text{ mi}}{200 \text{ mi}}$$

Next Day

c. On the first day, you walk 4 miles and burn 300 calories. The next day, you walk 3 miles and burn 225 calories.

First Day

$$\frac{4 \text{ mi}}{3 \text{ mi}} \stackrel{?}{=} \frac{300 \text{ cal}}{225 \text{ cal}}$$

Next Day

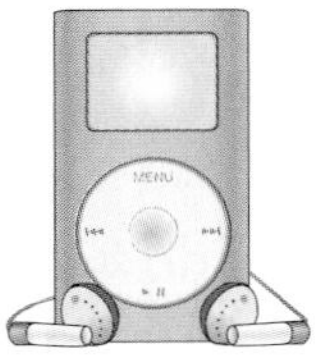

d. On the first day, you download 5 songs and pay $2.25. The next day, you download 4 songs and pay $2.00.

First Day

$$\frac{5 \text{ songs}}{4 \text{ songs}} \stackrel{?}{=} \frac{\$2.25}{\$2.00}$$

Next Day

2 ACTIVITY: Checking a Proportion

Work with a partner.

a. It is said that "one year in a dog's life is equivalent to seven years in a human's life." Explain why Newton thinks he has a score of 105 points. Did he solve the proportion correctly?

$$\frac{1 \text{ year}}{7 \text{ years}} \stackrel{?}{=} \frac{15 \text{ points}}{105 \text{ points}}$$

b. If Newton thinks his score is 98 points, how many points does he actually have? Explain your reasoning.

"I got 15 on my online test. That's 105 in dog points! Isn't that an A+?"

3 ACTIVITY: Determining Fairness

Work with a partner. Write a ratio for each sentence. If they are equal, then the answer is "It is fair." If they are not equal, then the answer is "It is not fair." Explain your reasoning.

a. You pay \$184 for 2 tickets to a concert. & I pay \$266 for 3 tickets to the same concert. → Is this fair?

b. You get 75 points for answering 15 questions correctly. & I get 70 points for answering 14 questions correctly. → Is this fair?

c. You trade 24 football cards for 15 baseball cards. & I trade 20 football cards for 32 baseball cards. → Is this fair?

What Is Your Answer?

4. Find a recipe for something you like to eat. Then show how two of the ingredient amounts are proportional when you double or triple the recipe.

5. **IN YOUR OWN WORDS** How can proportions help you decide when things are "fair"? Give an example.

Practice Use what you discovered about proportions to complete Exercises 17–22 on page 216.

Key Vocabulary
proportion, *p. 214*
proportional, *p. 214*
cross products, *p. 215*

Key Idea

Proportions

Words A **proportion** is an equation stating that two ratios are equivalent. Two quantities that form a proportion are **proportional**.

Numbers $\frac{2}{3} = \frac{4}{6}$ The proportion is read "2 is to 3 as 4 is to 6."

EXAMPLE 1 Determining Whether Ratios Form a Proportion

Tell whether the ratios form a proportion.

a. $\frac{4}{10}$ and $\frac{10}{25}$

Compare the ratios in simplest form.

$$\frac{4}{10} = \frac{4 \div 2}{10 \div 2} = \frac{2}{5}$$

$$\frac{10}{25} = \frac{10 \div 5}{25 \div 5} = \frac{2}{5}$$

The ratios are equivalent.

So, $\frac{4}{10}$ and $\frac{10}{25}$ form a proportion.

b. $\frac{6}{4}$ and $\frac{8}{12}$

Compare the ratios in simplest form.

$$\frac{6}{4} = \frac{6 \div 2}{4 \div 2} = \frac{3}{2}$$

$$\frac{8}{12} = \frac{8 \div 4}{12 \div 4} = \frac{2}{3}$$

The ratios are not equivalent.

So, $\frac{6}{4}$ and $\frac{8}{12}$ do not form a proportion.

On Your Own

Now You're Ready
Exercises 5–16

Tell whether the ratios form a proportion.

1. $\frac{1}{2}, \frac{5}{10}$ **2.** $\frac{4}{6}, \frac{18}{24}$ **3.** $\frac{10}{3}, \frac{5}{6}$ **4.** $\frac{25}{20}, \frac{15}{12}$

Key Ideas

Cross Products

In the proportion $\frac{a}{b} = \frac{c}{d}$, the products $a \cdot d$ and $b \cdot c$ are called **cross products**.

Cross Products Property

Words The cross products of a proportion are equal.

Numbers

$$\frac{2}{3} = \frac{4}{6}$$

$$2 \cdot 6 = 3 \cdot 4$$

Algebra

$$\frac{a}{b} = \frac{c}{d}$$

$ad = bc$,
where $b \neq 0$ and $d \neq 0$

EXAMPLE 2 Identifying Proportional Relationships

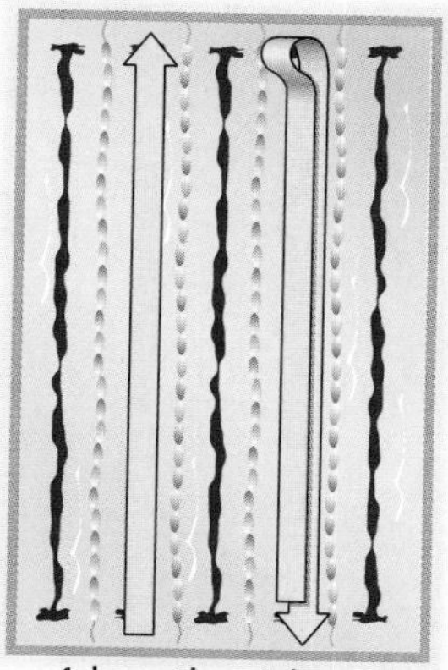

You swim your first 4 laps in 2.4 minutes. You complete 16 laps in 12 minutes. Is the number of laps proportional to your time?

Method 1: Compare unit rates.

$$\frac{2.4 \text{ min}}{4 \text{ laps}} = \frac{0.6 \text{ min}}{1 \text{ lap}} \quad (\div 4) \qquad \frac{12 \text{ min}}{16 \text{ laps}} = \frac{0.75 \text{ min}}{1 \text{ lap}} \quad (\div 16)$$

The unit rates are not equivalent.

∴ So, the number of laps is not proportional to the time.

Method 2: Use the Cross Products Property.

$\frac{2.4 \text{ min}}{4 \text{ laps}} \stackrel{?}{=} \frac{12 \text{ min}}{16 \text{ laps}}$ Test to see if the rates are equivalent.

$2.4 \cdot 16 \stackrel{?}{=} 4 \cdot 12$ Find the cross products.

$38.4 \neq 48$ The cross products are not equal.

∴ So, the number of laps is not proportional to the time.

On Your Own

Now You're Ready
Exercises 17–22

5. You read the first 20 pages of a book in 25 minutes. You read 36 pages in 45 minutes. Is the number of pages read proportional to your time?

5.3 Exercises

Check It Out
Help with Homework
BigIdeasMath.com

Vocabulary and Concept Check

1. **VOCABULARY** What does it mean for two ratios to form a proportion?
2. **VOCABULARY** What are two ways you can tell that two ratios form a proportion?
3. **OPEN-ENDED** Write two ratios that are equivalent to $\frac{3}{5}$.
4. **WHICH ONE DOESN'T BELONG?** Which ratio does *not* belong with the other three? Explain your reasoning.

$\frac{4}{10}$ $\frac{2}{5}$ $\frac{3}{5}$ $\frac{6}{15}$

Practice and Problem Solving

Tell whether the ratios form a proportion.

1 5. $\frac{1}{3}, \frac{7}{21}$

6. $\frac{1}{5}, \frac{6}{30}$

7. $\frac{3}{4}, \frac{24}{18}$

8. $\frac{2}{5}, \frac{40}{16}$

9. $\frac{48}{9}, \frac{16}{3}$

10. $\frac{18}{27}, \frac{33}{44}$

11. $\frac{7}{2}, \frac{16}{6}$

12. $\frac{12}{10}, \frac{14}{12}$

13. $\frac{27}{15}, \frac{18}{10}$

14. $\frac{4}{15}, \frac{15}{42}$

15. $\frac{76}{36}, \frac{19}{9}$

16. $\frac{49}{77}, \frac{38}{57}$

Tell whether the two rates form a proportion.

2 17. 7 inches in 9 hours; 42 inches in 54 hours

18. 12 players from 21 teams; 15 players from 24 teams

19. 440 calories in 4 servings; 300 calories in 3 servings

20. 120 units made in 5 days; 88 units made in 4 days

21. 66 wins in 82 games; 99 wins in 123 games

22. 68 hits in 172 at bats; 43 hits in 123 at bats

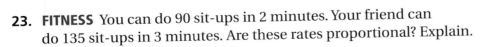

23. **FITNESS** You can do 90 sit-ups in 2 minutes. Your friend can do 135 sit-ups in 3 minutes. Are these rates proportional? Explain.

24. **HEARTBEAT** Find the heartbeat rates of you and your friend. Do these rates form a proportion? Explain.

	Heartbeats	Seconds
You	22	20
Friend	18	15

Tell whether the ratios form a proportion.

25. $\frac{3}{8}, \frac{31.5}{84}$

26. $\frac{14}{30}, \frac{75.6}{180}$

27. $\frac{2.5}{4}, \frac{7}{11.2}$

28. PAY RATE You earn \$56 walking your neighbor's dog for 8 hours. Your friend earns \$36 painting your neighbor's fence for 4 hours.

- **a.** What is your pay rate?
- **b.** What is your friend's pay rate?
- **c.** Are the pay rates equivalent? Explain.

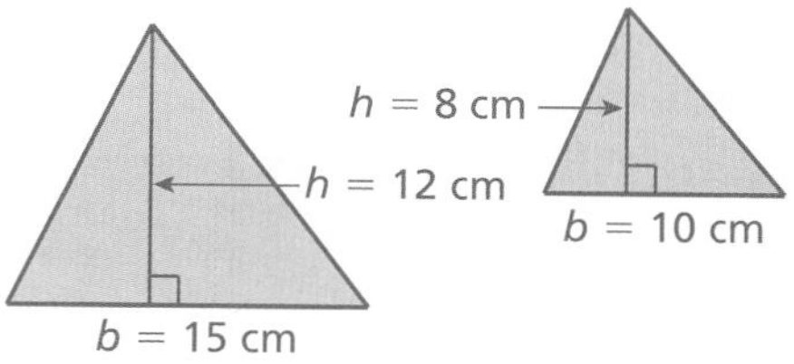

29. GEOMETRY Are the ratios of h to b in the two triangles proportional? Explain.

30. MUSIC You can buy 3 CDs for \$52.20 or 5 CDs for \$62.45. Are the rates proportional? Explain.

31. BASEBALL The table shows pitching statistics for four pitchers during the 2008 season.

- **a.** Which pitcher has the highest ratio of strikeouts to walks?
- **b.** Which of the pitchers have equivalent strikeout to walk ratios?

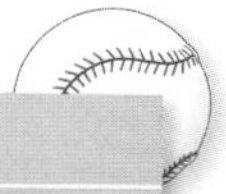

2008 Season		
Pitcher	**Strikeouts**	**Walks**
Pitcher 1	6	8
Pitcher 2	8	4
Pitcher 3	10	1
Pitcher 4	10	5

32. NAIL POLISH A specific shade of red nail polish requires 7 parts red to 2 parts yellow. A mixture contains 35 quarts of red and 8 quarts of yellow. How can you fix the mixture to make the correct shade of red?

33. COIN COLLECTION The ratio of quarters to dimes in a coin collection is 5 : 3. The same number of new quarters and dimes are added to the collection.

- **a.** Is the ratio of quarters to dimes still 5 : 3?
- **b.** If so, illustrate your answer with an example. If not, show why with a "counterexample."

34. Critical Thinking Ratio A is equivalent to ratio B. Ratio B is equivalent to ratio C. Is ratio A equivalent to ratio C? Explain.

Fair Game Review What you learned in previous grades & lessons

Add or subtract. *(Sections 1.2 and 1.3)*

35. $-28 + 15$

36. $-6 + (-11)$

37. $-10 - 8$

38. $-17 - (-14)$

39. MULTIPLE CHOICE Which fraction is not equivalent to $\frac{2}{6}$? *(Skills Review Handbook)*

Ⓐ $\frac{1}{3}$ Ⓑ $\frac{12}{36}$ Ⓒ $\frac{4}{12}$ Ⓓ $\frac{6}{9}$

5.4 Writing Proportions

STANDARDS OF LEARNING
7.4

Essential Question How can you write a proportion that solves a problem in real life?

1 ACTIVITY: Writing Proportions

Work with a partner. A rough rule for finding the correct bat length is "The bat length should be half of the batter's height." So, a 62-inch-tall batter uses a bat that is 31 inches long. Write a proportion to find the bat length for each given batter height.

a. 58 inches

b. 60 inches

c. 64 inches

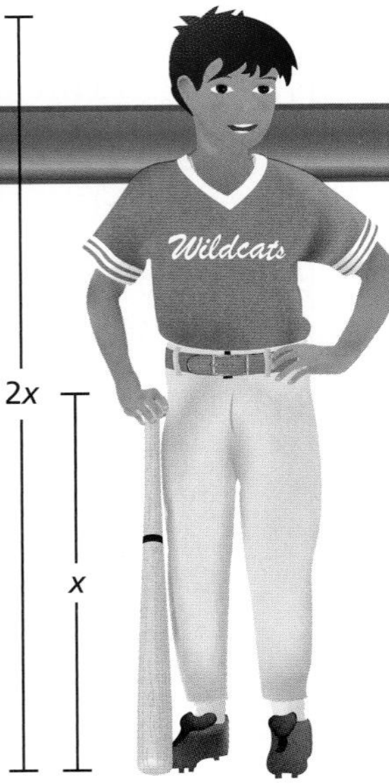

2 ACTIVITY: Bat Lengths

Work with a partner. Here is a more accurate table for determining the bat length for a batter. Find all of the batter heights for which the rough rule in Activity 1 is exact.

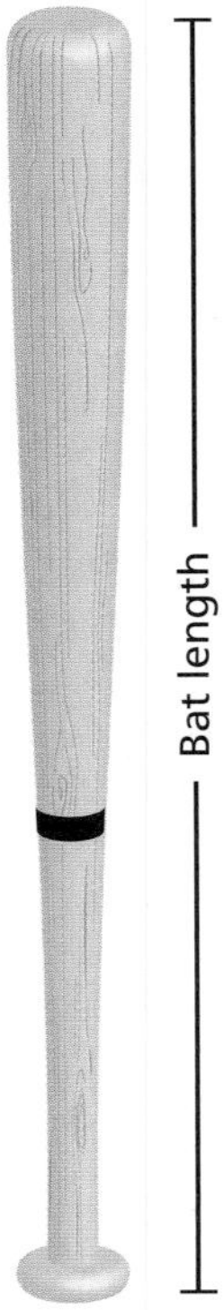

Weight of Batter (pounds)	Height of Batter (inches)							
	45–48	49–52	53–56	57–60	61–64	65–68	69–72	Over 72
Under 61	28	29	29					
61–70	28	29	30	30				
71–80	28	29	30	30	31			
81–90	29	29	30	30	31	32		
91–100	29	30	30	31	31	32		
101–110	29	30	30	31	31	32		
111–120	29	30	30	31	31	32		
121–130	29	30	30	31	32	33	33	
131–140	30	30	31	31	32	33	33	
141–150	30	30	31	31	32	33	33	
151–160	30	31	31	32	32	33	33	33
161–170		31	31	32	32	33	33	34
171–180				32	33	33	34	34
Over 180					33	33	34	34

3 ACTIVITY: Writing Proportions

Work with a partner. The batting average of a baseball player is the number of "hits" divided by the number of "at bats."

$$\text{Batting average} = \frac{\text{Hits } (H)}{\text{At bats } (A)}$$

A player whose batting average is 0.250 is said to be "batting 250."

$$\frac{20 \text{ hits}}{80 \text{ at bats}} = 0.250 = \frac{250 \text{ hits}}{1000 \text{ at bats}}$$

Actual hits → 20 hits

Actual at bats → 80 at bats

Batting average → 0.250

Batting 250 out of 1000 → $\frac{250 \text{ hits}}{1000 \text{ at bats}}$

Write a proportion to find how many hits H a player needs to achieve the given batting average. Then solve the proportion.

a. 50 times at bat; batting average is 0.200.

b. 84 times at bat; batting average is 0.250.

c. 80 times at bat; batting average is 0.350.

d. 1 time at bat; batting average is 1.000.

What Is Your Answer?

4. **IN YOUR OWN WORDS** How can you write a proportion that solves a problem in real life?

5. Two players have the same batting average.

	At Bats	Hits	Batting Average
Player 1	132	45	
Player 2	132	45	

Player 1 gets four hits in the next five at bats. Player 2 gets three hits in the next three at bats.

a. Who has the higher batting average?

b. Does this seem fair? Explain your reasoning.

Practice Use what you discovered about proportions to complete Exercises 4–7 on page 222.

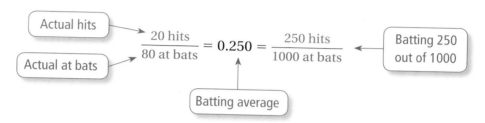

5.4 Lesson

Check It Out
Lesson Tutorials
BigIdeasMath.com

One way to write a proportion is to use a table.

	Last Month	This Month
Purchase	2 ringtones	3 ringtones
Total Cost	6 dollars	x dollars

Use the columns or the rows to write a proportion.

Use columns:

$$\frac{2 \text{ ringtones}}{6 \text{ dollars}} = \frac{3 \text{ ringtones}}{x \text{ dollars}}$$

Numerators have the same units.

Denominators have the same units.

Use rows:

$$\frac{2 \text{ ringtones}}{3 \text{ ringtones}} = \frac{6 \text{ dollars}}{x \text{ dollars}}$$

The units are the same on each side of the proportion.

EXAMPLE 1 Writing a Proportion

Black Bean Soup

1.5 cups black beans
0.5 cup salsa
2 cups water
1 tomato
2 teaspoons seasoning

A chef increases the amounts of ingredients in a recipe to make a proportional recipe. The new recipe has 6 cups of black beans. Write a proportion that gives the number x of tomatoes in the new recipe.

Organize the information in a table.

	Original Recipe	New Recipe
Black Beans	1.5 cups	6 cups
Tomatoes	1 tomato	x tomatoes

∴ One proportion is $\frac{1.5 \text{ cups beans}}{1 \text{ tomato}} = \frac{6 \text{ cups beans}}{x \text{ tomatoes}}$.

On Your Own

Now You're Ready
Exercises 8–11

1. In Example 1, write a different proportion that gives the number x of tomatoes in the new recipe.
2. In Example 1, write a proportion that gives the amount y of water in the new recipe.

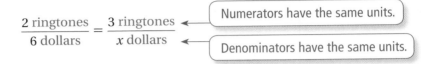

EXAMPLE 2 Solving Proportions Using Mental Math

Solve $\frac{3}{2} = \frac{x}{8}$.

Step 1: Think: The product of 2 and what number is 8?

$$\frac{3}{2} = \frac{x}{8}$$

$$2 \times ? = 8$$

Step 2: Because the product of 2 and 4 is 8, multiply the numerator by 4 to find x.

$$3 \times 4 = 12$$

$$\frac{3}{2} = \frac{x}{8}$$

$$2 \times 4 = 8$$

The solution is $x = 12$.

EXAMPLE 3 Solving Proportions Using Mental Math

In Example 1, how many tomatoes are in the new recipe?

Solve the proportion $\frac{1.5}{1} = \frac{6}{x}$. (cups black beans; tomatoes)

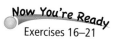

Step 1: Think: The product of 1.5 and what number is 6?

$$1.5 \times ? = 6$$

$$\frac{1.5}{1} = \frac{6}{x}$$

Step 2: Because the product of 1.5 and 4 is 6, multiply the denominator by 4 to find x.

$$1.5 \times 4 = 6$$

$$\frac{1.5}{1} = \frac{6}{x}$$

$$1 \times 4 = 4$$

So, there are 4 tomatoes in the new recipe.

On Your Own

Now You're Ready
Exercises 16–21

Solve the proportion.

3. $\frac{5}{8} = \frac{20}{d}$

4. $\frac{7}{z} = \frac{14}{10}$

5. $\frac{21}{24} = \frac{x}{8}$

6. A school has 950 students. The ratio of female students to all students is $\frac{48}{95}$. Write and solve a proportion to find the number f of students that are female.

5.4 Exercises

Vocabulary and Concept Check

1. **WRITING** Describe two ways you can use a table to write a proportion.

2. **WRITING** What is your first step when solving $\frac{x}{15} = \frac{3}{5}$? Explain.

3. **OPEN-ENDED** Write a proportion using an unknown value x and the ratio 5 : 6. Then solve it.

Practice and Problem Solving

Write a proportion to find how many points a student needs to score on the test to get the given score.

4. Test worth 50 points; test score of 40%
5. Test worth 50 points; test score of 78%
6. Test worth 80 points; test score of 80%
7. Test worth 150 points; test score of 96%

Use the table to write a proportion.

1 8.

	Game 1	Game 2
Points	12	18
Shots	14	w

9.

	May	June
Winners	n	34
Entries	85	170

10.

	Today	Yesterday
Miles	15	m
Hours	2.5	4

11.

	Race 1	Race 2
Meters	100	200
Seconds	x	22.4

12. **ERROR ANALYSIS** Describe and correct the error in writing the proportion.

	Monday	Tuesday
Dollars	2.08	d
Ounces	8	16

$$\frac{2.08}{16} = \frac{d}{8}$$

13. **T-SHIRTS** You can buy three T-shirts for \$24. Write a proportion that gives the cost c of buying seven T-shirts.

14. **COMPUTERS** A school requires two computers for every five students. Write a proportion that gives the number c of computers needed for 145 students.

15. **SWIM TEAM** The school team has 80 swimmers. The ratio of 6th grade swimmers to all swimmers is 5 : 16. Write a proportion that gives the number s of 6th grade swimmers.

Solve the proportion.

2 3 **16.** $\frac{1}{4} = \frac{z}{20}$

17. $\frac{3}{4} = \frac{12}{y}$

18. $\frac{35}{k} = \frac{7}{3}$

19. $\frac{15}{8} = \frac{45}{c}$

20. $\frac{b}{36} = \frac{5}{9}$

21. $\frac{1.4}{2.5} = \frac{g}{25}$

22. ORCHESTRA In an orchestra, the ratio of trombones to violas is 1 to 3.

a. There are nine violas. Write a proportion that gives the number t of trombones in the orchestra.

b. How many trombones are in the orchestra?

23. ATLANTIS Your science teacher has a 1 : 200 scale model of the Space Shuttle Atlantis. Which of the proportions can be used to find the actual length x of Atlantis? Explain.

$\frac{1}{200} = \frac{19.5}{x}$ $\quad$ $\frac{1}{200} = \frac{x}{19.5}$ $\quad$ $\frac{200}{19.5} = \frac{x}{1}$ $\quad$ $\frac{x}{200} = \frac{1}{19.5}$

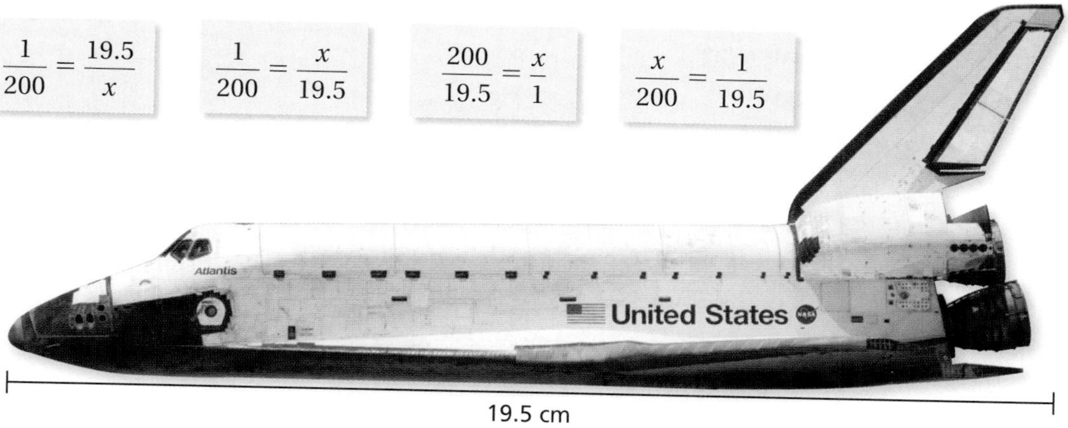

24. YOU BE THE TEACHER Your friend says "$48x = 6 \cdot 12$." Is your friend right? Explain.

Solve $\frac{6}{x} = \frac{12}{48}$.

25. Reasoning There are 180 white lockers in the school. There are 3 white lockers for every 5 blue lockers. How many lockers are in the school?

Fair Game Review What you learned in previous grades & lessons

Solve the equation. *(Section 3.5)*

26. $\frac{x}{6} = 25$

27. $8x = 72$

28. $150 = 2x$

29. $35 = \frac{x}{4}$

30. MULTIPLE CHOICE Which is the slope of a line? *(Section 5.2)*

Ⓐ $\frac{\text{change in } y}{1}$ Ⓑ $\frac{\text{change in } x}{1}$ Ⓒ $\frac{\text{change in } x}{\text{change in } y}$ Ⓓ $\frac{\text{change in } y}{\text{change in } x}$

Solving Proportions

Essential Question How can you use ratio tables and cross products to solve proportions in science?

STANDARDS OF LEARNING
7.4

1 ACTIVITY: Solving a Proportion in Science

SCIENCE Scientists use *ratio tables* to determine the amount of a compound (like salt) that is dissolved in a solution. Work with a partner to show how scientists use cross products to determine the unknown quantity in a ratio.

a. Sample: Salt Water

Salt Water	1 L	3 L
Salt	250 g	x g

l liter 3 liter

$$\frac{3\,\text{L}}{1\,\text{L}} = \frac{x\,\text{g}}{250\,\text{g}}$$ Write proportion.

$$3 \cdot 250 = 1 \cdot x$$ Set cross products equal.

$$750 = x$$ Simplify.

So, there are 750 grams of salt in the 3-liter solution.

b. White Glue Solution

Water	$\frac{1}{2}$ cup	1 cup
White Glue	$\frac{1}{2}$ cup	x cups

c. Borax Solution

Borax	1 tsp	2 tsp
Water	1 cup	x cups

d. Slime (see recipe)

Borax Solution	$\frac{1}{2}$ cup	1 cup
White Glue Solution	y cups	x cups

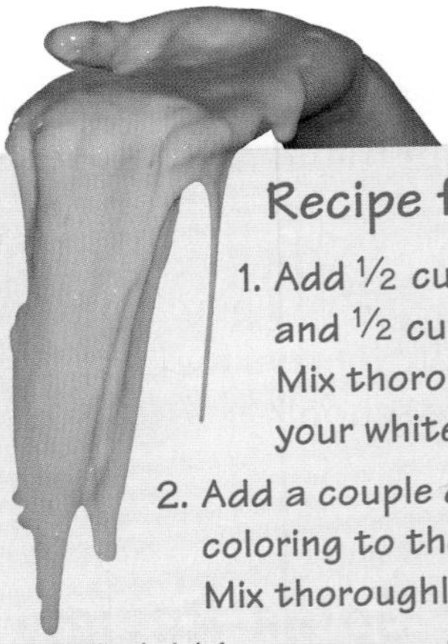

Recipe for SLIME

1. Add $\frac{1}{2}$ cup of water and $\frac{1}{2}$ cup white glue. Mix thoroughly. This is your white glue solution.
2. Add a couple drops of food coloring to the glue solution. Mix thoroughly.
3. Add 1 teaspoon of borax to 1 cup of water. Mix thoroughly. This is your borax solution (about 1 cup).
4. Pour the borax solution and the glue solution into a separate bowl.
5. Place the slime that forms in a plastic bag and squeeze the mixture repeatedly to mix it up.

2 ACTIVITY: The Game of Criss Cross

Preparation:

- Cut index cards to make 48 playing cards.
- Write each number on a card.

  1, 1, 1, 2, 2, 2, 3, 3, 3, 4, 4, 4, 5, 5, 5, 6, 6, 6, 7, 7,

  7, 8, 8, 8, 9, 9, 9, 10, 10, 10, 12, 12, 12, 13, 13,

  13, 14, 14, 14, 15, 15, 15, 16, 16, 16, 18, 20, 25
- Make a copy of the game board.

To Play:

- Play with a partner.
- Deal 8 cards to each player.
- Begin by drawing a card from the remaining cards. Use four of your cards to try to form a proportion.
- Lay the four cards on the game board. If you form a proportion, say "Criss Cross" and you earn 4 points. Place the four cards in a discard pile. Now it is your partner's turn.
- If you cannot form a proportion, then it is your partner's turn.
- When the original pile of cards is empty, shuffle the cards in the discard pile and start again.
- The first player to reach 20 points wins.

What Is Your Answer?

3. **IN YOUR OWN WORDS** How can you use ratio tables and cross products to solve proportions in science? Give an example.
4. **PUZZLE** Use each number once to form three proportions.

1	2	10	4	12	20
15	5	16	6	8	3

Practice Use what you discovered about solving proportions to complete Exercises 10–13 on page 228.

5.5 Lesson

Check It Out
Lesson Tutorials
BigIdeasMath.com

Key Idea

Solving Proportions

Method 1 Use mental math. *(Section 5.4)*

Method 2 Use the Multiplication Property of Equality. *(Section 5.5)*

Method 3 Use the Cross Products Property. *(Section 5.5)*

EXAMPLE 1 Solving Proportions Using Multiplication

Solve $\frac{5}{7} = \frac{x}{21}$.

$\frac{5}{7} = \frac{x}{21}$ Write the proportion.

$21 \cdot \frac{5}{7} = 21 \cdot \frac{x}{21}$ Multiply each side by 21.

$15 = x$ Simplify.

The solution is 15.

On Your Own

Now You're Ready
Exercises 4–9

Solve the proportion using multiplication.

1. $\frac{w}{6} = \frac{6}{9}$ **2.** $\frac{12}{10} = \frac{a}{15}$ **3.** $\frac{y}{6} = \frac{2}{4}$

EXAMPLE 2 Solving Proportions Using the Cross Products Property

Solve each proportion.

a. $\frac{x}{8} = \frac{7}{10}$

$x \cdot 10 = 8 \cdot 7$ Use the Cross Products Property.

$10x = 56$ Multiply.

$x = 5.6$ Divide.

The solution is 5.6.

b. $\frac{9}{y} = \frac{3}{17}$

$9 \cdot 17 = y \cdot 3$ Use the Cross Products Property.

$153 = 3y$ Multiply.

$51 = y$ Divide.

The solution is 51.

On Your Own

Now You're Ready
Exercises 10–21

Solve the proportion using the Cross Products Property.

4. $\frac{2}{7} = \frac{x}{28}$ **5.** $\frac{12}{5} = \frac{6}{y}$ **6.** $\frac{40}{z+1} = \frac{15}{6}$

EXAMPLE 3 Real-Life Application

The graph shows the toll y due on a turnpike for driving x miles. Your toll is \$7.50. How many *kilometers* did you drive?

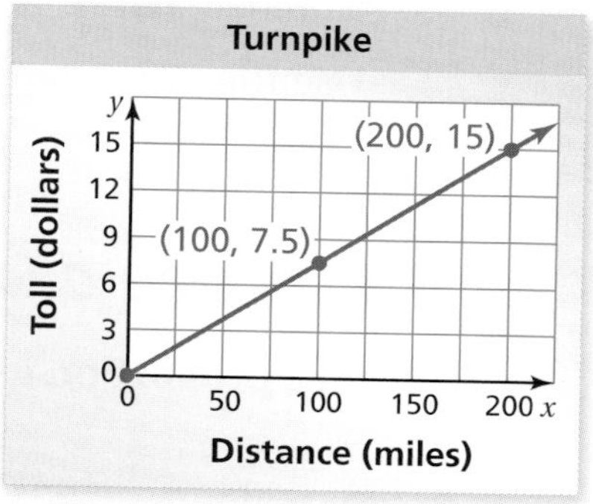

The point (100, 7.5) on the graph shows that the toll is \$7.50 for driving 100 miles. Convert 100 miles to kilometers.

Method 1: Convert using a ratio.

1 mi ≈ 1.6 km

$$100 \text{ mi} \times \frac{1.6 \text{ km}}{1 \text{ mi}} = 160 \text{ km}$$

So, you drove about 160 kilometers.

Method 2: Convert using a proportion.

Let x be the number of kilometers equivalent to 100 miles.

kilometers → $\frac{1.6}{1} = \frac{x}{100}$ ← kilometers
miles → ← miles

Write a proportion. Use 1.6 km ≈ 1 mi.

$$1.6 \cdot 100 = 1 \cdot x$$ Use Cross Products Property.

$$160 = x$$ Simplify.

So, you drove about 160 kilometers.

On Your Own

Now You're Ready
Exercises 28–30

Write and solve a proportion to complete the statement.

7. 7.5 in. ≈ ___ cm **8.** 100 g ≈ ___ oz

9. 2 L ≈ ___ qt **10.** 4 m ≈ ___ ft

5.5 Exercises

Check It Out
Help with Homework
BigIdeasMath.com

Vocabulary and Concept Check

1. **WRITING** What are three ways you can solve a proportion?

2. **OPEN-ENDED** Which way would you choose to solve $\frac{3}{x} = \frac{6}{14}$? Explain your reasoning.

3. **NUMBER SENSE** Does $\frac{x}{4} = \frac{15}{3}$ have the same solution as $\frac{x}{15} = \frac{4}{3}$? Use the Cross Products Property to explain your answer.

Practice and Problem Solving

Solve the proportion using multiplication.

1 4. $\frac{9}{5} = \frac{z}{20}$ 5. $\frac{h}{15} = \frac{16}{3}$ 6. $\frac{w}{4} = \frac{42}{24}$

7. $\frac{35}{28} = \frac{n}{12}$ 8. $\frac{7}{16} = \frac{x}{4}$ 9. $\frac{y}{9} = \frac{44}{54}$

Solve the proportion using the Cross Products Property.

2 10. $\frac{a}{6} = \frac{15}{2}$ 11. $\frac{10}{7} = \frac{8}{k}$ 12. $\frac{3}{4} = \frac{v}{14}$ 13. $\frac{5}{n} = \frac{16}{32}$

14. $\frac{36}{42} = \frac{24}{r}$ 15. $\frac{9}{10} = \frac{d}{6.4}$ 16. $\frac{x}{8} = \frac{3}{12}$ 17. $\frac{8}{m} = \frac{6}{15}$

18. $\frac{4}{24} = \frac{c}{36}$ 19. $\frac{20}{16} = \frac{d}{12}$ 20. $\frac{30}{20} = \frac{w}{14}$ 21. $\frac{2.4}{1.8} = \frac{7.2}{k}$

22. **ERROR ANALYSIS** Describe and correct the error in solving the proportion $\frac{m}{8} = \frac{15}{24}$.

✗

$$\frac{m}{8} = \frac{15}{24}$$
$$8 \cdot m = 24 \cdot 15$$
$$m = 45$$

23. **PENS** Forty-eight pens are packaged in four boxes. How many pens are packaged in nine boxes?

24. **PIZZA PARTY** How much does it cost to buy 10 medium pizzas?

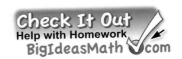

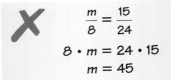

Solve the proportion.

25. $\frac{2x}{5} = \frac{9}{15}$ 26. $\frac{5}{2} = \frac{d-2}{4}$ 27. $\frac{4}{k+3} = \frac{8}{14}$

Write and solve a proportion to complete the statement.

3 **28.** 6 km ≈ ▢ mi **29.** 2.5 L ≈ ▢ gal **30.** 90 lb ≈ ▢ kg

31. TRUE OR FALSE? Tell whether the statement is *true* or *false*. Explain.

$$\text{If } \frac{a}{b} = \frac{2}{3}, \text{ then } \frac{3}{2} = \frac{b}{a}.$$

32. CLASS TRIP It costs $95 for 20 students to visit an aquarium. How much does it cost for 162 students?

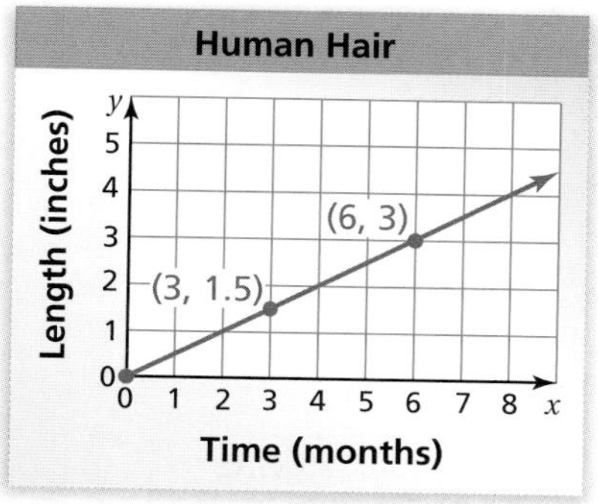

33. GRAVITY A person who weighs 120 pounds on Earth weighs 20 pounds on the moon. How much does a 93-pound person weigh on the moon?

34. HAIR The length of human hair is proportional to the number of months it has grown.

a. What is the hair length in *centimeters* after 6 months?

b. How long does it take hair to grow 8 inches?

c. Use a different method than the one in part (b) to find how long it takes hair to grow 20 inches.

35. CHEETAH Cheetahs are the fastest mammals in the world. They can reach speeds of 70 miles per hour.

a. At this speed, how long would it take a cheetah to run 17 miles?

b. **RESEARCH** Use the Internet or library to find how long a cheetah can maintain a speed of 70 miles per hour.

36. AUDIENCE There are 144 people in an audience. The ratio of adults to children is 5 to 3. How many are adults?

37. LAWN SEED Three pounds of lawn seed covers 1800 square feet. How many bags are needed to cover 8400 square feet?

38. Critical Thinking Consider the proportions $m = \frac{1}{2}$ and $k = \frac{1}{4}$. What is the ratio $\frac{m}{k}$? Explain your reasoning.

Fair Game Review What you learned in previous grades & lessons

Solve the equation. Check your solution. *(Section 3.4)*

39. $90 + x = 170$ **40.** $-15 = m + 75$ **41.** $n - 130 = -250$

42. MULTIPLE CHOICE Which function rule is shown by the table? *(Section 4.3)*

Ⓐ $y = -2x$ Ⓑ $y = -\frac{x}{2}$

Ⓒ $y = x - 3$ Ⓓ $y = x + 3$

Input, x	2	4	6	8
Output, y	−1	−2	−3	−4

5.3–5.5 Quiz

Check It Out
Progress Check
BigIdeasMath.com

Tell whether the ratios form a proportion. *(Section 5.3)*

1. $\frac{1}{8}, \frac{4}{32}$

2. $\frac{2}{3}, \frac{10}{30}$

3. $\frac{7}{4}, \frac{28}{16}$

Tell whether the two rates form a proportion. *(Section 5.3)*

4. 75 miles in 3 hours; 140 miles in 4 hours

5. 12 gallons in 4 minutes; 21 gallons in 7 minutes

6. 150 steps in 50 feet; 72 steps in 24 feet

7. 3 rotations in 675 days; 2 rotations in 730 days

Use the table to write a proportion. *(Section 5.4)*

8.

	Monday	Tuesday
Dollars	42	56
Hours	6	h

9.

	Series 1	Series 2
Games	g	6
Wins	4	3

Solve the proportion. *(Section 5.5)*

10. $\frac{7}{n} = \frac{42}{48}$

11. $\frac{x}{2} = \frac{40}{16}$

12. $\frac{3}{11} = \frac{27}{z}$

13. GAMING You advance 3 levels in 15 minutes. Your friend advances 5 levels in 20 minutes. Are these rates proportional? Explain. *(Section 5.3)*

14. CLASS TIME You spend 150 minutes in three classes. Write and solve a proportion to find how many minutes you spend in five classes. *(Section 5.4)*

15. CONCERT A benefit concert with three performers lasts 8 hours. At this rate, how many hours is a concert with four performers? *(Section 5.5)*

16. INVENTORY A store sells bags of tortilla chips and jars of salsa. *(Section 5.5)*

a. The ratio of bags of chips to jars of salsa is 7 to 4. There are 175 bags of chips. How many jars of salsa are there?

b. What is the cost of 5 bags of chips?

5 Chapter Review

Review Key Vocabulary

ratio, *p. 200*
rate, *p. 200*
unit rate, *p. 200*
slope, *p. 206*
proportion, *p. 214*
proportional, *p. 214*
cross products, *p. 215*

Review Examples and Exercises

5.1 Ratios and Rates *(pp. 198–203)*

There are 15 orangutans and 25 gorillas in a nature preserve. One of the orangutans swings 75 feet in 15 seconds on a rope.

a. Find the ratio of orangutans to gorillas.

b. How fast is the orangutan swinging?

a. $\dfrac{\text{orangutans}}{\text{gorillas}} = \dfrac{15}{25} = \dfrac{3}{5}$

∴ The ratio of orangutans to gorillas is $\dfrac{3}{5}$.

b. $75 \text{ feet in } 15 \text{ seconds} = \dfrac{75 \text{ ft}}{15 \text{ sec}}$

$$= \frac{75 \text{ ft} \div 15}{15 \text{ sec} \div 15}$$

$$= \frac{5 \text{ ft}}{1 \text{ sec}}$$

∴ The orangutan is swinging 5 feet per second.

Exercises

Find the unit rate.

1. 289 miles on 10 gallons

2. 975 revolutions in 3 minutes

3.

Servings	2	4	6	8
Calories	240	480	720	960

4.

Hours	2	3	4	5
Tips (dollars)	14	21	28	35

5.2 Slope (pp. 204–209)

The graph shows the number of visits your website received over the past 6 months. Find and interpret the slope.

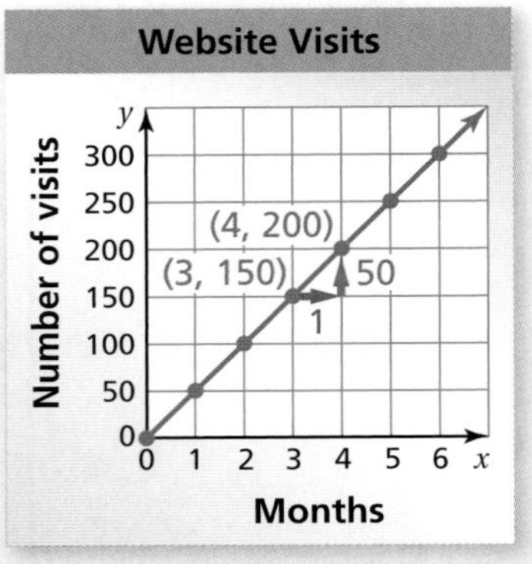

Choose any two points to find the slope of the line.

$$\text{slope} = \frac{\text{change in } y}{\text{change in } x}$$

$$= \frac{50}{1}$$ ← visits / ← months

$$= 50$$

∴ The slope of the line is 50. So, the number of visits increased by 50 each month.

Exercises

Find the slope of the line.

5.

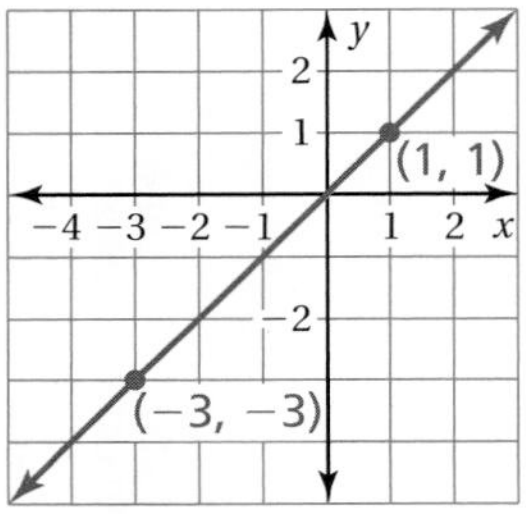

6.

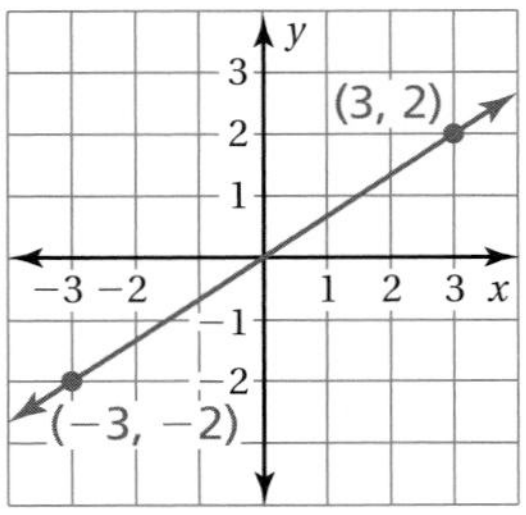

7.

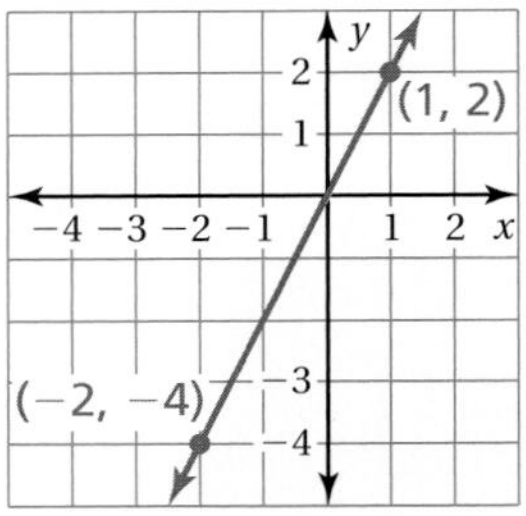

5.3 Proportions (pp. 212–217)

Tell whether the ratios $\frac{9}{12}$ and $\frac{6}{8}$ form a proportion.

$$\frac{9}{12} = \frac{9 \div 3}{12 \div 3} = \frac{3}{4}$$

$$\frac{6}{8} = \frac{6 \div 2}{8 \div 2} = \frac{3}{4}$$

The ratios are equivalent.

∴ So, $\frac{9}{12}$ and $\frac{6}{8}$ form a proportion.

Exercises

Tell whether the ratios form a proportion.

8. $\frac{4}{9}, \frac{2}{3}$ **9.** $\frac{12}{22}, \frac{18}{33}$ **10.** $\frac{8}{50}, \frac{4}{10}$ **11.** $\frac{32}{40}, \frac{12}{15}$

5.4 Writing Proportions (pp. 218–223)

Write a proportion that gives the number r of returns on Saturday.

	Friday	Saturday
Sales	40	85
Returns	32	r

One proportion is $\frac{40 \text{ sales}}{32 \text{ returns}} = \frac{85 \text{ sales}}{r \text{ returns}}$.

Exercises

Use the table to write a proportion.

12.

	Game 1	Game 2
Penalties	6	8
Minutes	16	m

13.

	Concert 1	Concert 2
Songs	15	18
Hours	2.5	h

14. SANITIZER You can buy three bottles of hand sanitizer for $9. Write a proportion that gives the cost c of buying seven bottles of hand sanitizer.

15. DANCE A school dance requires one chaperone for every twelve students. Write a proportion that gives the number c of chaperones needed for 84 students.

5.5 Solving Proportions (pp. 224–229)

Solve $\frac{15}{2} = \frac{30}{y}$.

$15 \cdot y = 2 \cdot 30$ Use the Cross Products Property.

$15y = 60$ Multiply.

$y = 4$ Divide.

The solution is 4.

Exercises

Solve the proportion.

16. $\frac{x}{4} = \frac{2}{5}$ **17.** $\frac{5}{12} = \frac{y}{15}$ **18.** $\frac{z}{7} = \frac{3}{16}$

19. $\frac{8}{20} = \frac{6}{w}$ **20.** $\frac{s+1}{4} = \frac{4}{8}$ **21.** $\frac{6}{16} = \frac{3}{t-2}$

5 Chapter Test

Check It Out
Test Practice
BigIdeasMath.com

Write the ratio as a fraction in simplest form.

1. 34 cars : 26 trucks
2. 3 feet to 9 feet

Find the unit rate.

3. 84 miles in 12 days
4. \$3.20 for 8 ounces

Graph the line that passes through the two points. Then find the slope of the line.

5. (15, 9), (−5, −3)
6. (2, 9), (4, 18)

Tell whether the ratios form a proportion.

7. $\frac{1}{9}, \frac{6}{54}$
8. $\frac{9}{12}, \frac{8}{72}$

Use the table to write a proportion.

9.

	Monday	Tuesday
Gallons	6	8
Miles	180	m

10.

	Thursday	Friday
Classes	6	c
Hours	8	4

Solve the proportion.

11. $\frac{x}{8} = \frac{9}{4}$
12. $\frac{17}{3} = \frac{y}{6}$

Write and solve a proportion to complete the statement.

13. 5 L ≈ ▢ qt
14. 56 lb ≈ ▢ kg

15. **MOVIE TICKETS** Five movie tickets cost \$36.25. What is the cost of eight movie tickets?

16. **CROSSWALK** The graph shows the number of cycles of a crosswalk signal during the day and during the night.
   a. Compare the steepness of the lines. What does this mean in the context of the problem?
   b. Find and interpret the slope of each line.

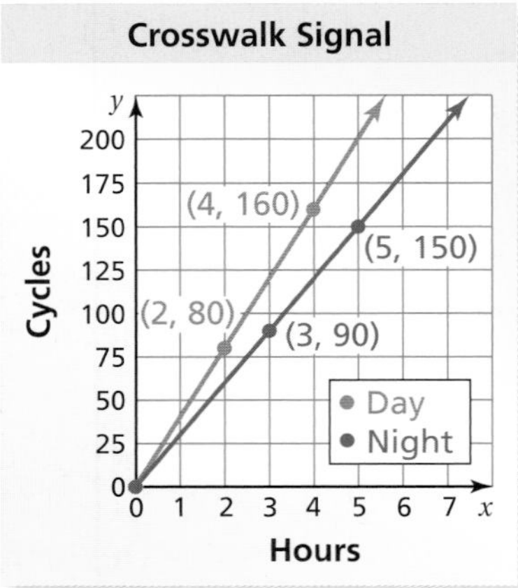

17. **GLAZE** A specific shade of green glaze requires 5 parts blue to 3 parts yellow. A glaze mixture contains 25 quarts of blue and 9 quarts of yellow. How can you fix the mixture to make the specific shade of green glaze?

5 Standardized Test Practice

Test-Taking Strategy

Read Question Before Answering

What is NOT the ratio of human years to dog years?
A 1/7 B 1:7 C 1 to 7 D 7
Newton the senior citizen

"Be sure to read the question before choosing your answer. You may find a word that changes the meaning."

1. The school store sells 4 pencils for \$0.50. At that rate, what would be the cost of 10 pencils?

A. \$1.10 **C.** \$2.00

B. \$1.25 **D.** \$5.00

2. Which expressions do *not* have a value of 3?

I. $|3|$ II. $|-3|$

III. $-|3|$ IV. $-|-3|$

F. I and II **H.** II and IV

G. I and III **I.** III and IV

3. What is the value of y in the equation below when $x = 12$ and $k = 3$?

$$xy = k$$

4. Use the coordinate plane to answer the question below.

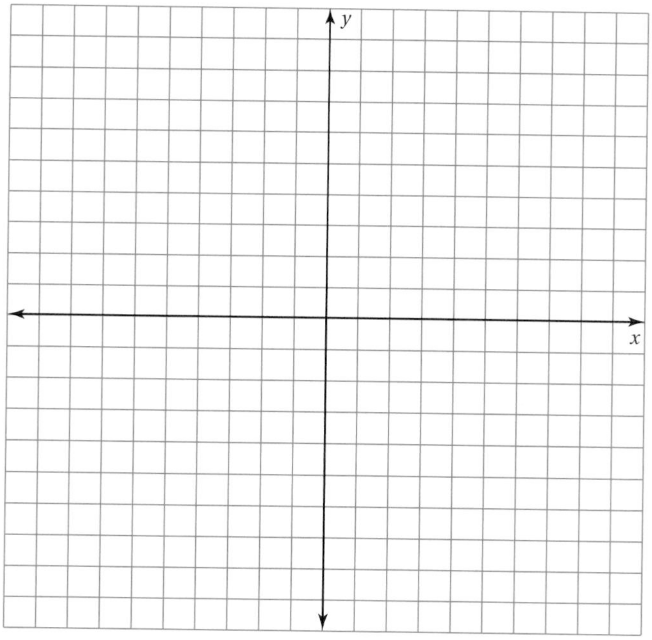

A line contains both the point (0, 5) and the point (5, 0). Which of the following points is also on this line?

A. (0, −5) **C.** (5, 5)

B. (3, 3) **D.** (7, −2)

5. The scores from a diving competition are shown in the line plot below.

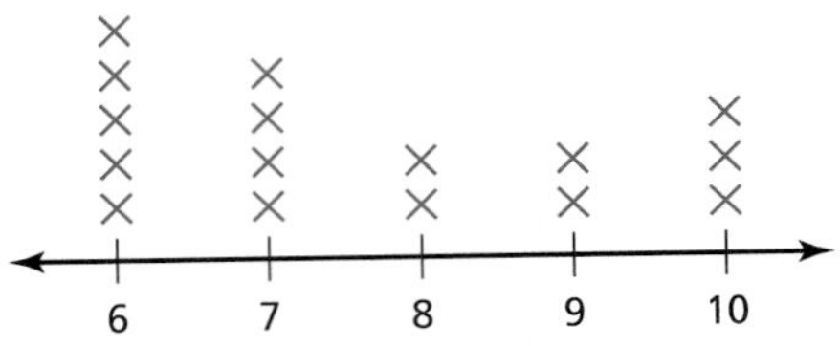

What is the median score?

F. 7

G. 6

H. 5

I. 4

6. Meli was solving the equation in the box below.

$$-\frac{1}{2}(4x - 10) = -16$$
$$-2x - 5 = -16$$
$$-2x - 5 + 5 = -16 + 5$$
$$-2x = -11$$
$$\frac{-2x}{-2} = \frac{-11}{-2}$$
$$x = \frac{11}{2}$$

What should Meli do to correct the error that she made?

A. Distribute the $-\frac{1}{2}$ to get $-2x + 5$.

B. Distribute the $-\frac{1}{2}$ to get $2x - 5$.

C. Divide -11 by -2 to get $-\frac{11}{2}$.

D. Add 2 to -11 to get -9.

7. What is the value of the expression below when $n = -8$ and $p = -4$?

$$-9n - p$$

F. 76

G. 68

H. -68

I. -76

8. How many millimeters are equivalent to 20 inches? (Use 1 millimeter $\approx$ 0.04 inch.)

9. If 5 dogs share equally a bag of dog treats, each dog gets 24 treats. Suppose 8 dogs share equally the bag of treats. How many treats does each dog get?

A. 3

B. 15

C. 21

D. 38

10. The figure below consists of a rectangle and a right triangle.

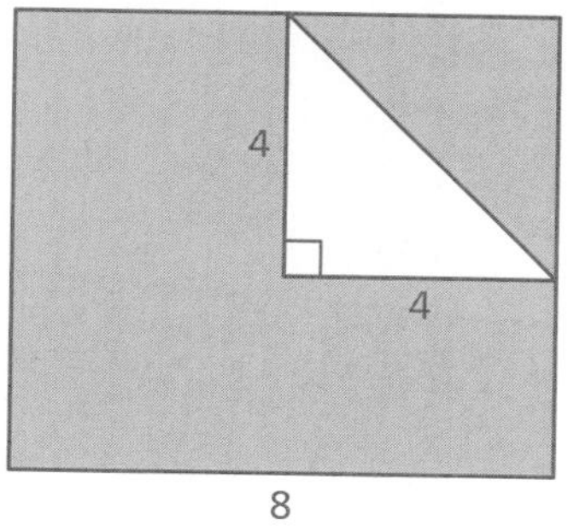

What is the area of the shaded region?

F. 23 units^2

G. 40 units^2

H. 48 units^2

I. 60 units^2

11. You can mow 800 square feet of lawn in 15 minutes. At this rate, how many minutes will you take to mow a lawn that measures 6000 square feet?

Think Solve Explain

Part A Write a proportion to represent the problem. Use m to represent the number of minutes. Explain your reasoning.

Part B Solve the proportion you wrote in Part A and use it to answer the problem. Show your work.

12. What number belongs in the box to make the equation true?

$$5\frac{7}{8} = \square \cdot \left(-\frac{1}{4}\right)$$

A. $-23\frac{1}{2}$

B. $-1\frac{15}{32}$

C. $6\frac{1}{8}$

D. $23\frac{1}{2}$

6 Similarity

"Hey Descartes, did you know that your kite is not actually a kite?"
Oh, go fly a rhombus.
"That's right. My kite is a real kite. You aren't flying a kite, you are flying a rhombus."

2nd
3rd
1st
"I'm at 3rd base. You are running to 1st base and Fluffy is running to 2nd base."
Throw to the adjacent angle. Heh, heh, heh.
"Should I throw the ball to 2nd to get Fluffy out or throw it to 1st to get you out?"

What You Learned Before

"These clouds are making me hungry."

Finding Perimeters

Example 1 **Find the perimeter.**

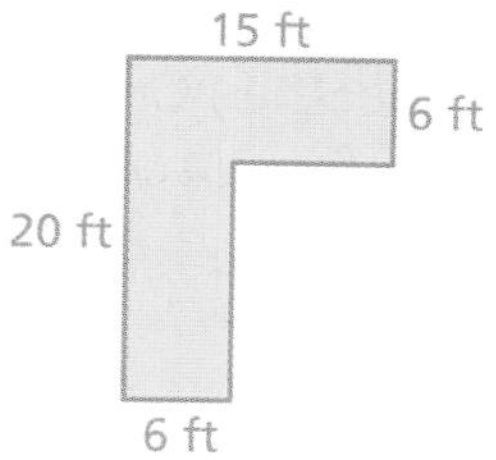

$P = 15 + 6 + 9 + 14 + 6 + 20$

$= 70$ ft

Example 2 **Find the circumference.**

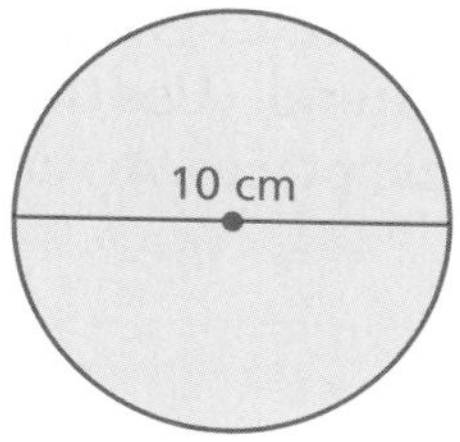

$C = 2\pi r$

$= 2 \cdot 3.14 \cdot 5$ ← $r = \frac{d}{2} = \frac{10}{2} = 5$

$= 31.4$ cm

Try It Yourself

Find the perimeter.

1.

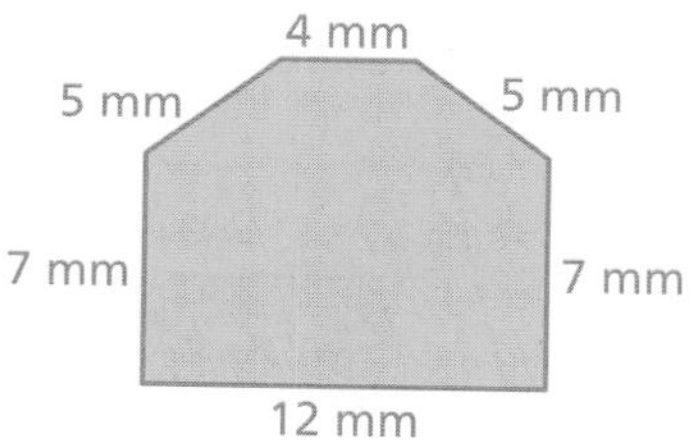

2.

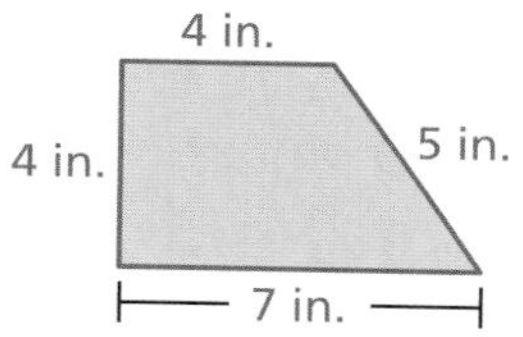

3.

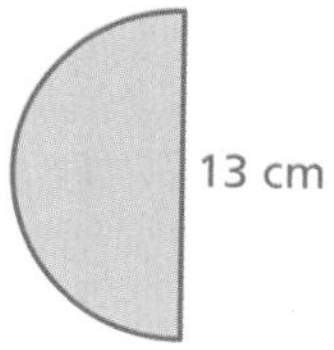

Solving Proportions

Example 3 **Solve the proportion.**

a. $\frac{x}{32} = \frac{3}{4}$

$\frac{x}{32} = \frac{3}{4}$	Write the proportion.	$\frac{3x}{20} = \frac{3}{5}$
$4x = 96$	Use the Cross Products Property.	$15x = 60$
$x = 24$	Solve for *x*.	$x = 4$

b. $\frac{3x}{20} = \frac{3}{5}$

Try It Yourself

Solve the proportion.

4. $\frac{2}{7} = \frac{x}{21}$ **5.** $\frac{3}{4} = \frac{3y}{8}$ **6.** $\frac{3}{14} = \frac{9}{y}$ **7.** $\frac{8}{9x} = \frac{2}{9}$

6.1 Quadrilaterals

STANDARDS OF LEARNING
7.7

Essential Question What are the properties of the angles of special types of quadrilaterals?

1 ACTIVITY: Classifying Quadrilaterals

Work with a partner. Trace the quadrilaterals. Label each quadrilateral as a square, rectangle, parallelogram, trapezoid, rhombus, kite, or simply as a quadrilateral. (Use the name that is the most specific.) Cut the shapes to form 12 puzzle pieces. Arrange the pieces so that they form a square.

2 ACTIVITY: Properties of Quadrilaterals

Work with a partner. Use a protractor to measure the angles of each puzzle piece in Activity 1. Label each angle. Talk about any properties that you observe. Then answer each of the following.

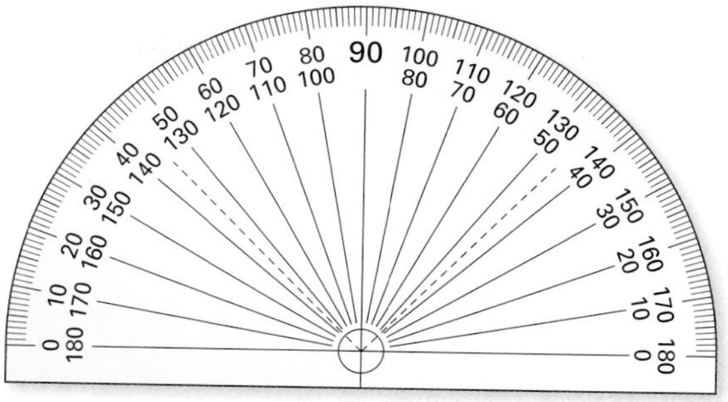

a. Describe a property of the measures of the angles of a square.

b. Describe a property of the measures of the angles of a rectangle.

c. Describe a property of the measures of opposite angles of a parallelogram.

d. Describe a property of the measures of adjacent angles of a parallelogram.

e. The Greek word *isos* means equal. The Greek word *skelos* means leg. Write a definition for isosceles trapezoid. Then, describe a property of the measures of the angles of an isosceles trapezoid.

f. Describe a property of the measures of the angles of a kite.

g. Write a definition for right trapezoid. Describe a property of the measures of the two nonright angles of a right trapezoid.

What Is Your Answer?

3. **IN YOUR OWN WORDS** What are the properties of the angles of special types of quadrilaterals? Organize your work in a table.

4. Describe how knowing properties of the angles of special types of quadrilaterals is important in a construction career.

Practice Use what you learned about quadrilaterals to complete Exercises 3–5 on page 244.

6.1 Lesson

Check It Out
Lesson Tutorials
BigIdeasMath.com

A quadrilateral is a polygon with four sides. The diagram shows properties of different types of quadrilaterals and how they are related. When identifying a quadrilateral, use the name that is most specific.

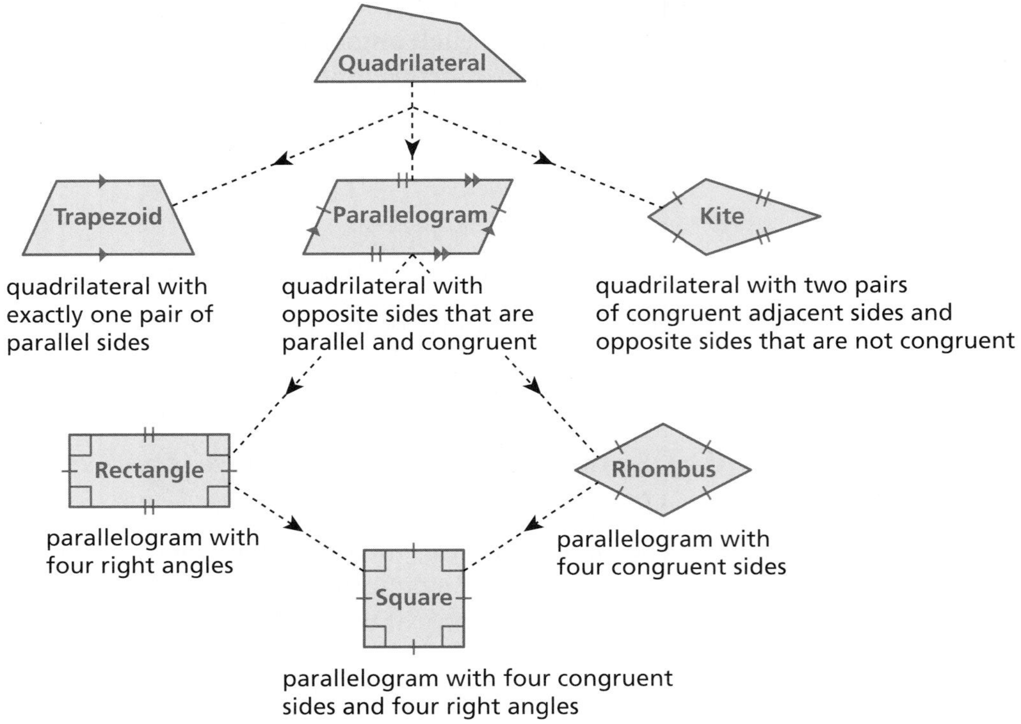

EXAMPLE 1 Standardized Test Practice

Which name *cannot* be used to identify the figure?

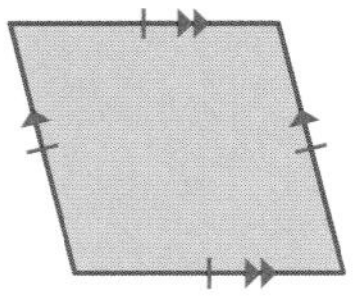

Ⓐ kite Ⓑ parallelogram

Ⓒ quadrilateral Ⓓ rhombus

The figure has four sides, so it is a quadrilateral. Opposite sides are parallel and congruent, so it is also a parallelogram. All four sides are congruent, so it is also a rhombus.

Because opposite sides are congruent, it is not a kite.

So, the correct answer is Ⓐ.

On Your Own

Now You're Ready
Exercises 3–8

Identify the quadrilateral.

1.

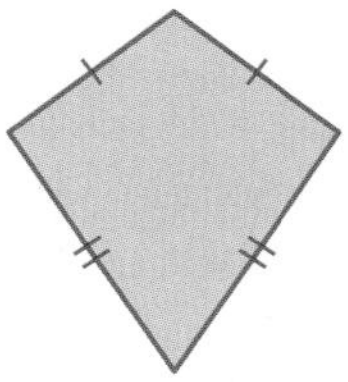

2.

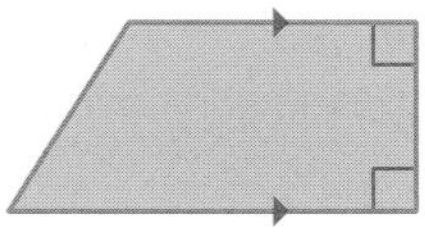

3.

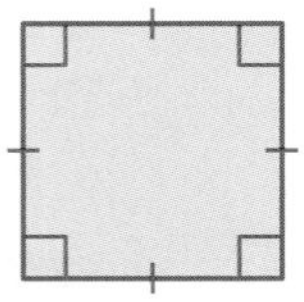

Key Ideas

Angles of Parallelograms

Opposite angles are congruent.

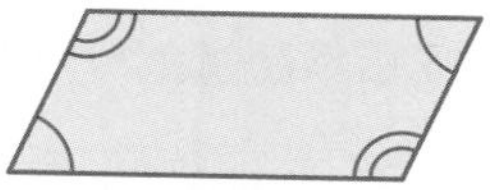

Angles of Kites

Exactly one pair of opposite angles is congruent. This pair is formed by the noncongruent sides.

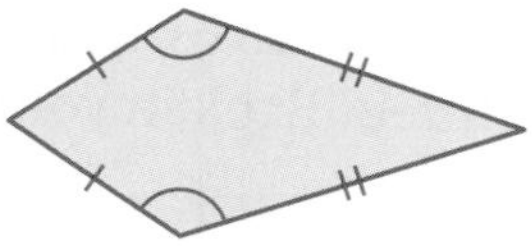

EXAMPLE 2 Finding Angle Measures of a Quadrilateral

Find the missing angle measures.

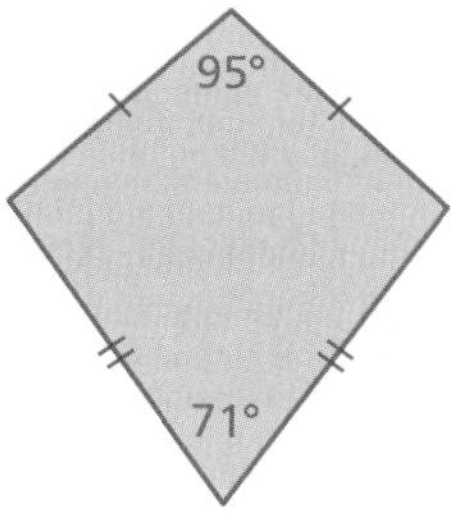

The quadrilateral is a kite because it has two pairs of congruent adjacent sides and opposite sides that are not congruent.

Remember

The sum of the angle measures of a quadrilateral is 360°.

The missing angles are formed by noncongruent sides. So, the missing angles are congruent. Let x° be the measure of each angle.

$$
\begin{aligned}
\text{Sum of angle measures} &= 360^\circ \\
95 + 71 + x + x &= 360 && \text{Write an equation.} \\
166 + 2x &= 360 && \text{Combine like terms.} \\
-166 \quad\quad &\quad -166 && \text{Subtract 166 from each side.} \\
2x &= 194 && \text{Simplify.} \\
\frac{2x}{2} &= \frac{194}{2} && \text{Divide each side by 2.} \\
x &= 97 && \text{Simplify.}
\end{aligned}
$$

∴ The measure of each missing angle in the kite is 97°.

On Your Own

Now You're Ready
Exercises 9–11

Find the missing angle measures.

4.

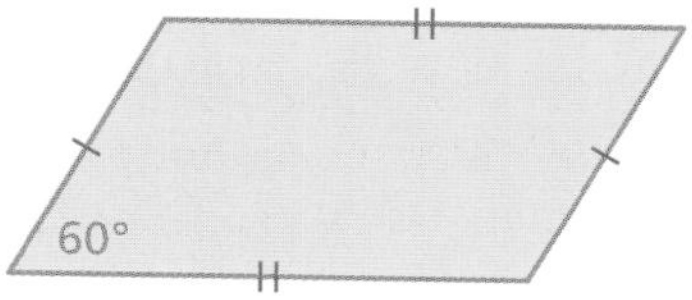

5.

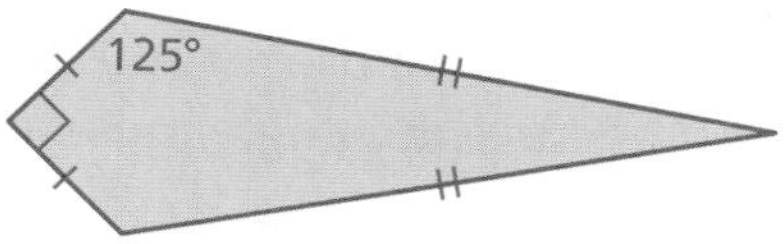

6.1 Exercises

Check It Out
Help with Homework
BigIdeasMath.com

Vocabulary and Concept Check

1. **REASONING** How do you determine the most specific name of a quadrilateral?
2. **WHICH ONE DOESN'T BELONG?** Which one does *not* belong with the other three? Explain your reasoning.

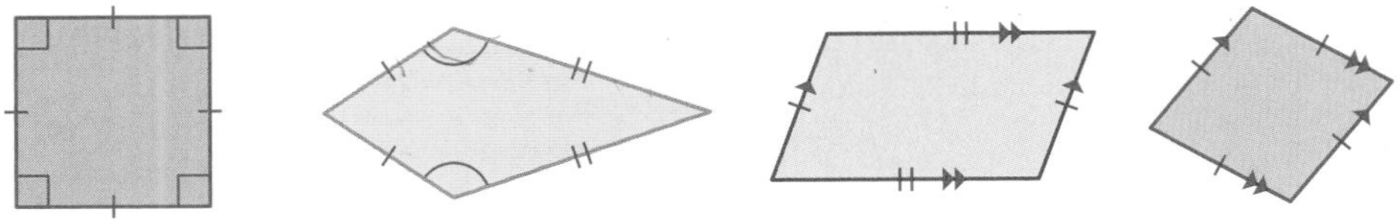

Practice and Problem Solving

Identify the quadrilateral.

1 3.
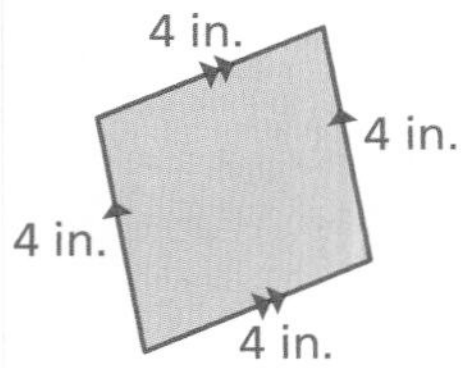

4.
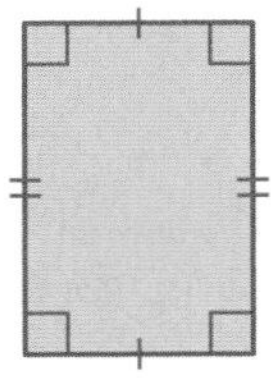

5.
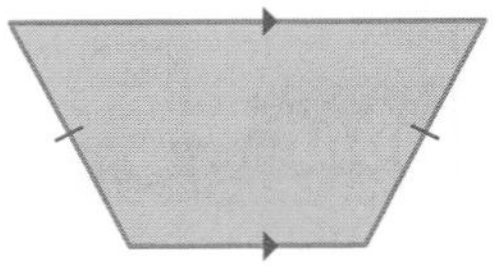

6.
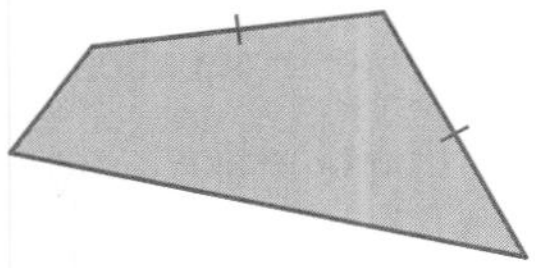

7.
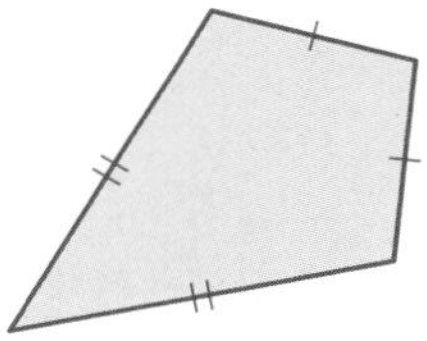

8.
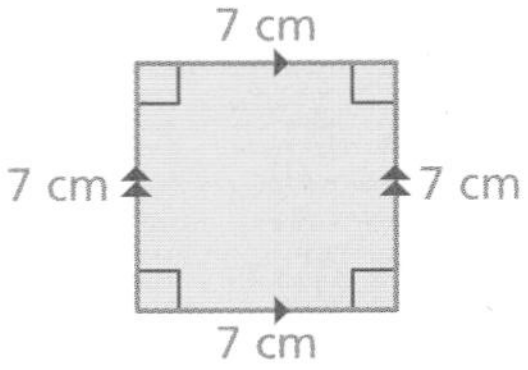

Find the missing angle measure(s).

2 9.
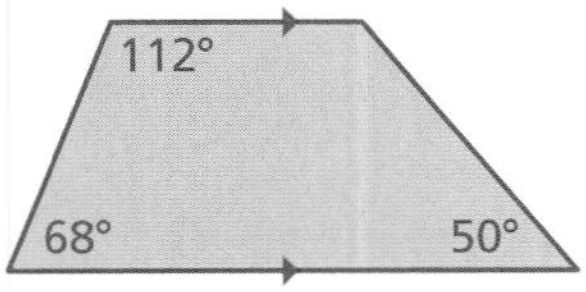

10.
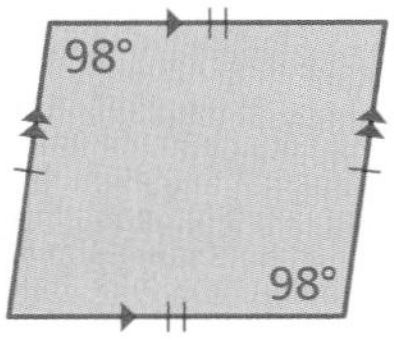

11.
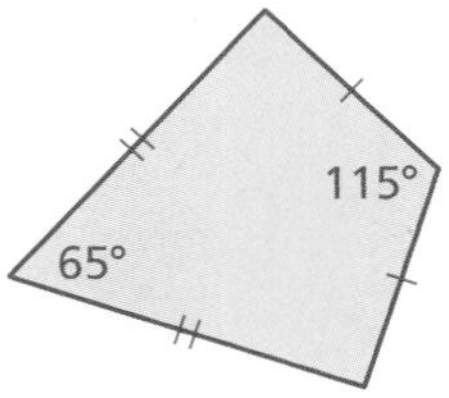

12. **ERROR ANALYSIS** Describe and correct the error in finding the missing angle measures.

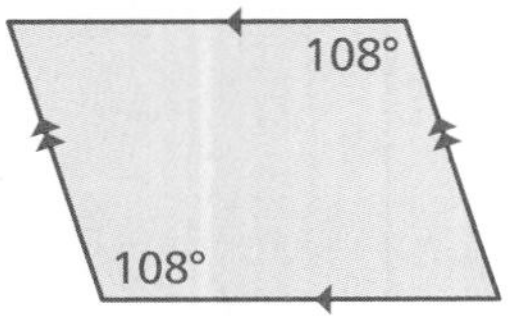

✗

$108 + 108 + x = 360$
$216 + x = 360$
$x = 144$

The measure of each missing angle is 144°.

13. SEWING A skirt pattern is in the shape of an isosceles trapezoid. The angles at the top are congruent and the angles at the bottom are congruent. What is the angle measure of the bottom right corner?

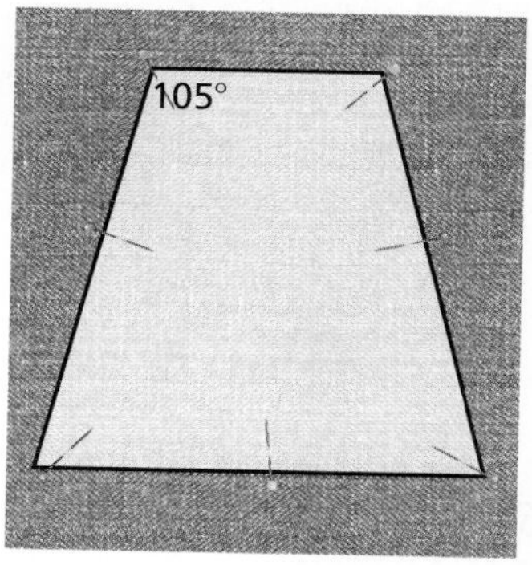

14. REASONING The measures of two adjacent angles of a kite are each 110°. Find the missing angle measures. Explain.

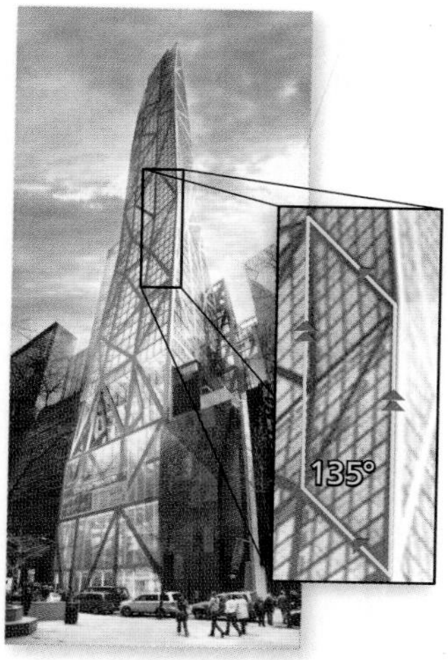

15. ARCHITECTURE Jean Nouvel, a French architect, designed the building shown.

a. Identify the yellow quadrilateral.

b. Find the missing angle measures.

16. CRITICAL THINKING Draw a Venn diagram that shows the relationship of the quadrilaterals in the lesson.

17. **Geometry** A quadrilateral is *concave* if at least one line segment connecting any two vertices lies outside the quadrilateral. Is the sum of the angle measures of any concave quadrilateral equal to 360°? Draw and measure the angles of at least 3 concave quadrilaterals to support your reasoning.

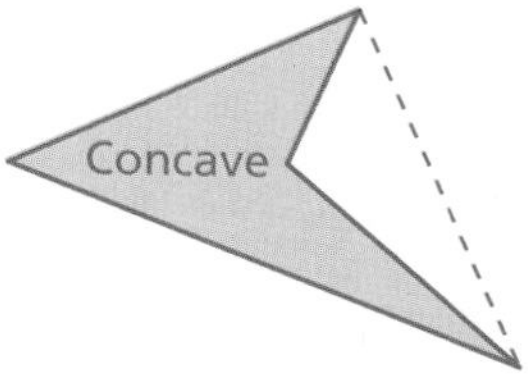

Fair Game Review What you learned in previous grades & lessons

Tell whether the ratios form a proportion. *(Section 5.3)*

18. $\frac{3}{4}, \frac{15}{25}$ **19.** $\frac{2}{3}, \frac{24}{36}$ **20.** $\frac{7}{35}, \frac{13}{65}$ **21.** $\frac{48}{60}, \frac{39}{54}$

22. MULTIPLE CHOICE Which graph represents the function $y = -2x + 1$? *(Section 4.4)*

Ⓐ
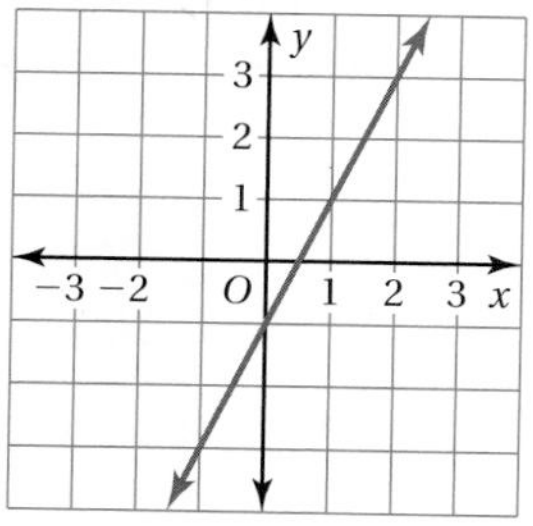

Ⓑ
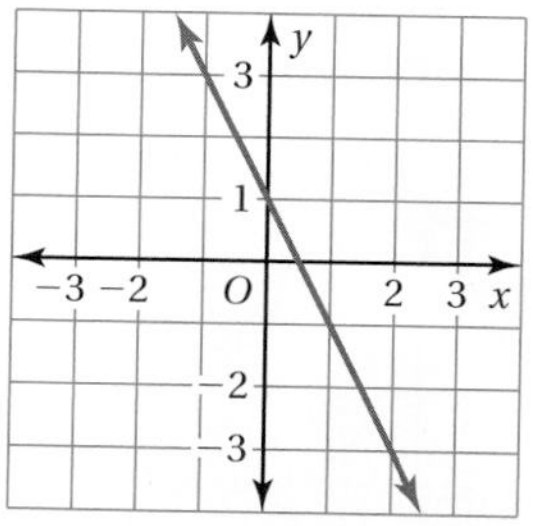

Ⓒ
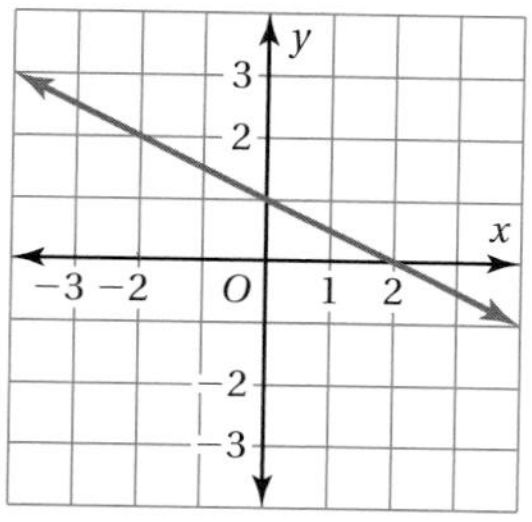

Ⓓ
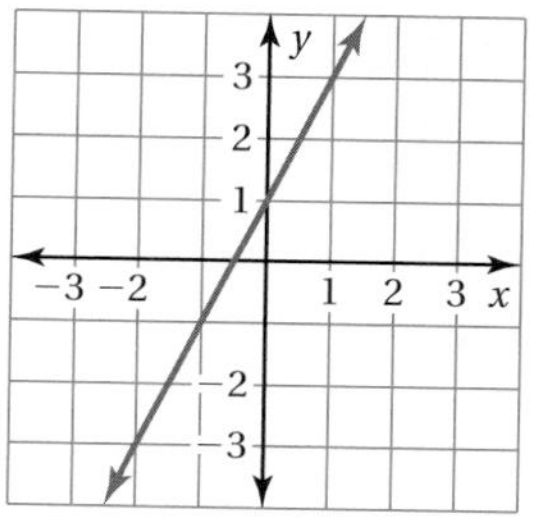

6.2 Identifying Similar Figures

STANDARDS OF LEARNING
7.6

Essential Question How can you use proportions to help make decisions in art, design, and magazine layouts?

Original Photograph

In a computer art program, when you click and drag on a side of a photograph, you distort it.

But when you click and drag on a corner of the photograph, it remains proportional to the original.

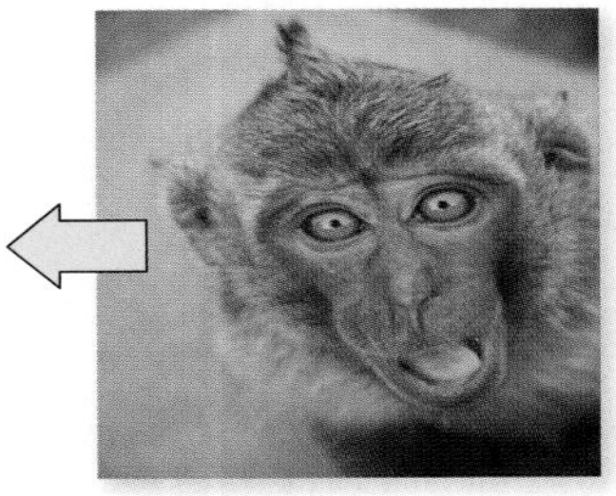

Distorted

Distorted

Proportional

1 ACTIVITY: Reducing Photographs

Work with a partner. You are trying to reduce the photograph to the indicated size for a nature magazine. Can you reduce the photograph to the indicated size without distorting or cropping? Explain your reasoning.

a.

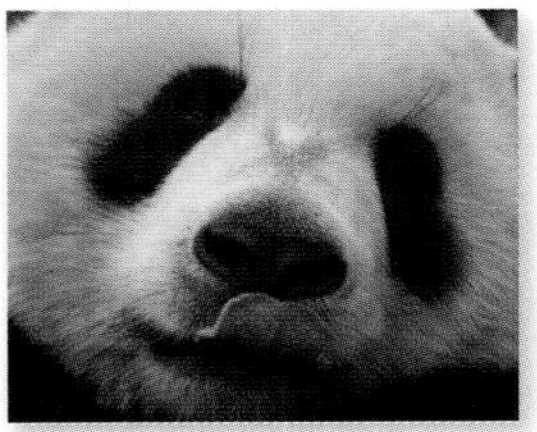

5 in.

6 in.

4 in.

5 in.

b.

5 in.

5 in.

4 in.

4 in.

c.

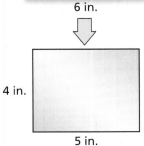

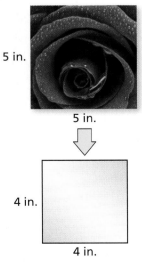

6 in.

8 in.

3 in.

4 in.

2 ACTIVITY: Proportional Designs

Work with a partner.

a. Tell whether the new designs are proportional to the original design. Explain your reasoning.

Original **Design 1** **Design 2**

b. Draw two designs that are proportional to the given design. Make one bigger and one smaller. Label the sides of the designs with their lengths.

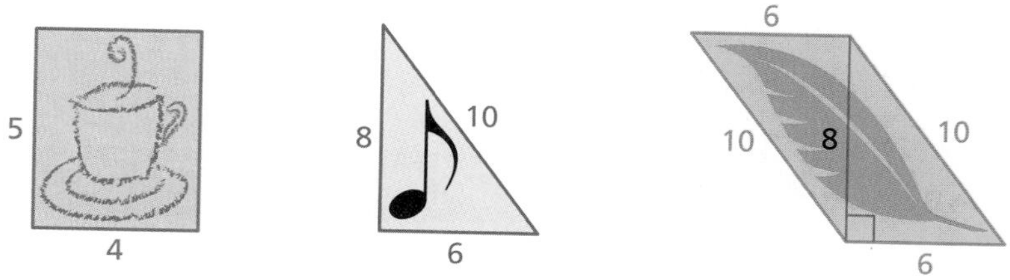

What Is Your Answer?

3. **IN YOUR OWN WORDS** How can you use proportions to help make decisions in art, design, and magazine layouts? Give two examples.

4. **a.** Use a computer art program to draw two rectangles that are proportional to each other.

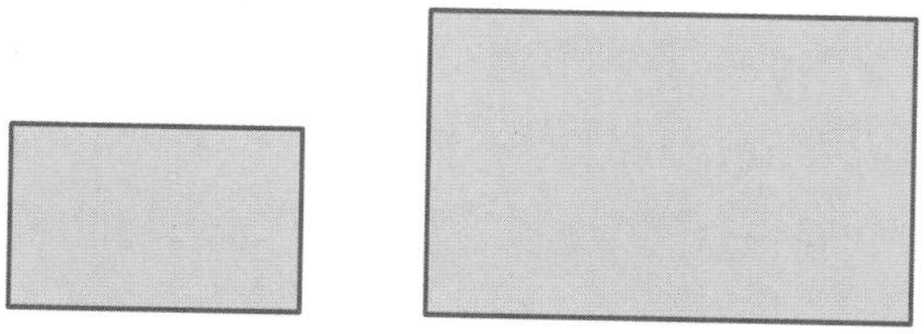

You've got to hand it to him. He's right.

"I love this statue. It seems similar to a big statue I saw in New York."

b. Print the two rectangles on the same piece of paper.

c. Use a centimeter ruler to measure the length and width of each rectangle.

d. Find the following ratios. What can you conclude?

$$\frac{\text{Length of Larger}}{\text{Length of Smaller}} \qquad \frac{\text{Width of Larger}}{\text{Width of Smaller}}$$

Practice Use what you learned about similar figures to complete Exercises 9 and 10 on page 250.

6.2 Lesson

Check It Out
Lesson Tutorials
BigIdeasMath.com

Key Vocabulary
similar figures, *p. 248*
corresponding angles, *p. 248*
corresponding sides, *p. 248*

Key Idea

Similar Figures

Figures that have the same shape but not necessarily the same size are called **similar figures**. The triangles below are similar.

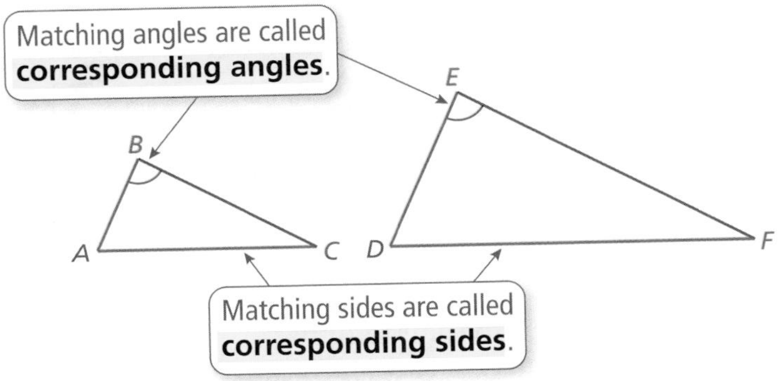

EXAMPLE 1 Naming Corresponding Parts

The trapezoids are similar. (a) Name the corresponding angles. (b) Name the corresponding sides.

Reading
$\angle A$ is read as "angle A."

a. Corresponding angles:

$\angle A$ and $\angle P$

$\angle B$ and $\angle Q$

$\angle C$ and $\angle R$

$\angle D$ and $\angle S$

b. Corresponding sides:

Side AB and Side PQ

Side BC and Side QR

Side CD and Side RS

Side AD and Side PS

On Your Own

Now You're Ready
Exercises 5 and 6

1. The figures are similar.
   a. Name the corresponding angles.
   b. Name the corresponding sides.

Multi-Language Glossary at BigIdeasMath.com.

Key Idea

Reading

The symbol ~ means "is similar to."
The symbol ≅ means "is congruent to."

Common Error

When writing a similarity statement, make sure to list the vertices of the figures in the correct order.

Identifying Similar Figures

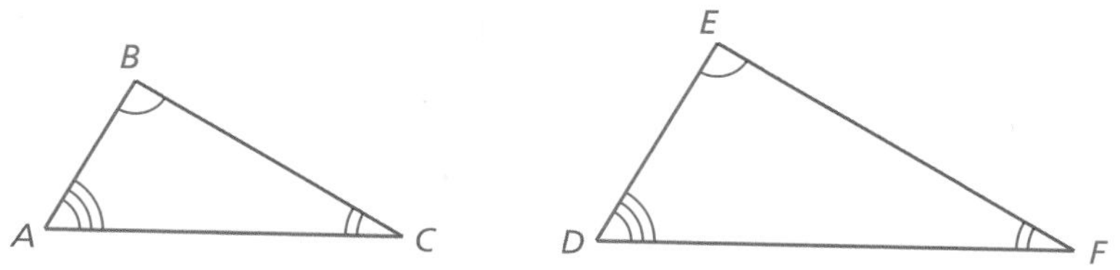

Triangle *ABC* is similar to triangle *DEF*.

Words Two figures are similar if

- corresponding side lengths are proportional, and
- corresponding angles are congruent.

Symbols

Side Lengths	*Angles*	*Figures*
$\frac{AB}{DE} = \frac{BC}{EF} = \frac{AC}{DF}$	$\angle A \cong \angle D$ $\angle B \cong \angle E$ $\angle C \cong \angle F$	$\triangle ABC \sim \triangle DEF$

EXAMPLE 2 Identifying Similar Figures

Which rectangle is similar to Rectangle A?

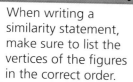

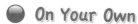

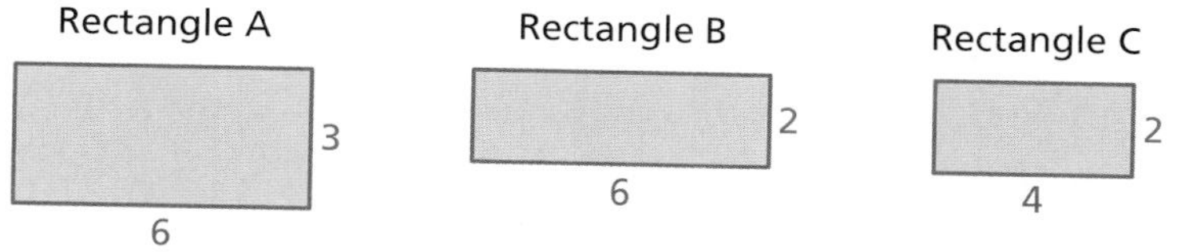

Each figure is a rectangle. So, corresponding angles are congruent. Check to see if corresponding side lengths are proportional.

Rectangle A and Rectangle B

$$\frac{\text{Length of A}}{\text{Length of B}} = \frac{6}{6} = 1 \qquad \frac{\text{Width of A}}{\text{Width of B}} = \frac{3}{2}$$ Not proportional

Rectangle A and Rectangle C

$$\frac{\text{Length of A}}{\text{Length of C}} = \frac{6}{4} = \frac{3}{2} \qquad \frac{\text{Width of A}}{\text{Width of C}} = \frac{3}{2}$$ Proportional

∴ So, Rectangle C is similar to Rectangle A.

On Your Own

Now You're Ready
Exercises 7–12

2. Rectangle D is 3 units long and 1 unit wide. Which rectangle in Example 2 is similar to Rectangle D?

6.2 Exercises

Check It Out
Help with Homework
BigIdeasMath.com

Vocabulary and Concept Check

1. **VOCABULARY** How are corresponding angles of two similar figures related?
2. **VOCABULARY** How are corresponding side lengths of two similar figures related?
3. **OPEN-ENDED** Give examples of two real-world objects whose shapes are similar.
4. **CRITICAL THINKING** Are two figures that have the same size and shape similar? Explain.

Practice and Problem Solving

Name the corresponding angles and the corresponding sides of the similar figures.

1 5.

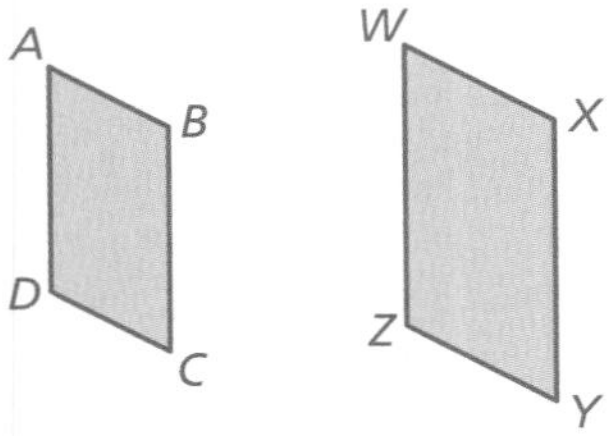

6.

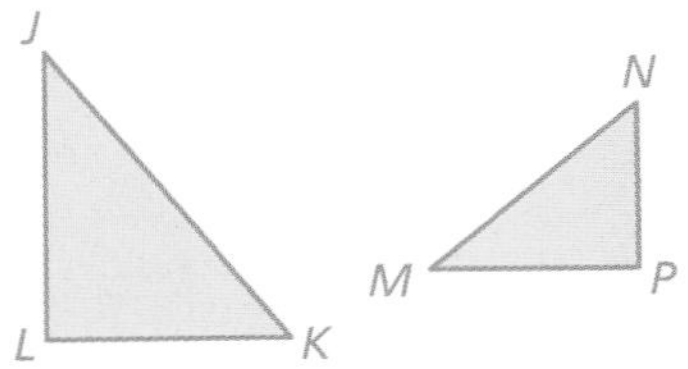

In a coordinate plane, draw the figures with the given vertices. Which figures are similar? Explain your reasoning.

2 7. Triangle A: (0, 0), (3, 0), (0, 3)
Triangle B: (0, 0), (5, 0), (0, 5)
Triangle C: (0, 0), (3, 0), (0, 6)

8. Rectangle A: (0, 0), (4, 0), (4, 2), (0, 2)
Rectangle B: (0, 0), (−6, 0), (−6, 3), (0, 3)
Rectangle C: (0, 0), (4, 0), (4, 2), (0, 2)

Tell whether the two figures are similar. Explain your reasoning.

9.

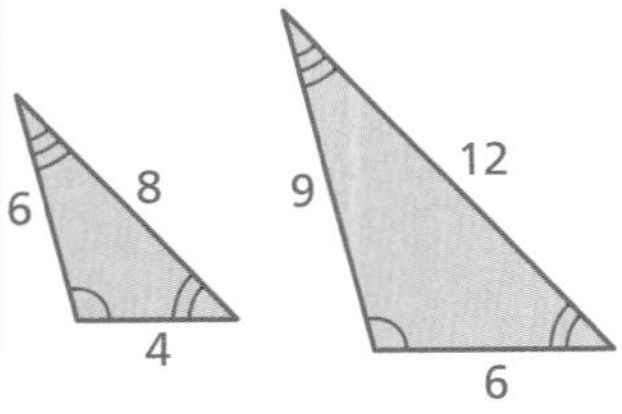

10.

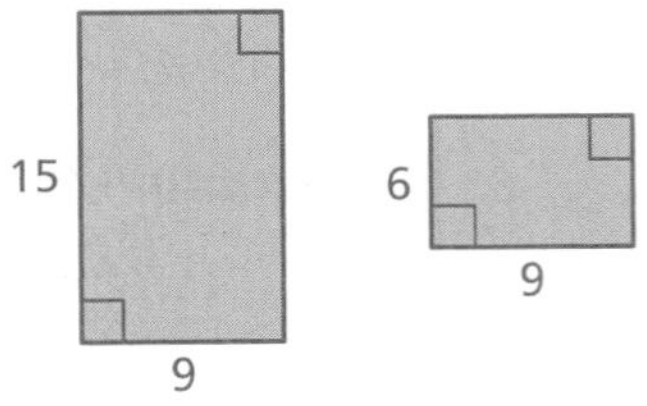

11. **MEXICO** A Mexican flag is 63 inches long and 36 inches high. Is the drawing at the right similar to the Mexican flag?

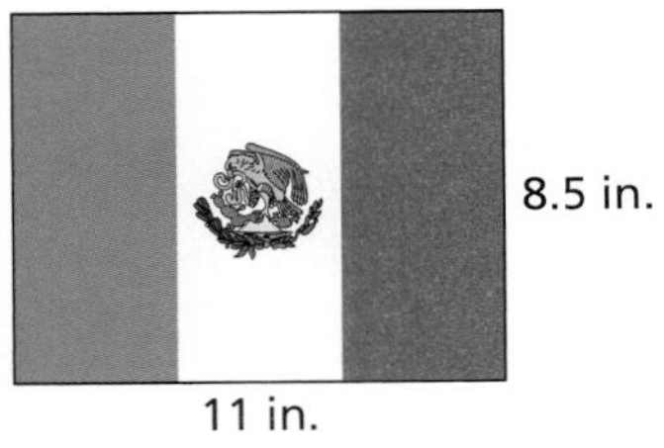

12. **DESKS** A student's rectangular desk is 30 inches long and 18 inches wide. The teacher's rectangular desk is 60 inches long and 36 inches wide. Are the desks similar?

The two triangles are similar. Find the measure of the angle.

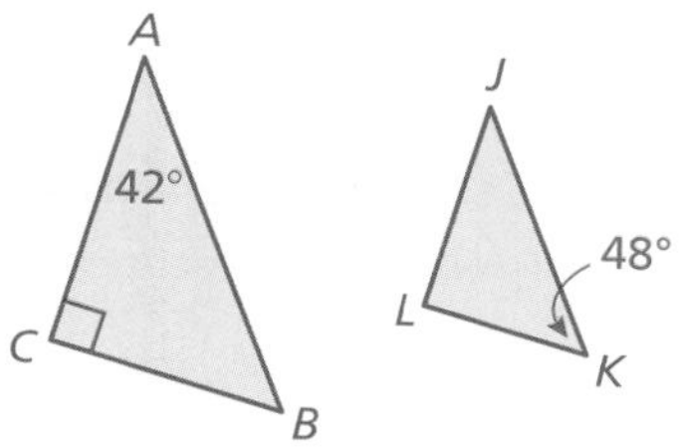

13. $\angle B$ **14.** $\angle L$ **15.** $\angle J$

16. REASONING Given $\triangle FGH \sim \triangle QRT$, name the corresponding angles and the corresponding sides.

17. PHOTOS You want to buy only photos that are similar rectangles. Which of the photo sizes should you buy?

Photo Size
4 in. × 5 in.
5 in. × 7 in.
8 in. × 12 in.
11 in. × 14 in.
18 in. × 27 in.

18. CRITICAL THINKING Are the following figures *always, sometimes,* or *never* similar? Explain.

a. Two triangles **b.** Two squares

c. Two rectangles **d.** A square and a triangle

19. CRITICAL THINKING Can you draw two quadrilaterals each having two 130° angles and two 50° angles that are *not* similar? Justify your answer.

20. SIGN All of the angle measures in the sign are 90°.

a. Each side length is increased by 20%. Is the new sign similar to the original?

b. Each side length is increased by 6 inches. Is the new sign similar to the original?

21. GEOMETRY Use a ruler to draw two different isosceles triangles similar to the one shown. Measure the heights of each triangle to the nearest centimeter.

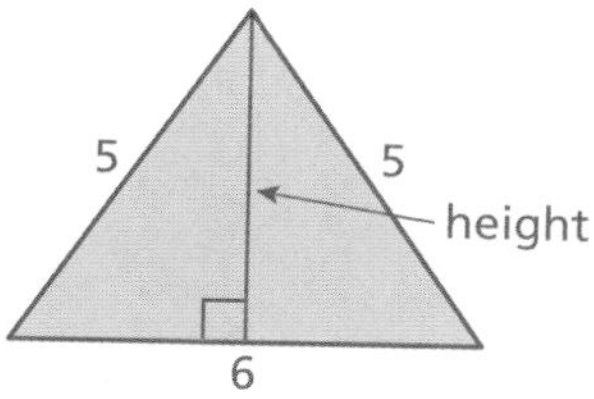

a. Is the ratio of the corresponding heights proportional to the ratio of the corresponding side lengths?

b. Do you think this is true for all similar triangles? Explain.

22. Critical Thinking Given $\triangle ABC \sim \triangle DEF$ and $\triangle DEF \sim \triangle JKL$, is $\triangle ABC \sim \triangle JKL$? Give an example or non-example.

Fair Game Review What you learned in previous grades & lessons

Simplify. *(Section 3.3)*

23. $\left(\frac{4}{9}\right)^2$ **24.** $\left(\frac{3}{8}\right)^2$ **25.** $\left(\frac{7}{4}\right)^2$ **26.** $\left(\frac{6.5}{2}\right)^2$

27. MULTIPLE CHOICE Which function has the steepest graph? *(Section 4.5)*

Ⓐ $y = \frac{x}{3}$ Ⓑ $y = \frac{3}{2}x$ Ⓒ $y = \frac{3}{4}x$ Ⓓ $y = \frac{4}{3}x$

6.3 Perimeters and Areas of Similar Figures

STANDARDS OF LEARNING
7.6

Essential Question How do changes in dimensions of similar geometric figures affect the perimeters and areas of the figures?

1 ACTIVITY: Comparing Perimeters and Areas

Work with a partner. Use pattern blocks to make a figure whose dimensions are 2, 3, and 4 times greater than those of the original figure. Find the perimeter *P* and area *A* of each larger figure.

a. Sample: Square

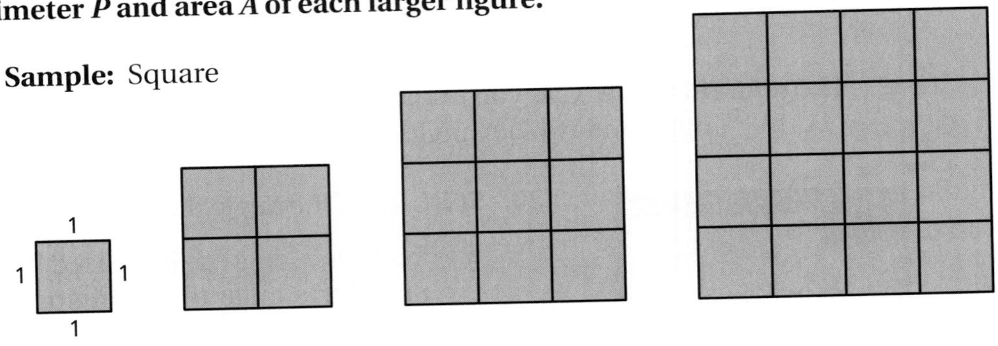

$P = 4$	$P = 8$	$P = 12$	$P = 16$
$A = 1$	$A = 4$	$A = 9$	$A = 16$

b. Triangle

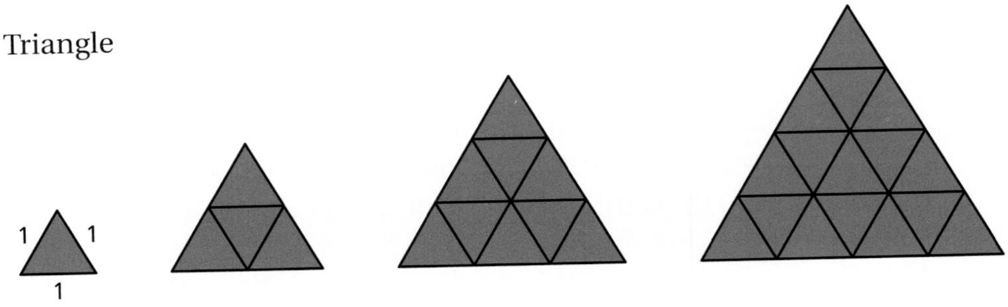

$P = 3$	$P = 6$	$P =$	$P =$
$A = B$	$A = 4B$	$A =$	$A =$

c. Rectangle

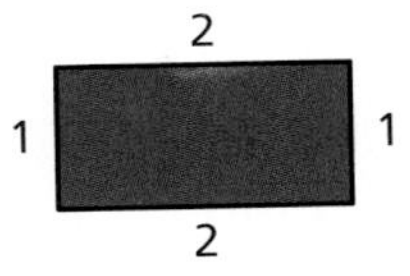

$P = 6$

$A = 2$

d. Parallelogram

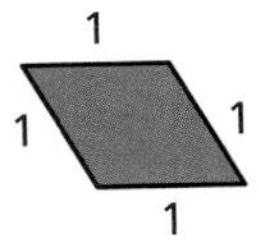

$P = 4$

$A = C$

2 ACTIVITY: Finding Patterns for Perimeters

Work with a partner. Copy and complete the table for the perimeters of the figures in Activity 1. Describe the pattern.

	Figure	Original Side Lengths	Double Side Lengths	Triple Side Lengths	Quadruple Side Lengths
Perimeters		$P = 4$	$P = 8$	$P = 12$	$P = 16$
	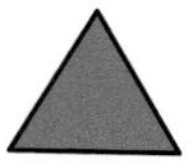	$P = 3$	$P = 6$		
	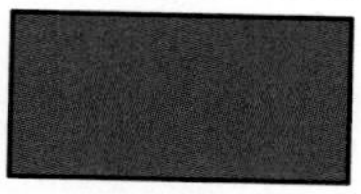	$P = 6$			
	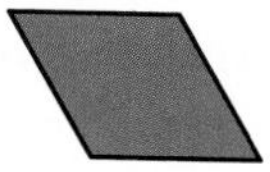	$P = 4$			

3 ACTIVITY: Finding Patterns for Areas

Work with a partner. Copy and complete the table for the areas of the figures in Activity 1. Describe the pattern.

	Figure	Original Side Lengths	Double Side Lengths	Triple Side Lengths	Quadruple Side Lengths
Areas		$A = 1$	$A = 4$	$A = 9$	$A = 16$
	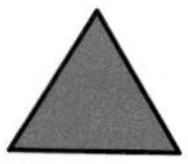	$A = B$	$A = 4B$		
	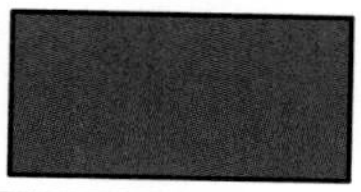	$A = 2$			
	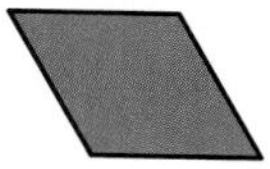	$A = C$			

What Is Your Answer?

4. **IN YOUR OWN WORDS** How do changes in dimensions of similar geometric figures affect the perimeters and areas of the figures?

Practice Use what you learned about perimeters and areas of similar figures to complete Exercises 8–11 on page 256.

6.3 Lesson

Check It Out
Lesson Tutorials
BigIdeasMath.com

Key Idea

Perimeters of Similar Figures

If two figures are similar, then the ratio of their perimeters is equal to the ratio of their corresponding side lengths.

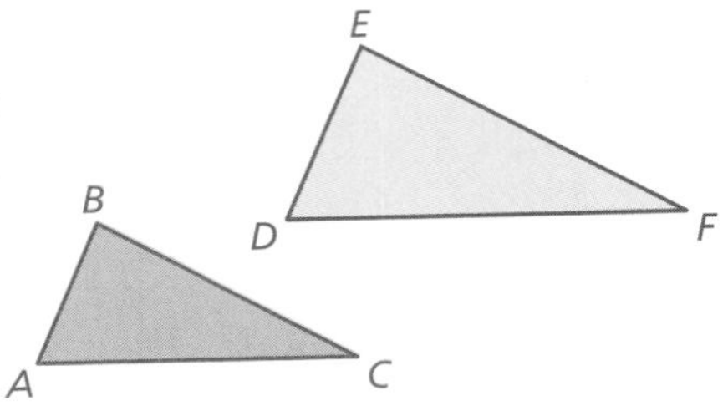

$$\frac{\text{Perimeter of } \triangle ABC}{\text{Perimeter of } \triangle DEF} = \frac{AB}{DE} = \frac{BC}{EF} = \frac{AC}{DF}$$

EXAMPLE 1 Finding Ratios of Perimeters

Find the ratio (red to blue) of the perimeters of the similar rectangles.

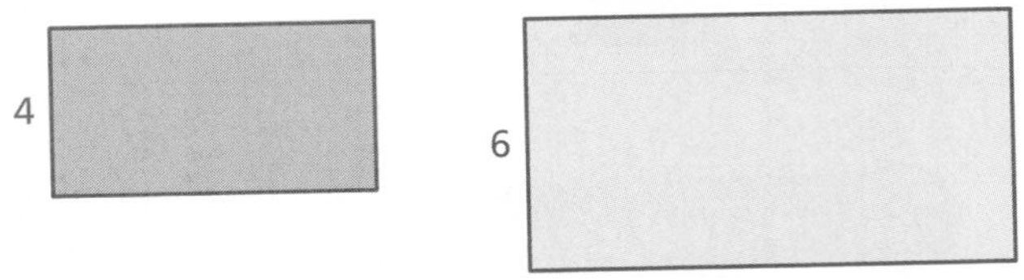

$$\frac{\text{Perimeter of red rectangle}}{\text{Perimeter of blue rectangle}} = \frac{4}{6} = \frac{2}{3}$$

∴ The ratio of the perimeters is $\frac{2}{3}$.

On Your Own

1. The height of Figure A is 9 feet. The height of a similar Figure B is 15 feet. What is the ratio of the perimeter of A to the perimeter of B?

Key Idea

Areas of Similar Figures

If two figures are similar, then the ratio of their areas is equal to the *square* of the ratio of their corresponding side lengths.

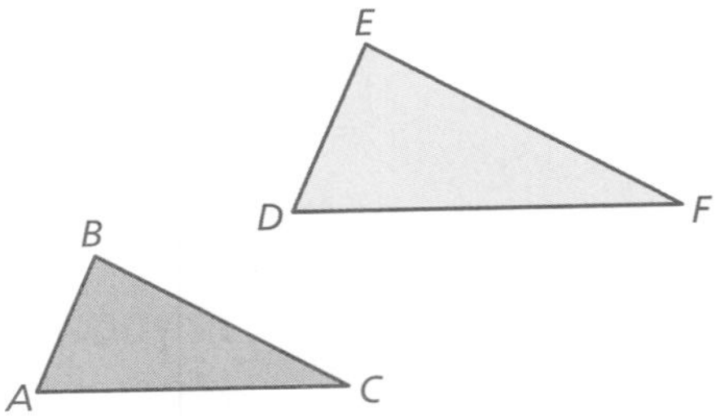

$$\frac{\text{Area of } \triangle ABC}{\text{Area of } \triangle DEF} = \left(\frac{AB}{DE}\right)^2 = \left(\frac{BC}{EF}\right)^2 = \left(\frac{AC}{DF}\right)^2$$

EXAMPLE 2 Finding Ratios of Areas

Find the ratio (red to blue) of the areas of the similar triangles.

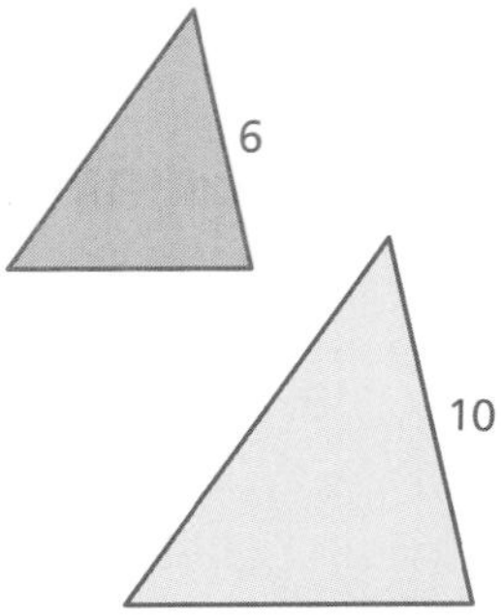

$$\frac{\text{Area of red triangle}}{\text{Area of blue triangle}} = \left(\frac{6}{10}\right)^2$$

$$= \left(\frac{3}{5}\right)^2 = \frac{9}{25}$$

∴ The ratio of the areas is $\frac{9}{25}$.

EXAMPLE 3 Real-Life Application

You place a picture on a page of a photo album. The page and the picture are similar rectangles.

a. How many times greater is the area of the page than the area of the picture?

b. The area of the picture is 45 square inches. What is the area of the page?

a. Find the ratio of the area of the page to the area of the picture.

$$\frac{\text{Area of page}}{\text{Area of picture}} = \left(\frac{\text{length of page}}{\text{length of picture}}\right)^2$$

$$= \left(\frac{8}{6}\right)^2 = \left(\frac{4}{3}\right)^2 = \frac{16}{9}$$

∴ The area of the page is $\frac{16}{9}$ times greater than the area of the picture.

b. Multiply the area of the picture by $\frac{16}{9}$.

$$45 \cdot \frac{16}{9} = 80$$

∴ The area of the page is 80 square inches.

On Your Own

Now You're Ready
Exercises 4–13

2. The base of Triangle P is 8 meters. The base of a similar Triangle Q is 7 meters. What is the ratio of the area of P to the area of Q?

3. In Example 3, the perimeter of the picture is 27 inches. What is the perimeter of the page?

6.3 Exercises

Check It Out
Help with Homework
BigIdeasMath.com

Vocabulary and Concept Check

1. **WRITING** How are the perimeters of two similar figures related?
2. **WRITING** How are the areas of two similar figures related?
3. **VOCABULARY** Rectangle *ABCD* is similar to Rectangle *WXYZ*. The area of *ABCD* is 30 square inches. Explain how to find the area of *WXYZ*.

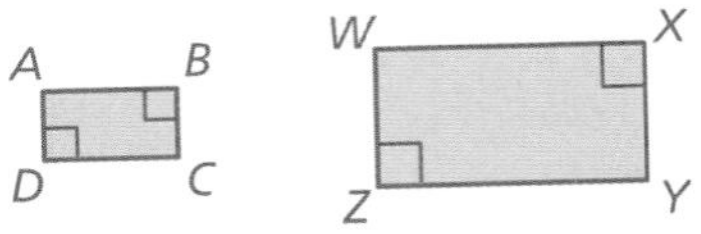

$\frac{AD}{WZ} = \frac{1}{2}$ $\frac{AB}{WX} = \frac{1}{2}$

Practice and Problem Solving

The two figures are similar. Find the ratios (red to blue) of the perimeters and of the areas.

1 2 **4.**

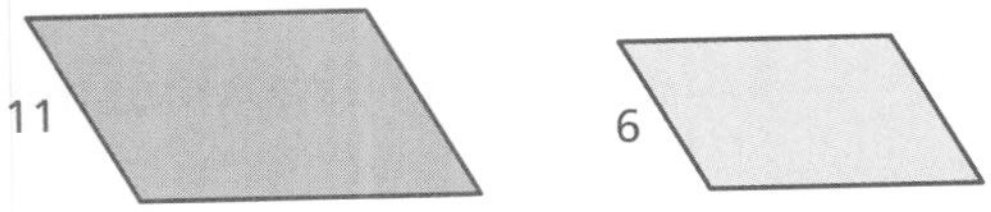

5.

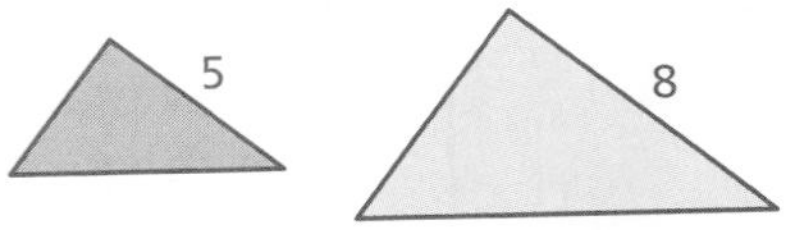

6.

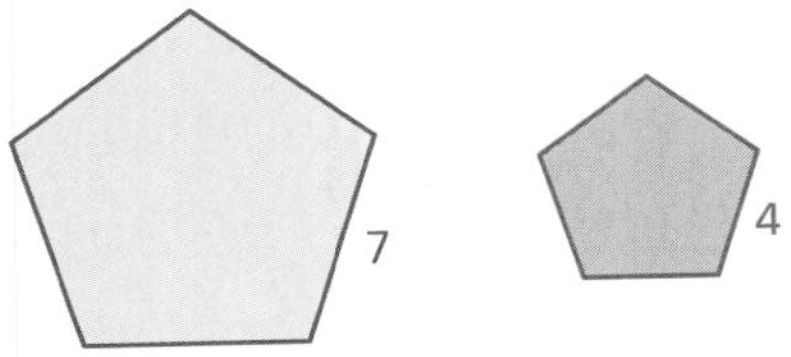

7.

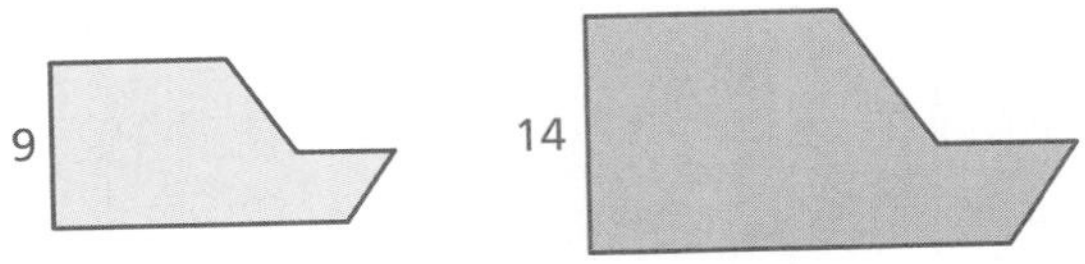

8. How does doubling the side lengths of a triangle affect its perimeter?
9. How does tripling the side lengths of a triangle affect its perimeter?
10. How does doubling the side lengths of a rectangle affect its area?
11. How does quadrupling the side lengths of a rectangle affect its area?

12. **FOOSBALL** The playing surfaces of two foosball tables are similar. The ratio of the corresponding side lengths is 10 : 7. What is the ratio of the areas?
13. **LAPTOP** The ratio of the corresponding side lengths of two similar computer screens is 13 : 15. The perimeter of the smaller screen is 39 inches. What is the perimeter of the larger screen?

Triangle *ABC* is similar to Triangle *DEF*. Tell whether the statement is *true* or *false*. Explain your reasoning.

14. $\frac{\text{Perimeter of } \triangle ABC}{\text{Perimeter of } \triangle DEF} = \frac{AB}{DE}$

15. $\frac{\text{Area of } \triangle ABC}{\text{Area of } \triangle DEF} = \frac{AB}{DE}$

16. FABRIC The cost of the fabric is $1.31. What would you expect to pay for a similar piece of fabric that is 18 inches by 42 inches?

17. AMUSEMENT PARK A model of a merry-go-round and the merry-go-round are similar.

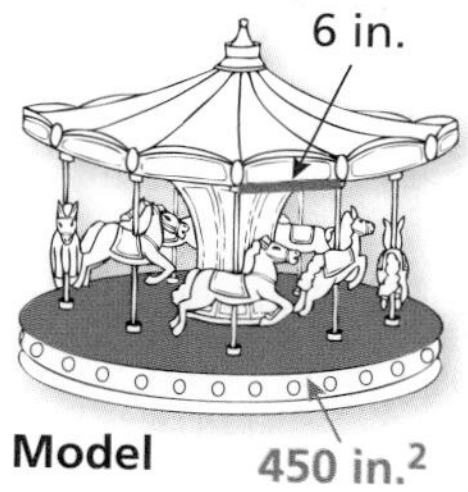

a. How many times greater is the base area of the actual merry-go-round than the base area of the model? Explain.

b. What is the base area of the actual merry-go-round in square inches?

c. What is the base area of the actual merry-go-round in square feet?

18. CRITICAL THINKING The circumference of Circle K is π. The circumference of Circle L is 4π.

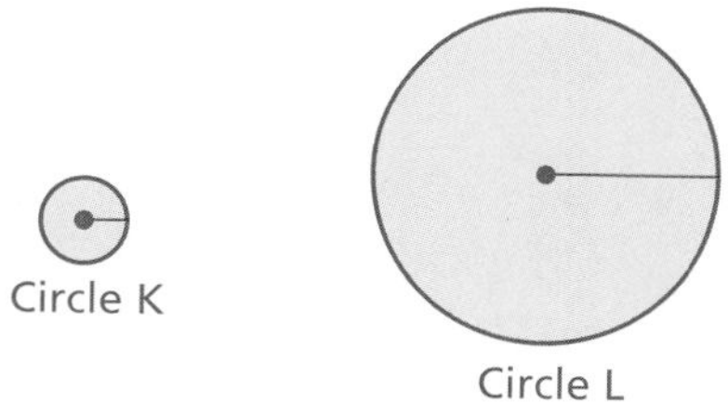

a. What is the ratio of their circumferences? of their radii? of their areas?

b. What do you notice?

19. GEOMETRY Rhombus A is similar to Rhombus B. What is the ratio (A to B) of the corresponding side lengths?

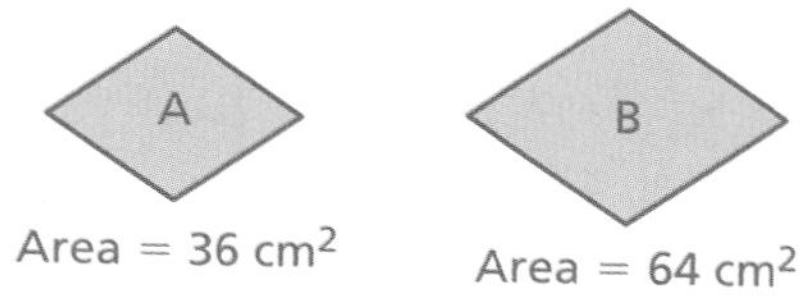

20. Geometry A triangle with an area of 10 square meters has a base of 4 meters. A similar triangle has an area of 90 square meters. What is the *height* of the larger triangle?

Fair Game Review What you learned in previous grades & lessons

Write the next three terms of the geometric sequence. *(Section 4.7)*

21. 2, 8, 32, 128, . . .

22. $-32, -8, -2, -\frac{1}{2}, \ldots$

23. 297, −99, 33, −11, . . .

24. MULTIPLE CHOICE A runner completes an 800-meter race in 2 minutes 40 seconds. What is the runner's speed? *(Section 5.1)*

Ⓐ $\frac{3 \text{ sec}}{10 \text{ m}}$ Ⓑ $\frac{160 \text{ sec}}{1 \text{ m}}$ Ⓒ $\frac{5 \text{ m}}{1 \text{ sec}}$ Ⓓ $\frac{10 \text{ m}}{3 \text{ sec}}$

6 Study Help

Check It Out
Graphic Organizer
BigIdeasMath.com

You can use an **example and non-example chart** to list examples and non-examples of a vocabulary word or term. Here is an example and non-example chart for similar figures.

Similar Figures

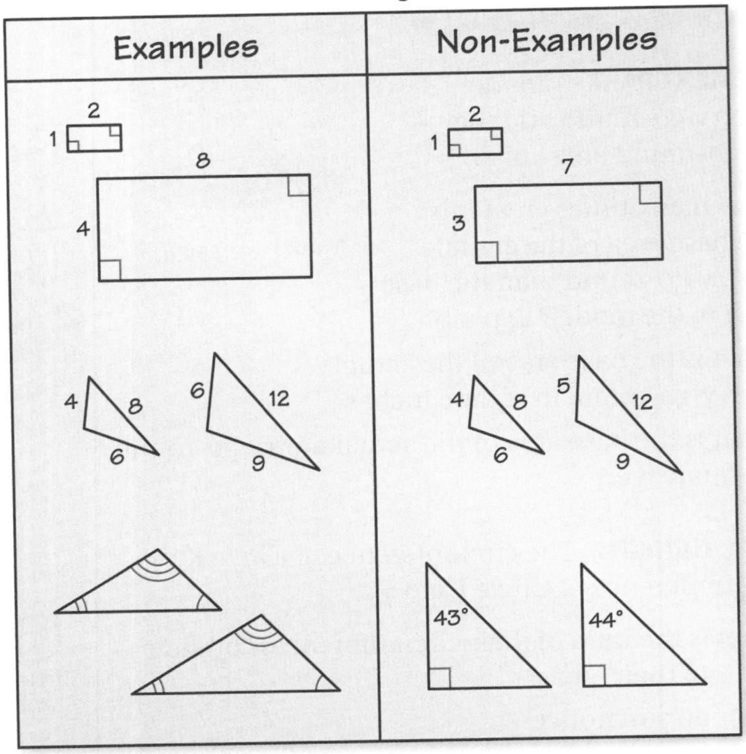

On Your Own

Make an example and non-example chart to help you study these topics.

1. quadrilaterals
2. corresponding angles
3. corresponding sides
4. perimeters of similar figures
5. areas of similar figures

After you complete this chapter, make example and non-example charts for the following topics.

6. unknown measures in similar figures
7. scale factor

"I'm using an example and non-example chart for a talk on cat hygiene."

6.1–6.3 Quiz

Check It Out
Progress Check
BigIdeasMath.com

Identify the quadrilateral. *(Section 6.1)*

1.

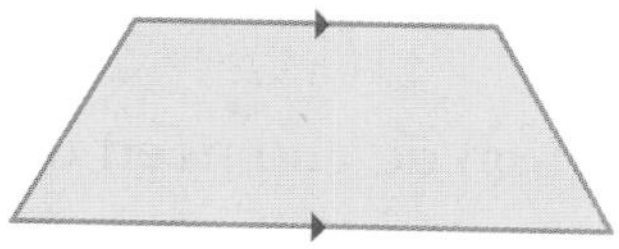

2.

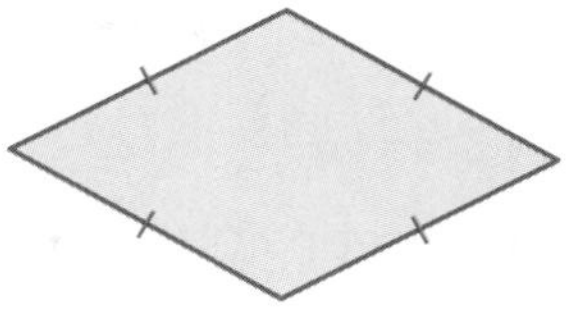

Find the missing angle measures. *(Section 6.1)*

3.

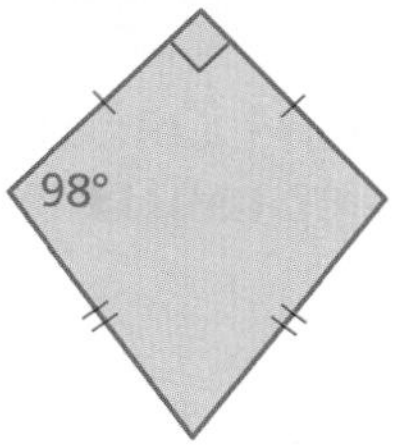

4.

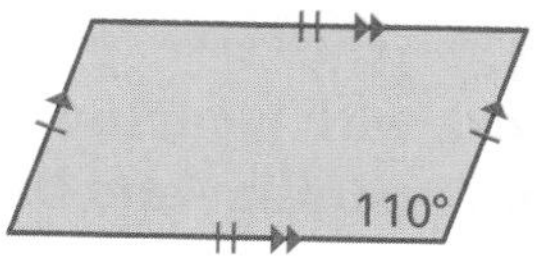

5. Tell whether the two rectangles are similar. Explain your reasoning. *(Section 6.2)*

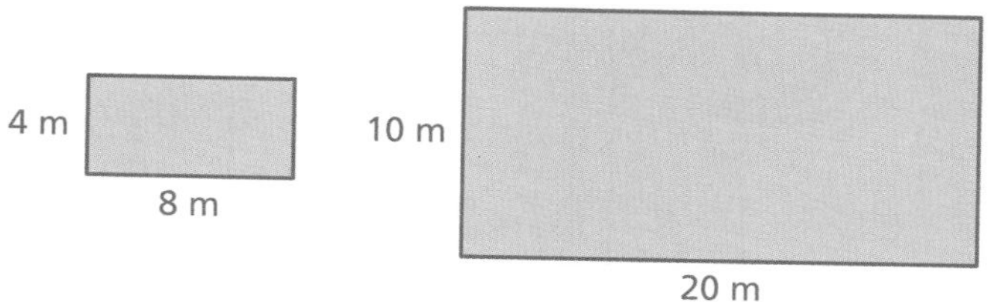

The two figures are similar. Find the ratios (red to blue) of the perimeters and of the areas. *(Section 6.3)*

6.

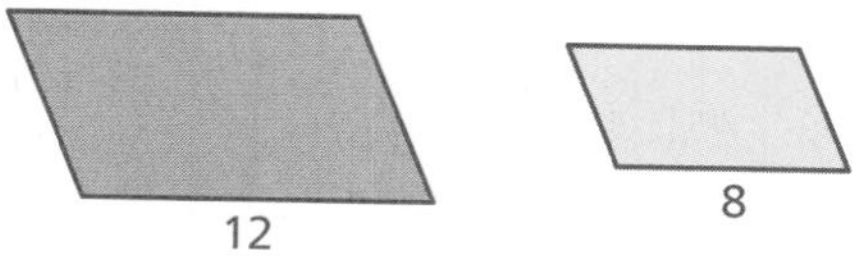

7.

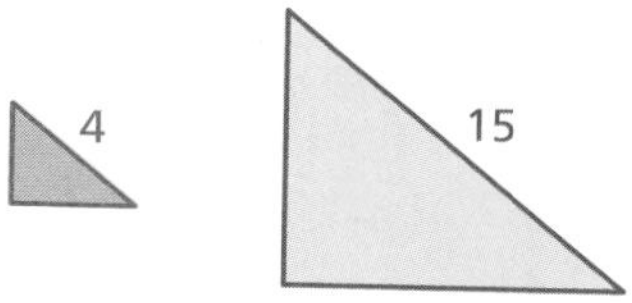

8. POSTERS The ratio of the corresponding side lengths of two similar posters is 7 in. : 20 in. What is the ratio of the perimeters? *(Section 6.3)*

9. PENDANT Find the missing angle measures of the kite-shaped pendant. *(Section 6.1)*

10. TENNIS COURT The tennis courts for singles and doubles matches are different sizes. Are the courts similar? Explain. *(Section 6.2)*

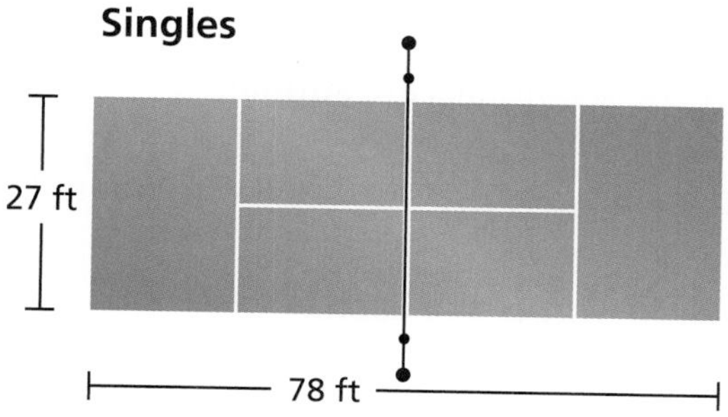

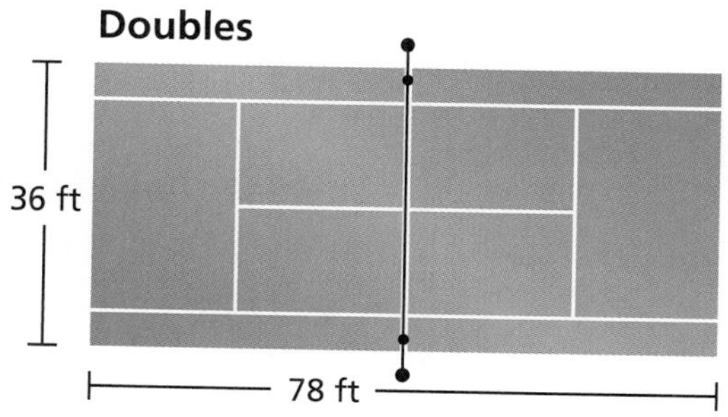

6.4 Finding Unknown Measures in Similar Figures

STANDARDS OF LEARNING
7.6

Essential Question What information do you need to know to find the dimensions of a figure that is similar to another figure?

1 ACTIVITY: Drawing and Labeling Similar Figures

Work with a partner. You are given the red rectangle. Find a blue rectangle that is similar and has one side from $(-1, -6)$ to $(5, -6)$. Label the vertices.

a. Sample:

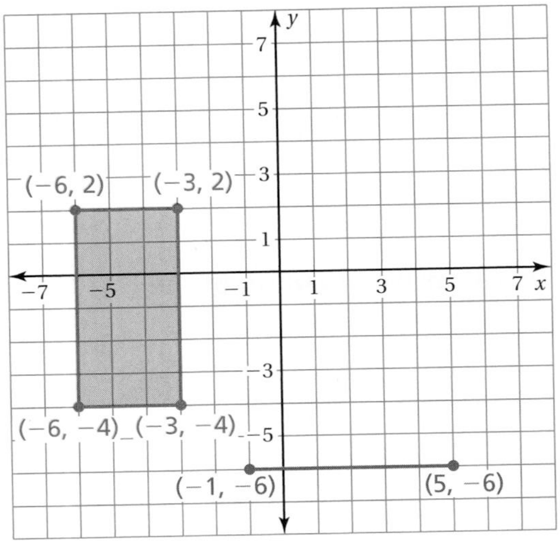

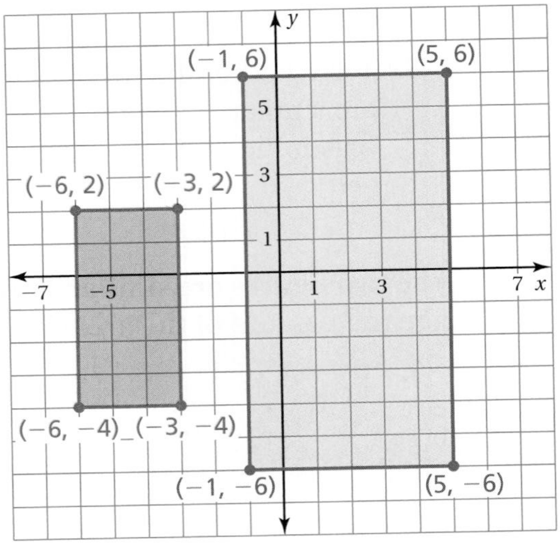

You can see that the two rectangles are similar by showing that ratios of corresponding sides are equal.

$$\frac{\text{Red Length}}{\text{Blue Length}} \stackrel{?}{=} \frac{\text{Red Width}}{\text{Blue Width}}$$

$$\frac{\text{change in } y}{\text{change in } y} \stackrel{?}{=} \frac{\text{change in } x}{\text{change in } x}$$

$$\frac{6}{12} \stackrel{?}{=} \frac{3}{6}$$

$$\frac{1}{2} = \frac{1}{2}$$

The ratios are equal. So, the rectangles are similar.

b. There are three other blue rectangles that are similar to the red rectangle and have the given side.

- Draw each one. Label the vertices of each.
- Show that each is similar to the original red rectangle.

2 ACTIVITY: Reading a Map

Work with a partner.

a. The red and blue rectangles are similar. Find the length of the blue rectangle. Explain your reasoning.

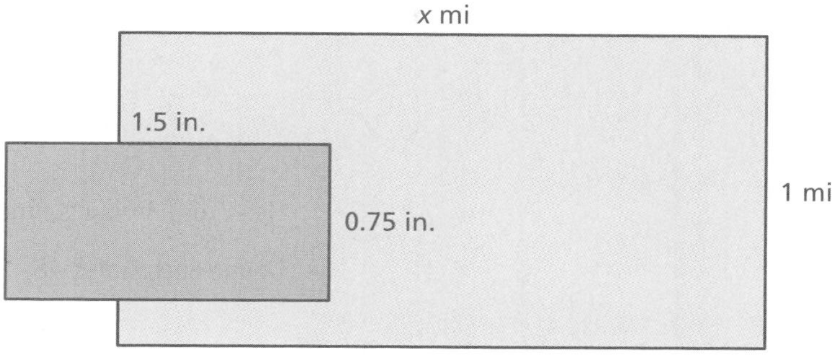

b. The distance marked by the vertical red line on the map is 1 mile. Find the distance marked by the horizontal red line. Explain your reasoning.

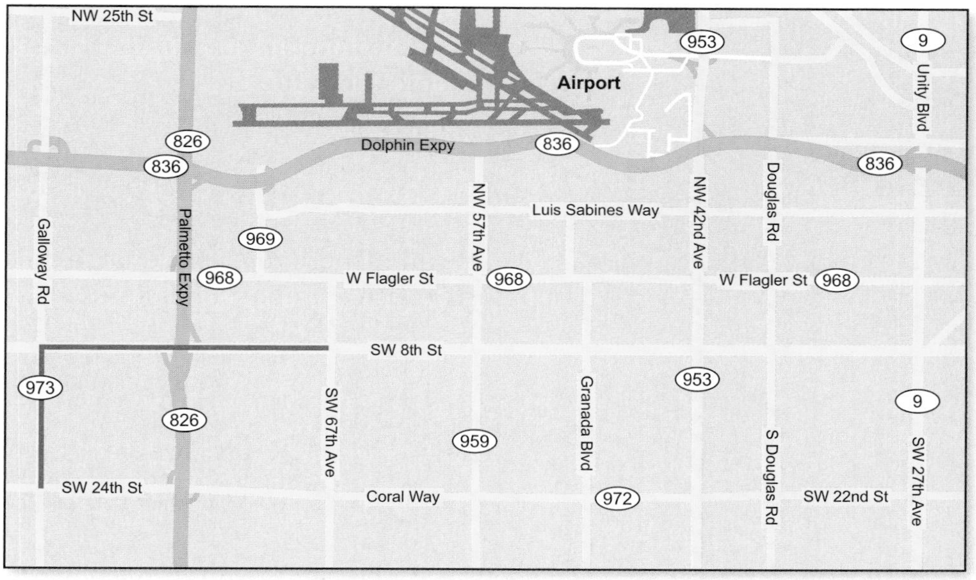

What Is Your Answer?

3. IN YOUR OWN WORDS What information do you need to know to find the dimensions of a figure that is similar to another figure? Give some examples using two rectangles.

4. When you know the length and width of one rectangle and the length of a similar rectangle, can you always find the missing width? Why or why not?

Practice Use what you learned about finding unknown measures in similar figures to complete Exercises 3 and 4 on page 264.

6.4 Lesson

Check It Out
Lesson Tutorials
BigIdeasMath.com

Key Vocabulary
indirect measurement, p. 263

EXAMPLE 1 Finding an Unknown Measure

The two triangles are similar. Find the value of x.

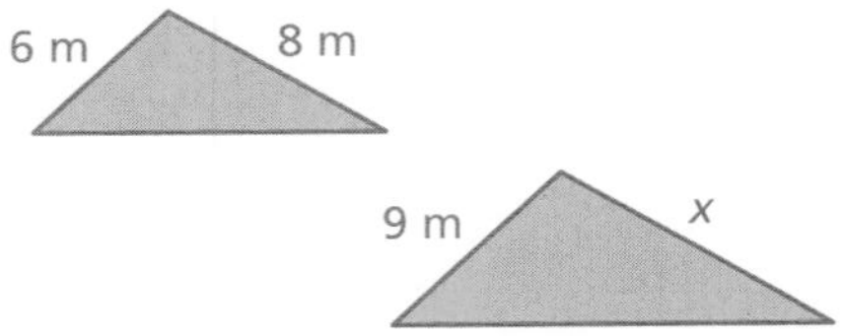

Corresponding side lengths are proportional. So, use a proportion to find x.

$\frac{6}{9} = \frac{8}{x}$ Write a proportion.

$6x = 72$ Use Cross Products Property.

$x = 12$ Divide each side by 6.

So, x is 12 meters.

EXAMPLE 2 Standardized Test Practice

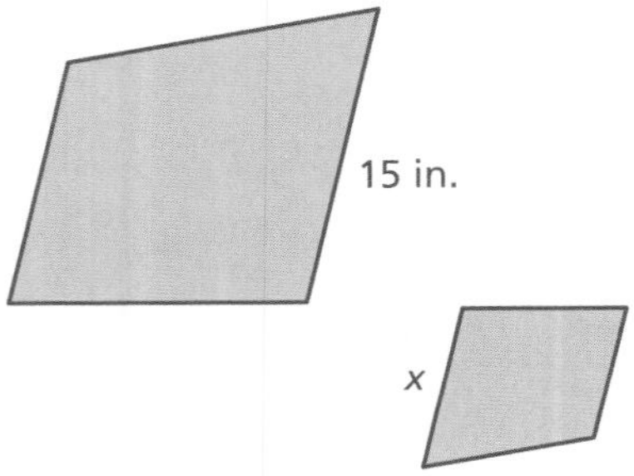

The two quadrilaterals are similar. The ratio of their perimeters is 12 : 5. Find the value of x.

Ⓐ 2.4 inches Ⓑ 4 inches

Ⓒ 6.25 inches Ⓓ 36 inches

The ratio of the perimeters is equal to the ratio of corresponding side lengths. So, use a proportion to find x.

$\frac{12}{5} = \frac{15}{x}$ Write a proportion.

$12x = 75$ Use Cross Products Property.

$x = 6.25$ Divide each side by 12.

So, x is 6.25 inches. The correct answer is Ⓒ.

On Your Own

Now You're Ready
Exercises 3–8

1. The two quadrilaterals are similar. The ratio of the perimeters is 3 : 4. Find the value of x.

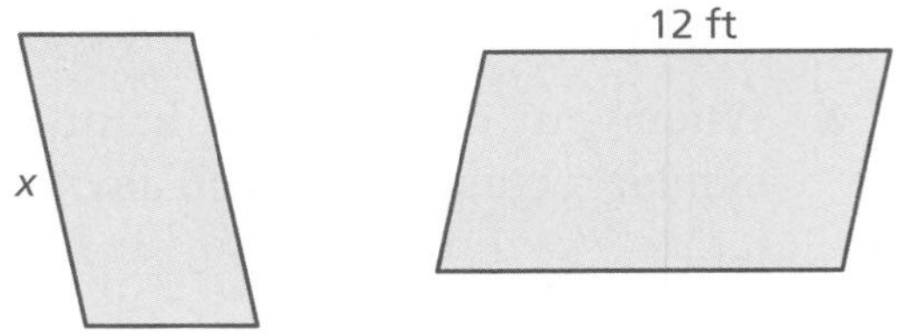

Multi-Language Glossary at BigIdeasMath.com.

Indirect measurement uses similar figures to find a missing measure that is difficult to find directly.

EXAMPLE 3 Using Indirect Measurement

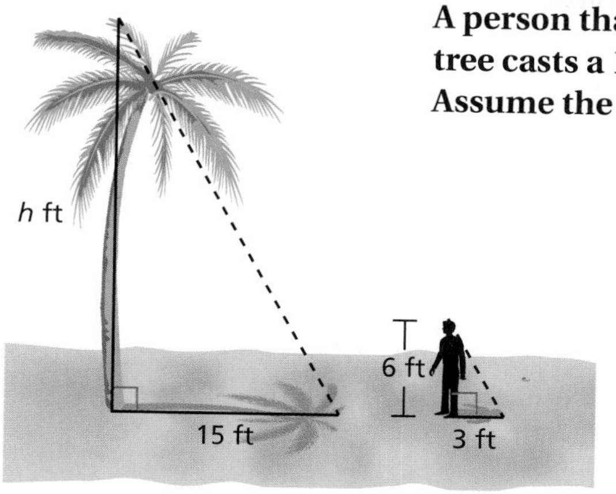
h ft
6 ft
15 ft
3 ft

A person that is 6 feet tall casts a 3-foot-long shadow. A nearby palm tree casts a 15-foot-long shadow. What is the height h of the palm tree? Assume the triangles are similar.

Corresponding side lengths are proportional.

$$\frac{h}{6} = \frac{15}{3}$$ Write a proportion.

$$6 \cdot \frac{h}{6} = \frac{15}{3} \cdot 6$$ Multiply each side by 6.

$$h = 30$$ Simplify.

∴ The palm tree is 30 feet tall.

On Your Own

Now You're Ready
Exercise 9

2. **WHAT IF?** Later in the day, the palm tree in Example 3 casts a 25-foot-long shadow. How long is the shadow of the person?

EXAMPLE 4 Using Proportions to Find Area

A swimming pool is similar in shape to a volleyball court. What is the area A of the pool?

18 yd

$$\frac{\text{Area of court}}{\text{Area of pool}} = \left(\frac{\text{width of court}}{\text{width of pool}}\right)^2$$

$$\frac{200}{A} = \left(\frac{10}{18}\right)^2$$ Substitute.

$$\frac{200}{A} = \frac{100}{324}$$ Simplify.

$$A = 648$$ Solve the proportion.

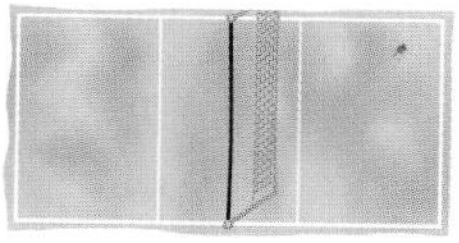
10 yd
Area = 200 yd²

∴ The area of the pool is 648 square yards.

On Your Own

3. The length of the volleyball court in Example 4 is 20 yards. What is the perimeter of the pool?

6.4 Exercises

Check It Out
Help with Homework
BigIdeasMath.com

Vocabulary and Concept Check

1. **REASONING** How can you use corresponding side lengths to find unknown measures in similar figures?

2. **CRITICAL THINKING** In which of the situations would you likely use indirect measurement? Explain your reasoning.

Finding the height of a statue	Finding the width of a doorway
Finding the width of a river	Finding the length of a lake

Practice and Problem Solving

The polygons are similar. Find the value of x.

① ② **3.**

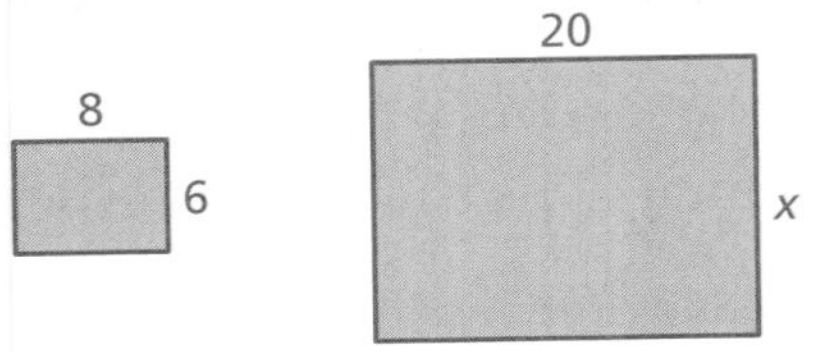

4.

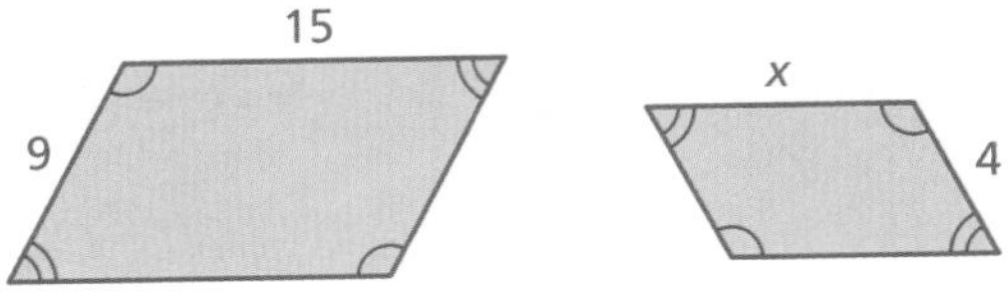

5.

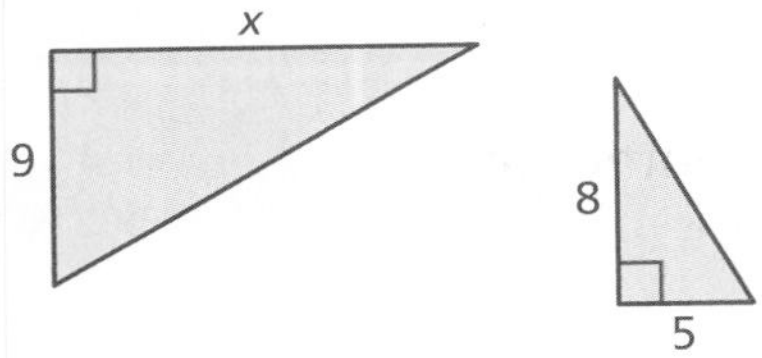

6.

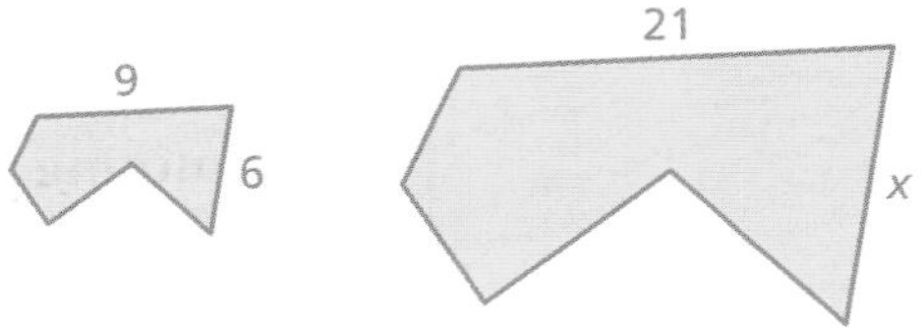

7. The ratio of the perimeters is $7:10$.

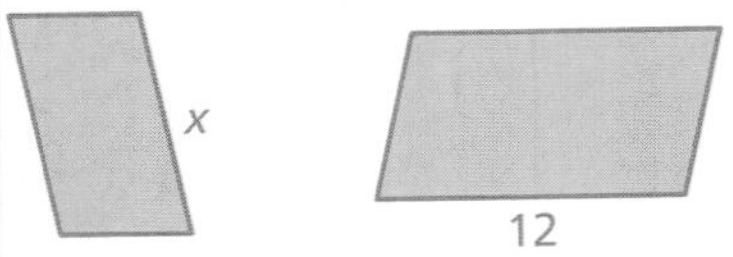

8. The ratio of the perimeters is $8:5$.

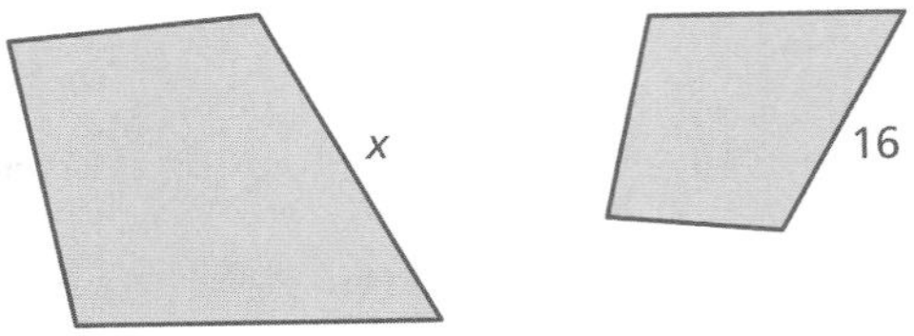

③ **9.** **FLAGPOLE** What is the height x of the flagpole? Assume the triangles are similar.

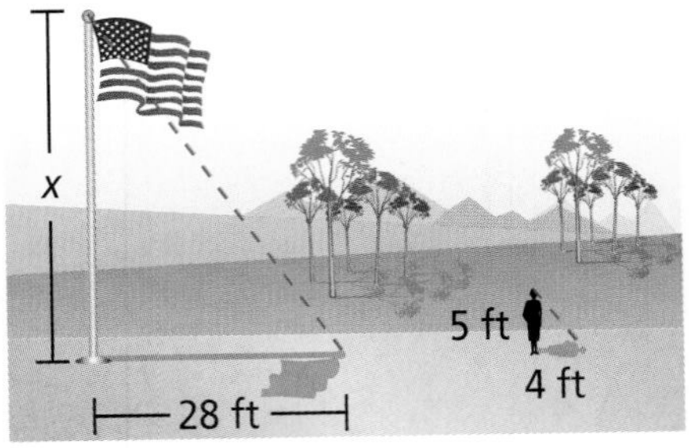

10. **CHEERLEADING** A rectangular school banner has a length of 44 inches and a perimeter of 156 inches. The cheerleaders make signs similar to the banner. The length of a sign is 11 inches. What is its perimeter?

11. **SQUARE** The ratio of the side length of Square A to the side length of Square B is 4 : 9. The side length of Square A is 12 yards. What is the perimeter of Square B?

12. **RIVER** Is the distance *QP* across the river greater than 100 meters? Explain.

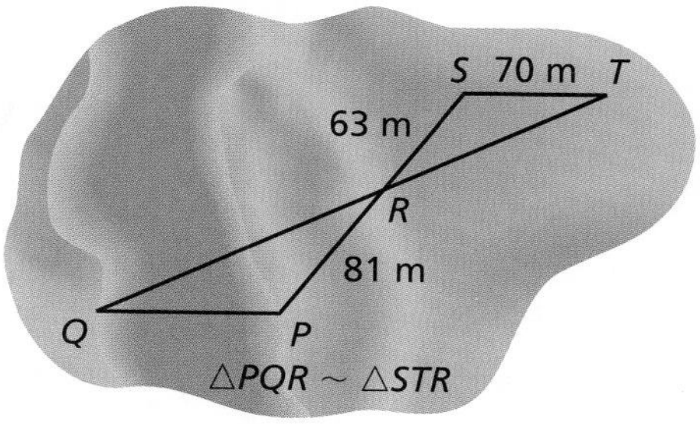

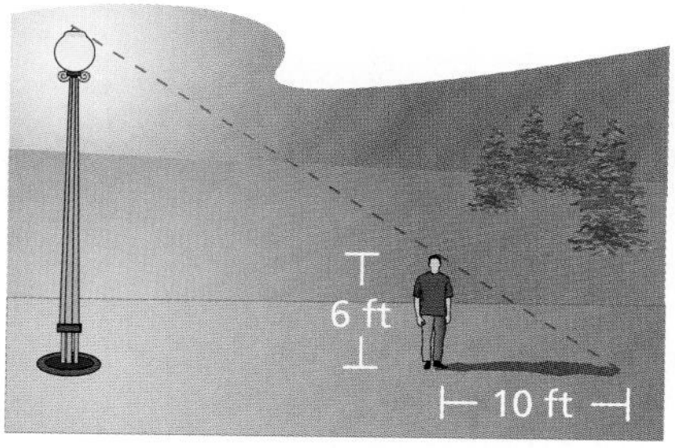

13. **STREET LIGHT** A person standing 20 feet from a street light casts a shadow as shown. How many times taller is the street light than the person? Assume the triangles are similar.

14. **AREA** A school playground is similar in shape to the community park. You can mow 250 square yards of grass in 15 minutes. How long would it take you to mow the grass on the playground?

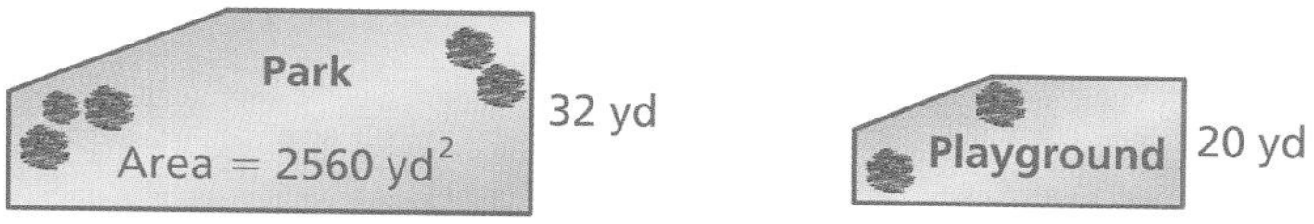

15. **Critical Thinking** Two bottles of fertilizer are needed to treat the flower garden shown. How many bottles are needed to treat a similar garden with a perimeter of 105 feet?

Fair Game Review What you learned in previous grades & lessons

Write and solve a proportion to complete the statement. *(Section 5.5)*

16. 4 mi ≈ ___ km **17.** 12.5 in. ≈ ___ cm **18.** 110 kg ≈ ___ lb

19. 6.2 km ≈ ___ mi **20.** 10 cm ≈ ___ in. **21.** 92.5 lb ≈ ___ kg

22. **MULTIPLE CHOICE** A recipe that makes 8 pints of salsa uses 22 tomatoes. Which proportion can you use to find the number n of tomatoes needed to make 12 pints of salsa? *(Section 5.4)*

Ⓐ $\frac{n}{8} = \frac{22}{12}$ Ⓑ $\frac{8}{22} = \frac{12}{n}$ Ⓒ $\frac{22}{n} = \frac{12}{8}$ Ⓓ $\frac{8}{22} = \frac{n}{12}$

6.5 Scale Drawings

STANDARDS OF LEARNING
7.4

Essential Question How can you use a scale drawing to estimate the cost of painting a room?

1 ACTIVITY: Making Scale Drawings

Work with a partner. You have decided that your classroom needs to be painted. Start by making a scale drawing of each of the four walls.

- **Measure each of the walls.**
- **Measure the locations and dimensions of parts that will *not* be painted.**
- **Decide on a scale for your drawings.**
- **Make a scale drawing of each of the walls.**

Sample: Wall #1

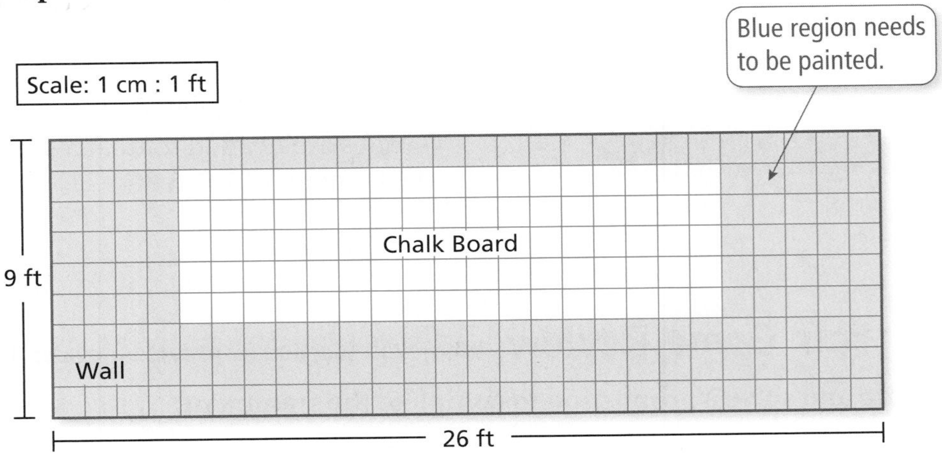

- **For each wall, find the area of the part that needs to be painted.**

	Dimensions	*Area*
Dimensions of the wall	9 ft by 26 ft	$9 \times 26 = 234$ sq ft
Dimensions of the part that will *not* be painted	5 ft by 17 ft	$5 \times 17 = 85$ sq ft
Area of painted part		149 sq ft

2 ACTIVITY: Using Scale Drawings

Work with a partner.

You are using a paint that covers 200 square feet per gallon. Each wall will need two coats of paint.

a. Find the total area of the walls from Activity 1 that needs to be painted.

b. Find the amount of paint you need to buy.

c. Estimate the total cost of painting your classroom.

Interior latex paint	$40 per gallon
Roller, pan, and brush set	$12

What Is Your Answer?

3. **IN YOUR OWN WORDS** How can you use a scale drawing to estimate the cost of painting a room?

4. Use a scale drawing to estimate the cost of painting another room, such as your bedroom or another room in your house.

5. Look at some maps in your school library or on the Internet. Make a list of the different scales used on the maps.

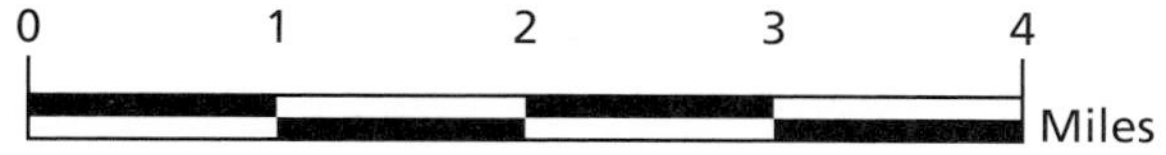

6. When you view a map on the Internet, how does the scale change when you zoom out? How does the scale change when you zoom in?

"I don't get it. According to this map, we only have to drive $8\frac{1}{2}$ inches."

Practice Use what you learned about scale drawings to complete Exercises 4–7 on page 270.

6.5 Lesson

Check It Out
Lesson Tutorials
BigIdeasMath.com

Key Vocabulary
scale drawing, *p. 268*
scale model, *p. 268*
scale, *p. 268*
scale factor, *p. 269*

Key Ideas

Scale Drawings and Models

A **scale drawing** is a proportional two-dimensional drawing of an object. A **scale model** is a proportional three-dimensional model of an object.

Scale

Measurements in scale drawings and models are proportional to the measurements of the actual object. The **scale** gives the ratio that compares the measurements of the drawing or model with the actual measurements.

$\frac{1 \text{ in.}}{10 \text{ mi}}$ ← drawing distance (1 in.); actual distance (10 mi)

1 in. : 10 mi (1 in. = drawing, 10 mi = actual)

Study Tip

Scales are written so that the drawing distance comes first in the ratio.

EXAMPLE 1 Finding an Actual Distance

What is the actual distance *d* between Cadillac and Detroit?

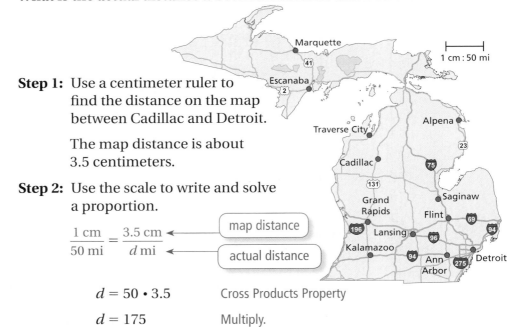

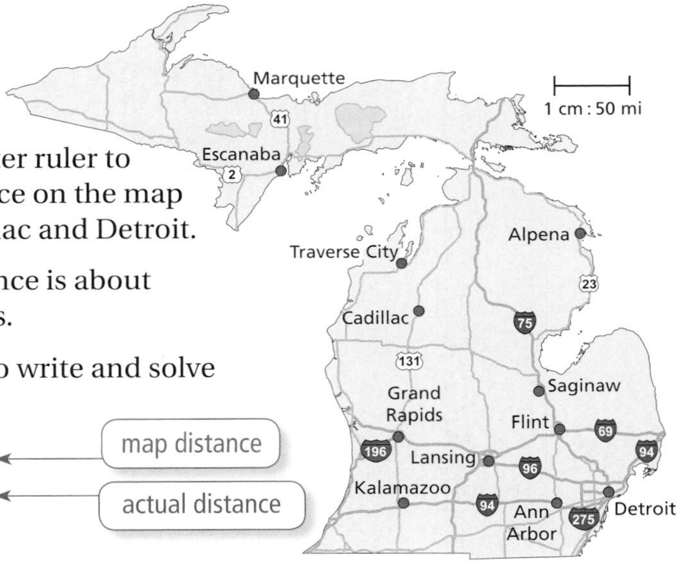

Step 1: Use a centimeter ruler to find the distance on the map between Cadillac and Detroit.

The map distance is about 3.5 centimeters.

Step 2: Use the scale to write and solve a proportion.

$$\frac{1 \text{ cm}}{50 \text{ mi}} = \frac{3.5 \text{ cm}}{d \text{ mi}}$$ ← map distance (numerators); actual distance (denominators)

$d = 50 \cdot 3.5$ Cross Products Property

$d = 175$ Multiply.

∴ The distance between Cadillac and Detroit is about 175 miles.

On Your Own

Now You're Ready
Exercises 8–11

1. What is the actual distance between Traverse City and Marquette?

Multi-Language Glossary at BigIdeasMath.com.

EXAMPLE 2 Standardized Test Practice

The liquid outer core of Earth is 2300 kilometers thick. A scale model of the layers of Earth has a scale of 1 in. : 500 km. How thick is the liquid outer core of the model?

Ⓐ 0.2 in. Ⓑ 4.6 in. Ⓒ 0.2 km Ⓓ 4.6 km

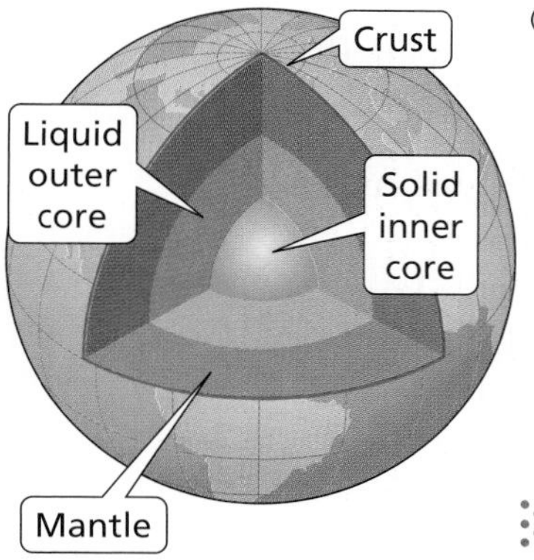

$$\frac{1\text{ in.}}{500\text{ km}} = \frac{x\text{ in.}}{2300\text{ km}}$$ ← model thickness (numerator), actual thickness (denominator)

$$\frac{1\text{ in.}}{500\text{ km}} \cdot 2300\text{ km} = \frac{x\text{ in.}}{2300\text{ km}} \cdot 2300\text{ km}$$ Multiply each side by 2300 km.

$$4.6 = x$$ Simplify.

∴ The liquid outer core of the model is 4.6 inches thick. The correct answer is Ⓑ.

On Your Own

2. The mantle of Earth is 2900 kilometers thick. How thick is the mantle of the model?

A scale can be written without units when the units are the same. A scale without units is called a **scale factor**.

EXAMPLE 3 Finding a Scale Factor

A scale drawing of a spider is 5 centimeters long. The actual spider is 10 millimeters long. (a) What is the scale of the drawing? (b) What is the scale factor of the drawing?

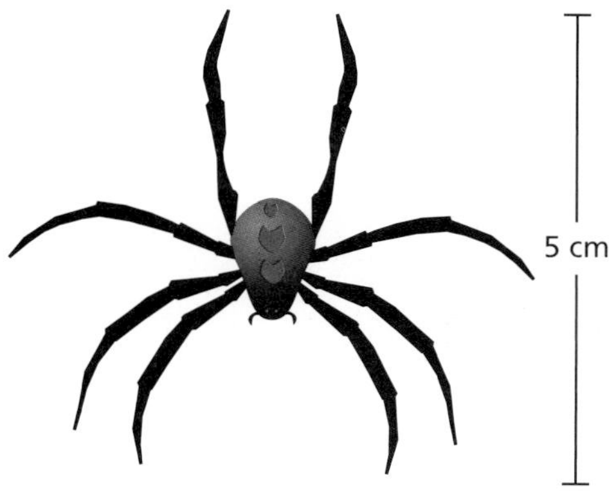

a. $\frac{\text{drawing length}}{\text{actual length}} = \frac{5\text{ cm}}{10\text{ mm}} = \frac{1\text{ cm}}{2\text{ mm}}$

∴ The scale is 1 cm : 2 mm.

b. Write the scale with the same units. Use the fact that 1 cm = 10 mm.

$$\text{scale factor} = \frac{1\text{ cm}}{2\text{ mm}} = \frac{10\text{ mm}}{2\text{ mm}} = \frac{5}{1}$$

∴ The scale factor is 5 : 1.

On Your Own

Now You're Ready
Exercises 12–16

3. A model has a scale of 1 mm : 20 cm. What is the scale factor of the model?

6.5 Exercises

Check It Out
Help with Homework
BigIdeasMath.com

Vocabulary and Concept Check

1. **VOCABULARY** Compare and contrast the terms *scale* and *scale factor*.
2. **CRITICAL THINKING** The scale of a drawing is 2 cm : 1 mm. Is the scale drawing *larger* or *smaller* than the actual object? Explain.
3. **REASONING** How would you find a scale factor of a drawing that shows a length of 4 inches when the actual object is 8 feet long?

Practice and Problem Solving

Use the drawing and a centimeter ruler.

4. What is the actual length of the flower garden?
5. What are the actual dimensions of the rose bed?
6. What are the actual perimeters of the perennial beds?
7. The area of the tulip bed is what percent of the area of the rose bed?

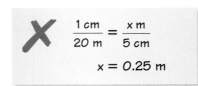

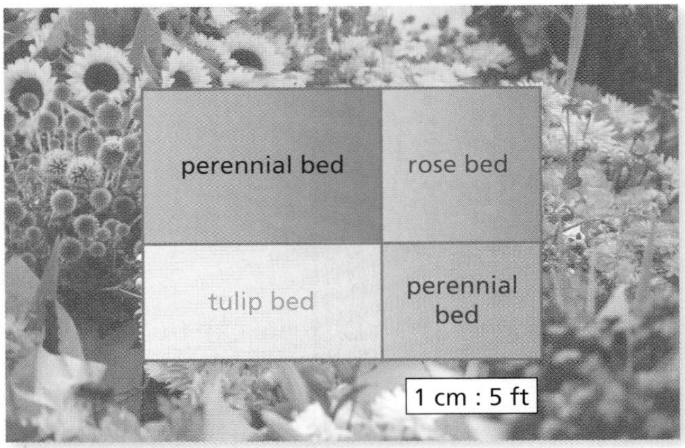

Use the map in Example 1 to find the actual distance between the cities.

(1) 8. Kalamazoo and Ann Arbor
9. Lansing and Flint
10. Grand Rapids and Escanaba
11. Saginaw and Alpena

Find the missing dimension. Use the scale factor 1 : 12.

	Item	Model	Actual
(2) (3) 12.	Mattress	Length: 6.25 in.	Length: ___ in.
13.	Corvette	Length: ___ in.	Length: 15 ft
14.	Water Tower	Depth: 32 cm	Depth: ___ m
15.	Wingspan	Width: 5.4 ft	Width: ___ yd
16.	Football Helmet	Diameter: ___ mm	Diameter: 21 cm

17. **ERROR ANALYSIS** A scale is 1 cm : 20 m. Describe and correct the error in finding the actual distance that corresponds to 5 cm.

✗ $\frac{1 \text{ cm}}{20 \text{ m}} = \frac{x \text{ m}}{5 \text{ cm}}$

$x = 0.25 \text{ m}$

Use a centimeter ruler to measure the segment shown. Find the scale of the drawing.

18.

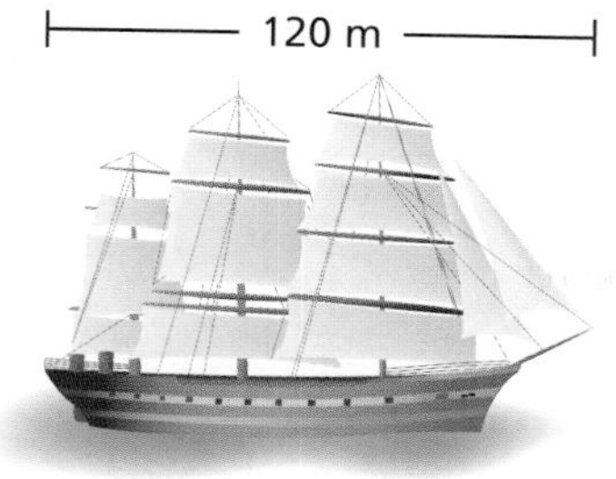

19.

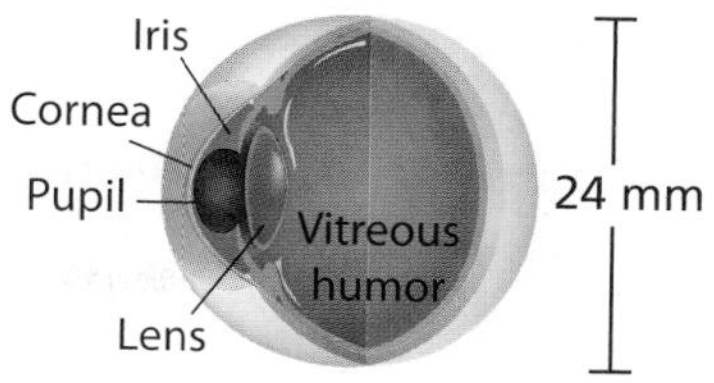

20. REASONING You know the length and width of a scale model. What additional information do you need to know to find the scale of the model?

21. OPEN-ENDED You are in charge of creating a billboard advertisement with the dimensions shown.

- **a.** Choose a product. Then design the billboard using words and a picture.
- **b.** What is the scale factor of your design?

Reduced drawing of blueprint

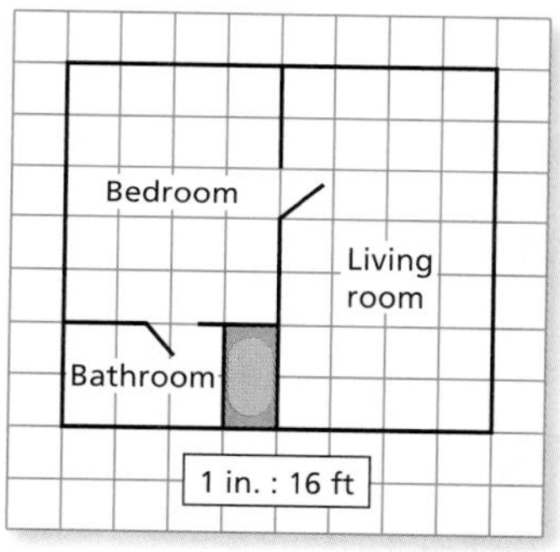

22. BLUEPRINT In a blueprint, each square has a side length of $\frac{1}{4}$ inch.

- **a.** Ceramic tile costs \$5 per square foot. How much would it cost to tile the bathroom?
- **b.** Carpet costs \$18 per square yard. How much would it cost to carpet the bedroom and living room?
- **c.** Which has a higher unit cost, the tile or the carpet? Explain.

23. REASONING You are making a scale model of the solar system. The radius of Earth is 6378 kilometers. The radius of the Sun is 695,500 kilometers. Is it reasonable to choose a baseball as a model of Earth? Explain your reasoning.

24. Critical Thinking A map on the Internet has a scale of 1 in. : 10 mi. You zoom out one level. The map has been reduced so that 2.5 inches on the old map appears as 1 inch on the new map. What is the scale of the new map?

Fair Game Review What you learned in previous grades & lessons

Plot and label the ordered pair in a coordinate plane. *(Skills Review Handbook)*

25. $A(-4, 3)$ **26.** $B(2, -6)$ **27.** $C(5, 1)$ **28.** $D(-3, -7)$

29. MULTIPLE CHOICE Which function describes the arithmetic sequence 3, 4, 5, 6, . . . ? *(Section 4.6)*

Ⓐ $y = n + 2$ Ⓑ $y = n - 2$ Ⓒ $y = 2n$ Ⓓ $y = n - 3$

6.4–6.5 Quiz

Check It Out
Progress Check
BigIdeasMath.com

The polygons are similar. Find the value of x. *(Section 6.4)*

1.

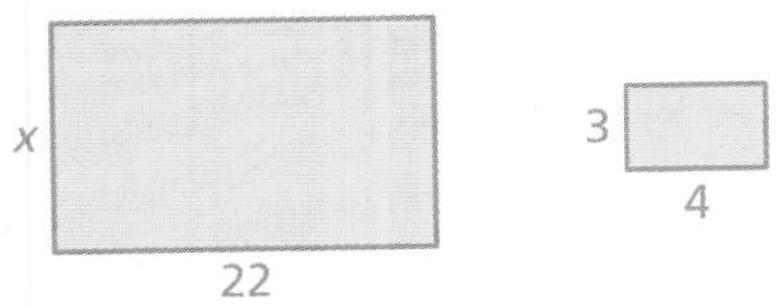

2.

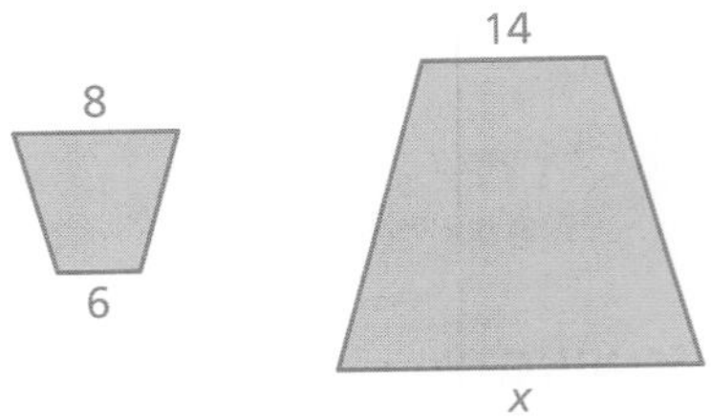

Find the missing dimension. Use the scale factor 1 : 20. *(Section 6.5)*

	Item	Model	Actual
3.	Basketball Player	Height: ▢ in.	Height: 90 in.
4.	Dinosaur	Length: 3.75 ft	Length: ▢ ft

5. DOLPHIN A dolphin in an aquarium is 12 feet long. A scale model of the dolphin is $3\frac{1}{2}$ inches long. What is the scale factor of the model? *(Section 6.5)*

6. STONEWALL JACKSON What is the height x of the statue? Assume the triangles are similar. *(Section 6.4)*

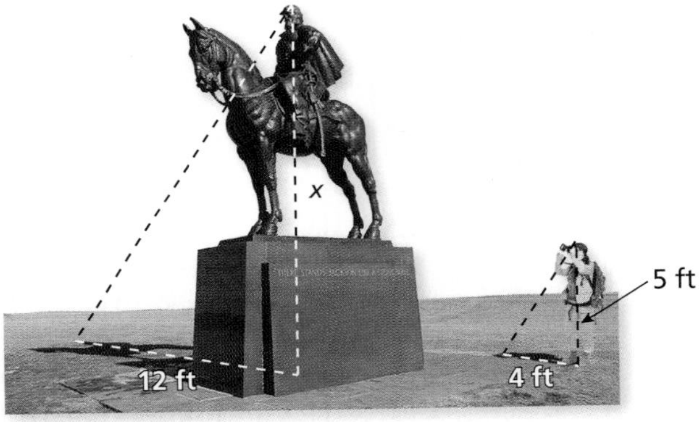

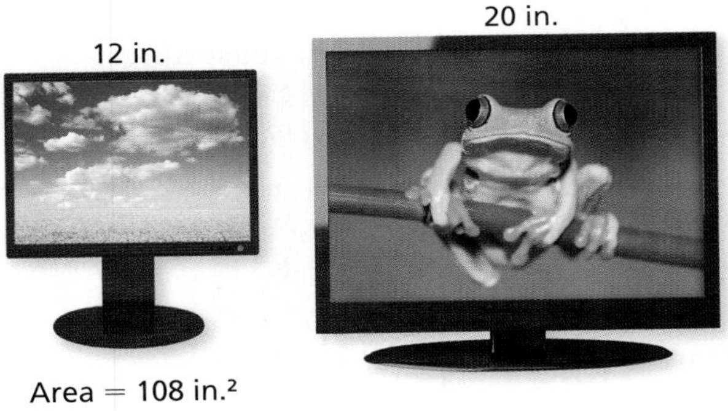

7. SCREENS The TV screen is similar to the computer screen. What is the area of the TV screen? *(Section 6.4)*

8. SOCCER A scale drawing of a soccer field is shown. The actual soccer field is 300 feet long. *(Section 6.5)*

a. What is the scale of the drawing?

b. What is the scale factor of the drawing?

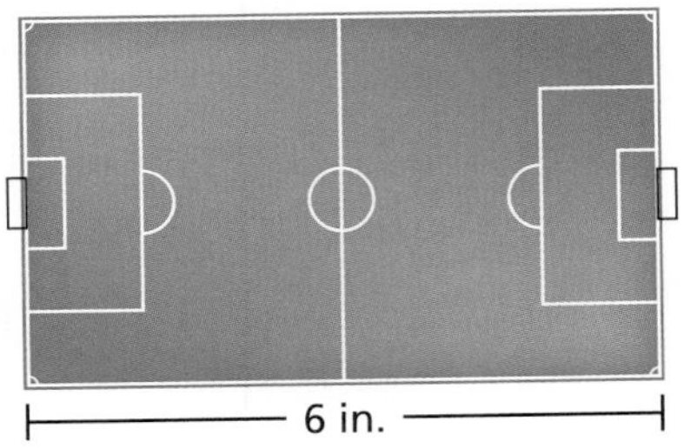

6 Chapter Review

Check It Out
Vocabulary Help
BigIdeasMath.com

Review Key Vocabulary

similar figures, *p. 248*
corresponding angles, *p. 248*
corresponding sides, *p. 248*
indirect measurement, *p. 263*
scale drawing, *p. 268*
scale model, *p. 268*
scale, *p. 268*
scale factor, *p. 269*

Review Examples and Exercises

6.1 Quadrilaterals *(pp. 240–245)*

Find the missing angle measures.

The missing angles are congruent because they are opposite angles of a parallelogram. Let $x°$ be the measure of each angle.

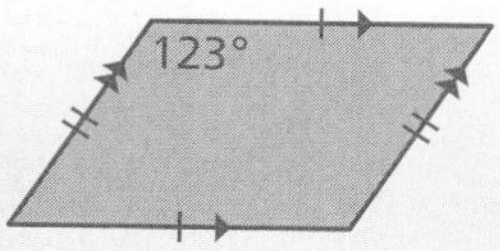

$$123 + 123 + x + x = 360$$
$$2x = 114$$
$$x = 57$$

So, the measure of each missing angle is 57°.

Exercises

Find the missing angle measure(s).

1.
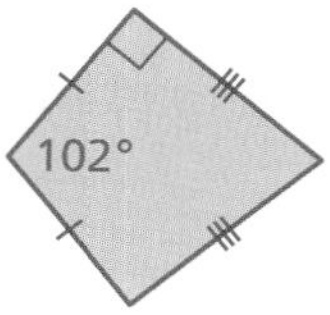

2.
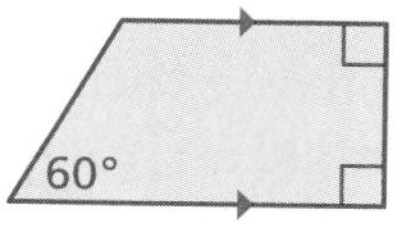

6.2 Identifying Similar Figures *(pp. 246–251)*

Is Rectangle A similar to Rectangle B?

Each figure is a rectangle. So, corresponding angles have the same measure. Check to see if corresponding side lengths are proportional.

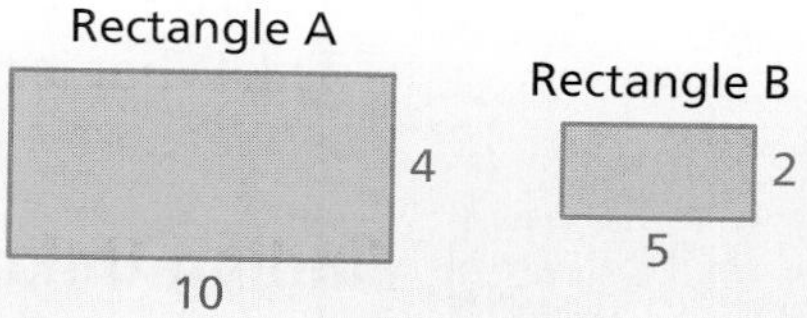

$\dfrac{\text{Length of A}}{\text{Length of B}} = \dfrac{10}{5} = 2$ $\quad \dfrac{\text{Width of A}}{\text{Width of B}} = \dfrac{4}{2} = 2$ $\quad$ Proportional

So, Rectangle A is similar to Rectangle B.

Exercises

Tell whether the two figures are similar. Explain your reasoning.

3.
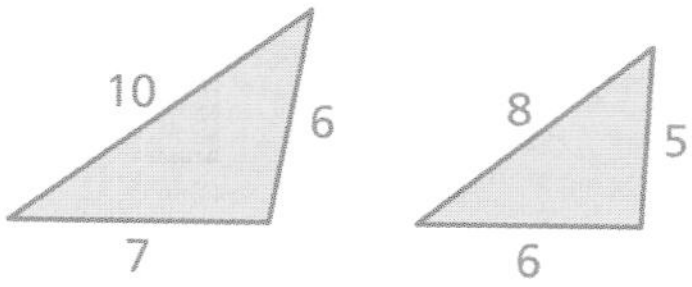

4.
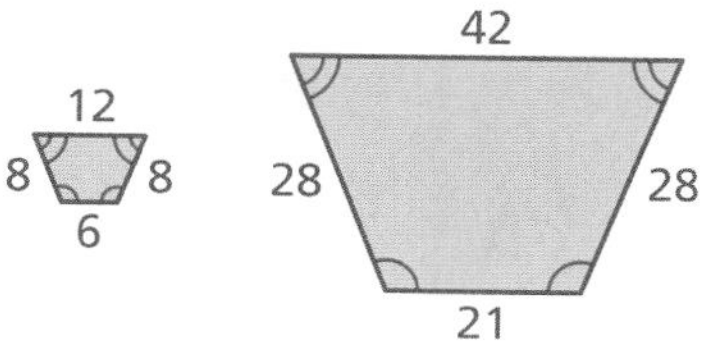

6.3 Perimeters and Areas of Similar Figures (pp. 252–257)

Find the ratio (red to blue) of the perimeters of the similar parallelograms.

$$\frac{\text{Perimeter of red parallelogram}}{\text{Perimeter of blue parallelogram}} = \frac{15}{9}$$

$$= \frac{5}{3}$$

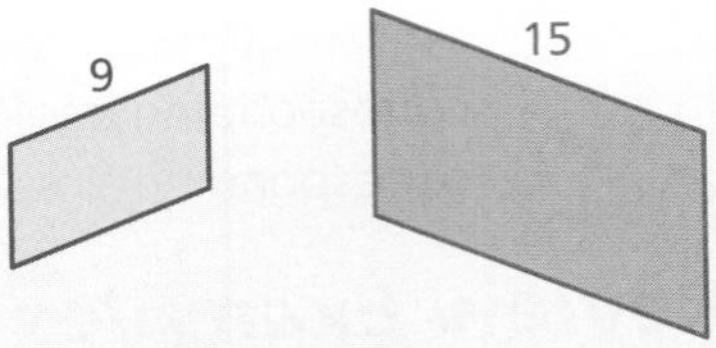

∴ The ratio of the perimeters is $\frac{5}{3}$.

Find the ratio (red to blue) of the areas of the similar figures.

$$\frac{\text{Area of red figure}}{\text{Area of blue figure}} = \left(\frac{3}{4}\right)^2$$

$$= \frac{9}{16}$$

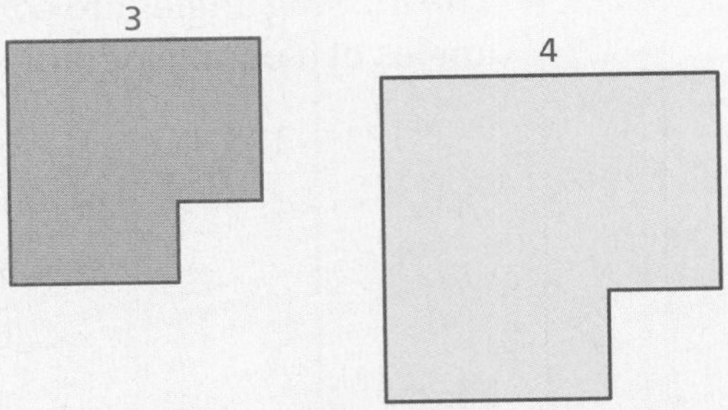

∴ The ratio of the areas is $\frac{9}{16}$.

Exercises

The two figures are similar. Find the ratios (red to blue) of the perimeters and of the areas.

5.

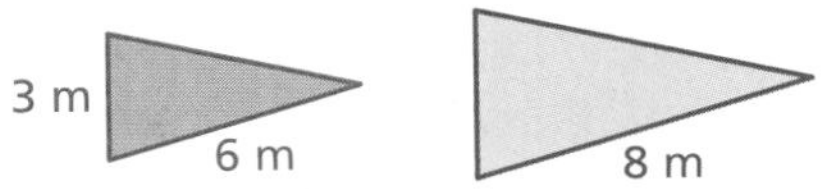

6.

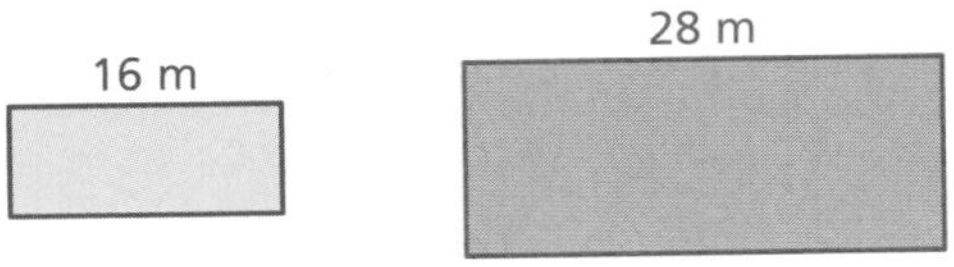

7. **PHOTOS** Two photos are similar. The ratio of the corresponding side lengths is 3 : 4. What is the ratio of their areas?

6.4 Finding Unknown Measures in Similar Figures (pp. 260–265)

The two rectangles are similar. Find the value of x.

Corresponding side lengths of similar figures are proportional. So, use a proportion to find x.

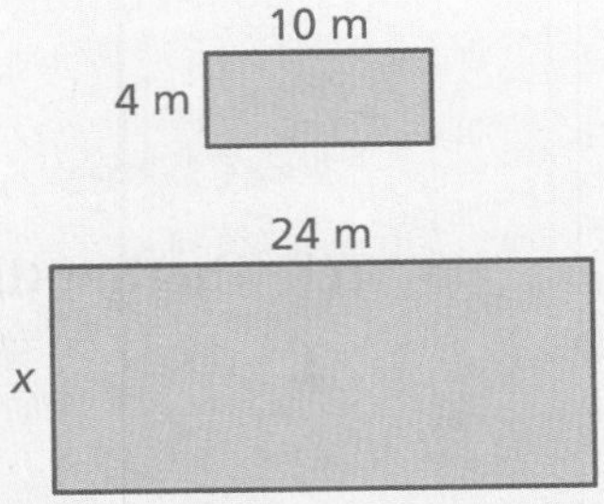

$\frac{10}{24} = \frac{4}{x}$	Write a proportion.
$10x = 96$	Use Cross Products Property.
$x = 9.6$	Divide each side by 10.

∴ So, x is 9.6 meters.

Exercises

The polygons are similar. Find the value of *x*.

8.

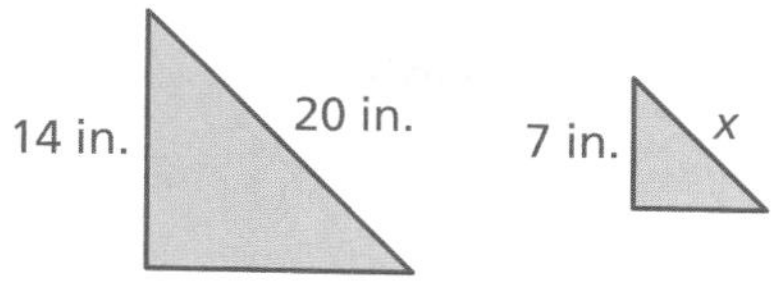

9.

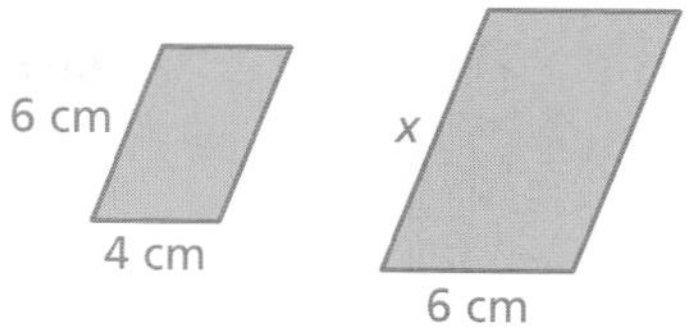

10.

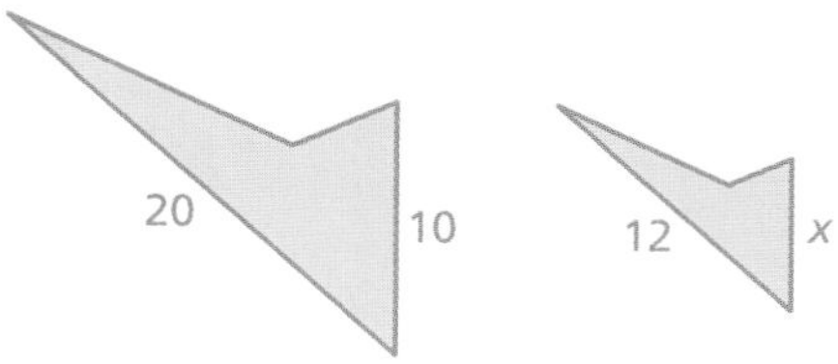

11.

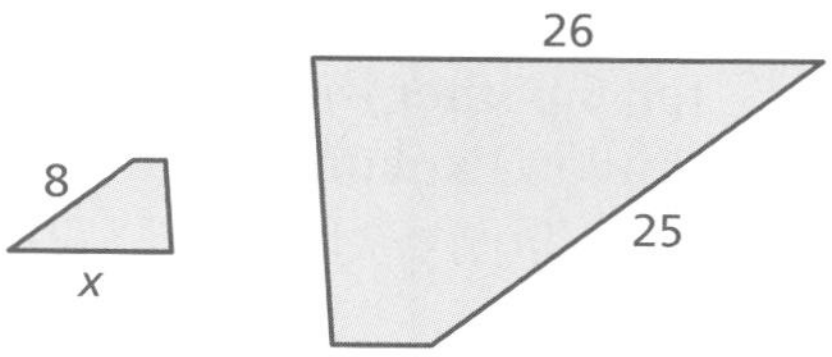

6.5 Scale Drawings *(pp. 266–271)*

A lighthouse is 160 feet tall. A scale model of the lighthouse has a scale of 1 in. : 8 ft. How tall is the model of the lighthouse?

$$\frac{1 \text{ in.}}{8 \text{ ft}} = \frac{x \text{ in.}}{160 \text{ ft}}$$

model height (x in.)

actual height (160 ft)

$$\frac{1 \text{ in.}}{8 \text{ ft}} \cdot 160 \text{ ft} = \frac{x \text{ in.}}{160 \text{ ft}} \cdot 160 \text{ ft}$$ Multiply each side by 160 ft.

$$20 = x$$ Simplify.

∴ The model of the lighthouse is 20 inches tall.

Exercises

Use a centimeter ruler to measure the segment shown. Find the scale of the drawing.

12.

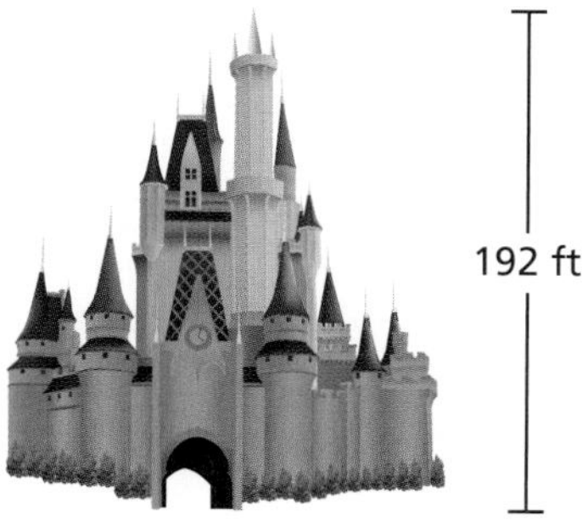

13.

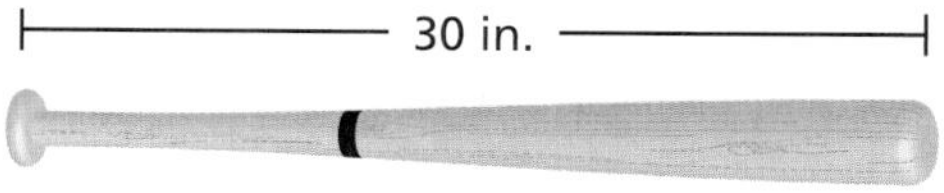

6 Chapter Test

Check It Out
Test Practice
BigIdeasMath.com

Find the missing angle measures.

1.

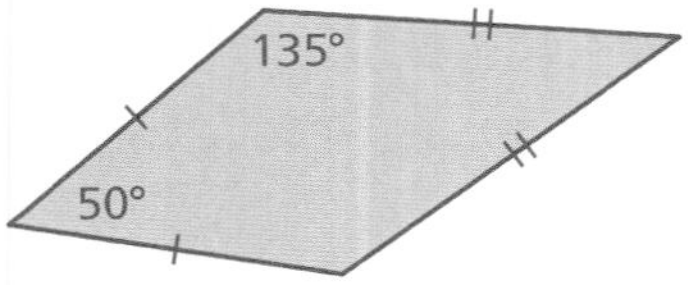

2.

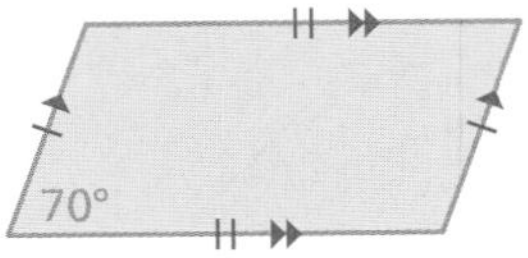

3. Tell whether the parallelograms are similar. Explain your reasoning.

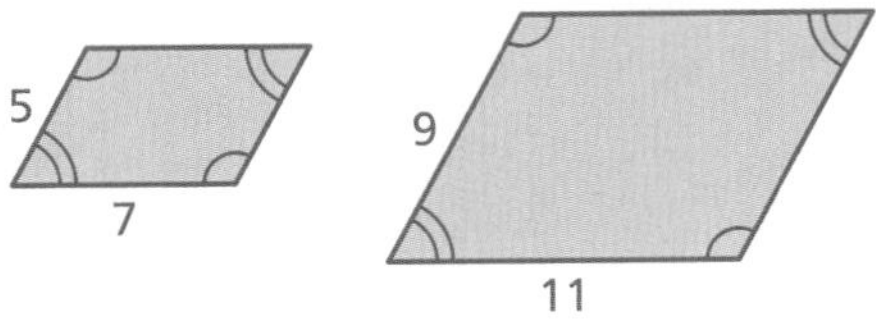

The two figures are similar. Find the ratios (red to blue) of the perimeters and of the areas.

4.

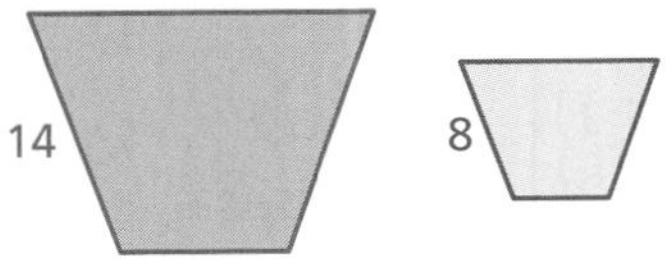

5.

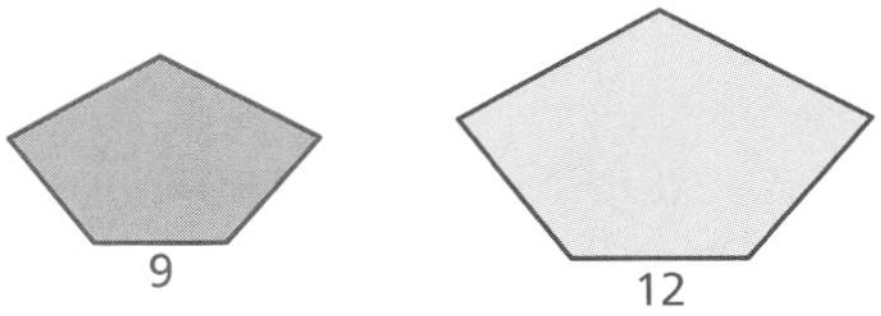

6. Use a centimeter ruler to measure the fish. Find the scale factor of the drawing.

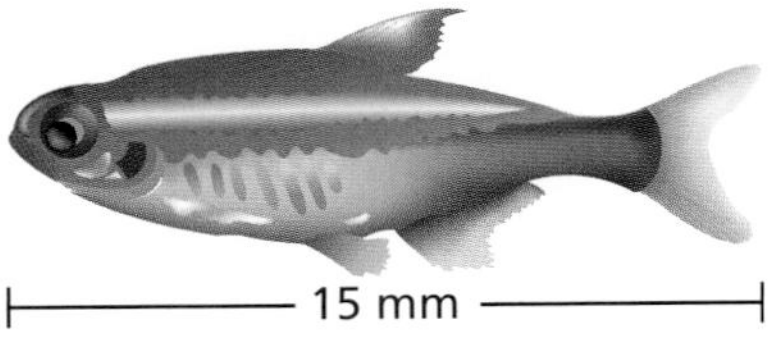

7. **SCREENS** A wide screen television measures 36 inches by 54 inches. A movie theater screen measures 42 feet by 63 feet. Are the screens similar? Explain.

8. **HOCKEY** An air hockey table and an ice hockey rink are similar. The ratio of their corresponding side lengths is 1 inch : 2 feet. What is the ratio of their areas?

9. **HEIGHT** You are five feet tall and cast a seven-foot eight-inch shadow. At the same time, a basketball hoop casts a 19-foot shadow. How tall is the basketball hoop? Assume the triangles are similar.

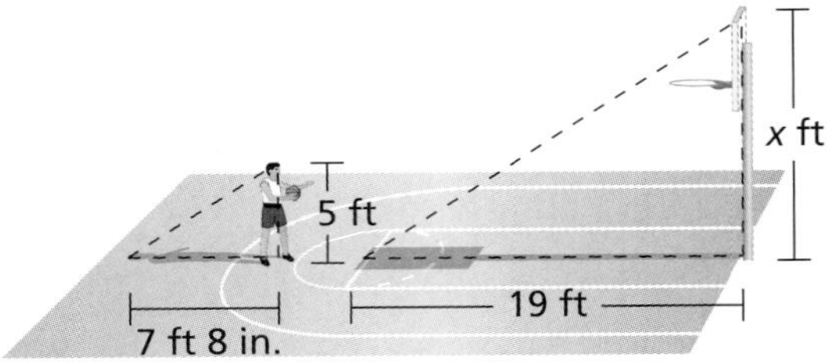

10. **CAD** An engineer is using computer-aided design (CAD) software to design a component for a space shuttle. The scale of the drawing is 1 cm : 5 ft. The actual length of the component is 12.5 feet. What is the length of the component in the drawing?

6 Standardized Test Practice

Test-Taking Strategy
Solve Problem Before Looking at Choices

Your paw has an area of 2 in.². A hyena's paw is twice as long. What is its area?
(A) 4 in.² (B) 6 in.² (C) 8 in.² (D) 10 in.²
4 times more area! Help!
CAT
HYENA

"Solve the problem before looking at the choices. You know area increases as the square of the scale. So, it's 8 in.²."

1. A set of data is shown below. Two of the data are missing.

8, 2, 10, 4, 8, 4, 8, 8, ____ , ____

The mean of the complete set of data is 6, and the median is 7. What are the two missing data?

A. 1 and 7 **C.** 4 and 4

B. 2 and 6 **D.** 6 and 6

2. What is the area of the shaded region in the figure below? (Use 3.14 for π.)

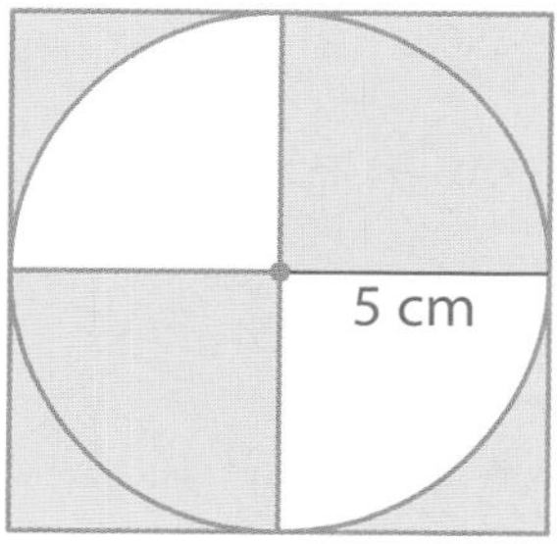

F. 21.5 cm^2 **H.** 80.375 cm^2

G. 60.75 cm^2 **I.** 84.3 cm^2

3. What is the solution to the proportion below?

$$\frac{8}{12} = \frac{x}{18}$$

4. You are building a scale model of a park that is planned for a city. The model uses the scale below.

1 centimeter = 2 meters

The park will have a rectangular reflecting pool with a length of 20 meters and a width of 12 meters. In your scale model, what will be the area of the reflecting pool?

A. 60 cm^2 **C.** 480 cm^2

B. 120 cm^2 **D.** 960 cm^2

5. In the figure, $\triangle EFG \sim \triangle HIJ$.

Which proportion is *not* necessarily correct for $\triangle EFG$ and $\triangle HIJ$?

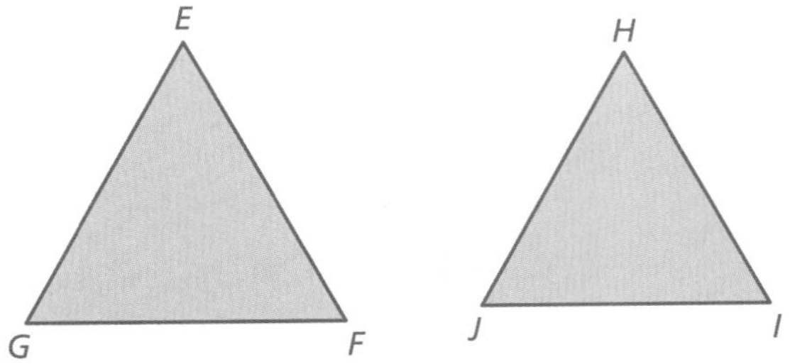

F. $\dfrac{EF}{FG} = \dfrac{HI}{IJ}$

G. $\dfrac{EG}{HI} = \dfrac{FG}{IJ}$

H. $\dfrac{GE}{EF} = \dfrac{JH}{HI}$

I. $\dfrac{EF}{HI} = \dfrac{GE}{JH}$

6. Brett was solving the equation in the box below.

$$\frac{c}{5} - (-15) = -35$$
$$\frac{c}{5} + 15 = -35$$
$$\frac{c}{5} + 15 - 15 = -35 - 15$$
$$\frac{c}{5} = -50$$
$$\frac{c}{5} = \frac{-50}{5}$$
$$c = -10$$

What should Brett do to correct the error that he made?

A. Subtract 15 from -35 to get -20.

B. Rewrite $\frac{c}{5} - (-15)$ as $\frac{c}{5} - 15$.

C. Multiply both sides of the equation by 5 to get $c = -250$.

D. Multiply both sides of the equation by -5 to get $c = 250$.

7. In the figure below, $\triangle ABC \sim \triangle DEF$.

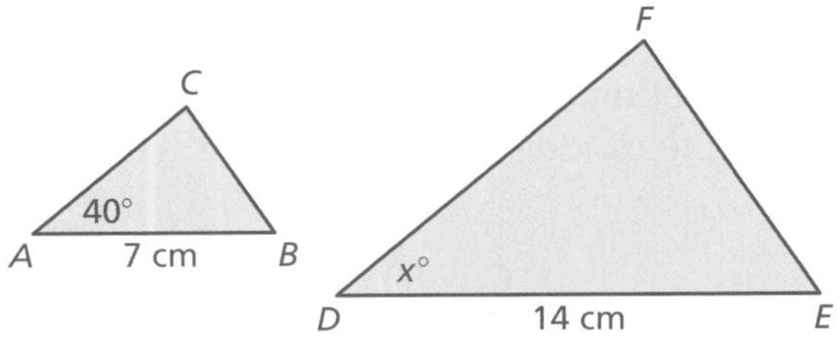

What is the value of x?

8. In the figure below, rectangle $EFGH \sim$ rectangle $IJKL$.

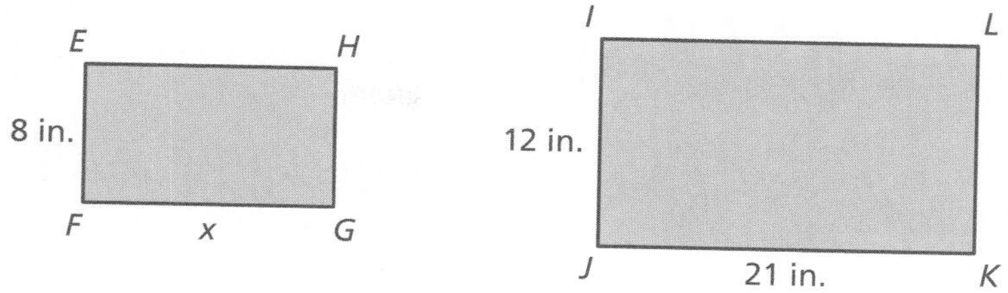

What is the value of x?

F. 14 in.

G. 15 in.

H. 16 in.

I. 17 in.

9. Two cubes are shown below.

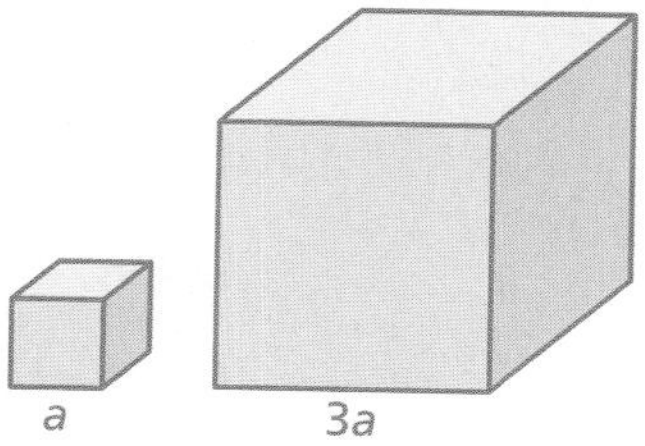

How many of the smaller cubes can be stacked to completely fill the larger cube?

A. 3

B. 9

C. 27

D. 54

10. A map of Donna's state has the following scale:

Think Solve Explain

$$\frac{1}{2} \text{ inch} = 10 \text{ miles}$$

Part A Donna measured the distance between her town and the state capitol on the map. Her measurement was $4\frac{1}{2}$ inches. Based on Donna's measurement, what is the actual distance, in miles, between her town and the state capitol? Show your work and explain your reasoning.

Part B Donna wants to mark her favorite campsite on the map. She knows that the campsite is 65 miles north of her town. What distance, in inches, on the map represents an actual distance of 65 miles? Show your work and explain your reasoning.

7 Transformations

"Just 2 more minutes. I'm almost done with my 'cat tessellation' painting."

"If you hold perfectly still..." "...each frame becomes a horizontal..." "...translation of the previous frame..."

What You Learned Before

Lines of Symmetry

How many lines of symmetry does the figure have?

Example 1

There is one way to fold the butterfly so that it coincides with itself.

∴ So, the butterfly has one line of symmetry.

Example 2

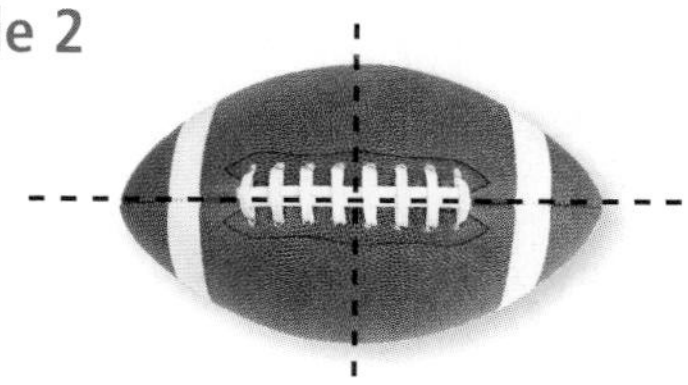

There are two ways to fold the football so that it coincides with itself.

∴ So, the football has two lines of symmetry.

Try It Yourself

How many lines of symmetry does the figure have?

1. 

2. 

3. 

Rotational Symmetry

Example 3 **Tell whether the figure has rotational symmetry.**

You can turn the figure 180° or less and produce an image that fits exactly on the original figure.

∴ So, the figure has rotational symmetry.

Try It Yourself

Tell whether the figure has rotational symmetry.

4. 

5. 

6. 

7.1 Translations

STANDARDS OF LEARNING
7.8

Essential Question How can you use translations to make a tessellation?

Share Your Work at... My.BigIdeasMath.com

When you slide a tile it is called a **translation**. When tiles can be used to cover a floor with no empty spaces, the collection of tiles is called a *tessellation*.

1 ACTIVITY: Describing Tessellations

Work with a partner. Can you make the pattern by using a translation of single tiles that are all of the same shape and design? If so, show how.

a. Sample:

Tile Pattern

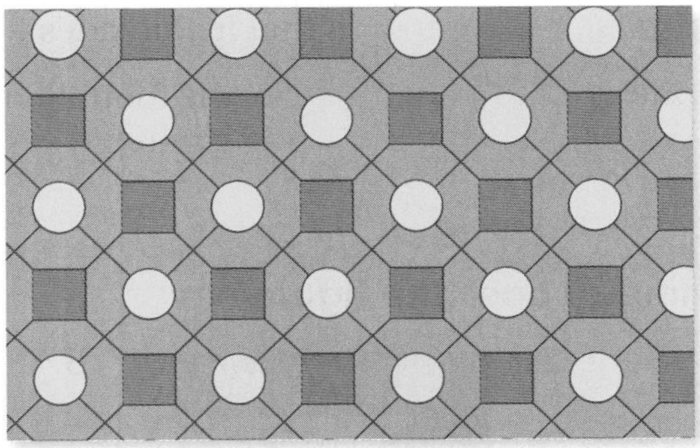

Single Tiles

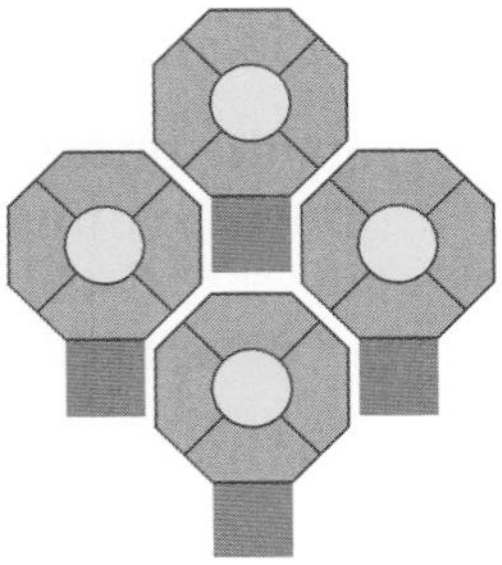

b.

c.

d.

e.

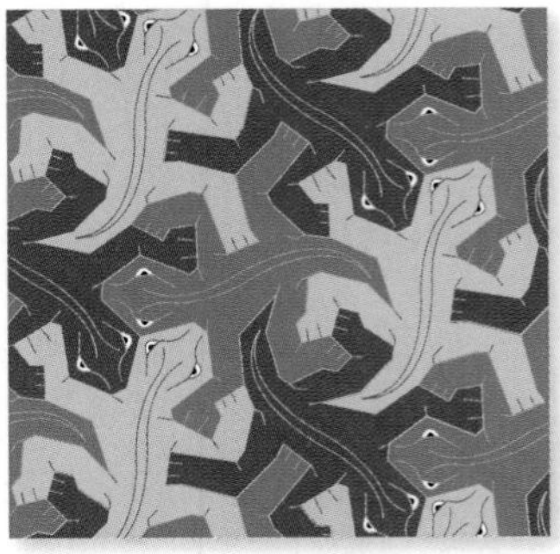

2 ACTIVITY: Tessellations and Basic Shapes

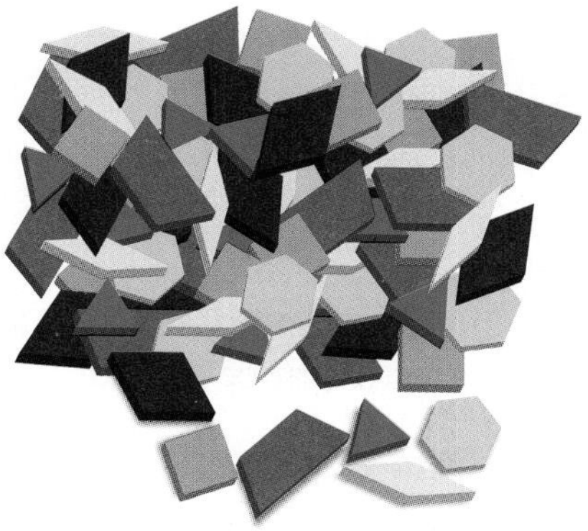

Work with a partner.

a. Which pattern blocks can you use to make a tessellation?

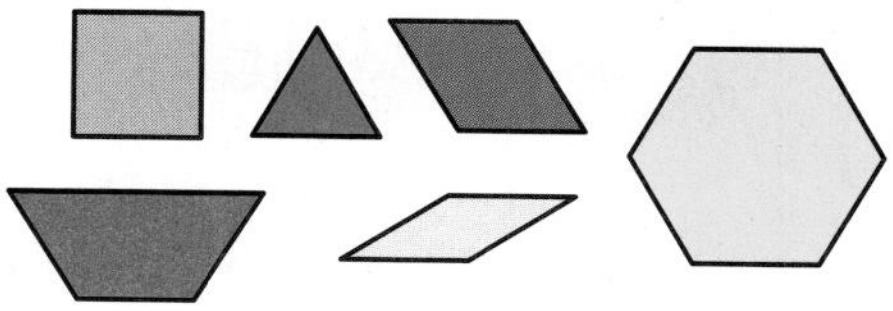

b. For each one that works, draw the tessellation.

c. Can you make the tessellation using only translation, or do you have to rotate or flip the pattern blocks?

3 ACTIVITY: Designing Tessellations

Work with a partner. Design your own tessellation. Use one of the basic shapes from Activity 2.

Sample:

Start with a square.

Cut a design out of one side.

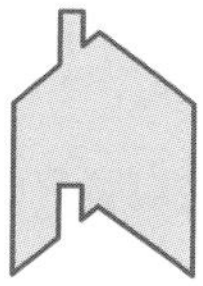

Tape it to the other side to make your pattern.

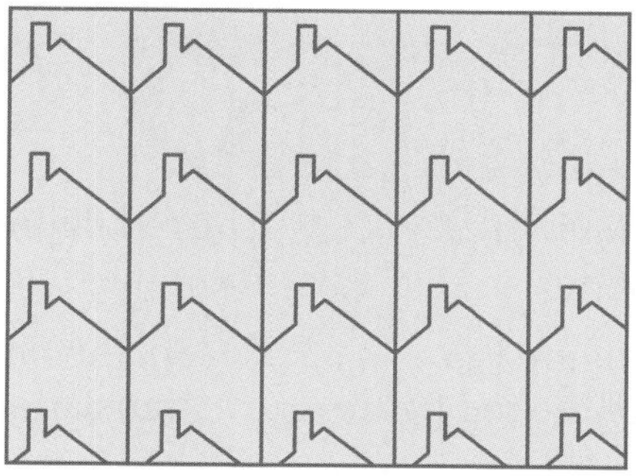

Use the pattern and translations to make your tessellation.

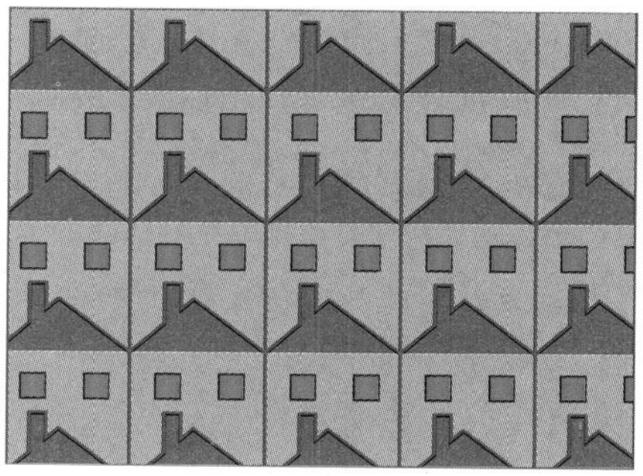

Color the tessellation.

What Is Your Answer?

4. **IN YOUR OWN WORDS** How can you use translations to make a tessellation? Give an example.

5. Draw any parallelogram. Does it tessellate? Is it true that any parallelogram can be translated to make a tessellation? Explain why.

Practice Use what you learned about translations to complete Exercises 4–6 on page 286.

7.1 Lesson

Check It Out
Lesson Tutorials
BigIdeasMath.com

Key Vocabulary
transformation, *p. 284*
image, *p. 284*
translation, *p. 284*

A **transformation** changes a figure into another figure. The new figure is called the **image**.

Key Idea

Translations

A **translation** is a transformation in which a figure *slides* but does not turn. Every point of the figure moves the same distance and in the same direction.

The original figure and its image have the same size and shape.

EXAMPLE 1 Identifying a Translation

Tell whether the blue figure is a translation of the red figure.

a.

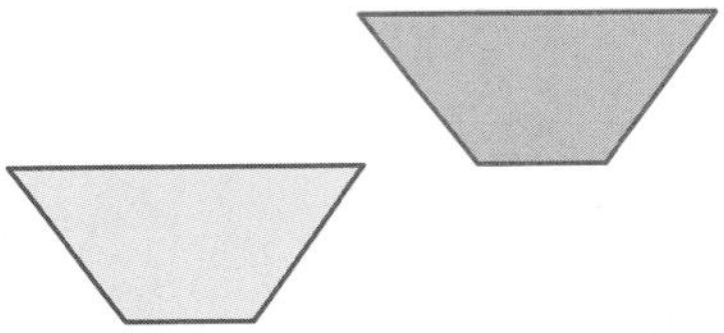

The red figure *slides* to form the blue figure.

∴ So, the blue figure is a translation of the red figure.

b.

The red figure *turns* to form the blue figure.

∴ So, the blue figure is *not* a translation of the red figure.

On Your Own

Now You're Ready
Exercises 4–9

Tell whether the blue figure is a translation of the red figure. Explain.

1.

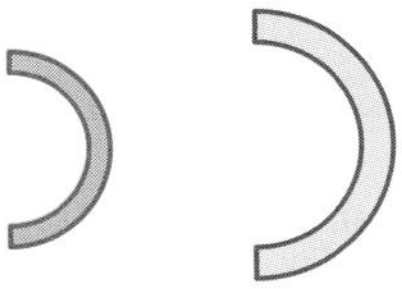

2.

3.

4.

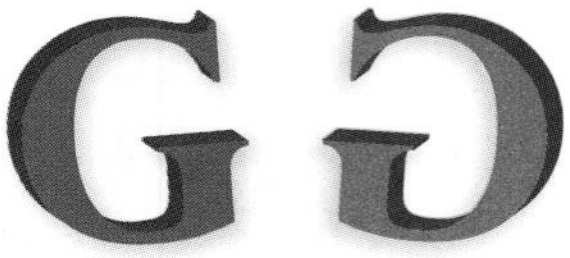

Multi-Language Glossary at BigIdeasMath.com.

EXAMPLE 2 Translating a Figure

Translate the red triangle 3 units right and 3 units down. What are the coordinates of the image?

> **Reading**
>
> A' is read "A prime." Use *prime* symbols when naming an image.
>
> $A \longrightarrow A'$
>
> $B \longrightarrow B'$
>
> $C \longrightarrow C'$

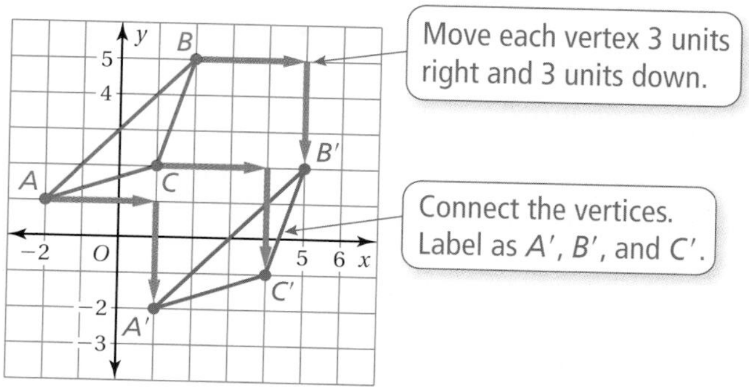

The coordinates of the image are $A'(1, -2)$, $B'(5, 2)$, and $C'(4, -1)$.

On Your Own

Now You're Ready
Exercises 10 and 11

5. The red triangle is translated 4 units left and 2 units up. What are the coordinates of the image?

EXAMPLE 3 Translating a Figure

The vertices of a square are $A(1, -2)$, $B(3, -2)$, $C(3, -4)$, and $D(1, -4)$. Draw the figure and its image after a translation 4 units left and 6 units up.

Subtract 4 from each x-coordinate.

Add 6 to each y-coordinate.

Vertices of $ABCD$	$(x - 4, y + 6)$	Vertices of $A'B'C'D'$
$A(1, -2)$	$(1 - 4, -2 + 6)$	$A'(-3, 4)$
$B(3, -2)$	$(3 - 4, -2 + 6)$	$B'(-1, 4)$
$C(3, -4)$	$(3 - 4, -4 + 6)$	$C'(-1, 2)$
$D(1, -4)$	$(1 - 4, -4 + 6)$	$D'(-3, 2)$

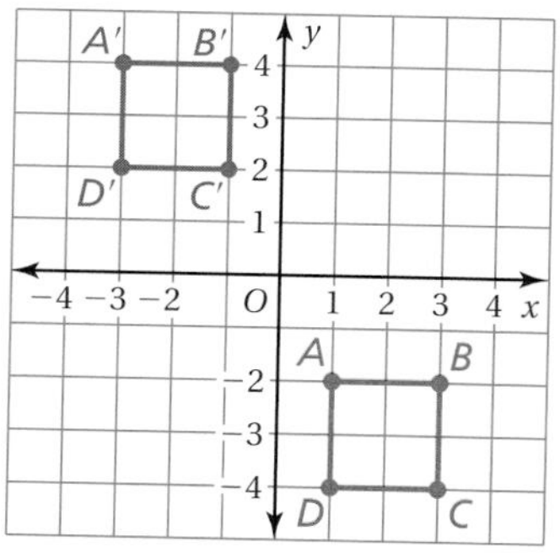

The figure and its image are shown at the right.

On Your Own

Now You're Ready
Exercises 12–15

6. The vertices of a triangle are $A(-2, -2)$, $B(0, 2)$, and $C(3, 0)$. Draw the figure and its image after a translation 1 unit left and 2 units up.

7.1 Exercises

Check It Out
Help with Homework
BigIdeasMath.com

Vocabulary and Concept Check

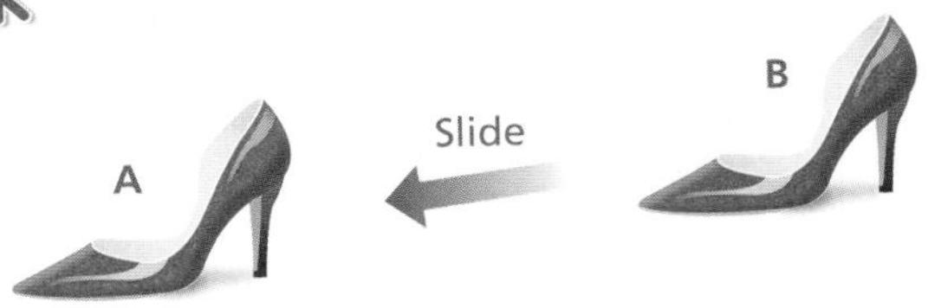

1. **VOCABULARY** Which figure is the image?
2. **VOCABULARY** How do you translate a figure in a coordinate plane?
3. **CRITICAL THINKING** Can you translate the letters in the word TOKYO to form the word KYOTO? Explain.

Practice and Problem Solving

Tell whether the blue figure is a translation of the red figure.

(1) **4.**

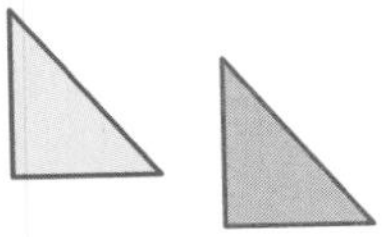

5.

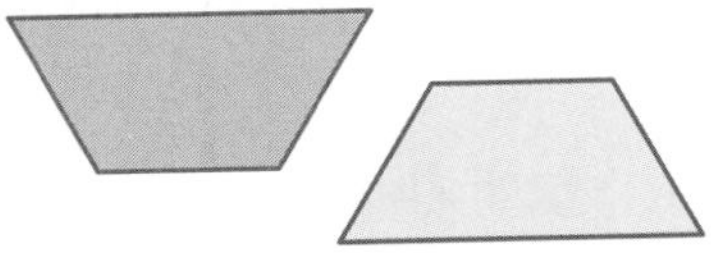

6.

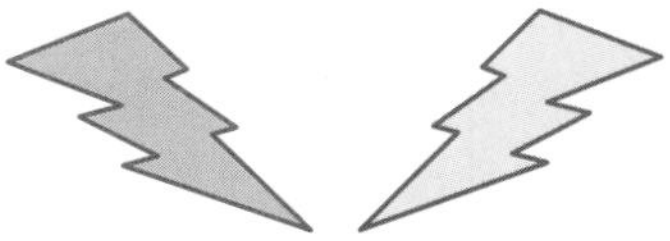

7.

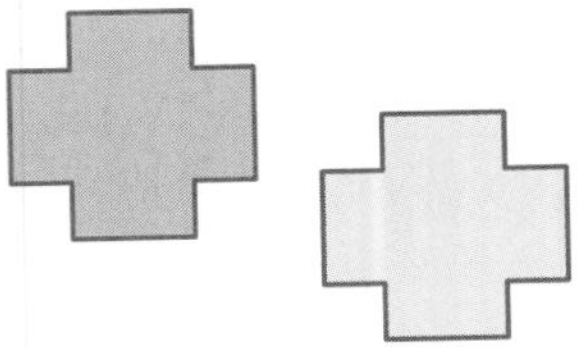

8.

9.

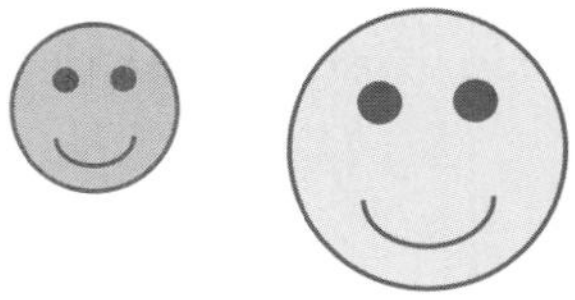

(2) **10.** Translate the triangle 4 units right and 3 units down. What are the coordinates of the image?

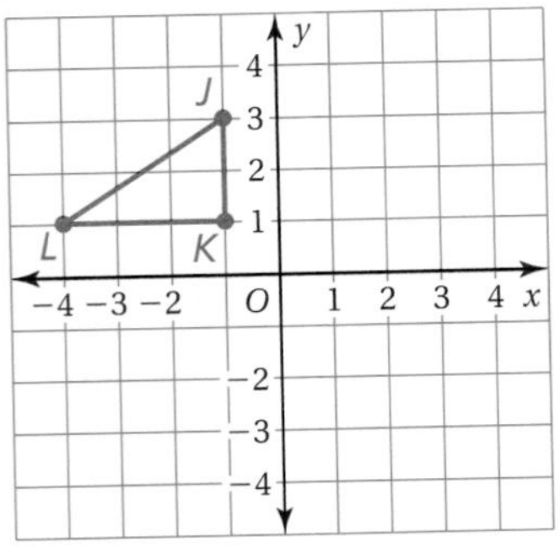

11. Translate the figure 2 units left and 4 units down. What are the coordinates of the image?

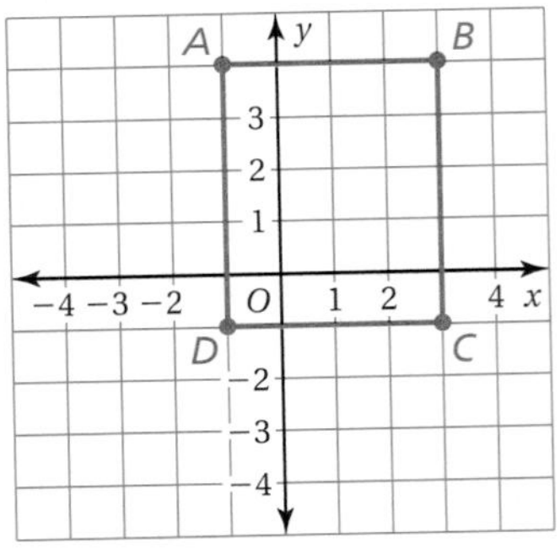

The vertices of a triangle are $L(0, 1)$, $M(1, -2)$, and $N(-2, 1)$. Draw the figure and its image after the translation.

(3) **12.** 6 units up

13. 5 units right

14. 2 units right and 3 units up

15. 3 units left and 4 units down

16. ICONS You can click and drag an icon on a computer screen. Is this an example of a translation? Explain.

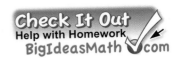

Describe the translation of the point to its image.

17. $(3, -2) \longrightarrow (1, 0)$

18. $(-8, -4) \longrightarrow (-3, 5)$

Describe the translation from the red figure to the blue figure.

19.

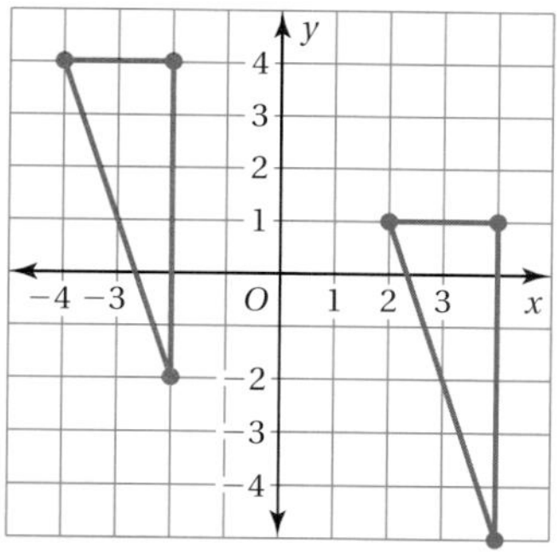

20.

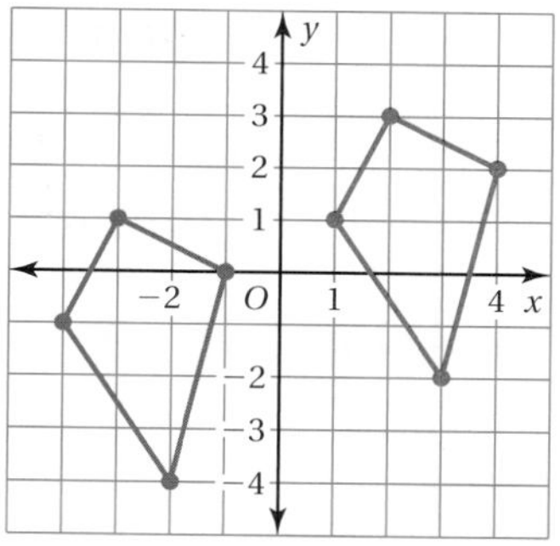

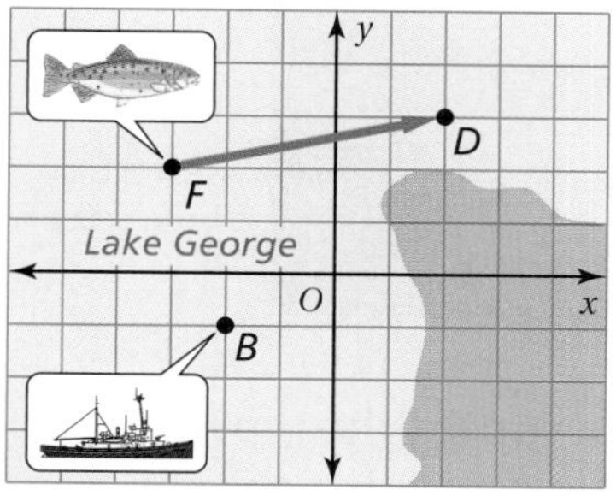

21. **FISHING** A school of fish translates from point F to point D.

a. Describe the translation of the school of fish.

b. Can the fishing boat make the same translation? Explain.

c. Describe a translation the fishing boat could make to get to point D.

22. **REASONING** A triangle is translated 5 units right and 2 units up. Then the image is translated 3 units left and 8 units down. Write a translation of the original triangle to the ending position.

23. Critical Thinking In chess, a knight can move only in an L-shape pattern:

- *two* vertical squares then *one* horizontal square;
- *two* horizontal squares then *one* vertical square;
- *one* vertical square then *two* horizontal squares; or
- *one* horizontal square then *two* vertical squares.

Write a series of translations to move the knight from g8 to g5.

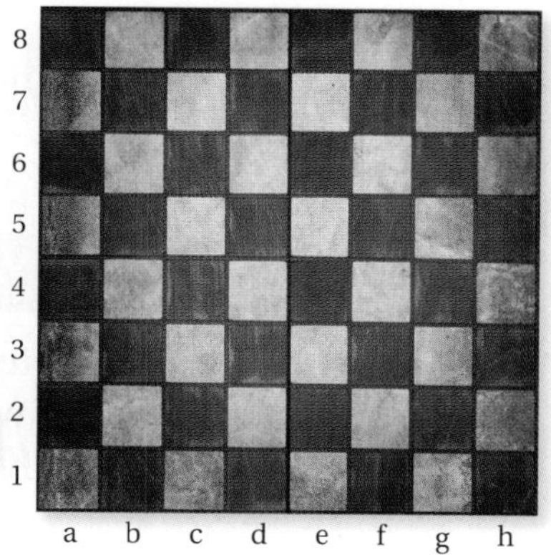

Fair Game Review What you learned in previous grades & lessons

Tell whether each figure can be folded in half so that one side matches the other. *(Skills Review Handbook)*

24.

25.

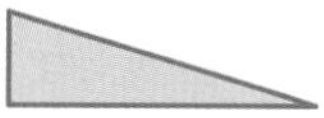

26.

27.

28. **MULTIPLE CHOICE** A 5-pound bag of potatoes costs \$2.95. Which equation can be used to find the cost c per pound? *(Section 2.2)*

Ⓐ $2.95c = 5$ Ⓑ $5c = 2.95$ Ⓒ $c \div 5 = 2.95$ Ⓓ $5 \div c = 2.95$

7.2 Reflections

STANDARDS OF LEARNING
7.8

Share Your Work at… My.BigIdeasMath.com

Essential Question

How can you use reflections to classify a frieze pattern?

The Meaning of a Word ● Reflection

When you look at a mountain by a lake, you can see the **reflection**, or mirror image, of the mountain in the lake.

If you fold the photo on its axis, the mountain and its reflection will align.

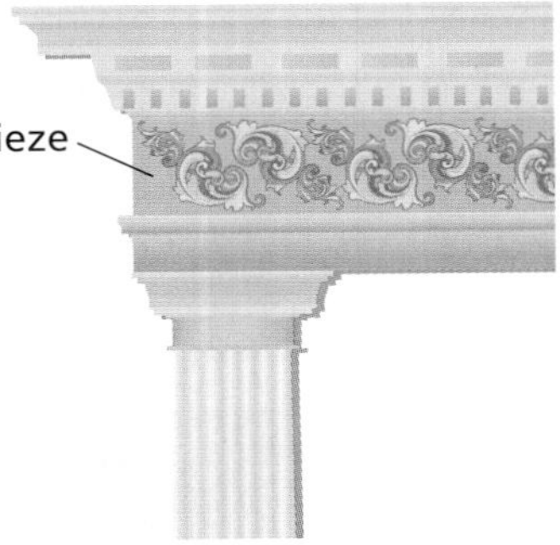

A *frieze* is a horizontal band that runs at the top of a building. A frieze is often decorated with a design that repeats.

- All frieze patterns are translations of themselves.
- Some frieze patterns are reflections of themselves.

1 EXAMPLE: Frieze Patterns

Is the frieze pattern a reflection of itself when folded horizontally? vertically?

- Fold (reflect) on horizontal axis. The pattern coincides.

- Fold (reflect) on vertical axis. The pattern coincides.

∴ This frieze pattern is a reflection of itself when folded horizontally *and* vertically.

2 ACTIVITY: Frieze Patterns and Reflections

Work with a partner. Is the frieze pattern a reflection of itself when folded *horizontally*, *vertically*, or *neither*?

a.

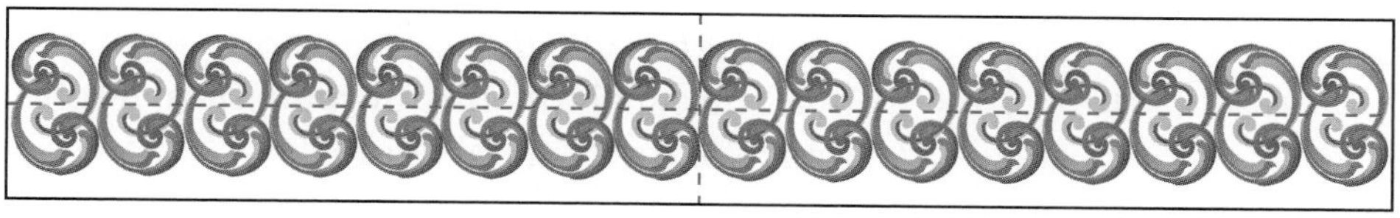

b.

c.

d.

e.

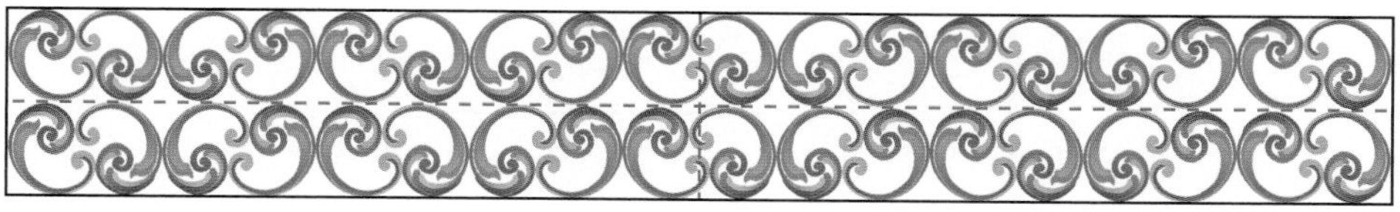

f.

What Is Your Answer?

3. Draw a frieze pattern that is a reflection of itself when folded horizontally.
4. Draw a frieze pattern that is a reflection of itself when folded vertically.
5. Draw a frieze pattern that is not a reflection of itself when folded horizontally or vertically.
6. **IN YOUR OWN WORDS** How can you use reflections to classify a frieze pattern?

Practice Use what you learned about reflections to complete Exercises 4–6 on page 292.

7.2 Lesson

Check It Out
Lesson Tutorials
BigIdeasMath.com

Key Vocabulary
reflection, *p. 290*
line of reflection, *p. 290*

Key Idea

Reflections

A **reflection**, or *flip*, is a transformation in which a figure is reflected in a line called the **line of reflection**. A reflection creates a mirror image of the original figure.

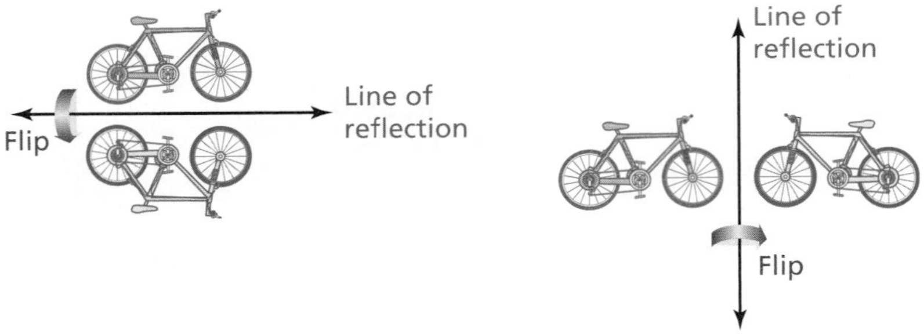

The original figure and its image have the same size and shape.

EXAMPLE 1 Identifying a Reflection

Tell whether the blue figure is a reflection of the red figure.

a.

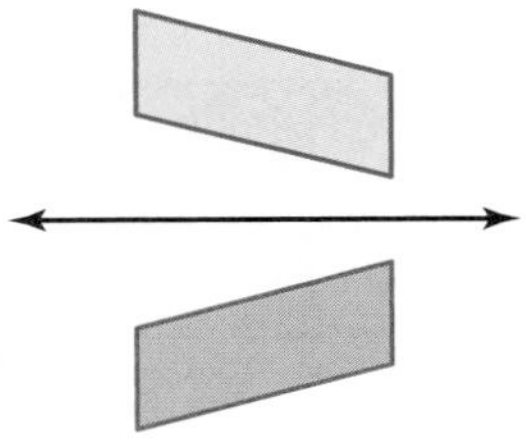

The red figure can be *flipped* to form the blue figure.

∴ So, the blue figure is a reflection of the red figure.

b.

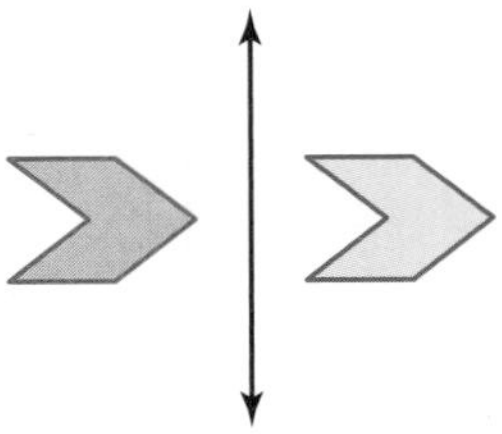

If the red figure were *flipped*, it would point to the left.

∴ So, the blue figure is *not* a reflection of the red figure.

On Your Own

Now You're Ready
Exercises 4–9

Tell whether the blue figure is a reflection of the red figure. Explain.

1.

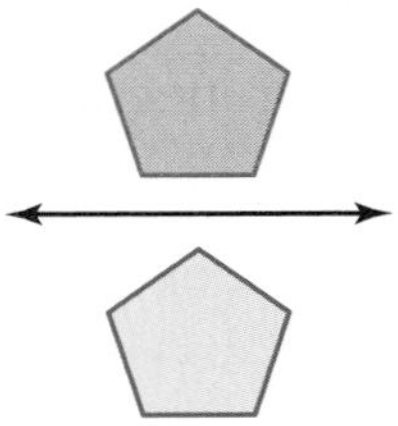

2.

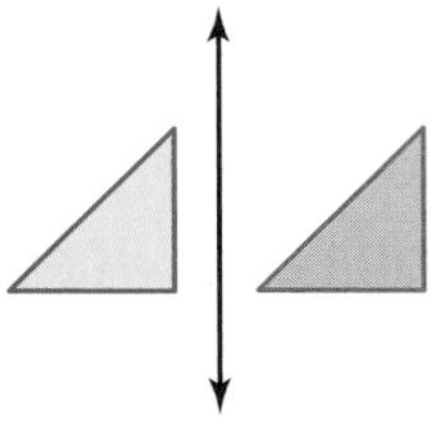

3.

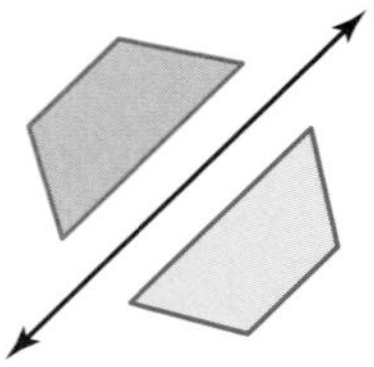

Multi-Language Glossary at BigIdeasMath.com.

EXAMPLE 2 Reflecting a Figure in the x-axis

The vertices of a triangle are $A(-1, 1)$, $B(-1, 3)$, and $C(6, 3)$. Draw this triangle and its reflection in the x-axis. What are the coordinates of the image?

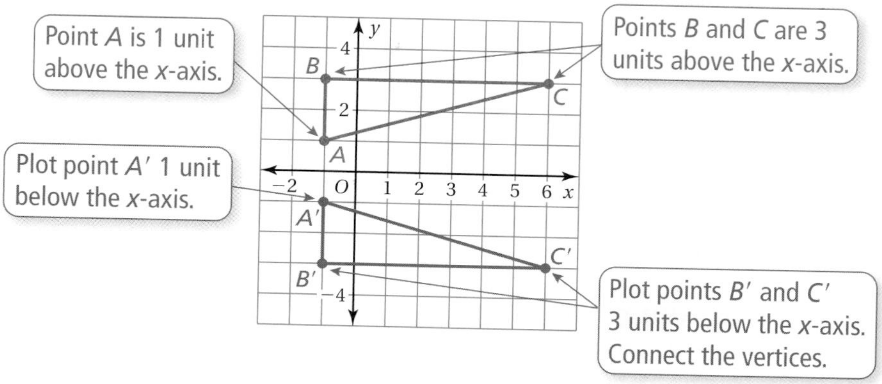

Point A is 1 unit above the x-axis.
Plot point A′ 1 unit below the x-axis.
Points B and C are 3 units above the x-axis.
Plot points B′ and C′ 3 units below the x-axis. Connect the vertices.

∴ The coordinates of the image are $A'(-1, -1)$, $B'(-1, -3)$, and $C'(6, -3)$.

EXAMPLE 3 Reflecting a Figure in the y-axis

The vertices of a quadrilateral are $P(-2, 5)$, $Q(-1, -1)$, $R(-4, 2)$, and $S(-4, 4)$. Draw this quadrilateral and its reflection in the y-axis. What are the coordinates of the image?

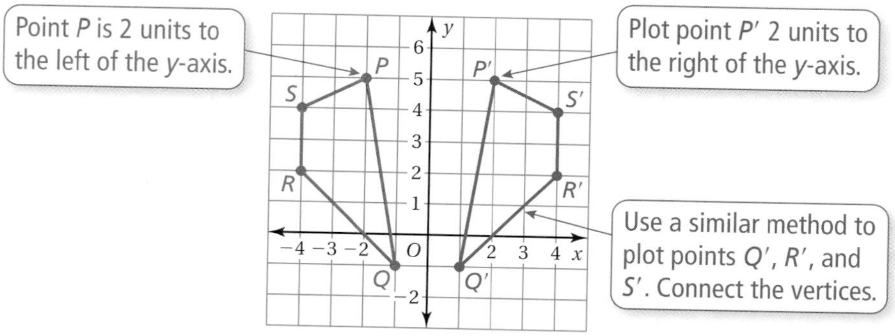

Point P is 2 units to the left of the y-axis.
Plot point P′ 2 units to the right of the y-axis.
Use a similar method to plot points Q′, R′, and S′. Connect the vertices.

∴ The coordinates of the image are $P'(2, 5)$, $Q'(1, -1)$, $R'(4, 2)$, and $S'(4, 4)$.

On Your Own

Now You're Ready
Exercises 10–17

4. The vertices of a rectangle are $A(-4, -3)$, $B(-4, -1)$, $C(-1, -1)$, and $D(-1, -3)$.
   a. Draw the rectangle and its reflection in the x-axis.
   b. Draw the rectangle and its reflection in the y-axis.
   c. Are the images in parts (a) and (b) the same size and shape? Explain.

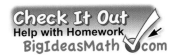

7.2 Exercises

Check It Out
Help with Homework
BigIdeasMath.com

Vocabulary and Concept Check

1. **WHICH ONE DOESN'T BELONG?** Which transformation does *not* belong with the other three? Explain your reasoning.

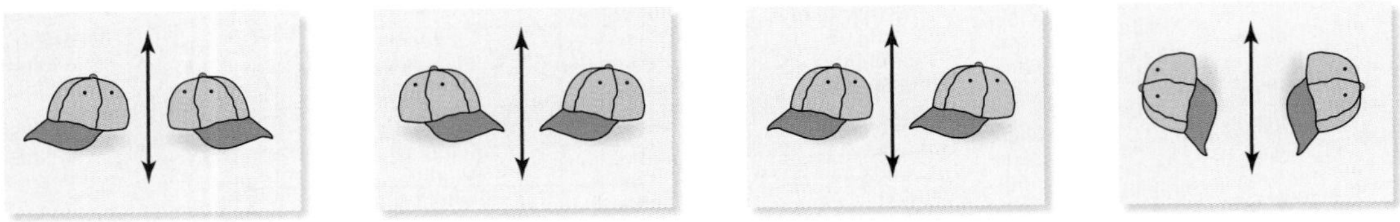

2. **WRITING** How can you tell when one figure is a reflection of another figure?

3. **REASONING** A figure lies entirely in Quadrant I. The figure is reflected in the x-axis. In which quadrant is the image?

Practice and Problem Solving

Tell whether the blue figure is a reflection of the red figure.

(1) 4.

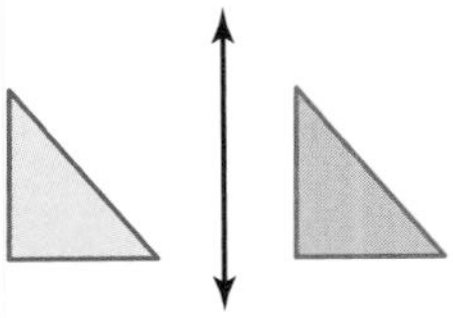

5.

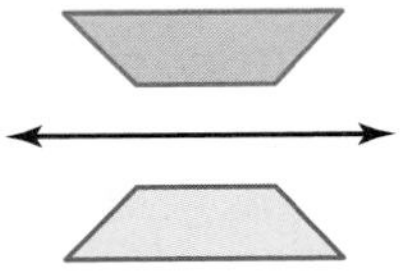

6.

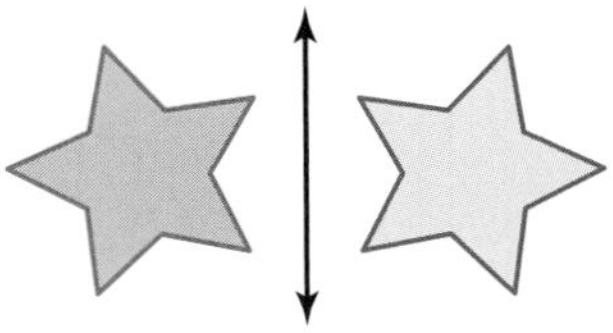

7.

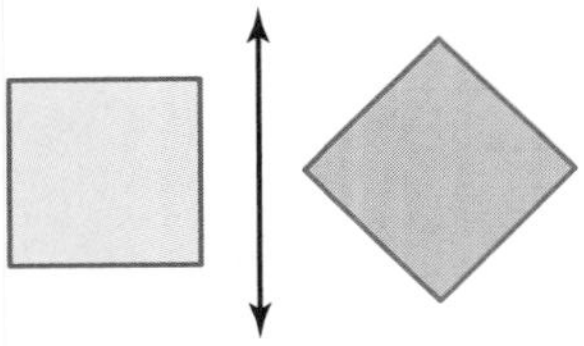

8.

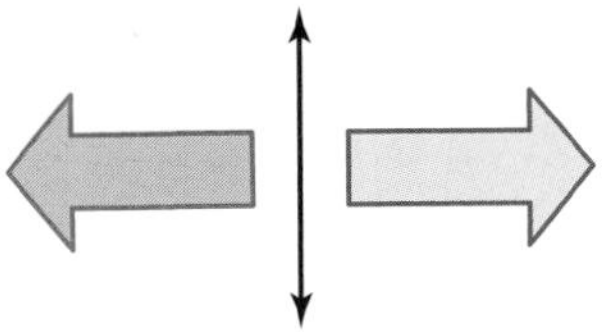

9.

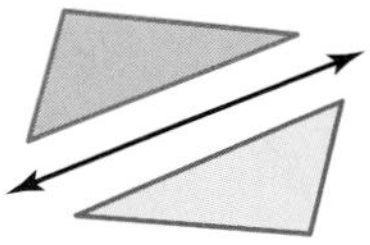

Find the coordinates of the figure after reflecting in the x-axis.

(2) 10. $A(1, 4)$, $B(4, 4)$, $C(1, 3)$

11. $M(-2, 1)$, $N(0, 3)$, $P(2, 2)$

12. $H(2, -2)$, $J(4, -1)$, $K(6, -3)$, $L(5, -4)$

13. $D(-2, -1)$, $E(0, -1)$, $F(-2, -5)$, $G(0, -5)$

Find the coordinates of the figure after reflecting in the y-axis.

(3) 14. $Q(-4, 2)$, $R(-2, 4)$, $S(-1, 1)$

15. $T(4, -2)$, $U(4, 2)$, $V(6, -2)$

16. $W(2, -1)$, $X(5, -1)$, $Y(5, -4)$, $Z(2, -4)$

17. $J(2, 2)$, $K(7, 4)$, $L(9, -2)$, $M(3, -1)$

18. **ALPHABET** Which letters look the same when reflected in the line?

A B C D E F G H I J K L M N O P Q R S T U V W X Y Z

The coordinates of a point and its image are given. Is the reflection in the *x-axis* or *y-axis*?

19. $(2, -2) \longrightarrow (2, 2)$

20. $(-4, 1) \longrightarrow (4, 1)$

21. $(-2, -5) \longrightarrow (2, -5)$

22. $(-3, -4) \longrightarrow (-3, 4)$

23. Translate the triangle 1 unit right and 5 units down. Then reflect the image in the y-axis.

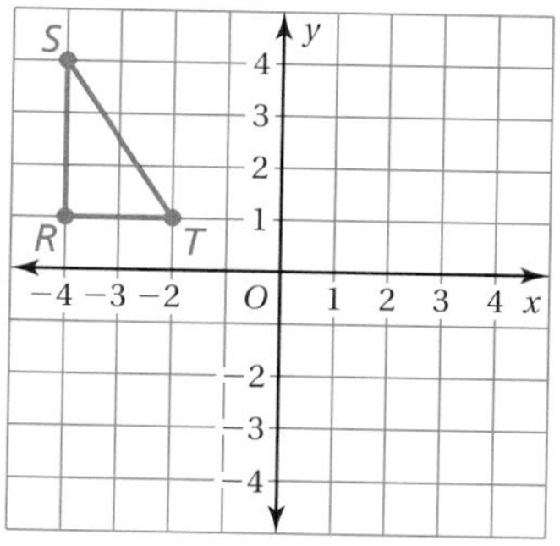

24. PROJECT Use a computer drawing program to create photographs of people by copying one side of the person's face and reflecting it in a vertical line. Does the person look normal or very different?

25. MIRROR IMAGE One of the faces shown is an exact reflection of itself. Which one is it? How can you tell?

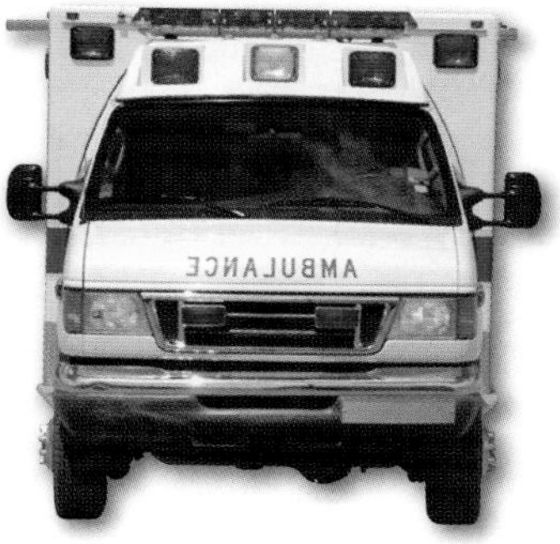

26. EMERGENCY VEHICLE Hold a mirror to the left side of the photo of the vehicle.

a. What word do you see in the mirror?

b. Why do you think it is written that way on the front of the vehicle?

27. Critical Thinking Reflect the triangle in the line $y = x$. How are the x- and y-coordinates of the image related to the x- and y-coordinates of the original triangle?

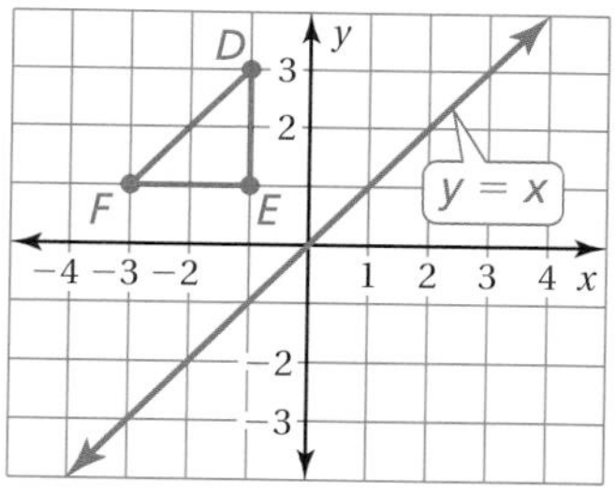

Fair Game Review What you learned in previous grades & lessons

Classify the angle as *acute, right, obtuse,* or *straight*. *(Skills Review Handbook)*

28. 

29. 

30. 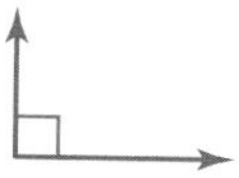

31. 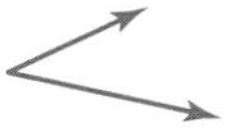

32. MULTIPLE CHOICE Which ratios form a proportion? *(Section 5.3)*

Ⓐ $\frac{1}{7}, \frac{3}{14}$ Ⓑ $\frac{2}{3}, \frac{8}{12}$ Ⓒ $\frac{21}{7}, \frac{63}{48}$ Ⓓ $\frac{18}{10}, \frac{24}{15}$

7 Study Help

Check It Out
Graphic Organizer
BigIdeasMath.com

You can use a **summary triangle** to explain a concept. Here is an example of a summary triangle for translating a figure.

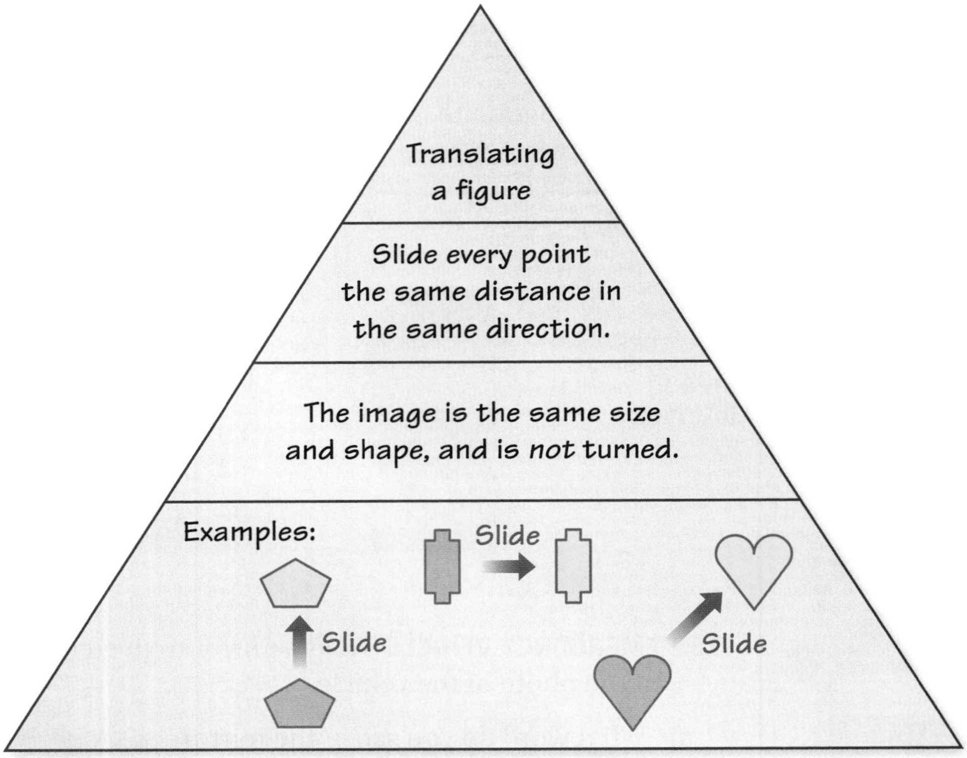

On Your Own

Make a summary triangle to help you study this topic.

1. reflecting a figure

After you complete this chapter, make summary triangles for the following topics.

2. rotating a figure
3. dilating a figure
4. transforming a figure

"I hope my owner sees my summary triangle. I just can't seem to learn 'roll over'."

7.1–7.2 Quiz

Check It Out
Progress Check
BigIdeasMath.com

Tell whether the blue figure is a translation of the red figure. *(Section 7.1)*

1.

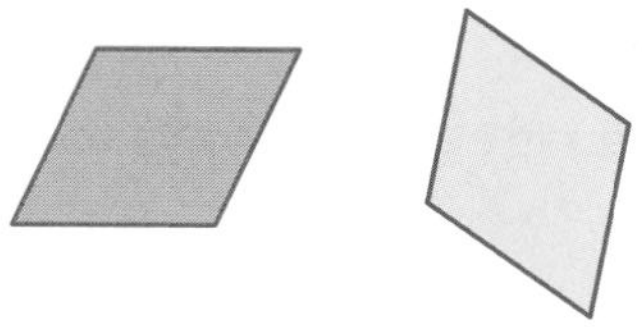

2.

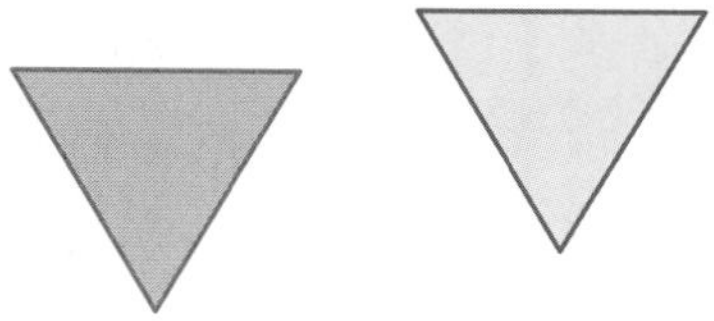

3. Translate the triangle 2 units left and 3 units down. What are the coordinates of the image? *(Section 7.1)*

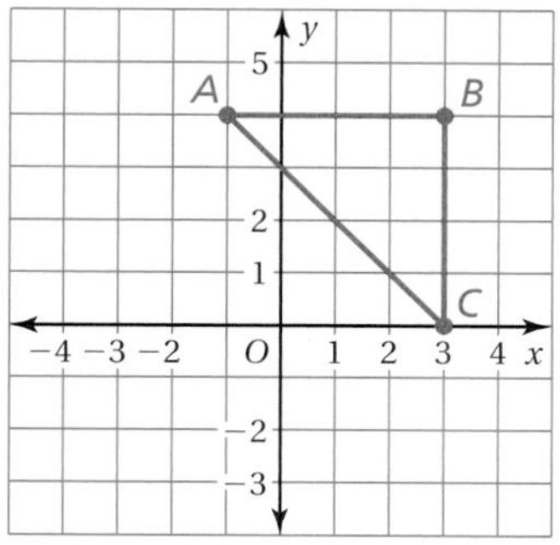

4. Translate the figure 2 units right and 4 units down. What are the coordinates of the image? *(Section 7.1)*

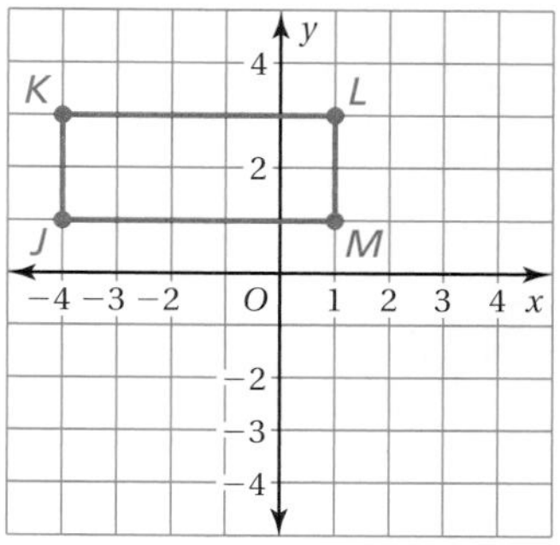

Tell whether the blue figure is a reflection of the red figure. *(Section 7.2)*

5.

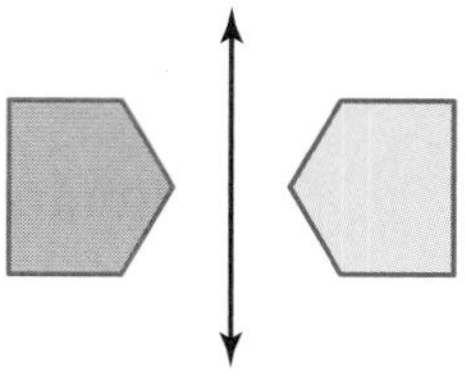

6.

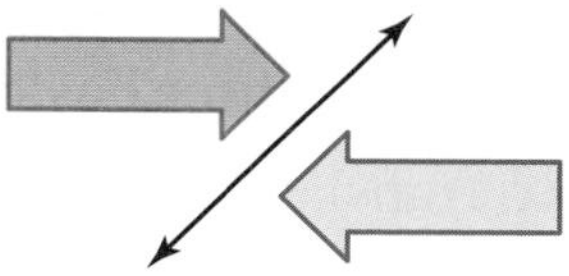

Find the coordinates of the figure after reflecting in the (a) x-axis and (b) y-axis. *(Section 7.2)*

7. $A(2, 0)$, $B(1, 5)$, $C(4, 3)$

8. $D(-5, -5)$, $E(-5, -1)$, $F(-2, -2)$, $G(-2, -5)$

9. AIRPLANE Describe a translation of the airplane from point A to point B. *(Section 7.1)*

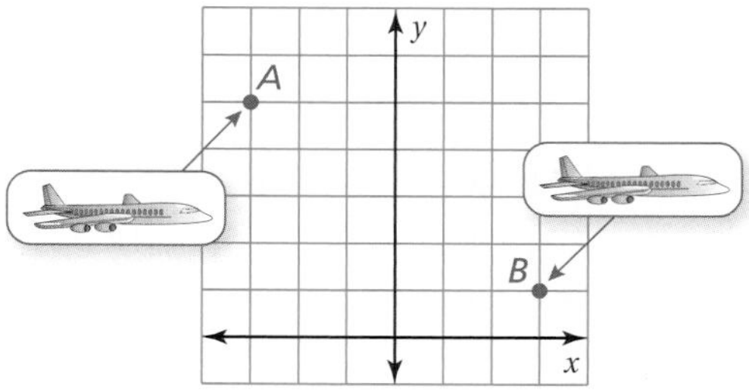

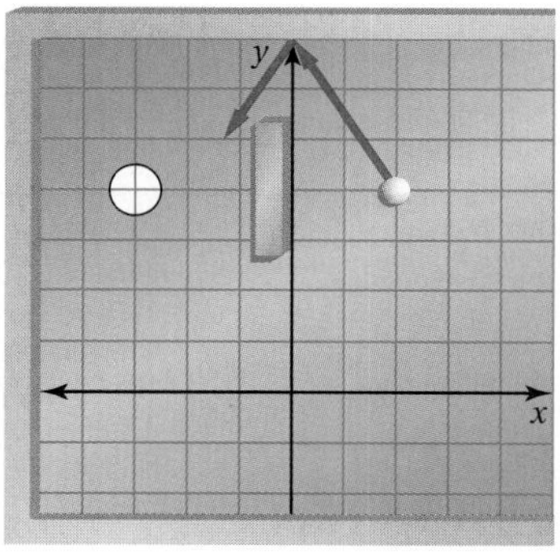

10. MINI GOLF You hit the golf ball along the red path so that its image will be a reflection in the y-axis. Does the golf ball land in the hole? Explain. *(Section 7.2)*

7.3 Rotations

STANDARDS OF LEARNING
7.8

Essential Question What are the three basic ways to move an object in a plane?

The Meaning of a Word ● Rotate

A bicycle wheel can **rotate** clockwise or counterclockwise.

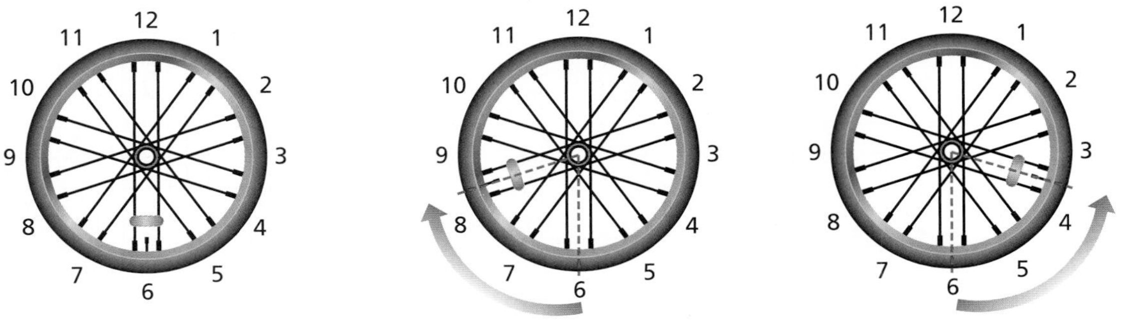

1 ACTIVITY: Three Basic Ways to Move Things

There are three basic ways to move objects on a flat surface.

1. Translate the object. **2.** Reflect the object. **3.** Rotate the object.

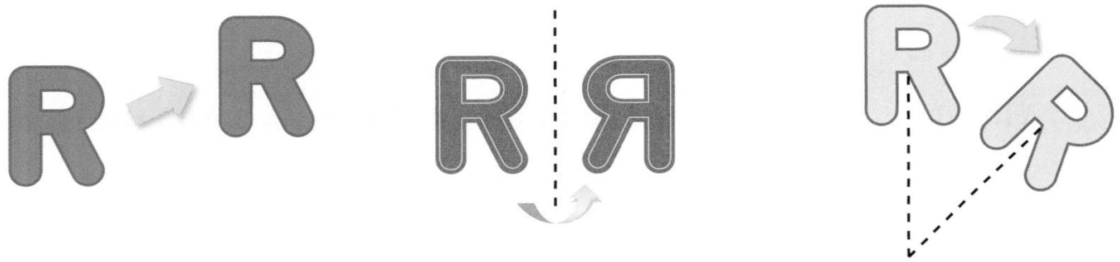

Work with a partner.

- Cut out a paper triangle that is the same size as the blue triangle shown.
- Decide how you can move the blue triangle to make each red triangle.
- Is each move a *translation*, a *reflection*, or a *rotation*?
- Draw four other red triangles in a coordinate plane. Describe how you can move the blue triangle to make each red triangle.

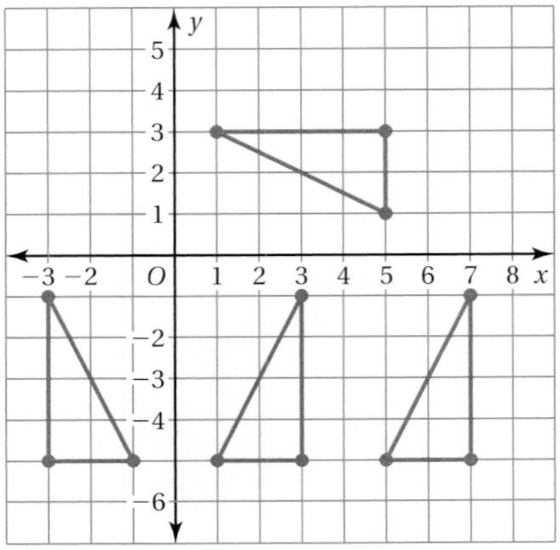

2 ACTIVITY: Tessellating a Plane

Work with a partner.

a. Describe how the figure labeled 1 in each diagram can be moved to make the other figures.

Triangles

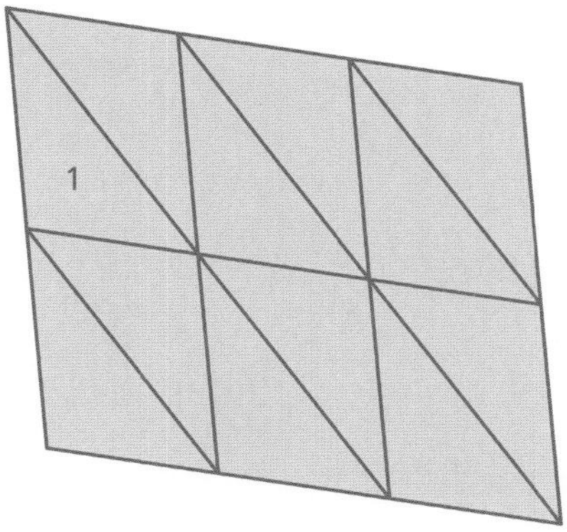

Quadrilaterals

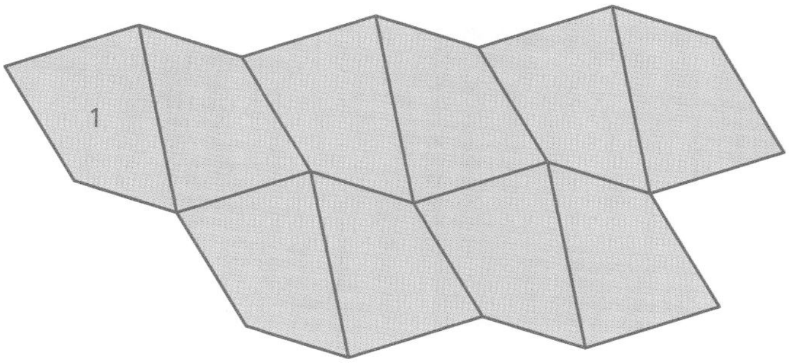

b. **EXPERIMENT** Will *any* triangle tessellate? Conduct an experiment to gather information to help form your conclusion. Draw a triangle. Cut it out. Then use it to trace other triangles so that you cover the plane with triangles that are all the same shape.

c. **EXPERIMENT** Will *any* quadrilateral tessellate? Conduct an experiment to gather information to help form your conclusion. Draw a quadrilateral. Cut it out. Then use it to trace other quadrilaterals so that you cover the plane with quadrilaterals that are all the same shape.

What Is Your Answer?

3. **IN YOUR OWN WORDS** What are the three basic ways to move an object in a plane? Draw an example of each.

"Dear Sub Shop: Why do you put the cheese on the subs so some parts have double coverage and some have none?"

"My suggestion is that you use the tessellation property of triangles for even cheese coverage."

Practice Use what you learned about rotations to complete Exercises 7–9 on page 300.

Key Vocabulary

rotation, *p. 298*
center of rotation, *p. 298*
angle of rotation, *p. 298*

Key Idea

Rotations

A **rotation**, or *turn*, is a transformation in which a figure is rotated about a point called the **center of rotation**. The number of degrees a figure rotates is the **angle of rotation**.

The original figure and its image have the same size and shape.

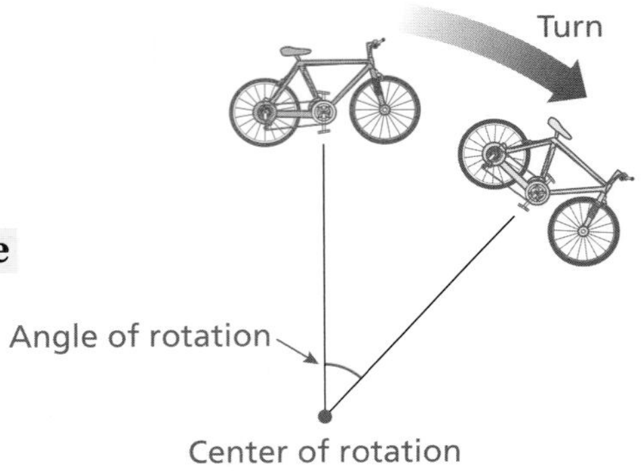

EXAMPLE 1 Standardized Test Practice

You must rotate the puzzle piece 270° clockwise about point *P* to fit it into a puzzle. Which piece fits in the puzzle as shown?

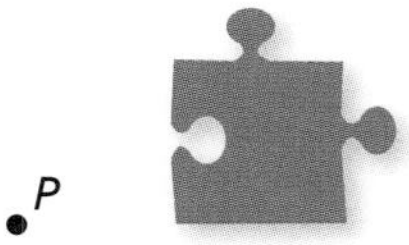

Ⓐ  Ⓑ  Ⓒ  Ⓓ 

Rotate the puzzle piece 270° clockwise about point *P*.

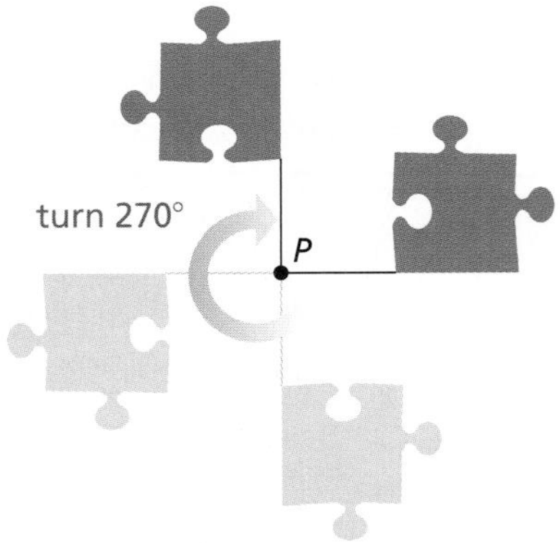

∴ The correct answer is Ⓒ.

Study Tip

When rotating figures, it may help to sketch the rotation in several steps, as shown in Example 1.

On Your Own

Now You're Ready
Exercises 7–12

1. Which piece is a 90° counterclockwise rotation about point *P*?
2. Is choice D a rotation of the original puzzle piece? If not, what kind of transformation does the image show?

EXAMPLE 2 Rotating a Figure

The vertices of a triangle are $X(-4, 2)$, $Y(-1, 4)$, and $Z(-1, 2)$. Rotate the triangle 180° clockwise about the origin. What are the coordinates of the image?

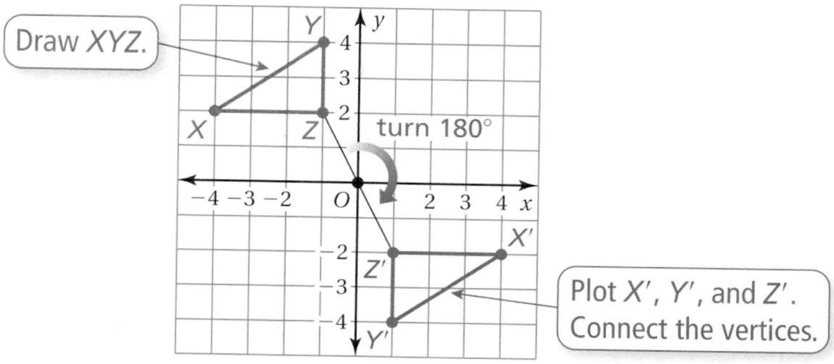

Draw XYZ.
turn 180°
Plot X′, Y′, and Z′. Connect the vertices.

The coordinates of the image are $X'(4, -2)$, $Y'(1, -4)$, and $Z'(1, -2)$.

EXAMPLE 3 Rotating a Figure

The vertices of a rectangle are $J(0, 0)$, $K(0, 5)$, $L(2, 5)$, and $M(2, 0)$. Rotate the rectangle 90° counterclockwise about the origin. What are the coordinates of the image?

Common Error

Be sure to pay attention to whether a rotation is clockwise or counterclockwise.

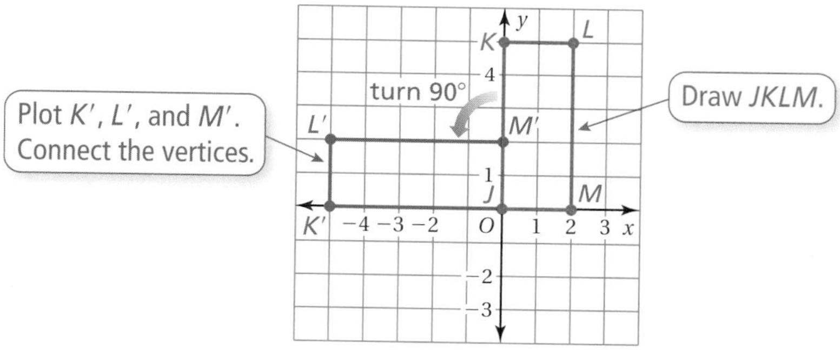

Plot K′, L′, and M′. Connect the vertices.
turn 90°
Draw JKLM.

The coordinates of the image are $J'(0, 0)$, $K'(-5, 0)$, $L'(-5, 2)$, and $M'(0, 2)$.

On Your Own

Now You're Ready
Exercises 13–16

3. A triangle has vertices $Q(4, 5)$, $R(4, 0)$, and $S(1, 0)$.

   a. Rotate the triangle 90° counterclockwise about the origin.

   b. Rotate the triangle 180° about the origin.

   c. Are the images in parts (a) and (b) the same size and shape? Explain.

7.3 Exercises

Check It Out
Help with Homework
BigIdeasMath.com

Vocabulary and Concept Check

1. **VOCABULARY** Identify the transformation shown.

a.
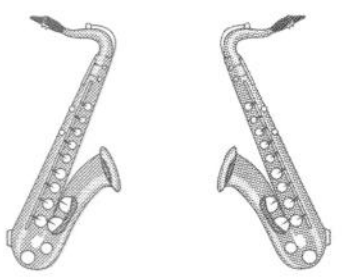

b.
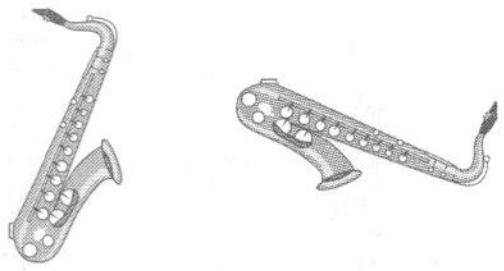

c.
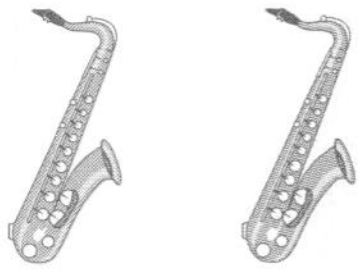

2. **VOCABULARY** What counterclockwise angle of rotation is equivalent to a 90° clockwise angle of rotation?

MENTAL MATH **A figure lies entirely in Quadrant II. In which quadrant will the figure lie after the given clockwise rotation about the origin?**

3. 90°
4. 180°
5. 270°
6. 360°

Practice and Problem Solving

Tell whether the blue figure is a rotation of the red figure about the origin. If so, give the angle and direction of rotation.

1 7.
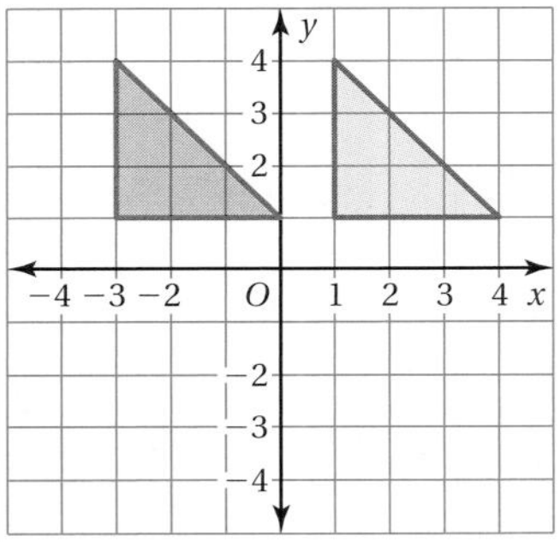

8.
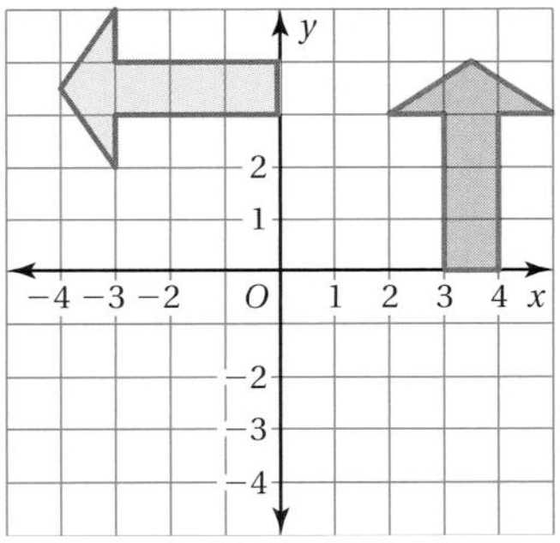

9.
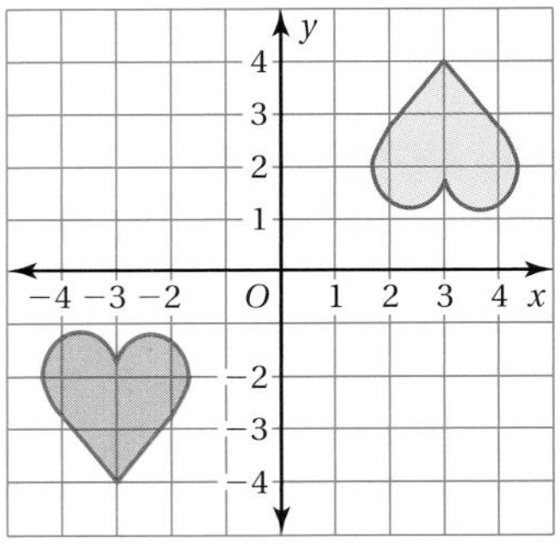

A figure has *rotational symmetry* if a rotation of 180° or less produces an image that fits exactly on the original figure. Explain why the figure has rotational symmetry.

10.

11.

12.
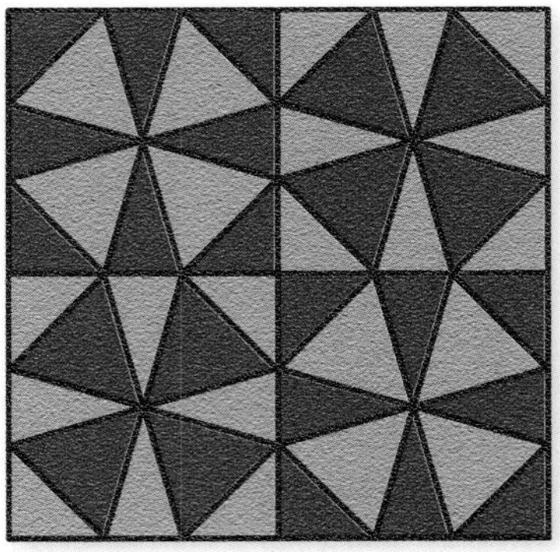

The vertices of a parallelogram are $A(-4, 1)$, $B(-3, 4)$, $C(-1, 4)$, and $D(-2, 1)$. Rotate the parallelogram about the origin as described. Find the coordinates of the image.

2 3 **13.** 90° counterclockwise

14. 270° clockwise

15. 180°

16. 270° counterclockwise

17. WRITING Why is it *not* necessary to use the words *clockwise* and *counterclockwise* when describing a rotation of 180°?

18. TRANSFORMATIONS You have learned three types of transformations: translations, reflections, and rotations.

a. Translate Triangle JKL four units left and four units up.

b. Reflect Triangle JKL in the x-axis.

c. Rotate Triangle JKL 180° about the origin.

d. Compare and contrast translations, reflections, and rotations.

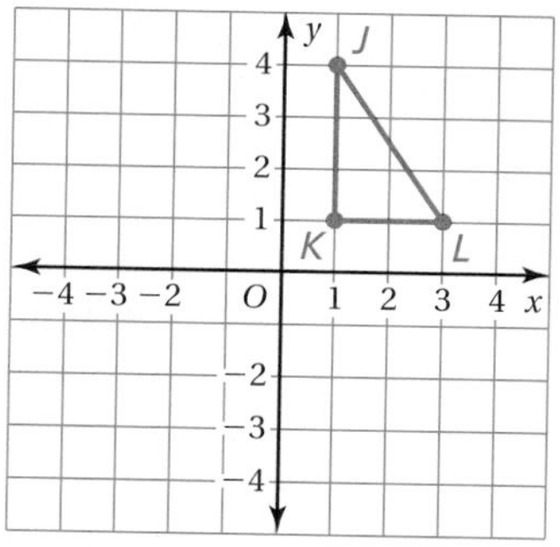

19. TREASURE MAP You want to find the treasure located on the map at ×. You are located at ●. The following transformations will lead you to the treasure, but they are not in the correct order. Find the correct order. Use each transformation exactly once.

- Rotate 180° about the origin.
- Reflect in the y-axis.
- Rotate 90° counterclockwise about the origin.
- Translate 1 unit right and 1 unit up.

20. Reasoning A triangle is rotated 90° counterclockwise about the origin. Its image is translated 1 unit left and 2 units down. The vertices of the final triangle are $(-5, -1)$, $(-2, 2)$, and $(-2, -1)$. What are the vertices of the original triangle?

Fair Game Review What you learned in previous grades & lessons

Identify the solid. *(Skills Review Handbook)*

21.

22.

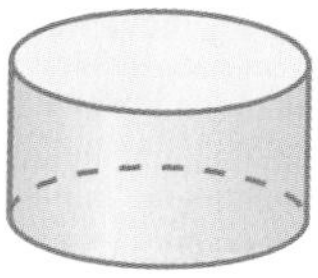

23. MULTIPLE CHOICE What is the value of $x - y$ when $x = -5$ and $y = -8$? *(Section 2.1)*

Ⓐ -13 Ⓑ -3 Ⓒ 3 Ⓓ 13

7.4 Dilations

STANDARDS OF LEARNING
7.8

Essential Question How can you enlarge or reduce a polygon in the coordinate plane?

The Meaning of a Word ● Dilate

When you have your eyes checked, the optometrist sometimes **dilates** one or both of the pupils of your eyes.

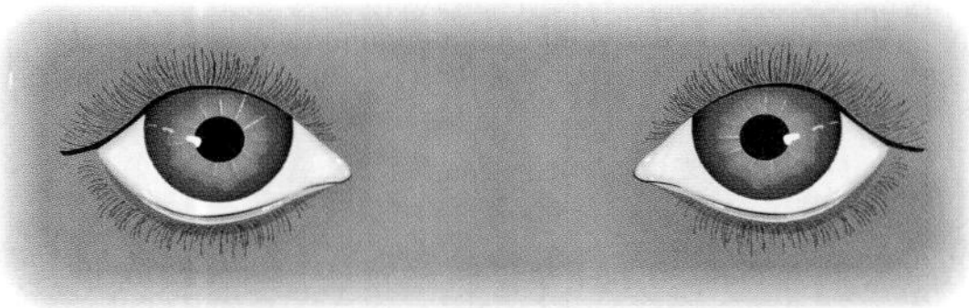

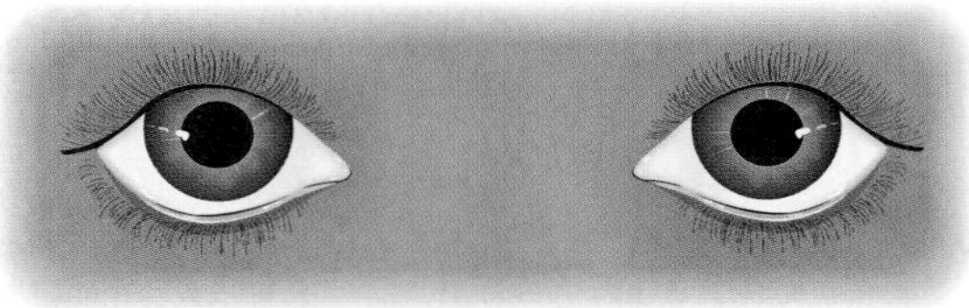

1 ACTIVITY: Comparing Triangles in a Coordinate Plane

Work with a partner. Write the coordinates of the vertices of the blue triangle. Then write the coordinates of the vertices of the red triangle.

a. How are the two sets of coordinates related?

b. How are the two triangles related? Explain your reasoning.

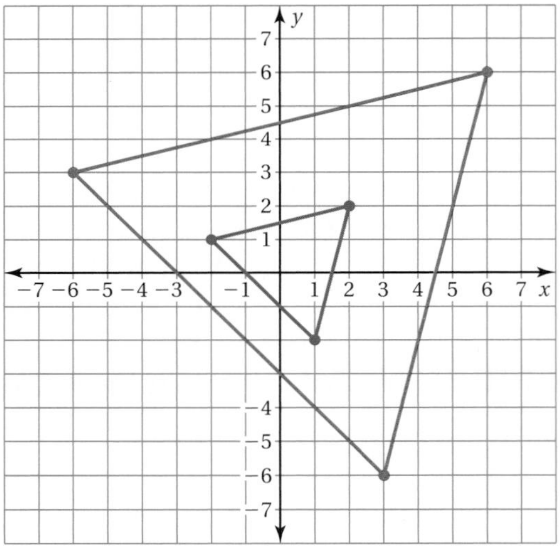

c. Draw a green triangle whose coordinates are twice the values of the corresponding coordinates of the blue triangle. How are the green and blue triangles related? Explain your reasoning.

d. How are the coordinates of the red and green triangles related? How are the two triangles related? Explain your reasoning.

2 ACTIVITY: Drawing Triangles in a Coordinate Plane

Work with a partner.

a. Draw the triangle whose vertices are (0, 2), (−2, 2), and (1, −2).

b. Multiply each coordinate of the vertices by 2 to obtain three new vertices. Draw the triangle given by the three new vertices. How are the two triangles related?

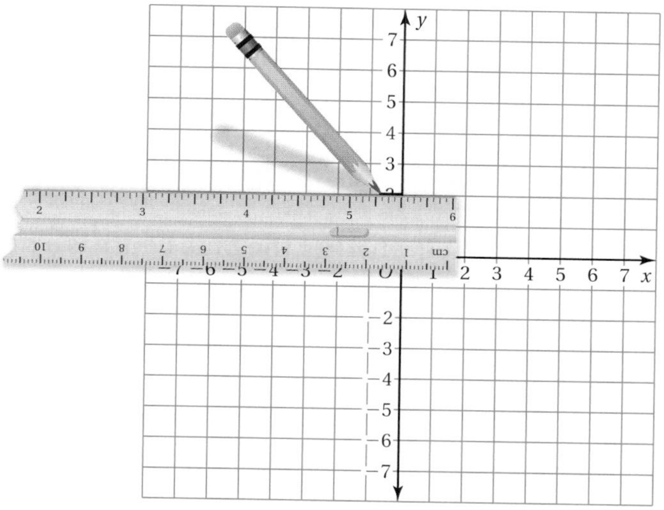

c. Repeat part (b) by multiplying by 3 instead of 2.

3 ACTIVITY: Summarizing Transformations

Work with a partner. Make a table that summarizes the relationships between the original figure and its image for the 4 types of transformations you studied in this chapter.

What Is Your Answer?

4. **IN YOUR OWN WORDS** How can you enlarge or reduce a polygon in the coordinate plane?

5. Describe how knowing how to enlarge or reduce figures in a technical drawing is important in a career such as drafting.

Practice Use what you learned about dilations to complete Exercises 4–6 on page 306.

7.4 Lesson

Check It Out
Lesson Tutorials
BigIdeasMath.com

Key Vocabulary
dilation, *p. 304*
center of dilation, *p. 304*
scale factor, *p. 304*

Key Idea

Dilations

A **dilation** is a transformation in which a figure is made larger or smaller with respect to a fixed point called the **center of dilation**.

The original figure and its image have the same shape but not the same size.

Center of dilation

EXAMPLE 1 Identifying a Dilation

Tell whether the blue figure is a dilation of the red figure.

a.

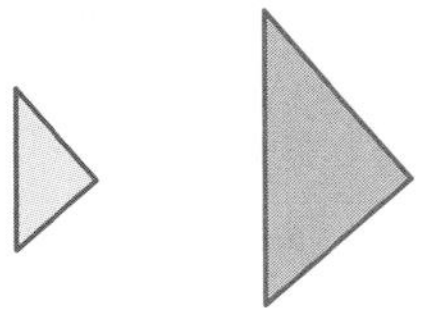

Lines connecting corresponding vertices meet at a point.

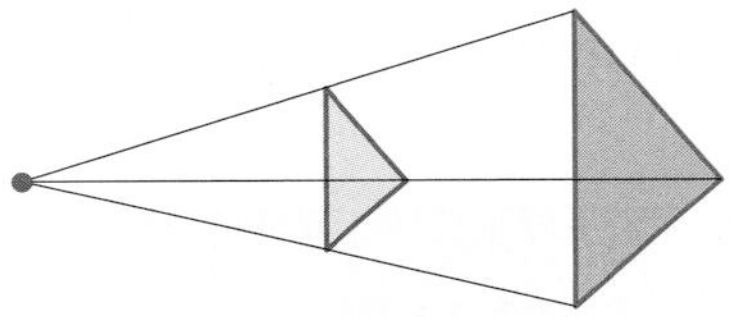

So, the blue figure is a dilation of the red figure.

b.

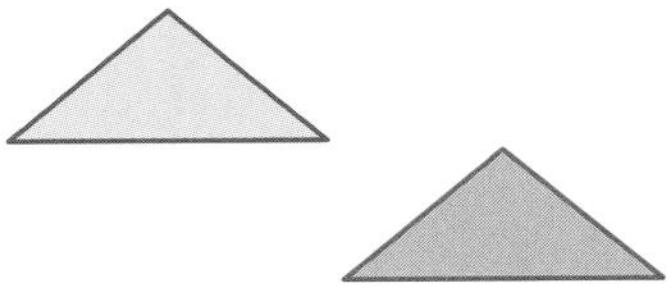

The figures have the same size and shape. The red figure *slides* to form the blue figure.

So, the blue figure is *not* a dilation of the red figure. It is a translation.

On Your Own

Now You're Ready
Exercises 7–12

Tell whether the blue figure is a dilation of the red figure. Explain.

1.

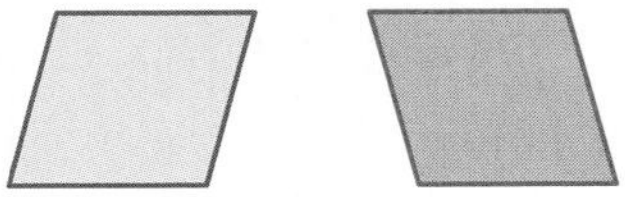

2.

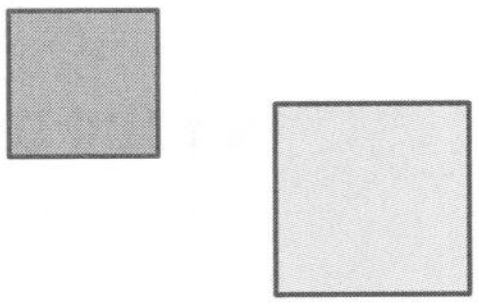

Because the image of a dilation has the same shape as the original figure, the two figures are similar. The ratio of the side lengths of the image to the corresponding side lengths of the original figure is the **scale factor** of the dilation.

To dilate a figure in the coordinate plane with respect to the origin, multiply the coordinates of each vertex by the scale factor k.

- When $k > 1$, the dilation is called an *enlargement*.
- When $k > 0$ and $k < 1$, the dilation is called a *reduction*.

Multi-Language Glossary at BigIdeasMath.com.

EXAMPLE 2 Dilating a Figure

Draw the image of Triangle *ABC* after a dilation with a scale factor of 3. Identify the type of dilation.

Multiply each *x*- and *y*-coordinate by the scale factor 3.

Vertices of *ABC*	$(x \cdot 3, y \cdot 3)$	Vertices of *A′B′C′*
$A(1, 3)$	$(1 \cdot 3, 3 \cdot 3)$	$A'(3, 9)$
$B(2, 3)$	$(2 \cdot 3, 3 \cdot 3)$	$B'(6, 9)$
$C(2, 1)$	$(2 \cdot 3, 1 \cdot 3)$	$C'(6, 3)$

Study Tip

You can check your answer by drawing a line from the origin through each vertex of the original figure. The vertices of the image should lie on these lines.

∴ The image is shown at the right. The dilation is an *enlargement* because the scale factor is greater than 1.

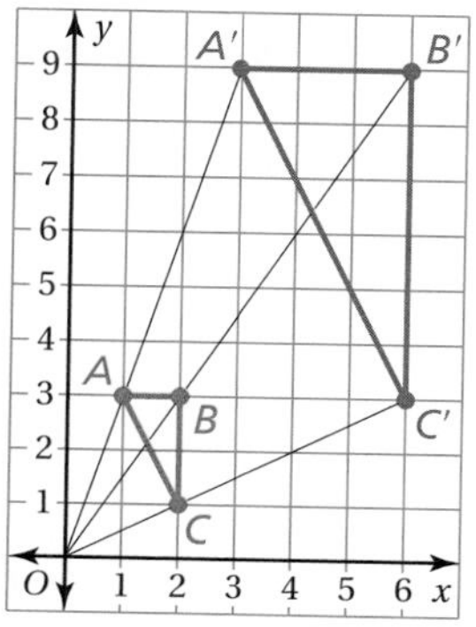

EXAMPLE 3 Dilating a Figure

Draw the image of Rectangle *WXYZ* after a dilation with a scale factor of 0.5. Identify the type of dilation.

Multiply each *x*- and *y*-coordinate by the scale factor 0.5.

Vertices of *WXYZ*	$(x \cdot 0.5, y \cdot 0.5)$	Vertices of *W′X′Y′Z′*
$W(-4, -6)$	$(-4 \cdot 0.5, -6 \cdot 0.5)$	$W'(-2, -3)$
$X(-4, 8)$	$(-4 \cdot 0.5, 8 \cdot 0.5)$	$X'(-2, 4)$
$Y(4, 8)$	$(4 \cdot 0.5, 8 \cdot 0.5)$	$Y'(2, 4)$
$Z(4, -6)$	$(4 \cdot 0.5, -6 \cdot 0.5)$	$Z'(2, -3)$

∴ The image is shown at the right. The dilation is a *reduction* because the scale factor is greater than 0 and less than 1.

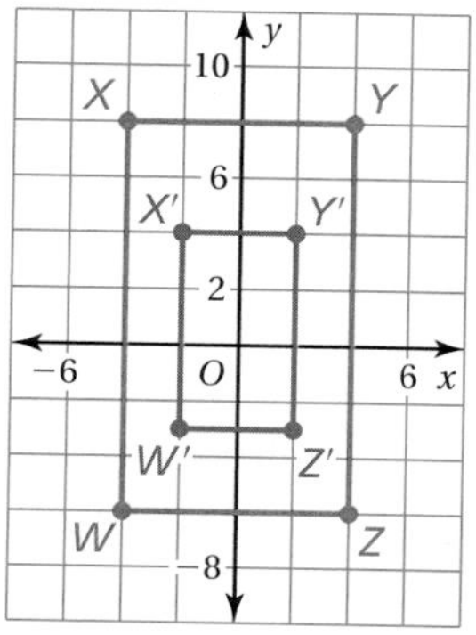

On Your Own

Now You're Ready
Exercises 13–18

3. Triangle *ABC* in Example 2 is dilated by a scale factor of 2. What are the coordinates of the image?
4. Rectangle *WXYZ* in Example 3 is dilated by a scale factor of $\frac{1}{4}$. What are the coordinates of the image?

7.4 Exercises

Check It Out
Help with Homework
BigIdeasMath.com

Vocabulary and Concept Check

1. **VOCABULARY** How is a dilation different from other transformations?
2. **VOCABULARY** For what values of scale factor k is a dilation called an enlargement? a reduction?
3. **REASONING** Which figure is *not* a dilation of the blue figure? Explain.

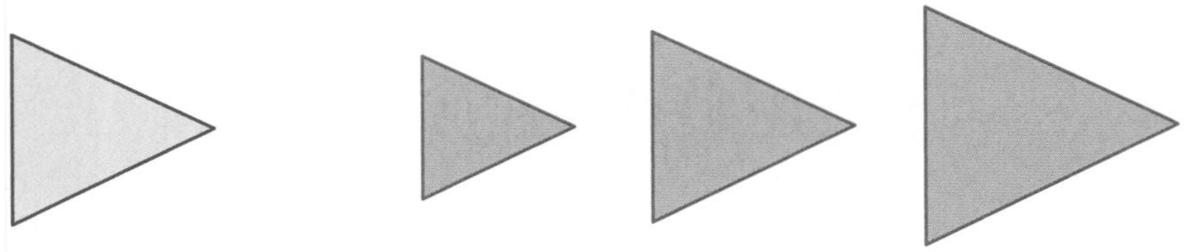

Practice and Problem Solving

Draw the triangle with the given vertices. Multiply each coordinate of the vertices by 3 and draw the new triangle. How are the two triangles related?

4. (0, 2), (3, 2), (3, 0)
5. (−1, 1), (−1, −2), (2, −2)
6. (−3, 2), (1, 2), (1, −4)

Tell whether the blue figure is a dilation of the red figure.

(1) 7.

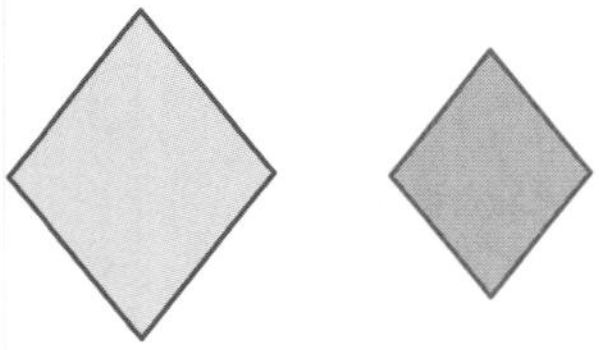

8.

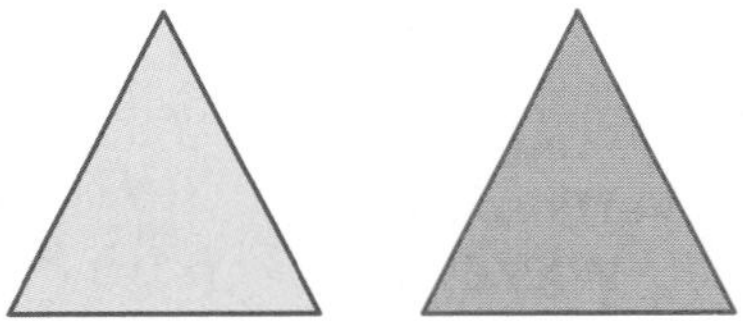

9.

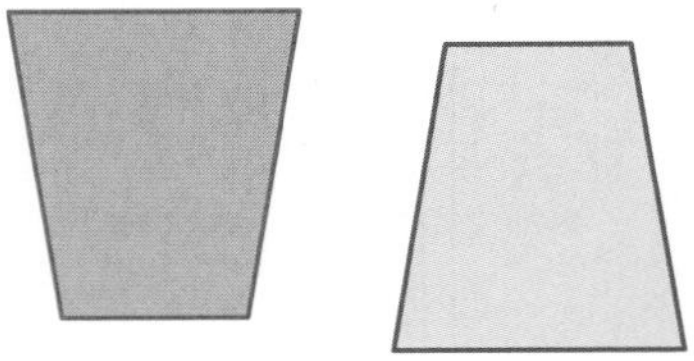

10.

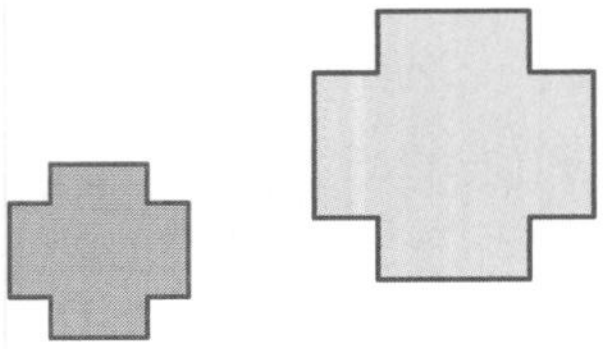

11.

12.

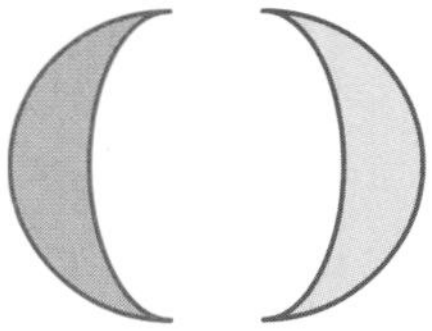

The vertices of a polygon are given. Draw the polygon and its image after a dilation with the given scale factor. Identify the type of dilation.

(2) (3) 13. $A(1, 1)$, $B(1, 4)$, $C(3, 1)$; $k = 4$

14. $D(0, 2)$, $E(6, 2)$, $F(6, 4)$; $k = 0.5$

15. $G(-2, -2)$, $H(-2, 6)$, $J(2, 6)$; $k = 0.25$

16. $M(2, 3)$, $N(5, 3)$, $P(5, 1)$; $k = 3$

17. $Q(-3, 0)$, $R(-3, 6)$, $T(4, 6)$, $U(4, 0)$; $k = \frac{1}{3}$

18. $V(-2, -2)$, $W(-2, 3)$, $X(5, 3)$, $Y(5, -2)$; $k = 5$

19. **ERROR ANALYSIS** Describe and correct the error in listing the coordinates of the image after a dilation with a scale factor of $\frac{1}{2}$.

Vertices of ABC	$(x \cdot 2, y \cdot 2)$	Vertices of $A'B'C'$
$A(2, 5)$	$(2 \cdot 2, 5 \cdot 2)$	$A'(4, 10)$
$B(2, 0)$	$(2 \cdot 2, 0 \cdot 2)$	$B'(4, 0)$
$C(4, 0)$	$(4 \cdot 2, 0 \cdot 2)$	$C'(8, 0)$

The blue figure is a dilation of the red figure. Identify the type of dilation and find the scale factor.

20.

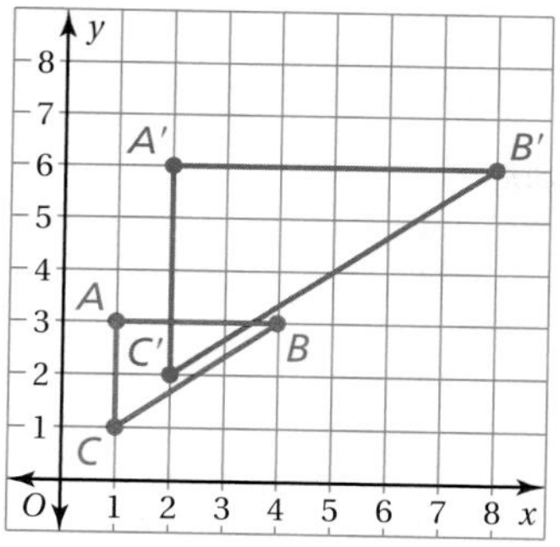

21.

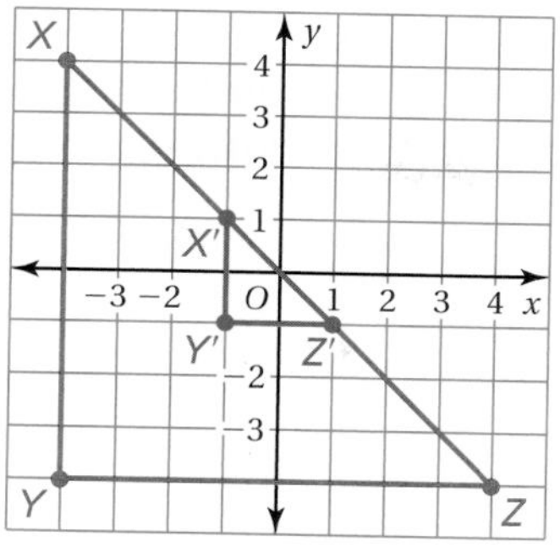

22.

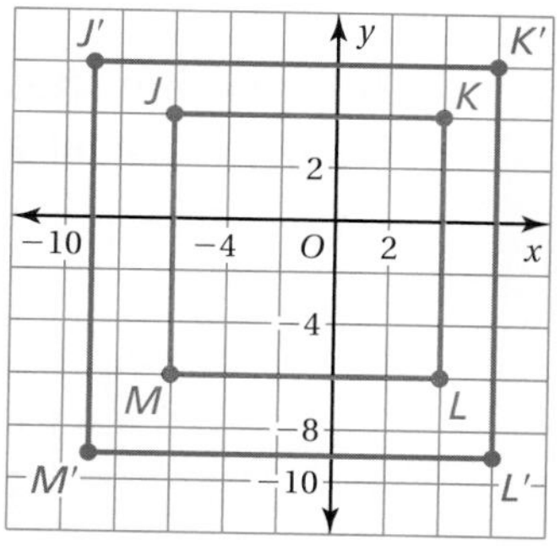

23. OPEN-ENDED Draw a rectangle on a coordinate plane. Choose a scale factor of 2, 3, 4, or 5, and dilate the rectangle. How many times greater is the area of the image than the area of the original rectangle?

24. SHADOW PUPPET You can use a flashlight and a shadow puppet (your hands) to project shadows on the wall.

a. Identify the type of dilation.

b. What does the flashlight represent?

c. The length of the ears on the shadow puppet is 3 inches. The length of the ears on the shadow is 4 inches. What is the scale factor?

d. Describe what happens as the shadow puppet moves closer to the flashlight. How does this affect the scale factor?

25. Critical Thinking A triangle is dilated using a scale factor of 3. The image is then dilated using a scale factor of $\frac{1}{2}$. What scale factor could you use to dilate the original triangle to get the final triangle? Explain.

Fair Game Review What you learned in previous grades & lessons

Evaluate the expression when $x = -4$, $y = 8$, and $z = -2$. *(Section 2.1)*

26. $3x - y$

27. $\frac{yz}{x}$

28. $xy - z^2$

29. $\frac{2z + x}{y}$

30. MULTIPLE CHOICE Which quadrilateral is *not* a parallelogram? *(Section 6.1)*

Ⓐ rhombus

Ⓑ trapezoid

Ⓒ square

Ⓓ rectangle

7.3–7.4 Quiz

Check It Out
Progress Check
BigIdeasMath.com

Tell whether the blue figure is a rotation of the red figure about the origin. If so, give the angle and direction of rotation. *(Section 7.3)*

1.

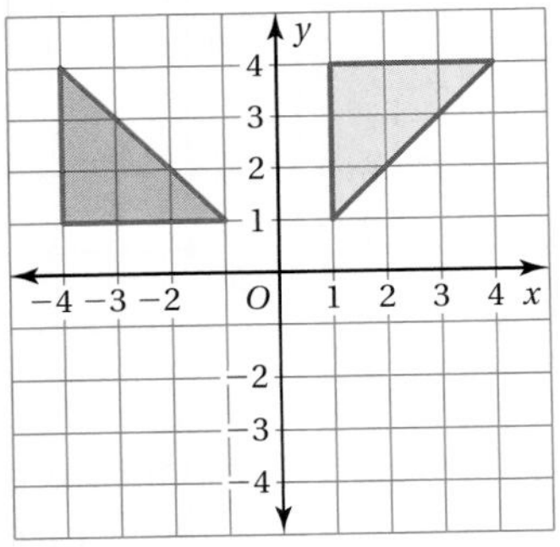

2.

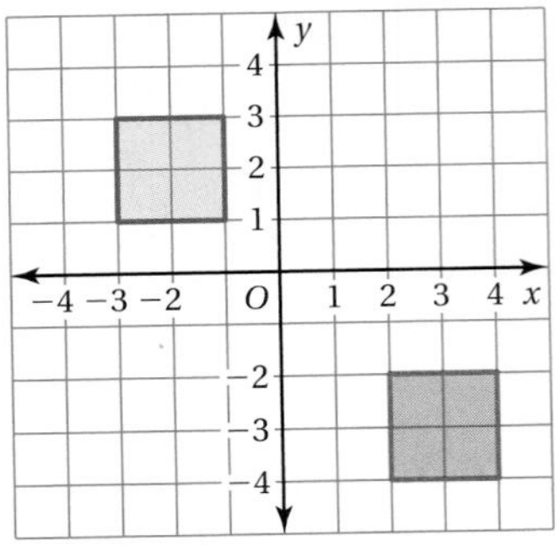

The vertices of a triangle are $A(-5, -2)$, $B(-3, -4)$, and $C(-3, -1)$. Rotate the triangle about the origin as described. Find the coordinates of the image. *(Section 7.3)*

3. 90° clockwise

4. 180°

Tell whether the blue figure is a dilation of the red figure. *(Section 7.4)*

5.

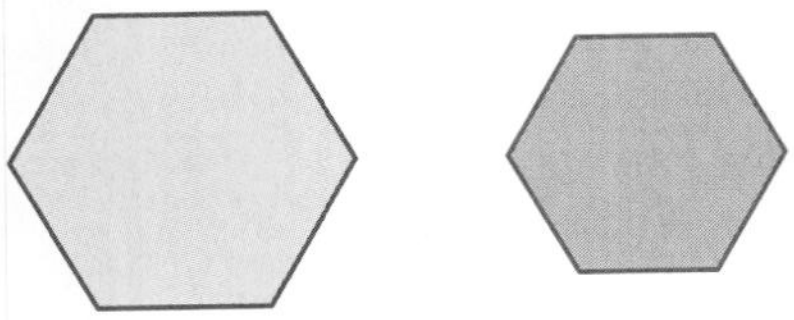

6.

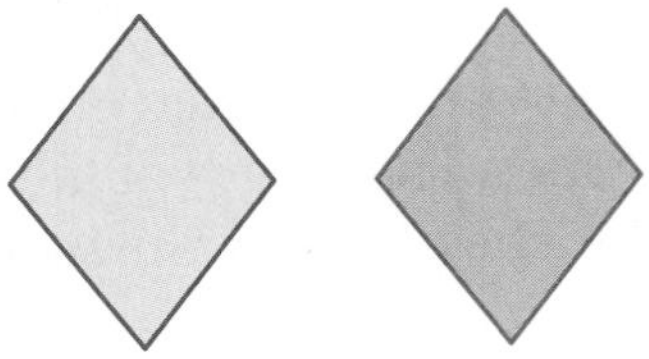

The blue figure is a dilation of the red figure. Identify the type of dilation and find the scale factor. *(Section 7.4)*

7.

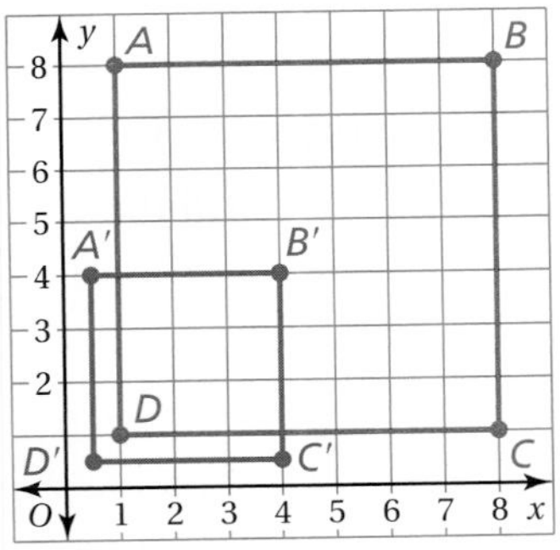

8.

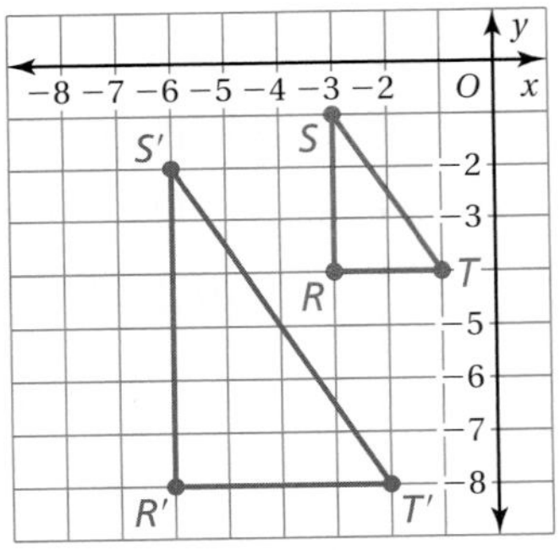

9. DRAWING The pivot point of a compass is at the origin. A circle is drawn starting at (3, 6). What point is the compass pencil on when the compass has rotated 270° counterclockwise? *(Section 7.3)*

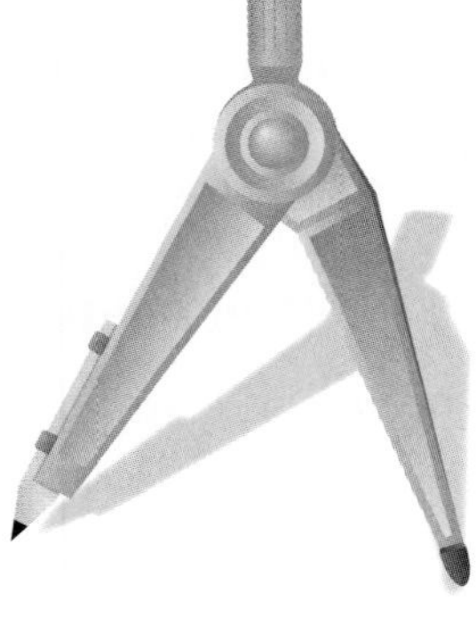

10. GEOMETRY The vertices of a rectangle are $A(2, 4)$, $B(5, 4)$, $C(5, -1)$, and $D(2, -1)$. Draw this rectangle and its image after a dilation with a scale factor of $\frac{1}{2}$. Identify the type of dilation. *(Section 7.4)*

Check It Out
Vocabulary Help
BigIdeasMath.com

Review Key Vocabulary

transformation, *p. 284*
image, *p. 284*
translation, *p. 284*
reflection, *p. 290*
line of reflection, *p. 290*
rotation, *p. 298*
center of rotation, *p. 298*
angle of rotation, *p. 298*
dilation, *p. 304*
center of dilation, *p. 304*
scale factor, *p. 304*

Review Examples and Exercises

7.1 Translations *(pp. 282–287)*

Translate the red triangle 4 units left and 1 unit down. What are the coordinates of the image?

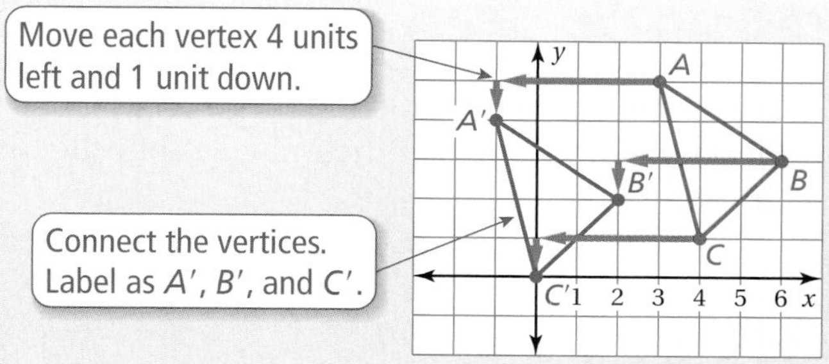

The coordinates of the image are $A'(-1, 4)$, $B'(2, 2)$, and $C'(0, 0)$.

Exercises

Tell whether the blue figure is a translation of the red figure.

1.

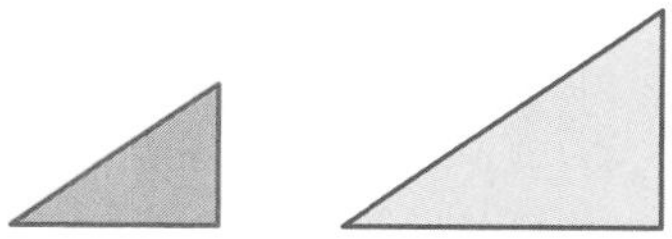

2.

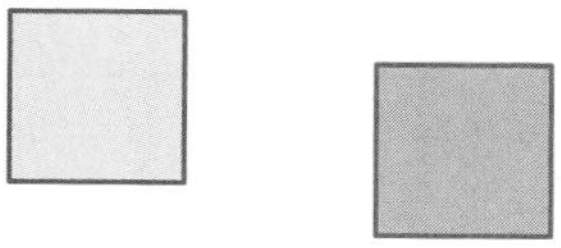

Translate the figure as described. What are the coordinates of the image?

3. 3 units left and 2 units down

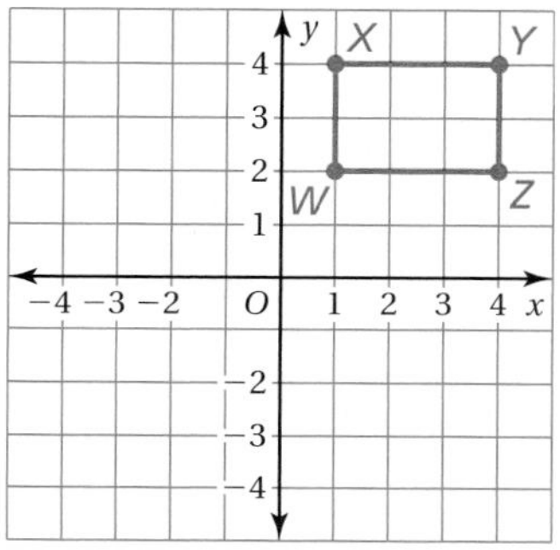

4. 5 units right and 1 unit up

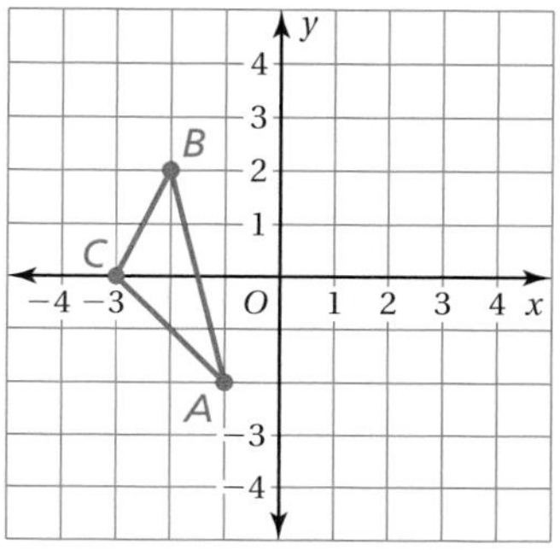

7.2 Reflections (pp. 288–293)

Tell whether the blue figure is a reflection of the red figure.

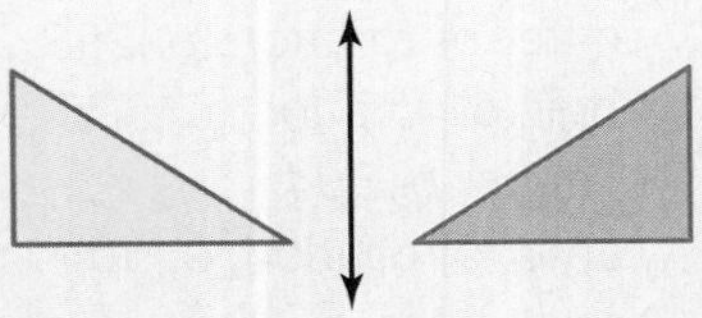

The red figure can be *flipped* to form the blue figure.

∴ So, the blue figure is a reflection of the red figure.

Exercises

Tell whether the blue figure is a reflection of the red figure.

5.

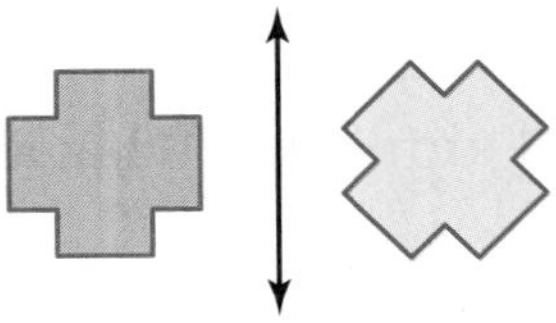

6.

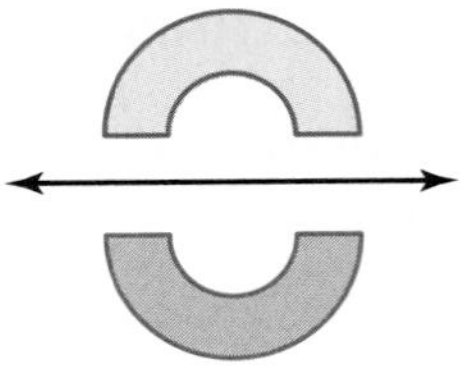

7. The vertices of a rectangle are $E(-1, 1)$, $F(-1, 3)$, $G(-5, 3)$, and $H(-5, 1)$. Find the coordinates of the image after reflecting in the (a) x-axis and (b) y-axis.

7.3 Rotations (pp. 296–301)

Tell whether the blue figure is a rotation of the red figure about the origin. If so, give the angle and direction of rotation.

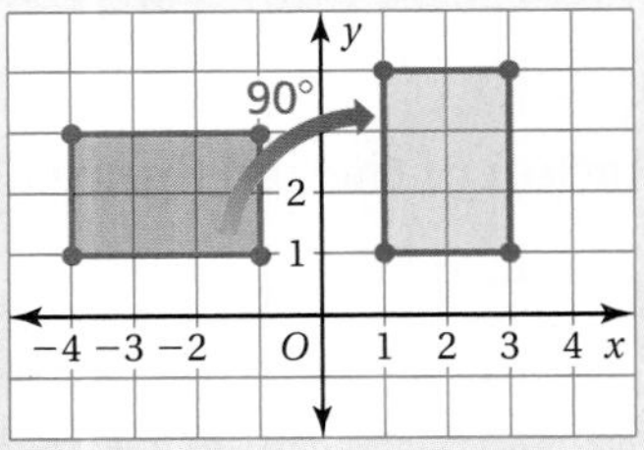

The red figure can be *turned* 90° clockwise about the origin to form the blue figure.

∴ So, the blue figure is a 90° clockwise rotation of the red figure.

Rotate the red triangle 90° counterclockwise about the origin. What are the coordinates of the image?

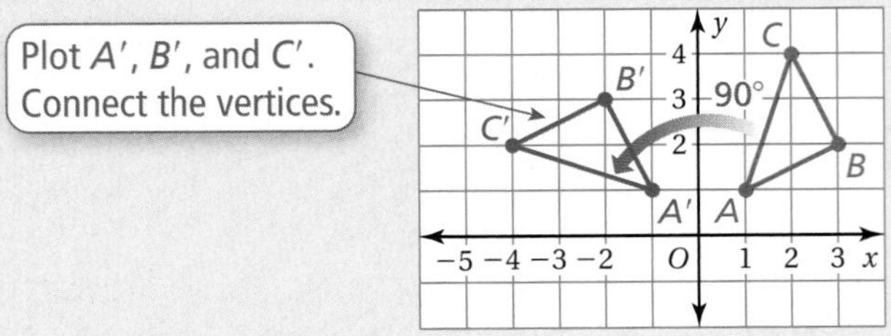

∴ The coordinates of the image are $A'(-1, 1)$, $B'(-2, 3)$, and $C'(-4, 2)$.

Exercises

Tell whether the blue figure is a rotation of the red figure about the origin. If so, give the angle and direction of rotation.

8.

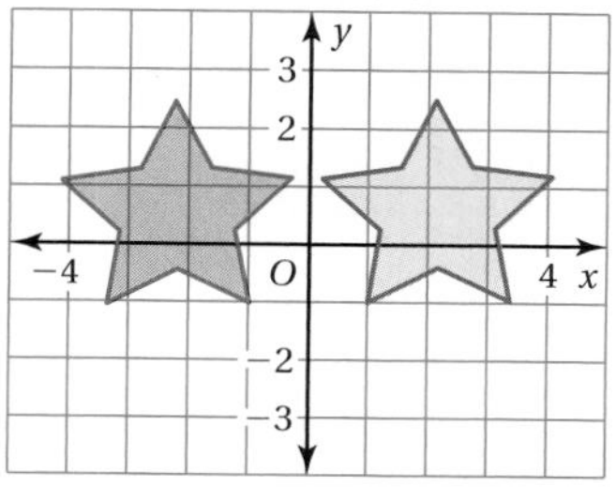

9.

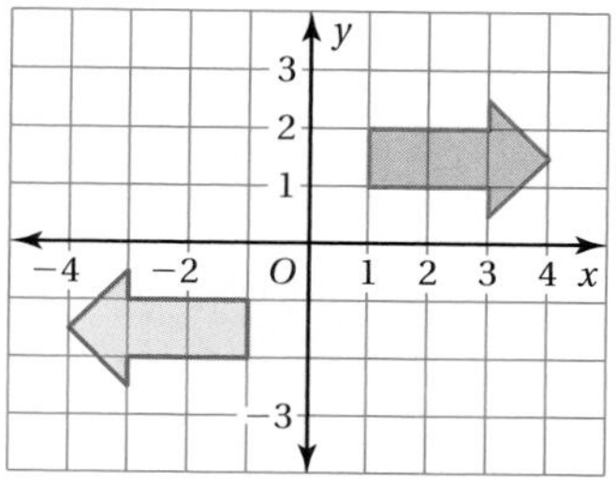

The vertices of a triangle are $A(-4, 2)$, $B(-2, 2)$, and $C(-3, 4)$. Rotate the triangle about the origin as described. Find the coordinates of the image.

10. 180°

11. 270° clockwise

7.4 Dilations *(pp. 302–307)*

Draw the image of Triangle *ABC* after a dilation with a scale factor of 2. Identify the type of dilation.

Multiply each *x*- and *y*-coordinate by the scale factor 2.

Vertices of *ABC*	$(x \cdot 2, y \cdot 2)$	Vertices of *A′B′C′*
$A(1, 1)$	$(1 \cdot 2, 1 \cdot 2)$	$A'(2, 2)$
$B(1, 2)$	$(1 \cdot 2, 2 \cdot 2)$	$B'(2, 4)$
$C(3, 2)$	$(3 \cdot 2, 2 \cdot 2)$	$C'(6, 4)$

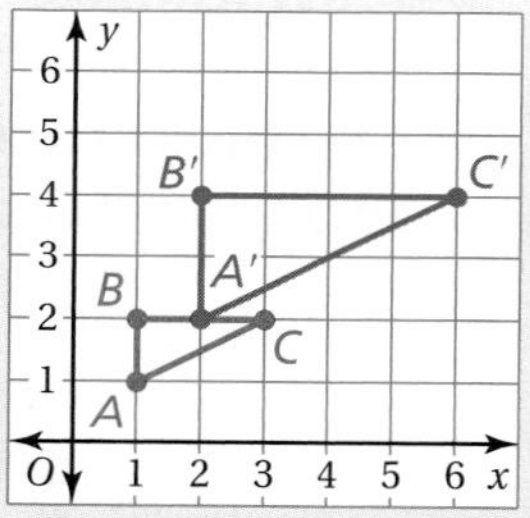

The image is shown at the right. The dilation is an *enlargement* because the scale factor is greater than 1.

Exercises

Tell whether the blue figure is a dilation of the red figure.

12.

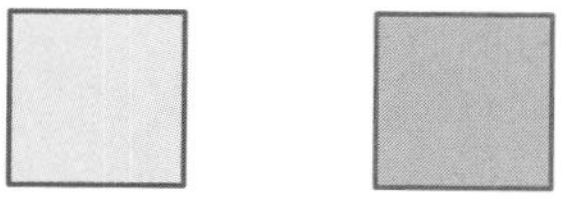

13.

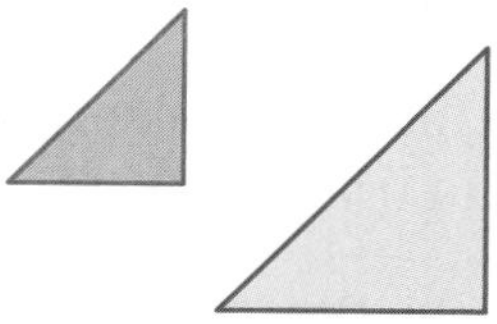

The vertices of a polygon are given. Draw the polygon and its image after a dilation with the given scale factor. Identify the type of dilation.

14. $P(-3, -2)$, $Q(-3, 0)$, $R(0, 0)$; $k = 4$

15. $B(3, 3)$, $C(3, 6)$, $D(6, 6)$, $E(6, 3)$; $k = \frac{1}{3}$

7 Chapter Test

Check It Out
Test Practice
BigIdeasMath.com

Tell whether the blue figure is a *translation*, *reflection*, *rotation*, or *dilation* of the red figure.

1. 

2. 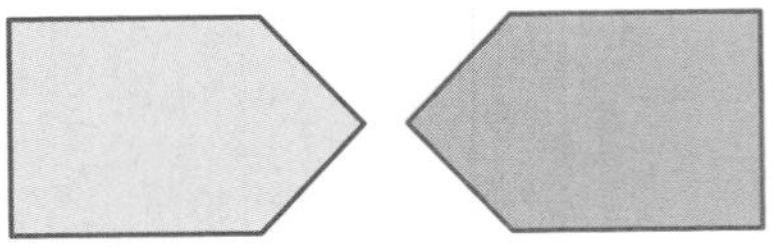

3. 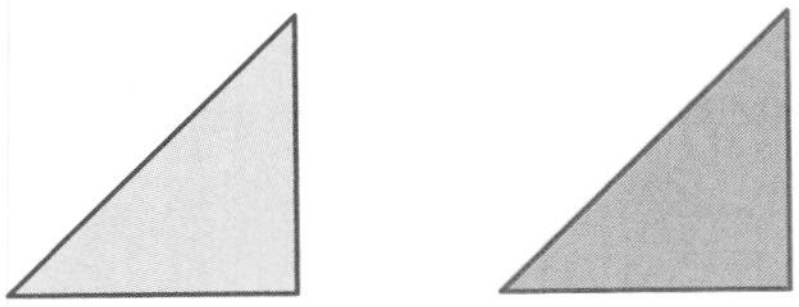

4. 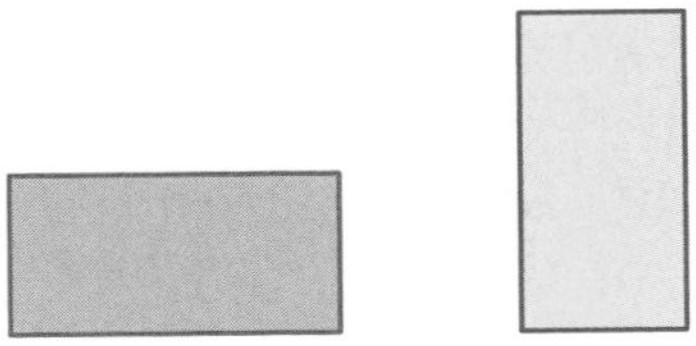

5. The vertices of a triangle are $A(2, 4)$, $B(2, 1)$, and $C(5, 1)$. Draw the triangle and its image after a translation of 1 unit left and 3 units down.

6. The vertices of a triangle are $A(2, 5)$, $B(1, 2)$, and $C(3, 1)$. Find the coordinates of the image after reflecting in the (a) x-axis and (b) y-axis.

The vertices of a triangle are $D(-3, -1)$, $E(-1, -4)$, and $F(-3, -4)$. Rotate the triangle about the origin as described. Find the coordinates of the image.

7. $180°$

8. $90°$ clockwise

The vertices of a rectangle are given. Draw the rectangle and its image after a dilation with the given scale factor. Identify the type of dilation.

9. $Q(-3, 1)$, $R(2, 1)$, $S(2, -2)$, $T(-3, -2)$; $k = 4$

10. $A(-6, -2)$, $B(4, -2)$, $C(4, 6)$, $D(-6, 6)$; $k = \frac{1}{2}$

11. **TRANSFORMATIONS** Several transformations are used to create the pattern.

    a. Describe the transformation of triangle *GLM* to triangle *DGH*.

    b. Describe the transformation of triangle *ALQ* to triangle *GLM*.

    c. Triangle *DFN* is a dilation of triangle *GHM*. Find the scale factor.

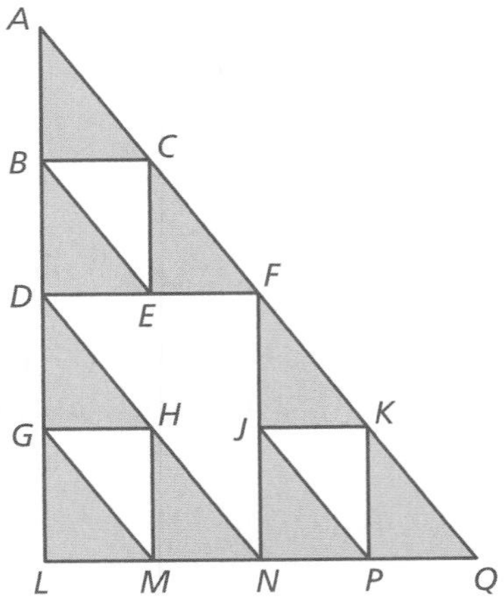

7 Standardized Test Practice

Test-Taking Strategy
After Answering Easy Questions, Relax

Which inequality best describes the annual cost x of owning a cat?
Vet Visits $546
Food $185
Boarding $119
Grooming $ 24
Treats/Toys $ 72
(A) x < $944 (B) x < $945
(C) x < $946 (D) x < $947
I'm worth it!

"After answering the easy questions, relax and try the harder ones. For this, x = $946. So, it's D."

1. What is the slope of the line shown in the graph below?

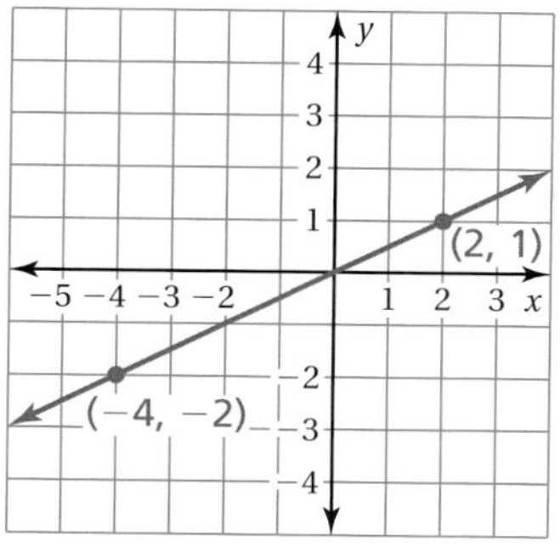

A. 2

B. $\frac{1}{6}$

C. $\frac{3}{2}$

D. $\frac{1}{2}$

2. A clockwise rotation of 90° is equivalent to a counterclockwise rotation of how many degrees?

3. Which parallelogram is similar to parallelogram *JKLM*? (Figures not drawn to scale.)

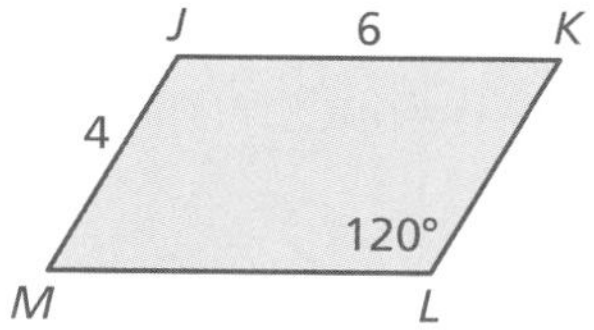

F.

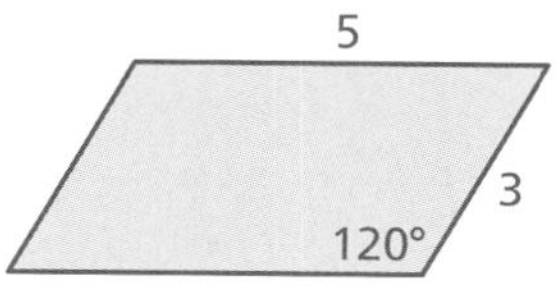

H.

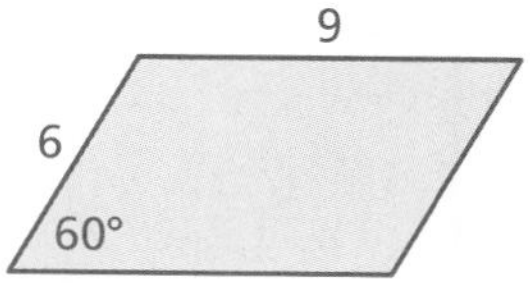

G.

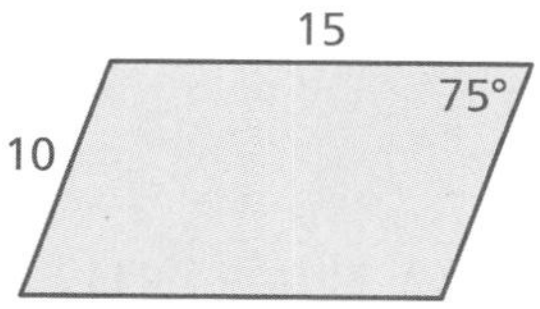

I.

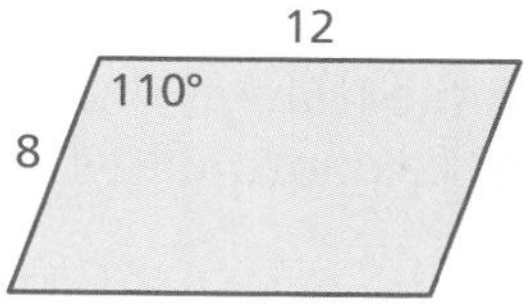

4. Dale was solving the equation in the box shown. What should Dale do to correct the error that he made?

$$-\frac{x}{3}+\frac{2}{5}=-\frac{7}{15}$$

$$-\frac{x}{3}+\frac{2}{5}-\frac{2}{5}=-\frac{7}{15}-\frac{2}{5}$$

$$-\frac{x}{3}=-\frac{13}{15}$$

$$3\cdot\left(-\frac{x}{3}\right)=3\cdot\left(-\frac{13}{15}\right)$$

$$x=-2\frac{3}{5}$$

A. Add $\frac{2}{5}$ to each side to get $-\frac{x}{3}=-\frac{1}{15}$.

B. Multiply each side by -3 to get $x+\frac{2}{5}=\frac{7}{5}$.

C. Multiply each side by -3 to get $x=2\frac{3}{5}$.

D. Subtract $\frac{2}{5}$ from each side to get $-\frac{x}{3}=-\frac{5}{10}$.

5. The table below describes the speed of four desktop printers.

Think Solve Explain

Printer	Description
Brand A	Prints 2 pages per second.
Brand B	Prints 1 page every 2 seconds.
Brand C	Prints 90 pages per minute.
Brand D	Prints 270 pages in 2 minutes.

Part A Write the speed of each printer as a unit rate in pages printed per second. Show your work.

Part B Which printer is the fastest? Explain your reasoning.

Part C How long does it take Brand C to print 27 pages?

6. A triangle is graphed in the coordinate plane below.

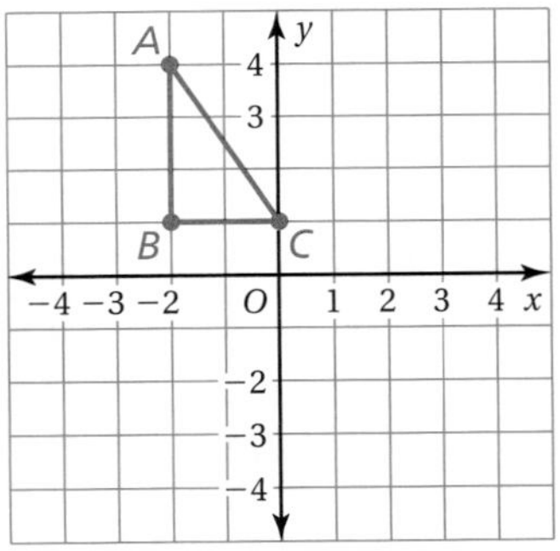

Translate the triangle 3 units right and 2 units down. What are the coordinates of the image?

F. $A'(1, 4), B'(1, 1), C'(3, 1)$

G. $A'(1, 2), B'(1, -1), C'(3, -1)$

H. $A'(-2, 2), B'(-2, -1), C'(0, -1)$

I. $A'(0, 1), B'(0, -2), C'(2, -2)$

7. Which transformation *flips* a figure?

A. translation

B. reflection

C. rotation

D. dilation

8. Jenny cut four congruent triangles off the corners of a rectangle to make an octagon, as shown below.

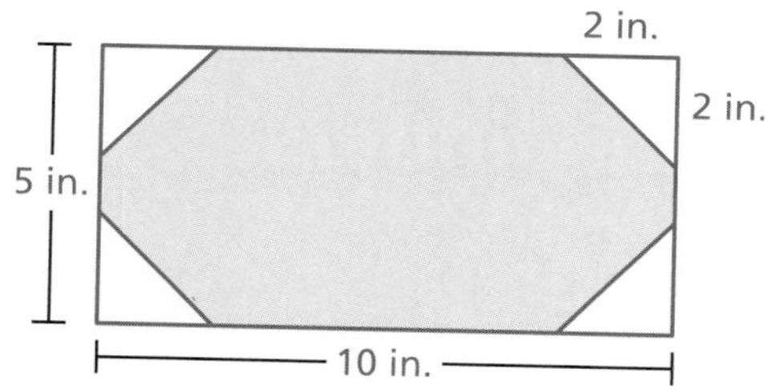

What is the area of the shaded region in square inches?

9. The vertices of a rectangle are $A(-4, 2)$, $B(3, 2)$, $C(3, -5)$, and $D(-4, -5)$. If the rectangle is dilated by a scale factor of 3, what will be the coordinates of vertex C'?

F. $(9, -15)$

G. $(-12, 6)$

H. $(-12, -15)$

I. $(9, 6)$

10. Manuel made a scale drawing of his tree house.

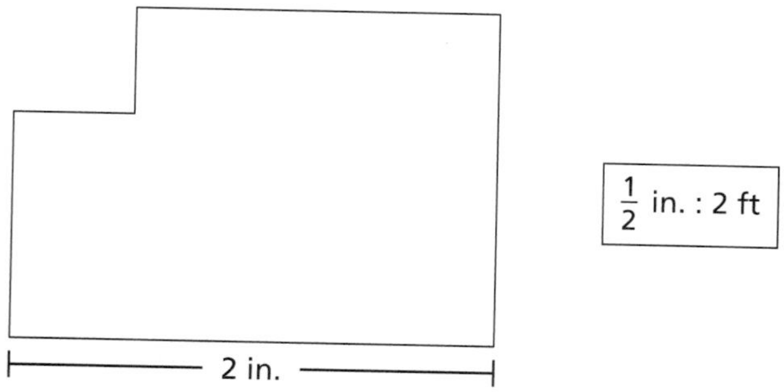

The width of the scale drawing of the tree house is 2 inches. What is the actual width, in feet, of the tree house?

A. 0.5

B. 2

C. 4

D. 8

8 Surface Area and Volume

8.1 Surface Areas of Rectangular Prisms

8.2 Volumes of Rectangular Prisms

8.3 Surface Areas of Cylinders

8.4 Volumes of Cylinders

"I was thinking that I want the Pagodal roof instead of the Swiss chalet roof for my new dog house."
Sometimes you are a true genius.
"Because PAGODAL rearranges to spell 'A DOG PAL.'"
"Take a deep breath and hold it."
Can I breathe now?
"Now, do you feel like your surface area or your volume is increasing more?"

What You Learned Before

"Descartes, how would you like it if I could double the height of your cat food can?"

Finding Areas of Squares and Rectangles

Example 1 **Find the area of the rectangle.**

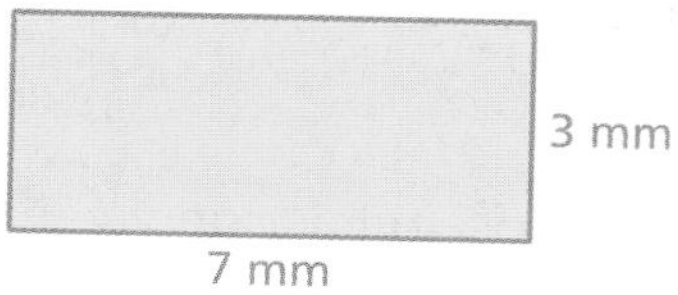

Area $= \ell w$	Write formula for area.
$= 7(3)$	Substitute 7 for ℓ and 3 for w.
$= 21$	Multiply.

The area of the rectangle is 21 square millimeters.

Try It Yourself

Find the area of the square or rectangle.

1.

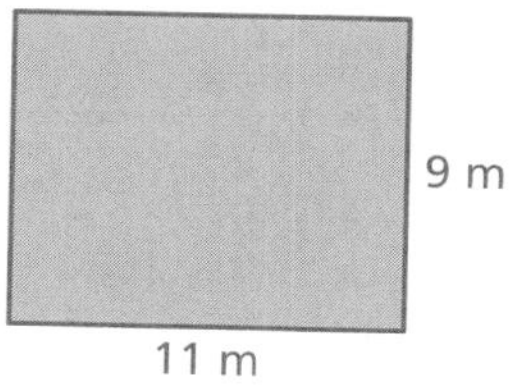

2.

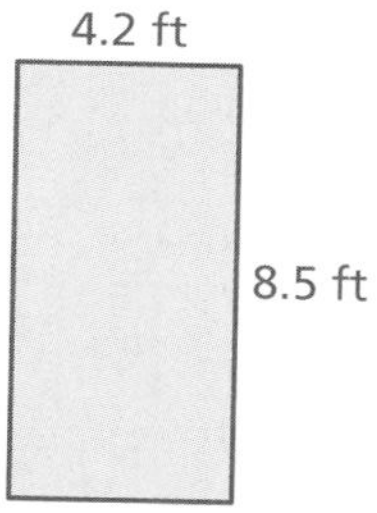

3.

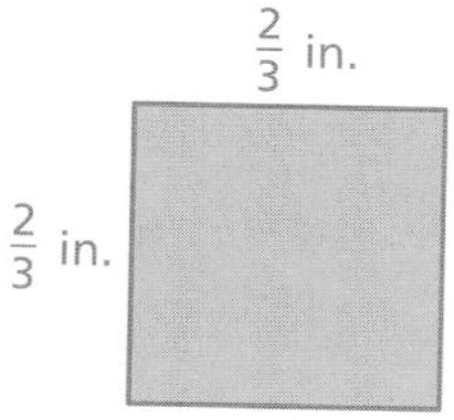

Finding Areas of Circles

Example 2 **Find the area of the circle. Use 3.14 for π.**

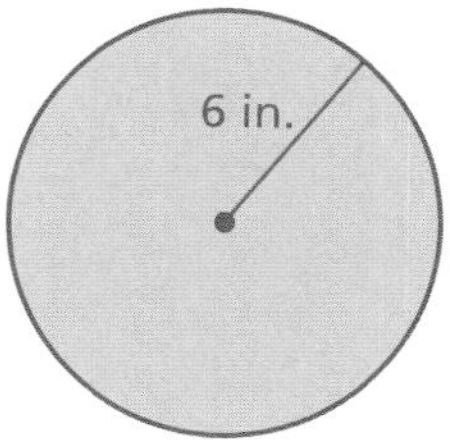

$A = \pi r^2$	Write formula for area.
$\approx 3.14 \cdot (6)^2$	Substitute 3.14 for π and 6 for r.
$= 3.14 \cdot 36$	Evaluate $(6)^2$.
$= 113.04$	Multiply.

The area is about 113.04 square inches.

Try It Yourself

Find the area of the circle. Use 3.14 or $\frac{22}{7}$ for π.

4.

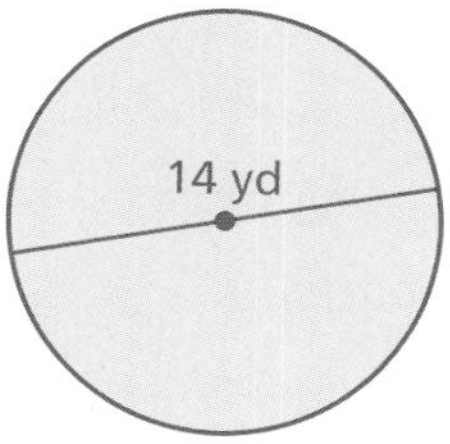

5.

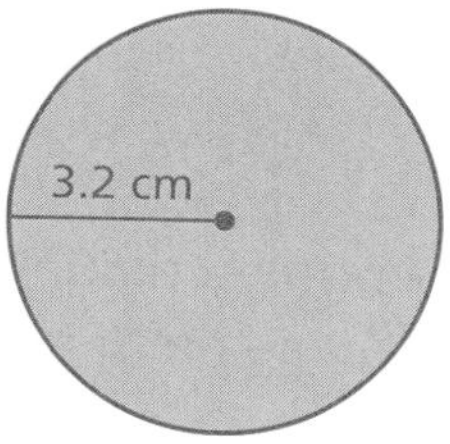

6.

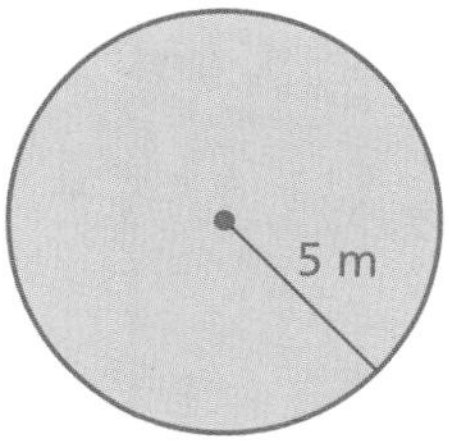

8.1 Surface Areas of Rectangular Prisms

STANDARDS OF LEARNING
7.5

Essential Question How does changing a dimension of a rectangular prism affect its surface area?

1 ACTIVITY: Comparing Surface Areas

Work with a partner.

a. Copy the net shown below on grid paper.

b. Double the width of the rectangular prism whose net is shown. Draw a net for the new prism.

c. Cut out both nets and fold them to form two prisms. Compare the two prisms visually. Estimate how much more surface area the larger prism has than the smaller prism.

d. Find the actual surface area of each prism. How close was your estimate?

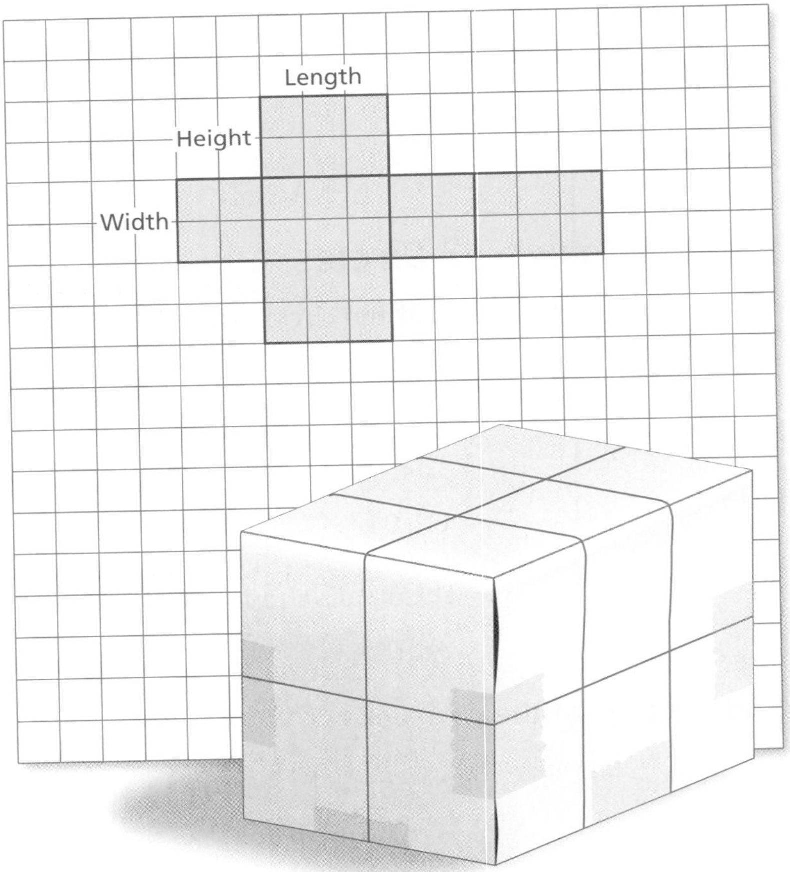

Note: Save your nets from Activities 1 and 2. You will use them again in Activity 8.2.

2 ACTIVITY: Comparing Surface Areas

Work with a partner.

a. Double all three dimensions of the original prism in Activity 1: the width, length, and height.

b. Calculate the surface area of the new prism. How many times greater is the surface area of the new prism?

c. Triple all three dimensions of the original prism in Activity 1.

d. Calculate the surface area of the new prism. How many times greater is the surface area of the new prism?

Original Prism

Prism with Dimensions Doubled

What Is Your Answer?

3. IN YOUR OWN WORDS How does changing a dimension of a rectangular prism affect its surface area?

4. Why is knowing the surface area of an object important in a career as an aeronautical engineer?

Practice Use what you learned about the surface areas of prisms to complete Exercises 4 and 5 on page 322.

8.1 Lesson

Check It Out
Lesson Tutorials
BigIdeasMath.com

Key Vocabulary
rectangular prism, *p. 320*

Remember
A prism has length, width, and height.

A **rectangular prism** is a three-dimensional figure that has 6 rectangular faces.

Key Idea

Surface Area of a Rectangular Prism

Words The surface area S of a rectangular prism is the sum of the areas of the bases and the lateral faces.

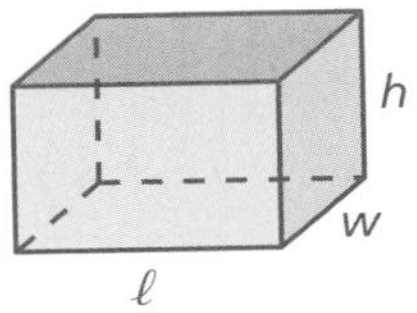

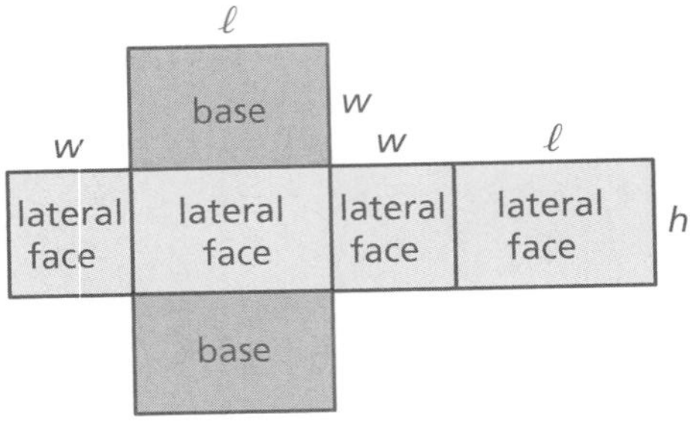

Algebra $S = 2\ell w + 2\ell h + 2wh$

Area of bases (points to $2\ell w$); Area of lateral faces (points to $2\ell h + 2wh$)

EXAMPLE 1 **Finding the Surface Area of a Rectangular Prism**

Find the surface area of the prism.

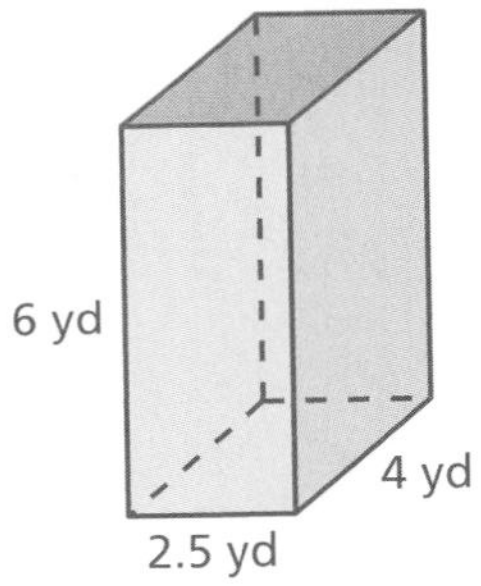

$$S = 2\ell w + 2\ell h + 2wh$$
$$= 2(4)(2.5) + 2(4)(6) + 2(2.5)(6)$$
$$= 20 + 48 + 30$$
$$= 98$$

∴ The surface area is 98 square yards.

On Your Own

Now You're Ready
Exercises 6–11

Find the surface area of the prism.

1.

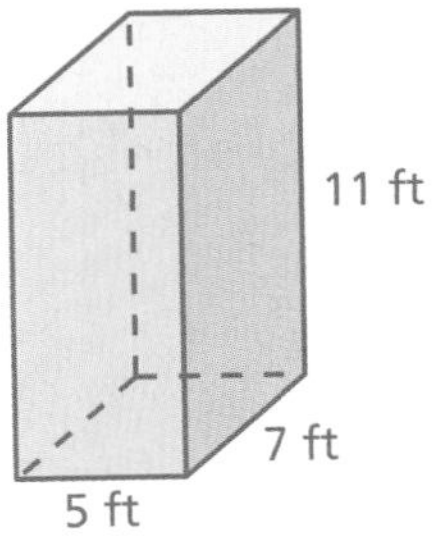

2.

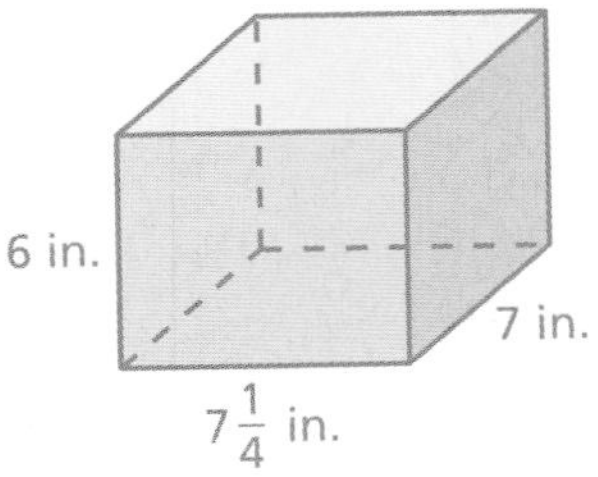

Multi-Language Glossary at BigIdeasMath.com.

EXAMPLE 2 Changing Dimensions of a Rectangular Prism

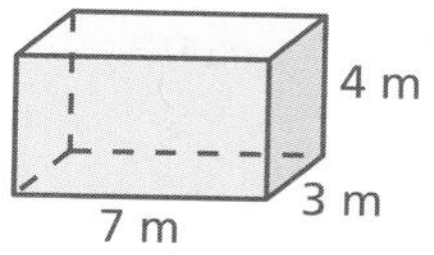

The dimensions of the red prism are twice the dimensions of the blue prism. How many times greater is the surface area of the red prism than the surface area of the blue prism?

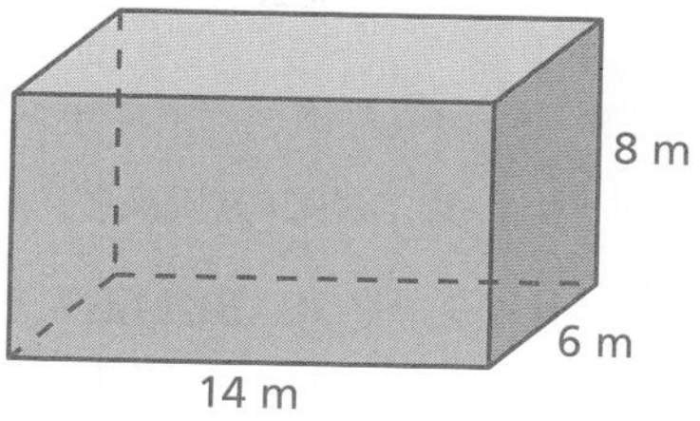

Find the surface area of each prism.

Blue Prism

$S = 2\ell w + 2\ell h + 2wh$

$= 2(7)(3) + 2(7)(4) + 2(3)(4)$

$= 42 + 56 + 24$

$= 122 \text{ m}^2$

Red Prism

$S = 2\ell w + 2\ell h + 2wh$

$= 2(14)(6) + 2(14)(8) + 2(6)(8)$

$= 168 + 224 + 96$

$= 488 \text{ m}^2$

The ratio of the surface areas is $\frac{488 \cancel{\text{m}^2}}{122 \cancel{\text{m}^2}} = 4$. So, the surface area of the red prism is 4 times greater than the surface area of the blue prism.

EXAMPLE 3 Real-Life Application

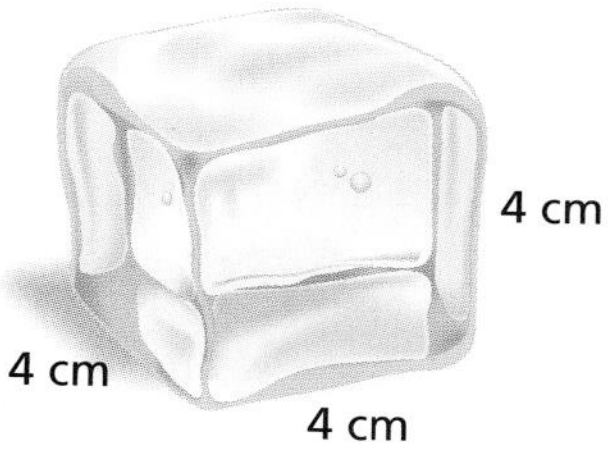

After the ice cube partially melts, its side lengths are one-half the original side lengths. Find the ratio of the surface areas. What can you conclude?

Find the surface area of the ice cube before and after melting.

Original

$S = 6s^2$

$= 6(4)^2$

$= 6(16) = 96 \text{ cm}^2$

Partially Melted

$S = 6s^2$

$= 6(2)^2$

$= 6(4) = 24 \text{ cm}^2$

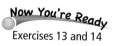

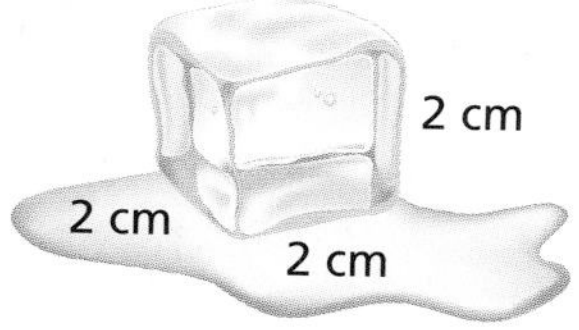

Remember

The formula for the surface area S of a cube with side length s is

$S = 6s^2$.

The ratio of the surface areas is $\frac{24 \cancel{\text{cm}^2}}{96 \cancel{\text{cm}^2}} = \frac{1}{4}$. So, the surface area of the partially melted ice cube is $\frac{1}{4}$ the original surface area.

On Your Own

Now You're Ready
Exercises 13 and 14

3. **WHAT IF?** The dimensions of a green prism are three times the dimensions of the blue prism in Example 2. How many times greater is the surface area of the green prism than the surface area of the blue prism?

4. **WHAT IF?** After the ice cube in Example 3 partially melts, its side lengths are one-fourth the original side lengths. Find the ratio of the surface areas. What can you conclude?

8.1 Exercises

Check It Out
Help with Homework
BigIdeasMath.com

Vocabulary and Concept Check

1. **VOCABULARY** What is a rectangular prism?
2. **VOCABULARY** Describe two ways to find the surface area of a rectangular prism.
3. **REASONING** Draw a net for the prism at the right.

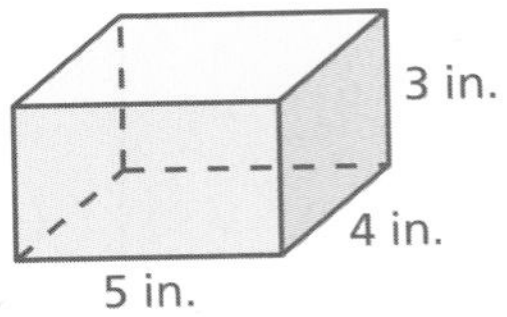

Practice and Problem Solving

Find the surface area of each prism. How many times greater is the surface area of the larger prism?

4. 

Original — Dimensions doubled

5. 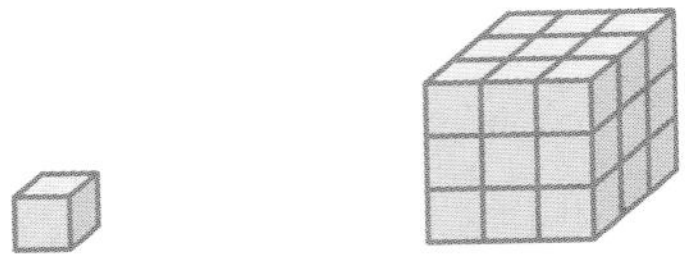

Original — Dimensions tripled

Find the surface area of the prism.

① 6. 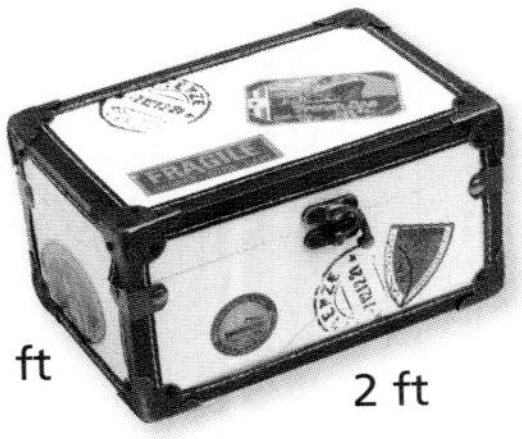

7. 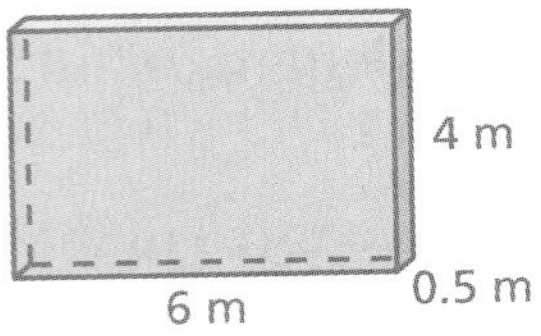

8. 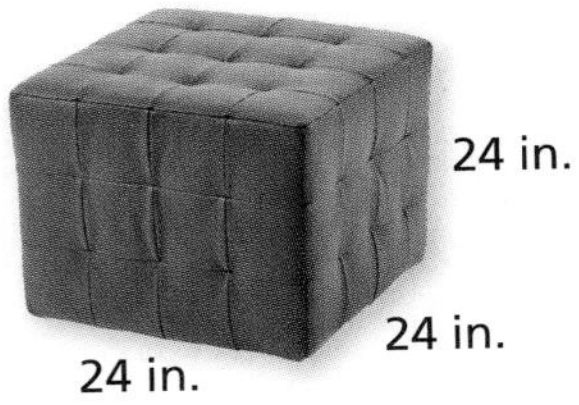

9. 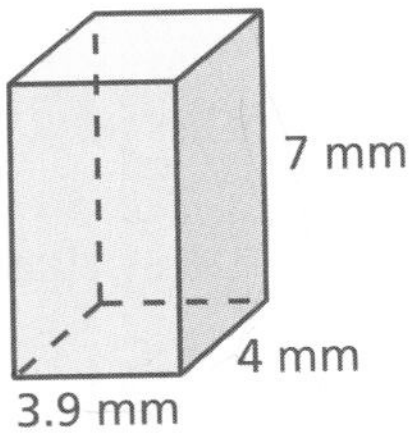

10. 

11. 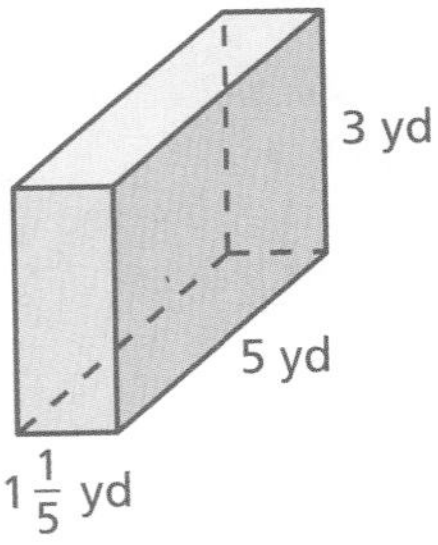

12. **ERROR ANALYSIS** Describe and correct the error in finding the surface area of the prism.

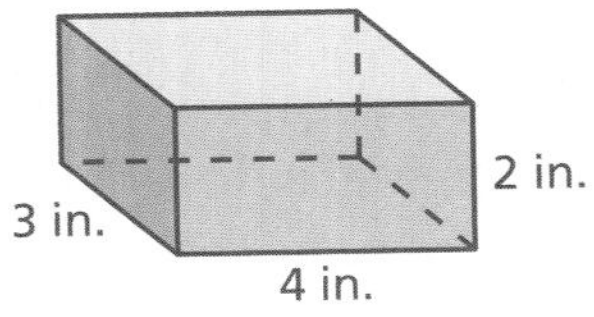

✗
$$S = 2(4)(3) + 2(4)(2) + 2(3)(4)$$
$$= 24 + 16 + 24$$
$$= 64 \text{ in.}^2$$

Compare the dimensions of the prisms. How many times greater is the surface area of the red prism than the surface area of the blue prism?

2 **13.**

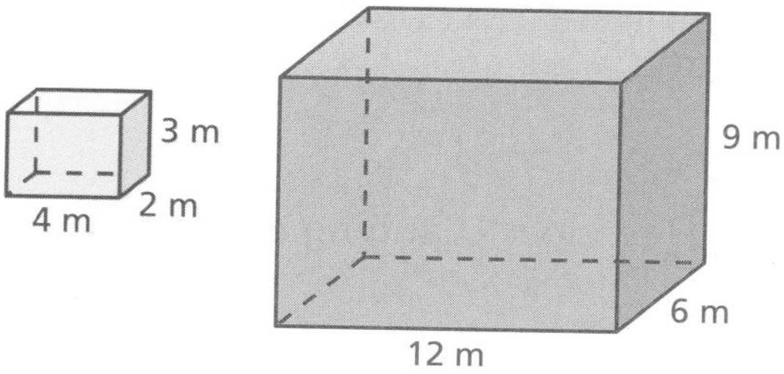

14.

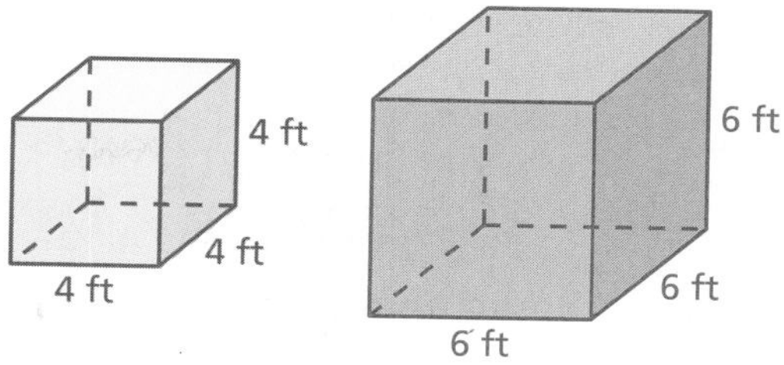

15. PEDESTALS You are painting the prize pedestals shown (including the bottoms). You need 0.5 pint of paint to paint the red pedestal.

a. The side lengths of the green pedestal are one-half the side lengths of the red pedestal. How much paint do you need to paint the green pedestal?

b. The side lengths of the blue pedestal are triple the side lengths of the green pedestal. How much paint do you need to paint the blue pedestal?

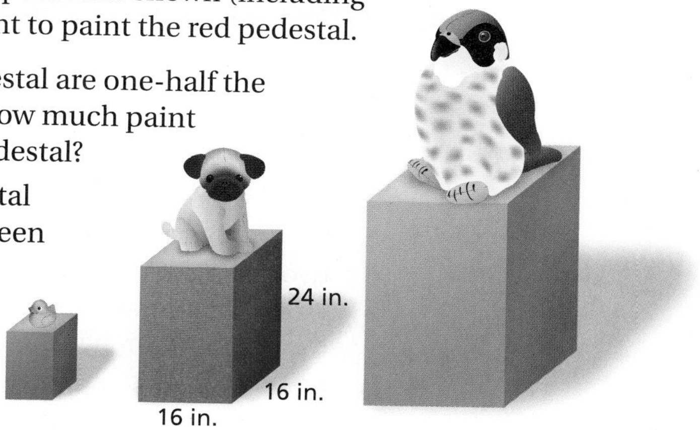

16. BREAD Fifty percent of the surface area of the bread is crust. What is the height h?

17. Critical Thinking You cut the foam cube into smaller cubes with side lengths that are one-fourth the original side lengths.

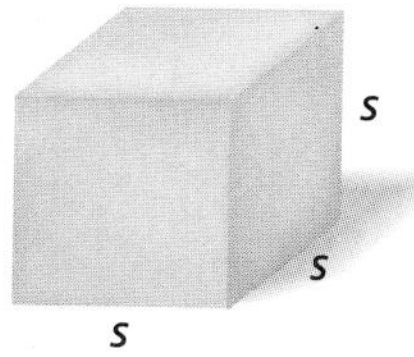

a. Find the number of smaller cubes.

b. Write an expression for the total surface area of the smaller cubes.

c. How many times greater is your answer in part (b) than the surface area of the original cube?

Fair Game Review What you learned in previous grades & lessons

The ratio of the corresponding side lengths of two similar figures is given. Find the ratios of the perimeters and of the areas. *(Section 6.3)*

18. 5 : 1

19. 3 : 7

20. 18 : 5

21. MULTIPLE CHOICE What is the circumference of the basketball? Use 3.14 for π. *(Skills Review Handbook)*

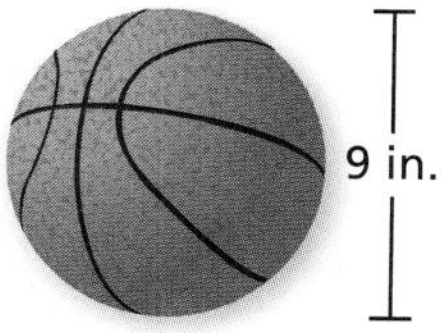

Ⓐ 14.13 in. Ⓑ 28.26 in.

Ⓒ 56.52 in. Ⓓ 254.34 in.

8.2 Volumes of Rectangular Prisms

STANDARDS OF LEARNING
7.5

Essential Question How does changing a dimension of a rectangular prism affect its volume?

1 ACTIVITY: Comparing Volumes

Work with a partner.

a. Copy the net shown below on grid paper.

b. Double the width of the rectangular prism whose net is shown. Draw a net for the new prism.

c. Cut out both nets and fold them to form two prisms. Compare the two prisms visually. Estimate how much more volume the larger prism has than the smaller prism.

d. Find the actual volume of each prism. How close was your estimate?

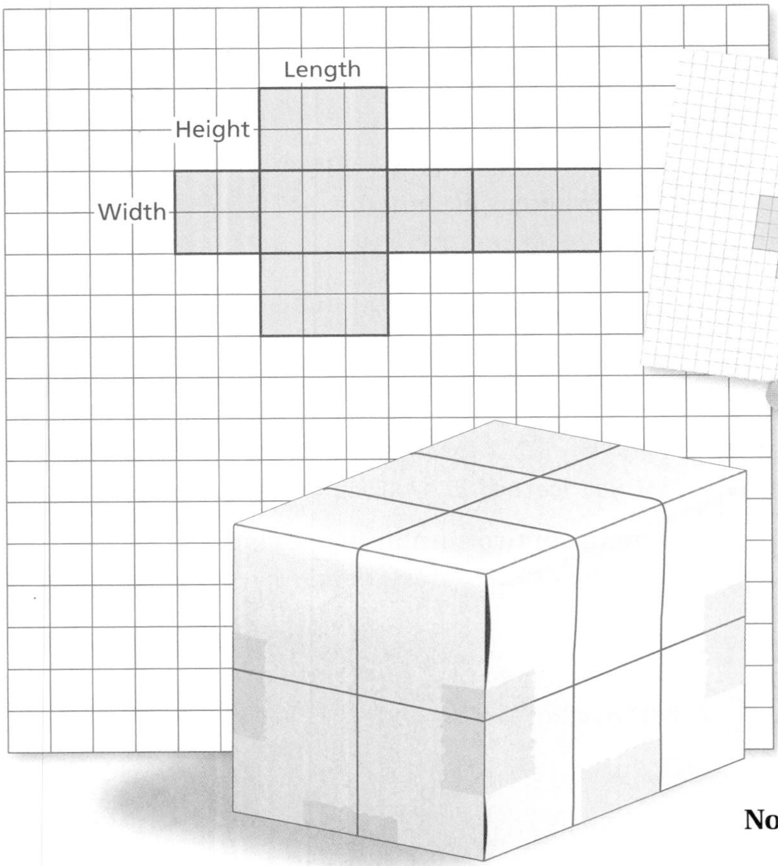

Note: You used these nets in Activities 1 and 2 in Section 8.1.

2 ACTIVITY: Comparing Volumes

Work with a partner.

a. Double all three dimensions of the original prism in Activity 1: the width, length, and height.

b. Calculate the volume of the new prism. How many times greater is the volume of the new prism?

c. Triple all three dimensions of the original prism in Activity 1.

d. Calculate the volume of the new prism. How many times greater is the volume of the new prism?

Original Prism

Prism with Dimensions Doubled

What Is Your Answer?

3. **IN YOUR OWN WORDS** How does changing a dimension of a rectangular prism affect its volume?

4. **BIG IDEAS SCIENCE** When designing an airplane, which of the following best describes your goal? Explain your reasoning.

a. Minimize surface area and maximize volume.

b. Minimize volume and maximize surface area.

Practice Use what you learned about volumes of rectangular prisms to complete Exercises 3 and 4 on page 328.

8.2 Lesson

Check It Out
Lesson Tutorials
BigIdeasMath.com

Key Vocabulary
volume, *p. 326*

The **volume** of a three-dimensional figure is a measure of the amount of space that it occupies. Volume is measured in cubic units.

Key Idea

Volume of a Rectangular Prism

Words The volume V of a rectangular prism is the product of the area of the base and the height of the prism.

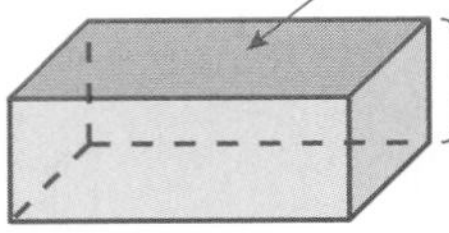

Algebra $V = Bh$

Area of base — Height of prism

EXAMPLE 1 Finding the Volume of a Rectangular Prism

Study Tip
The area of the base of a rectangular prism is the product of the length ℓ and the width w.
You can use $V = \ell wh$ to find the volume of a rectangular prism.

Find the volume of the prism.

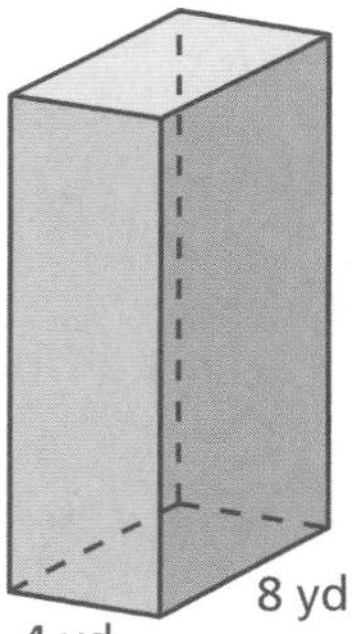

$V = Bh$	Write formula for volume.
$= 8(4) \cdot 12.5$	Substitute.
$= 32 \cdot 12.5$	Multiply.
$= 400$	Multiply.

∴ The volume is 400 cubic yards.

On Your Own

Now You're Ready
Exercises 5–10

Find the volume of the prism.

1.

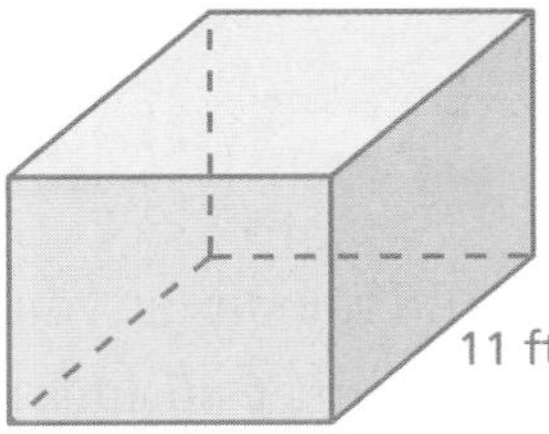

2.

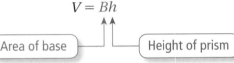

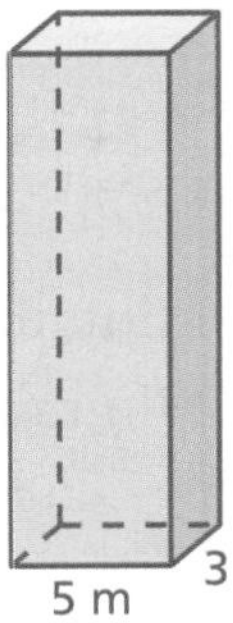

EXAMPLE 2 Changing One Dimension of a Rectangular Prism

Does doubling the height of the prism (a) double its volume? (b) double its surface area? Explain.

Prism A

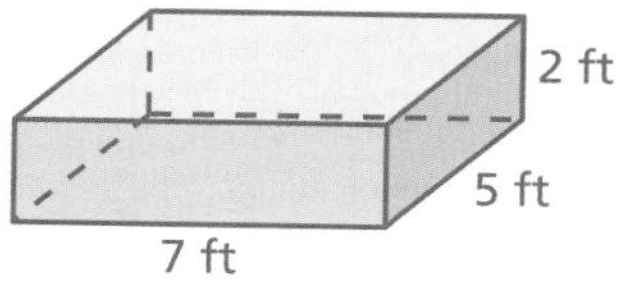

Prism B

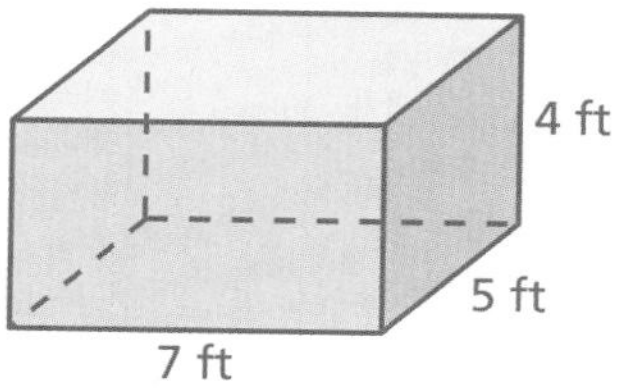

	Prism A	***Prism B***
a.	$V = Bh$	$V = Bh$
	$= 7(5) \cdot 2 = 70 \text{ ft}^3$	$= 7(5) \cdot 4 = 140 \text{ ft}^3$

The ratio of the volumes is $\frac{140 \text{ ft}^3}{70 \text{ ft}^3} = 2$. So, doubling the height of the prism doubles the volume.

b.	$S = 2\ell w + 2\ell h + 2wh$	$S = 2\ell w + 2\ell h + 2wh$
	$= 2(7)(5) + 2(7)(2) + 2(5)(2)$	$= 2(7)(5) + 2(7)(4) + 2(5)(4)$
	$= 70 + 28 + 20 = 118 \text{ ft}^2$	$= 70 + 56 + 40 = 166 \text{ ft}^2$

The ratio of the surface areas is $\frac{166 \text{ ft}^2}{118 \text{ ft}^2} \approx 1.4$. So, doubling the height of the prism does *not* double the surface area.

EXAMPLE 3 Real-Life Application

You can keep three angelfish in your aquarium. You buy an aquarium whose dimensions are two times greater. How many angelfish can you keep in your new aquarium?

Original aquarium

3 fish
$V = \ell wh$

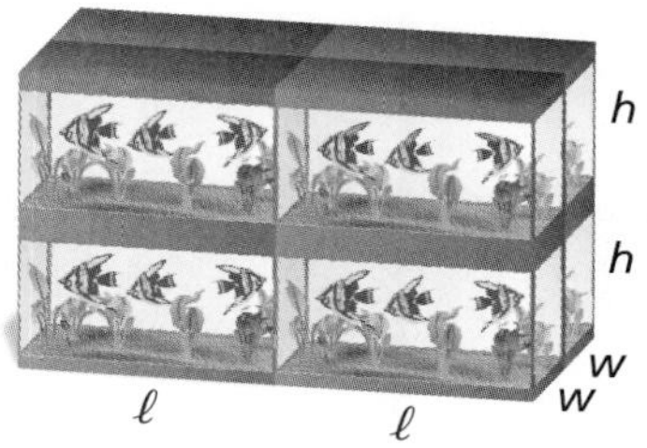

n fish
$V = 8 \cdot \ell wh$

The ratio of the volumes is $\frac{8 \cdot \ell wh}{\ell wh} = 8$. So, you can keep a maximum of $8 \cdot 3 = 24$ angelfish in your new aquarium.

On Your Own

Now You're Ready
Exercises 14–16

3. **WHAT IF?** In Example 2, does tripling the height (a) triple the volume? (b) triple the surface area? Explain.
4. **WHAT IF?** In Example 3, you buy an aquarium whose dimensions are three times greater. How many angelfish can you can keep in your new aquarium?

8.2 Exercises

Check It Out
Help with Homework
BigIdeasMath.com

Vocabulary and Concept Check

1. **VOCABULARY** What types of units are used to describe volume?
2. **CRITICAL THINKING** How are volume and surface area different?

Practice and Problem Solving

Find the volume of each prism. How many times greater is the volume of the larger prism?

3.

4.
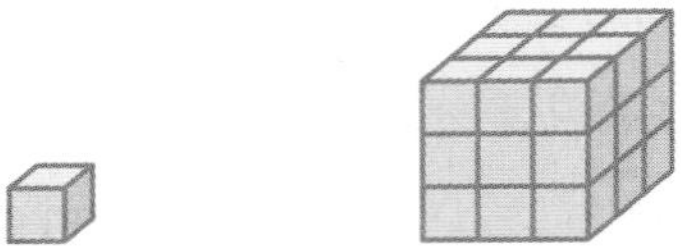

Find the volume of the prism.

1 5.
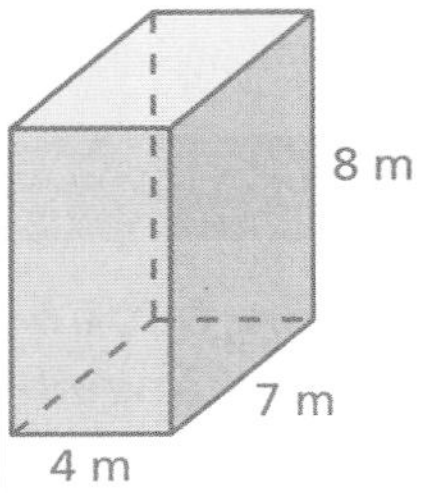

6.
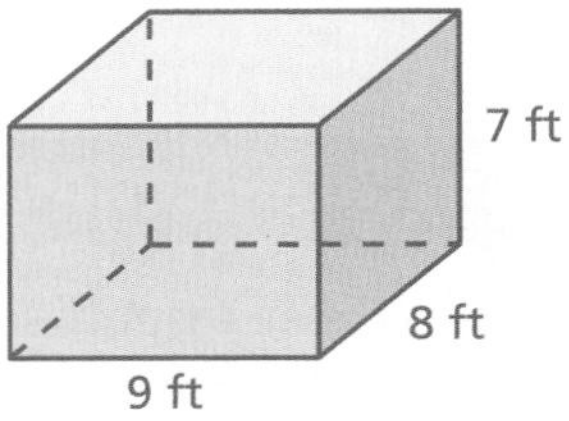

7.

8.
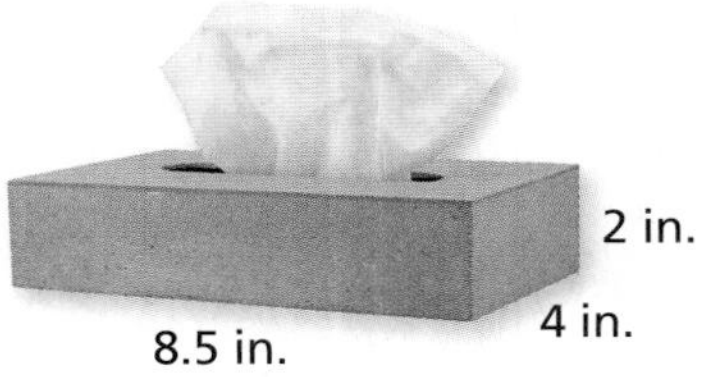

9.
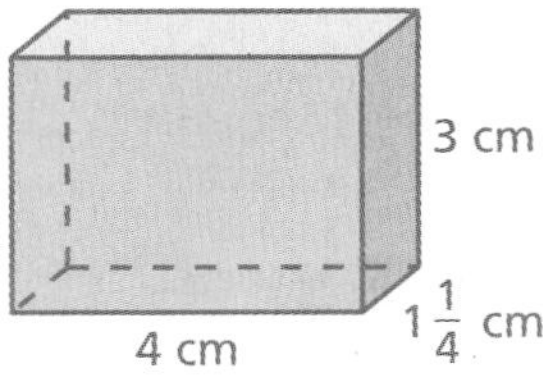

10.
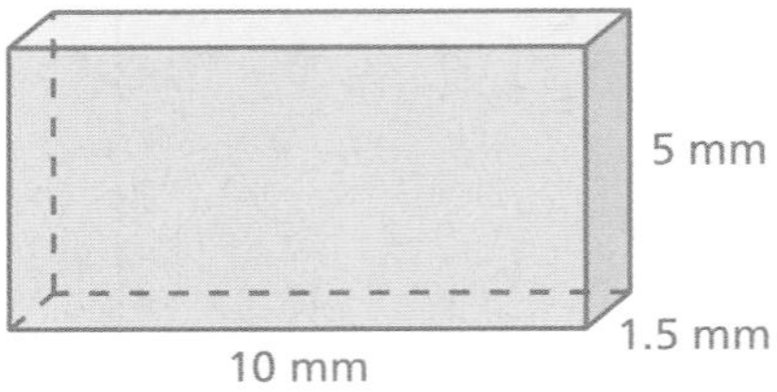

11. **ERROR ANALYSIS** Describe and correct the error in finding the volume of the prism.

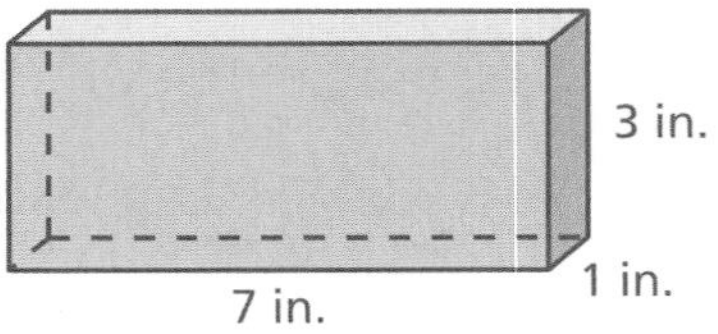

✗
$$V = Bh$$
$$= 7(1) \cdot 3$$
$$= 21 \text{ in.}^2$$

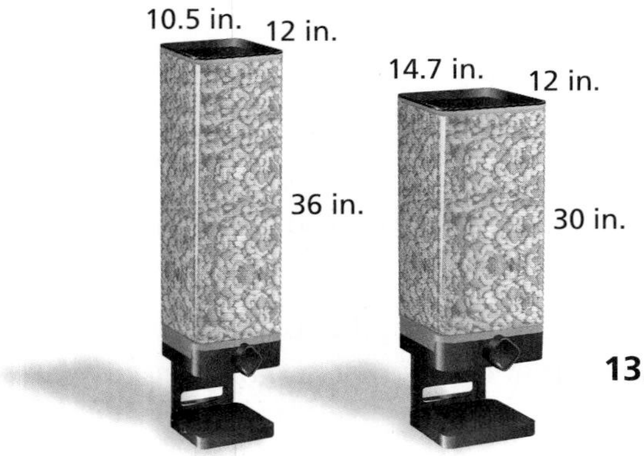

12. **MAIL** A box delivered in the mail is 7.5 inches by 2.6 inches by 9 inches. What is the volume of the box?

13. **BULK FOODS** Each bulk food dispenser is shaped like a rectangular prism. Which holds more food? Explain.

In Exercises 14 and 15, use the prism shown.

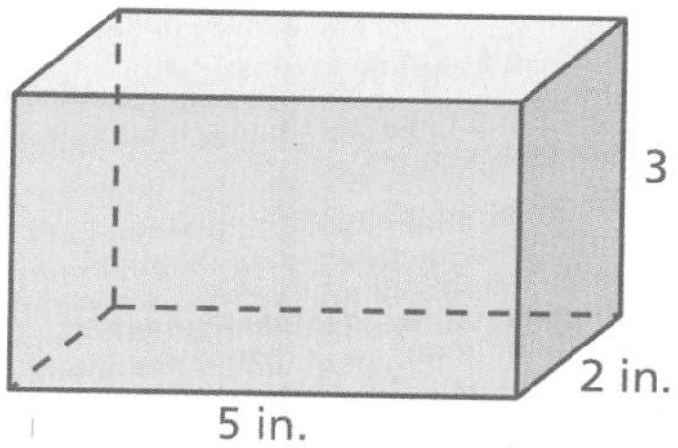

14. Double the width of the prism. How many times greater is the volume of the new prism than the volume of the original prism?

15. Triple the length of the prism. How many times greater is the volume of the new prism than the volume of the original prism?

16. **REASONING** Are your results in Exercises 14 and 15 true for any prism? Are these results true for surface area? Explain.

17. **SANDBOX** A sandbox at a playground is being rebuilt.

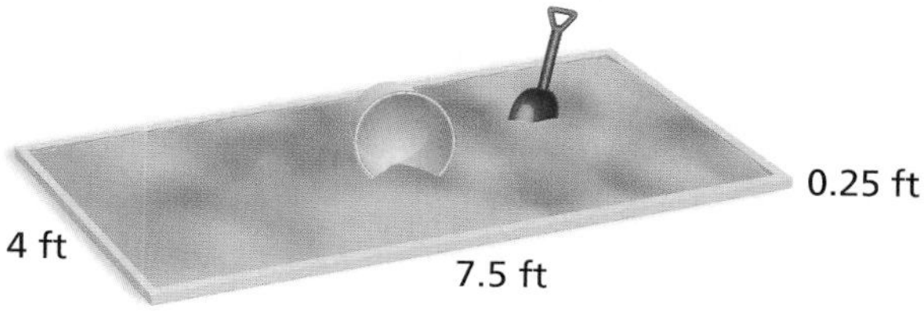

a. Find the volume of the sandbox.

b. After changing only one dimension, the volume of the new sandbox is 15 cubic feet. Describe how this could happen.

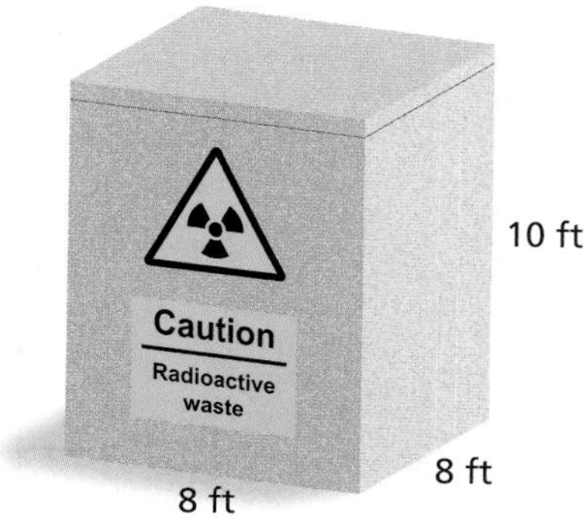

18. **WASTE** A radioactive waste storage container is shown.

a. Find the volume of the container.

b. The dimensions of a larger container are three times greater. Find the volume of the larger container.

c. It costs $256,000 to dispose of the waste in the smaller container. How much does it cost to dispose of the waste in the larger container?

19. Critical Thinking How does the given change affect the volume and surface area of a rectangular prism?

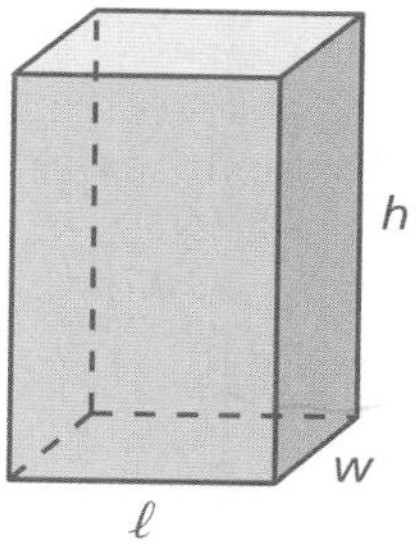

a. Height doubles

b. Height triples

c. All three dimensions double

d. All three dimensions triple

Fair Game Review What you learned in previous grades & lessons

Identify the transformation. *(Section 7.1, Section 7.2, and Section 7.3)*

20.

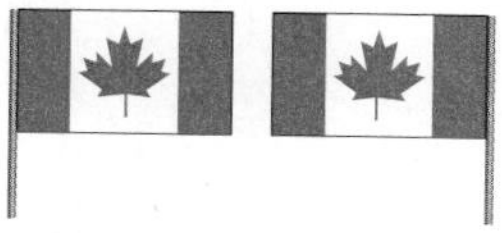

21.

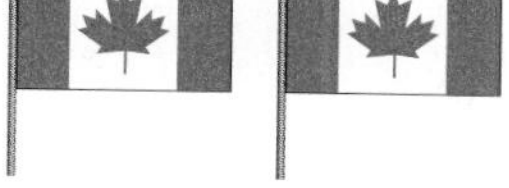

22.

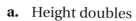

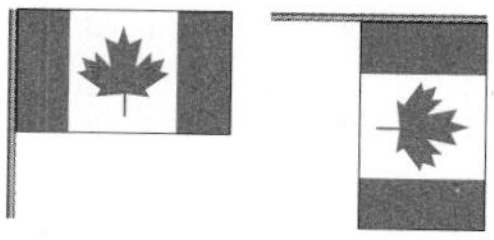

23. **MULTIPLE CHOICE** What is the solution of $\frac{x}{4} = \frac{22}{5}$? *(Section 5.5)*

Ⓐ 1 Ⓑ 17.6 Ⓒ 21 Ⓓ 25.4

You can use an **information frame** to help you organize and remember concepts. Here is an example of an information frame for surface areas of rectangular prisms.

Visual:

base

lateral face

lateral face

lateral face

base

lateral face

Words:

The surface area S of a rectangular prism is the sum of the areas of the bases and the lateral faces.

Surface Areas of Rectangular Prisms

Algebra:

$$S = 2\ell w + 2\ell h + 2wh$$

Area of bases

Area of lateral faces

Example:

$$S = 2\ell w + 2\ell h + 2wh$$
$$= 2(4)(3) + 2(4)(5) + 2(3)(5)$$
$$= 24 + 40 + 30$$
$$= 94 \text{ in.}^2$$

5 in.

4 in.

3 in.

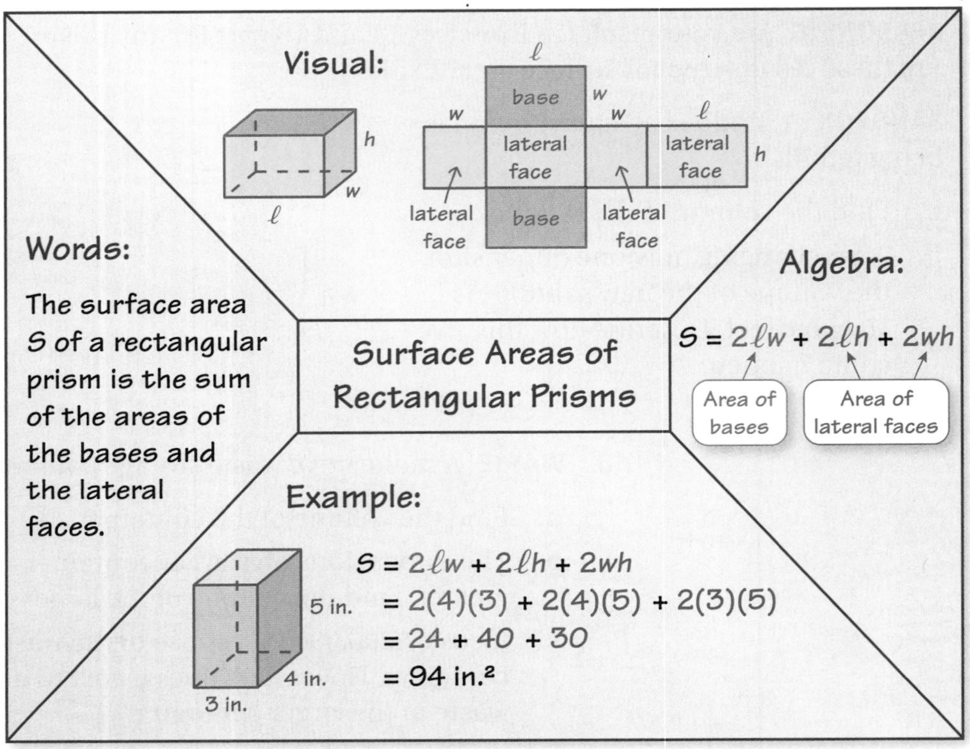

On Your Own

Make an information frame to help you study the topic.

1. volumes of rectangular prisms

After you complete this chapter, make information frames for the following topics.

2. surface areas of cylinders

3. volumes of cylinders

4. Pick three other topics that you studied earlier in this course. Make an information frame for each topic.

"I'm having trouble thinking of a good title for my information frame."

8.1–8.2 Quiz

Check It Out
Progress Check
BigIdeasMath.com

Find the surface area of the prism. *(Section 8.1)*

1.

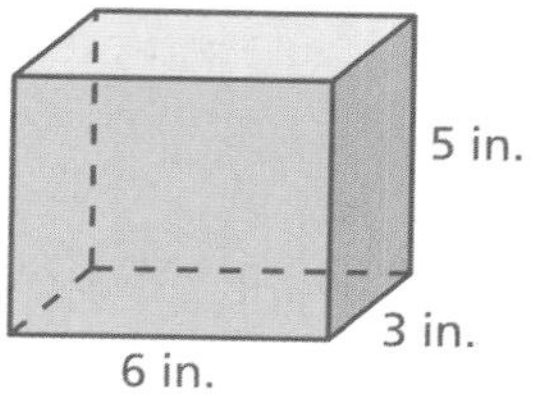

2.

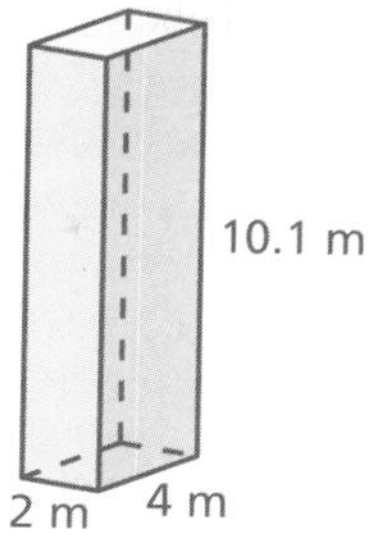

Find the volume of the prism. *(Section 8.2)*

3.

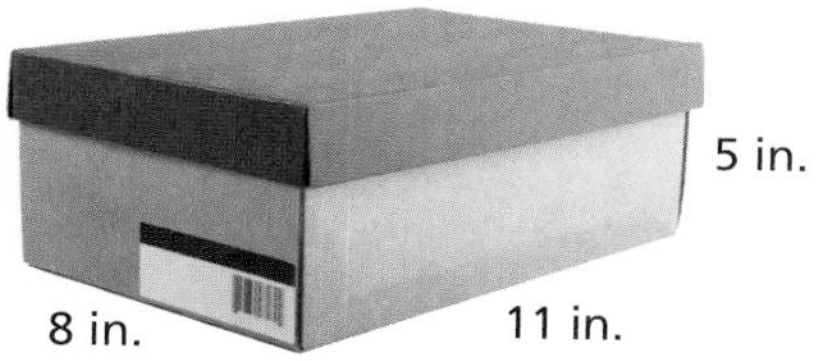

4.

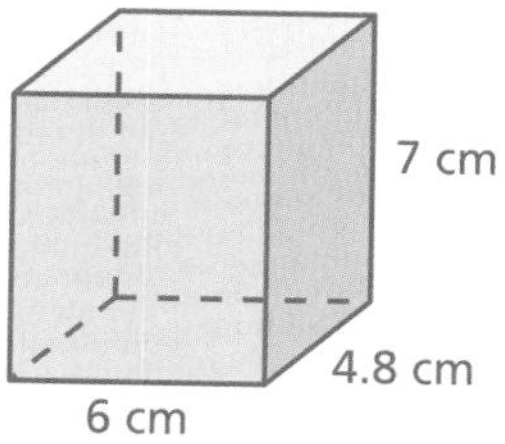

5.

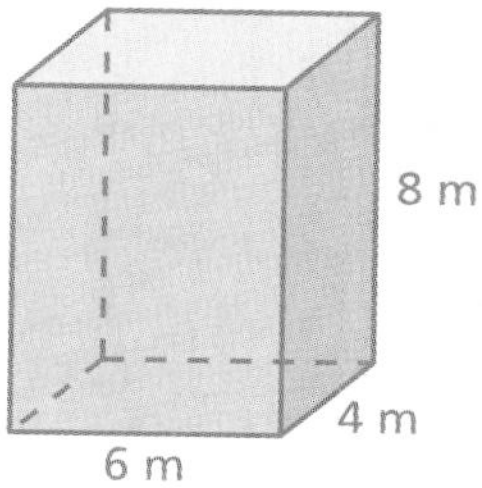

6.

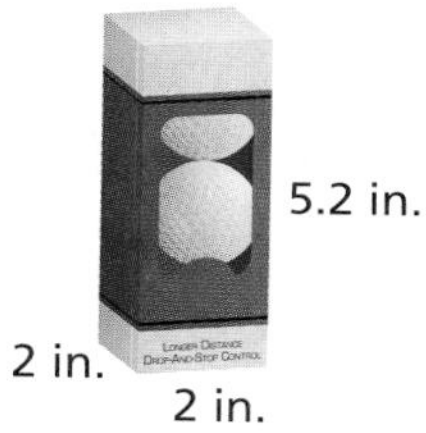

7. REASONING The volume of a prism is 200 cubic feet. What is the volume of the prism when its height is doubled? halved? *(Section 8.2)*

8. GEOMETRY How does doubling all three dimensions of the prism affect its volume? *(Section 8.2)*

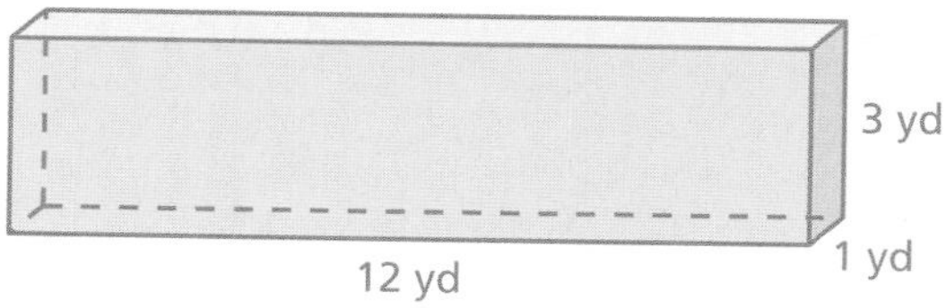

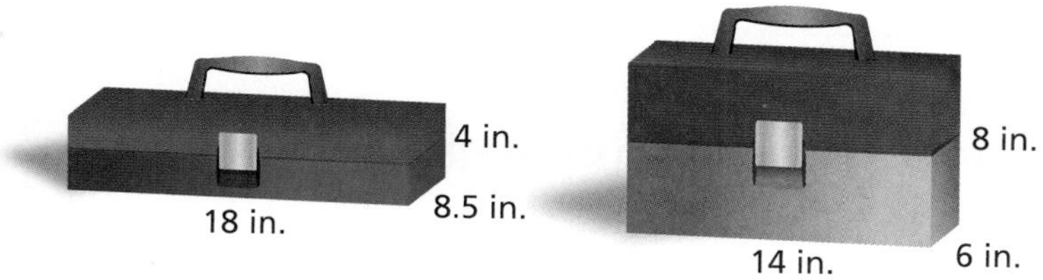

9. TACKLE BOX Which tackle box has the greater volume? Explain your reasoning. *(Section 8.2)*

10. WOODEN CHESTS All the faces of the wooden chest will be painted except for the bottom. *(Section 8.1)*

a. Find the area to be painted, in *square inches.*

b. The dimensions of a larger chest are two times greater. Find the area of the larger chest to be painted,in *square inches.* (Do not include the bottom.)

8.3 Surface Areas of Cylinders

STANDARDS OF LEARNING
7.5

Essential Question How can you find the surface area of a cylinder?

A **cylinder** is a solid that has two parallel, congruent circular bases.

The surface area of a cylinder is the sum of the areas of the bases and the lateral surface.

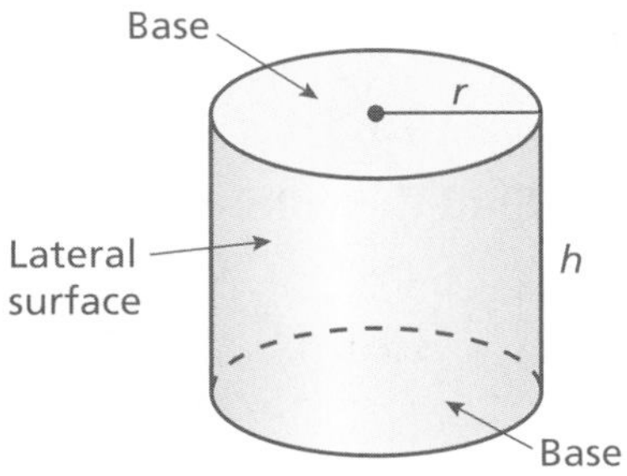

1 ACTIVITY: Finding Area

Work with a partner. Use a cardboard cylinder.

- **Talk about how you can find the area of the outside of the roll.**
- **Use a ruler to estimate the area of the outside of the roll.**
- **Cut the roll and press it out flat. Then find the area of the flattened cardboard. How close is your estimate to the actual area?**

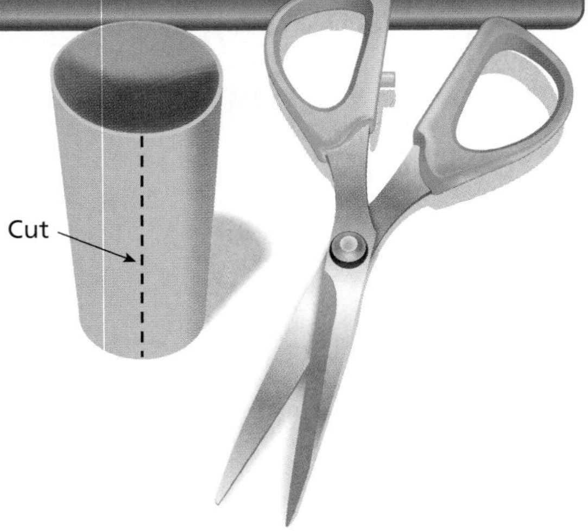

2 ACTIVITY: Finding Surface Area

Work with a partner.

- **Trace the top and bottom of a can on paper. Cut out the two shapes.**
- **Cut out a long paper rectangle. Make the width the same as the height of the can. Wrap the rectangle around the can. Cut off the excess paper so the edges just meet.**
- **Make a net for the can. Name the shapes in the net.**
- **How are the dimensions of the rectangle related to the dimensions of the can?**
- **Explain how to use the net to find the surface area of the can.**

3 ACTIVITY: Estimation

Work with a partner. From memory, estimate the dimensions of the real-life item in inches. Then use the dimensions to estimate the surface area of the item in square inches.

a.

b.

c.

d.

What Is Your Answer?

4. **IN YOUR OWN WORDS** How can you find the surface area of a cylinder? Give an example with your description. Include a drawing of the cylinder.

5. To eight decimal places, $\pi \approx 3.14159265$. Which of the following is closest to π?

   a. 3.14 **b.** $\frac{22}{7}$ **c.** $\frac{355}{113}$

"To approximate the irrational number $\pi \approx 3.141593$, I simply remember 1, 1, 3, 3, 5, 5."

"Then I compute the rational number $\frac{355}{113} \approx 3.141593$."

Practice Use what you learned about the surface area of a cylinder to complete Exercises 5–7 on page 336.

8.3 Lesson

Check It Out
Lesson Tutorials
BigIdeasMath.com

Key Vocabulary
cylinder, *p. 332*

The diagram reviews some important facts for circles.

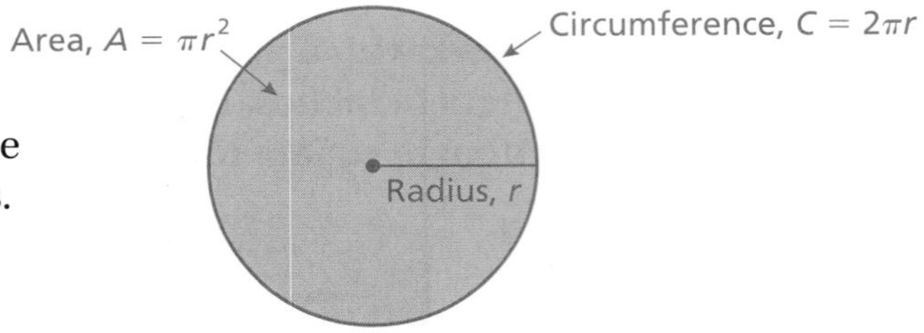

Key Idea

Surface Area of a Cylinder

Words The surface area S of a cylinder is the sum of the areas of the bases and the lateral surface.

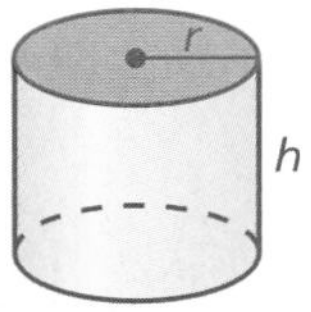

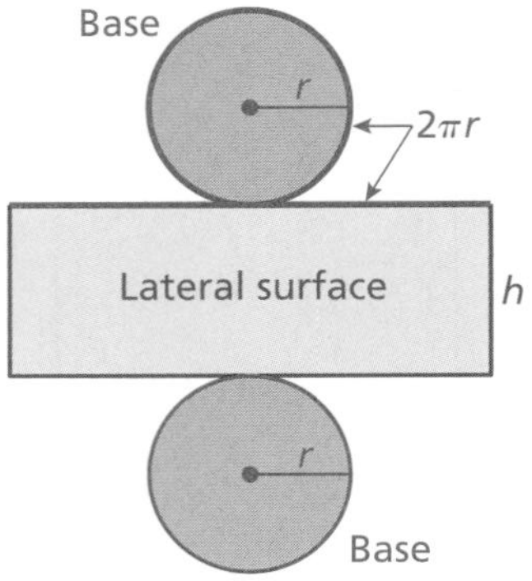

Algebra $S = 2\pi r^2 + 2\pi rh$

Area of bases (points to $2\pi r^2$)

Area of lateral surface (points to $2\pi rh$)

Remember

$\pi = \dfrac{\text{circumference}}{\text{diameter}}$

Pi can be approximated as 3.14 or $\dfrac{22}{7}$.

EXAMPLE 1 Finding the Surface Area of a Cylinder

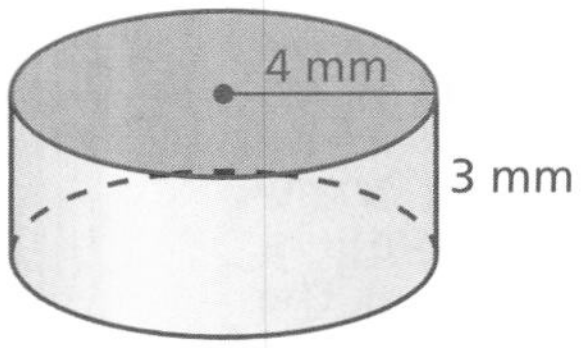

Find the surface area of the cylinder. Round your answer to the nearest tenth.

Draw a net.

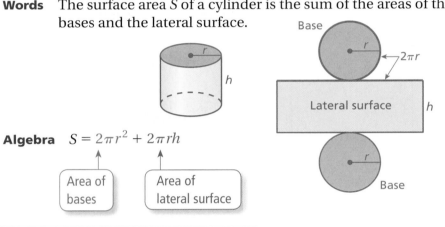

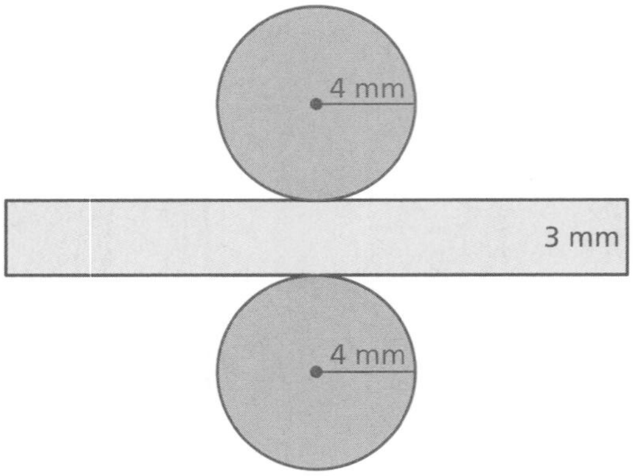

$S = 2\pi r^2 + 2\pi rh$

$= 2\pi(4)^2 + 2\pi(4)(3)$

$= 32\pi + 24\pi$

$= 56\pi \approx 175.8$

The surface area is about 175.8 square millimeters.

On Your Own

Now You're Ready
Exercises 8–10

1. A cylinder has a radius of 2 meters and a height of 5 meters. Find the surface area of the cylinder. Round your answer to the nearest tenth.

EXAMPLE 2 Finding Surface Area

How much paper is used for the label on the can of peas?

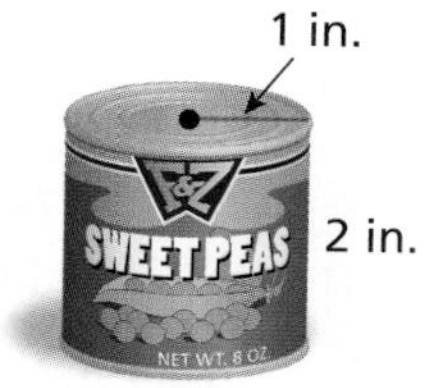

Find the *lateral* surface area of the cylinder.

$S = 2\pi rh$ ← Do not include the area of the bases in the formula.

$= 2\pi(1)(2)$ Substitute.

$= 4\pi \approx 12.56$ Multiply.

∴ About 12.56 square inches of paper is used for the label.

EXAMPLE 3 Real-Life Application

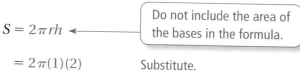

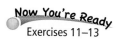

You earn \$0.01 for recycling the can in Example 2. How much can you expect to earn for recycling the tomato can? Assume that the recycle value is proportional to the surface area.

Find the surface area of each can.

Tomatoes

$S = 2\pi r^2 + 2\pi rh$

$= 2\pi(2)^2 + 2\pi(2)(5.5)$

$= 8\pi + 22\pi$

$= 30\pi$

Peas

$S = 2\pi r^2 + 2\pi rh$

$= 2\pi(1)^2 + 2\pi(1)(2)$

$= 2\pi + 4\pi$

$= 6\pi$

Use a proportion to find the recycle value x of the tomato can.

$$\frac{30\pi \text{ in.}^2}{x} = \frac{6\pi \text{ in.}^2}{\$0.01}$$ ← surface area / ← recycle value

$30\pi \cdot 0.01 = x \cdot 6\pi$ Use Cross Products Property.

$5 \cdot 0.01 = x$ Divide each side by 6π.

$0.05 = x$ Simplify.

∴ You can expect to earn \$0.05 for recycling the tomato can.

On Your Own

Now You're Ready
Exercises 11–13

2. **WHAT IF?** In Example 3, the height of the can of peas is doubled.
   a. Does the amount of paper used in the label double?
   b. Does the recycle value double? Explain.

8.3 Exercises

Check It Out
Help with Homework
BigIdeasMath.com

Vocabulary and Concept Check

1. **CRITICAL THINKING** Which part of the formula $S = 2\pi r^2 + 2\pi rh$ represents the lateral surface area of a cylinder?
2. **CRITICAL THINKING** Given the height and the circumference of the base of a cylinder, describe how to find the surface area of the entire cylinder.

Find the indicated area of the cylinder.

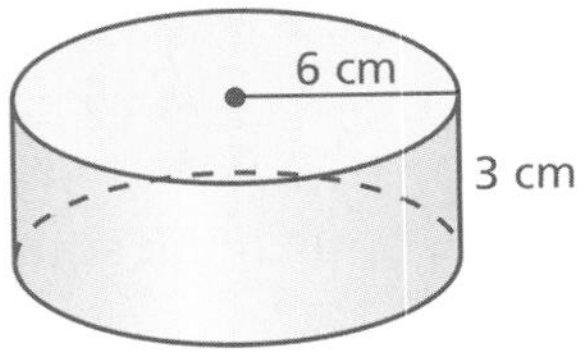

3. Area of a base
4. Surface area

Practice and Problem Solving

Make a net for the cylinder. Then find the surface area of the cylinder. Round your answer to the nearest tenth.

5.
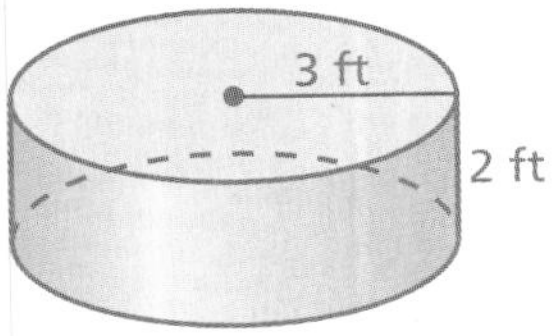

6.
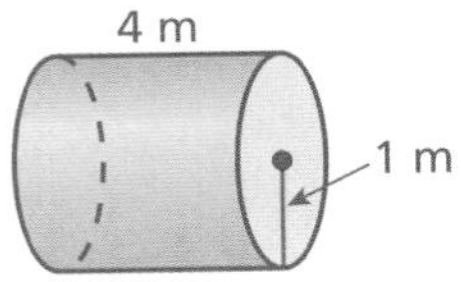

7.
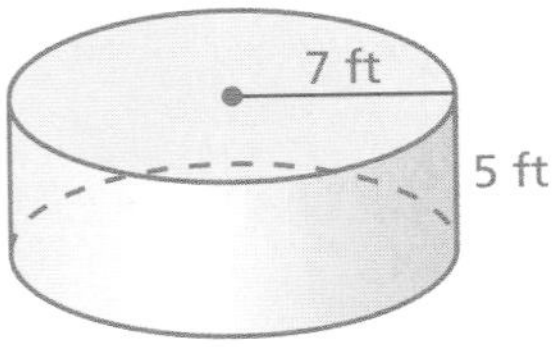

Find the surface area of the cylinder. Round your answer to the nearest tenth.

1 8.
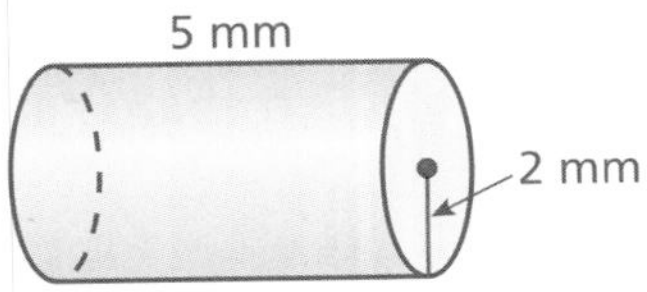

9.
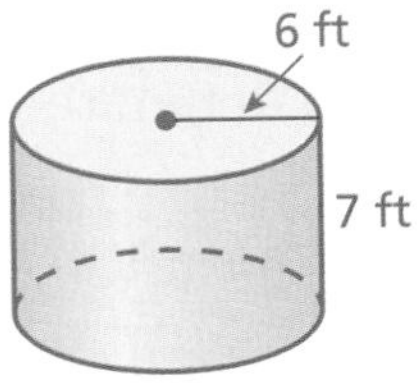

10.
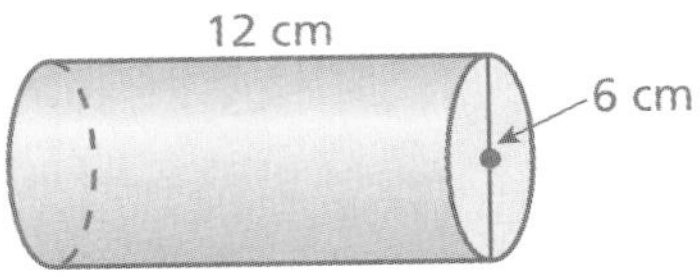

Find the lateral surface area of the cylinder. Round your answer to the nearest tenth.

2 11.
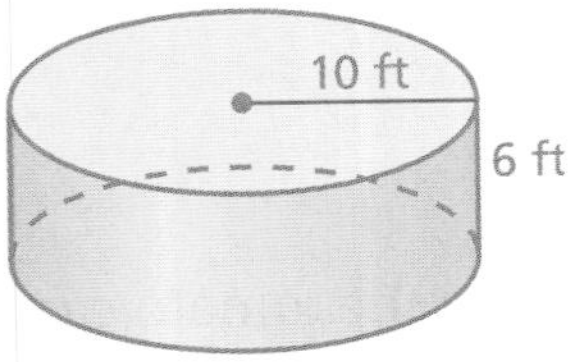

12.
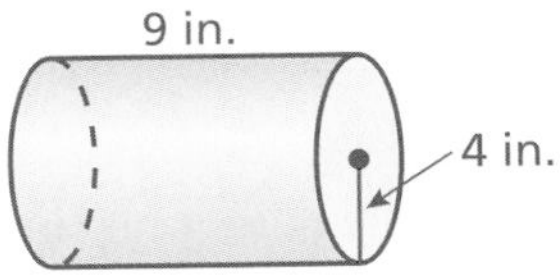

13.
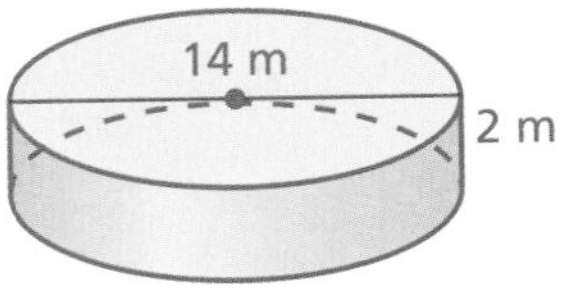

14. **TANKER** The truck's tank is a stainless steel cylinder. Find the surface area of the tank.

15. **ERROR ANALYSIS** Describe and correct the error in finding the surface area of the cylinder.

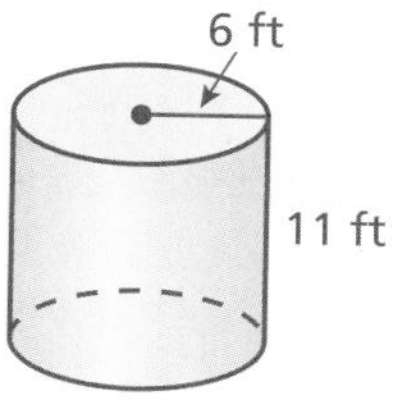

✗ $S = 2\pi rh$
$\approx 2\pi(6)(11)$
$= 132\pi \text{ ft}^2$

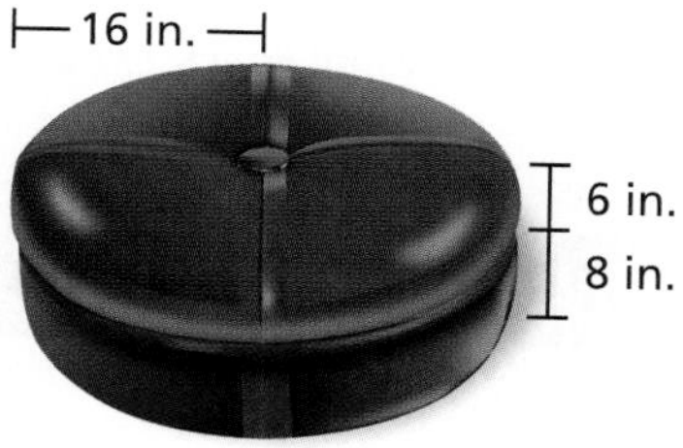

16. **OTTOMAN** What percent of the surface area of the ottoman is green (not including the bottom)?

17. **REASONING** You make two cylinders using 8.5-inch by 11-inch pieces of paper. One has a height of 8.5 inches and the other has a height of 11 inches. Without calculating, compare the surface areas of the cylinders.

18. **INSTRUMENT** A ganza is a percussion instrument used in samba music.

   a. Find the surface area of each of the two labeled ganzas.

   b. The weight of the smaller ganza is 1.1 pounds. Assume that the surface area is proportional to the weight. What is the weight of the larger ganza?

19. **BRIE CHEESE** The cut wedge represents one-eighth of the cheese.

   a. Find the surface area of the cheese before it is cut.

   b. Find the surface area of the remaining cheese after the wedge is removed. Did the surface area increase, decrease, or remain the same?

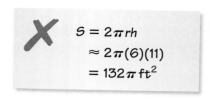

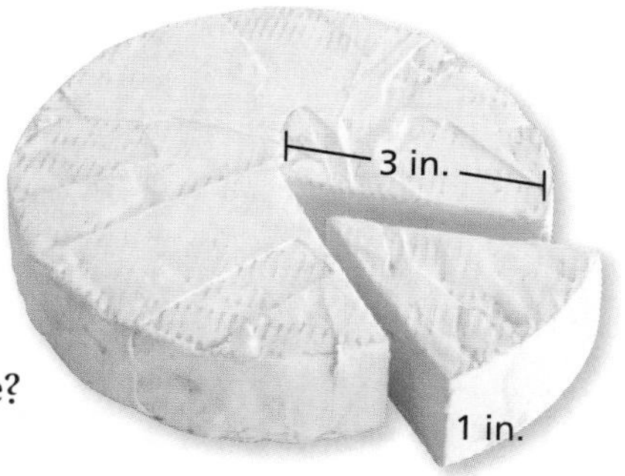

20. Critical Thinking The lateral surface area of a cylinder is 184 square centimeters. The radius is 9 centimeters. What is the surface area of the cylinder? Explain how you found your answer.

Fair Game Review What you learned in previous grades & lessons

Evaluate the expression. *(Skills Review Handbook)*

21. $\frac{1}{2}(26)(9)$

22. $\frac{1}{2}(8.24)(3) + 8.24$

23. $\frac{1}{2}(18.84)(3) + 28.26$

24. **MULTIPLE CHOICE** You can type 40 words in one minute. Which equation represents the number w of words you can type in m minutes? *(Section 4.2)*

Ⓐ $w = 40m$ Ⓑ $m = 40w$ Ⓒ $w = \frac{1}{40}m$ Ⓓ $w = 40m + 1$

8.4 Volumes of Cylinders

STANDARDS OF LEARNING
7.5

Essential Question How can you find the volume of a cylinder?

1 ACTIVITY: Finding a Formula Experimentally

Work with a partner.

a. Find the area of the face of a coin.

b. Find the volume of a stack of a dozen coins.

c. Generalize your results to find the volume of a cylinder.

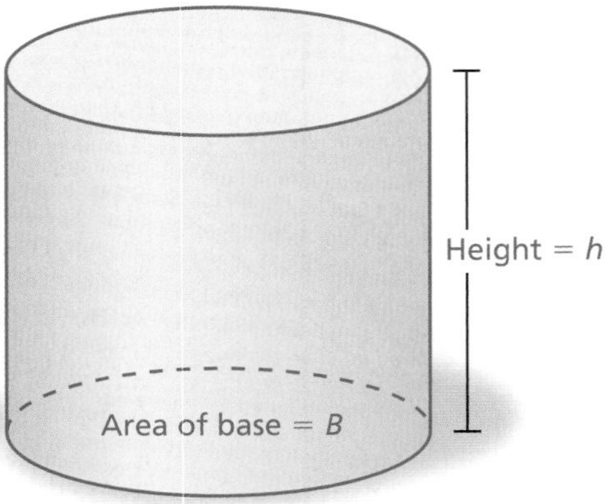

2 ACTIVITY: Making a Business Plan

Work with a partner. You are planning to make and sell 3 different sizes of cylindrical candles. You buy 1 cubic foot of candle wax for $20 to make 8 candles of each size.

a. Design the candles. What are the dimensions of each size?

b. You want to make a profit of $100. Decide on a price for each size.

c. Did you set the prices so that they are proportional to the volume of each size of candle? Why or why not?

3 ACTIVITY: Science Experiment

Work with a partner. Use the diagram to describe how you can find the volume of a small object.

4 ACTIVITY: Comparing Cylinders

Work with a partner.

a. Just by looking at the two cylinders, which one do you think has the greater volume? Explain your reasoning.

b. Find the volume of each cylinder. Was your prediction in part (a) correct? Explain your reasoning.

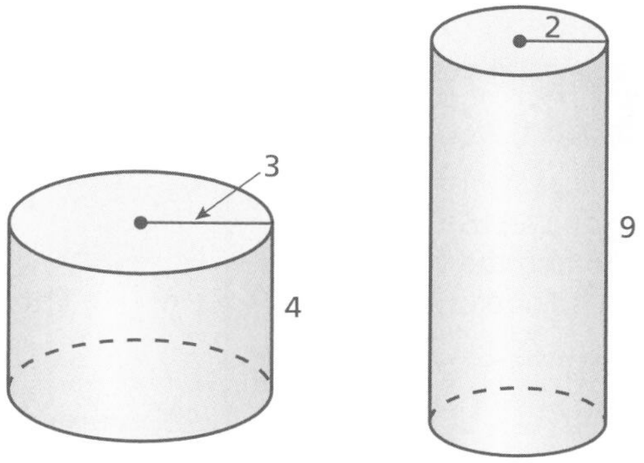

What Is Your Answer?

5. IN YOUR OWN WORDS How can you find the volume of a cylinder?

6. Compare your formula for the volume of a cylinder with the formula for the volume of a prism. How are they the same?

"Here's how I remember how to find the volume of any prism or cylinder."

"Base times tall, will fill 'em all."

Use what you learned about the volumes of cylinders to complete Exercises 3–5 on page 342.

8.4 Lesson

Check It Out
Lesson Tutorials
BigIdeasMath.com

Key Idea

Volume of a Cylinder

Words The volume V of a cylinder is the product of the area of the base and the height of the cylinder.

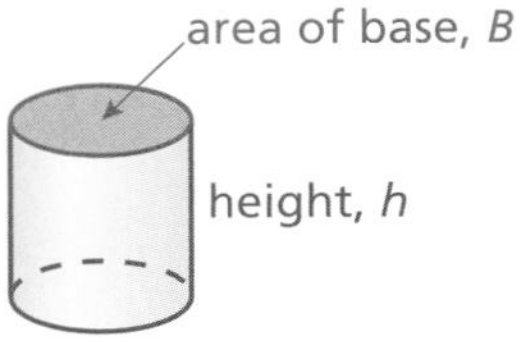

Algebra $V = Bh$

Area of base (B) — Height of cylinder (h)

EXAMPLE 1 Finding the Volume of a Cylinder

Find the volume of the cylinder. Round your answer to the nearest tenth.

Study Tip

Because $B = \pi r^2$, you can use $V = \pi r^2 h$ to find the volume of a cylinder.

$V = Bh$	Write formula for volume.
$= \pi(3)^2(6)$	Substitute.
$= 54\pi \approx 169.6$	Simplify.

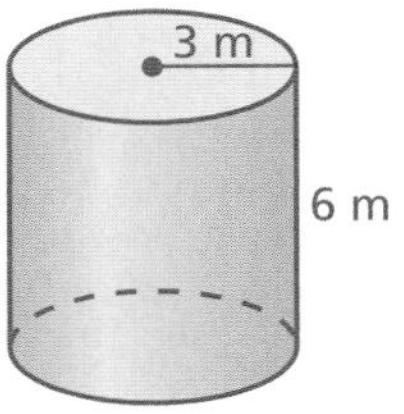

∴ The volume is about 169.6 cubic meters.

EXAMPLE 2 Finding the Height of a Cylinder

Find the height of the cylinder. Round your answer to the nearest whole number.

The diameter is 10 inches. So, the radius is 5 inches.

$V = Bh$	Write formula for volume.
$314 = \pi(5)^2(h)$	Substitute.
$314 = 25\pi h$	Simplify.
$4 \approx h$	Divide each side by 25π.

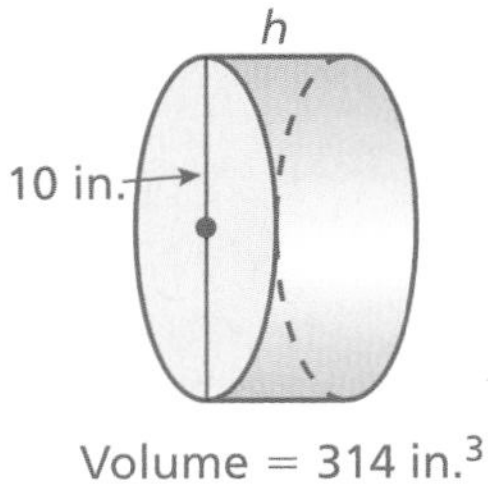

∴ The height is about 4 inches.

On Your Own

Now You're Ready
Exercises 3–11 and 13–15

Find the volume V or height h of the cylinder. Round your answer to the nearest tenth.

1.

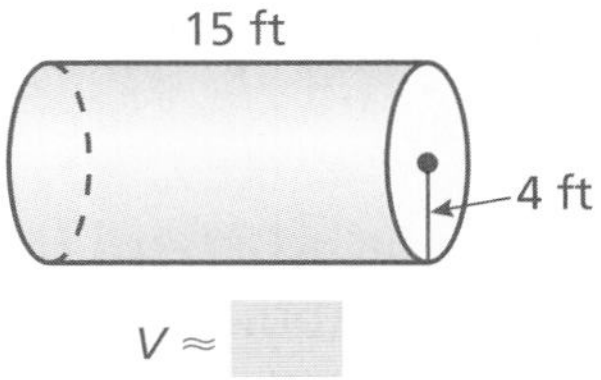

2.

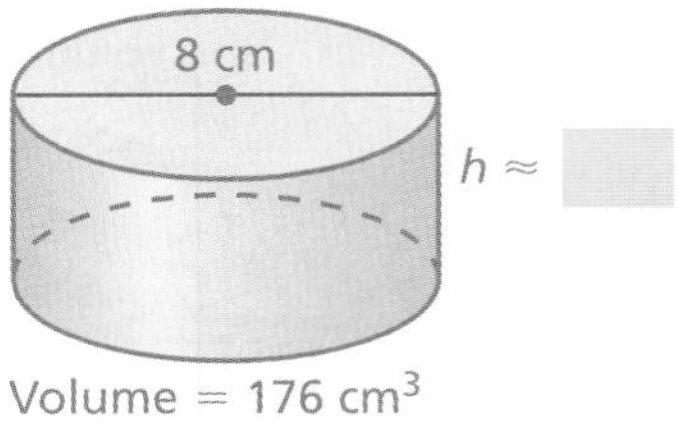

EXAMPLE 3 Real-Life Application

How much salsa is missing from the jar?

The missing salsa fills a cylinder with a height of $10 - 4 = 6$ centimeters and a radius of 5 centimeters.

$V = Bh$	Write formula for volume.
$= \pi(5)^2(6)$	Substitute.
$= 150\pi \approx 471$	Simplify.

∴ About 471 cubic centimeters of salsa are missing from the jar.

EXAMPLE 4 Standardized Test Practice

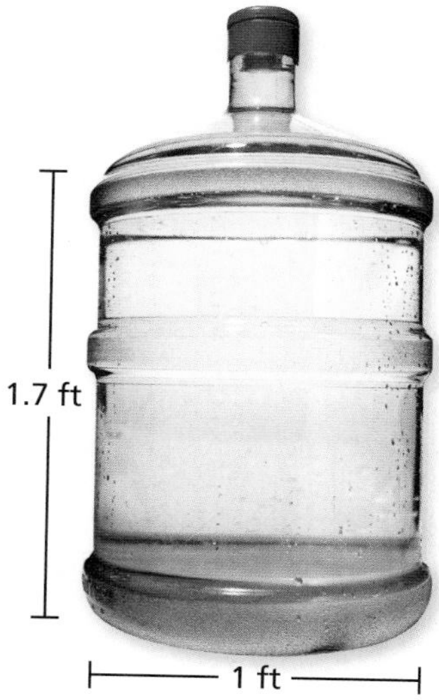

About how many gallons of water does the water cooler bottle contain? ($1\ \text{ft}^3 \approx 7.5$ gal)

Ⓐ 5.3 gal Ⓑ 10 gal Ⓒ 17 gal Ⓓ 40 gal

Find the volume of the cylinder. The diameter is 1 foot. So, the radius is 0.5 foot.

$V = Bh$	Write formula for volume.
$= \pi(0.5)^2(1.7)$	Substitute.
$= 0.425\pi \approx 1.3345$	Simplify.

So, the cylinder contains about 1.3345 cubic feet of water. To find the number of gallons it contains, multiply by $\frac{7.5\ \text{gal}}{1\ \text{ft}^3}$.

$$1.3345\ \cancel{\text{ft}^3} \times \frac{7.5\ \text{gal}}{1\ \cancel{\text{ft}^3}} \approx 10\ \text{gal}$$

∴ The water cooler bottle contains about 10 gallons of water. The correct answer is Ⓑ.

On Your Own

Now You're Ready
Exercise 12

3. **WHAT IF?** In Example 3, the height of the salsa in the jar is 5 centimeters. How much salsa is missing from the jar?
4. A cylindrical water tower has a diameter of 15 meters and a height of 5 meters. About how many gallons of water can the tower contain? ($1\ \text{m}^3 \approx 264$ gal)

8.4 Exercises

Check It Out
Help with Homework
BigIdeasMath.com

Vocabulary and Concept Check

1. **DIFFERENT WORDS, SAME QUESTION** Which is different? Find "both" answers.

   How much does it take to fill the cylinder?

   What is the capacity of the cylinder?

   How much does it take to cover the cylinder?

   How much does the cylinder contain?

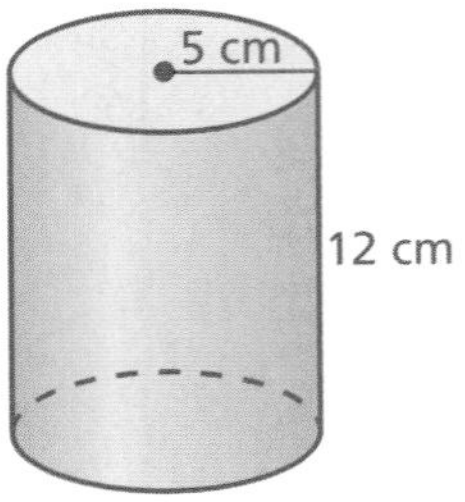

2. **REASONING** Without calculating, which of the solids has the greater volume? Explain.

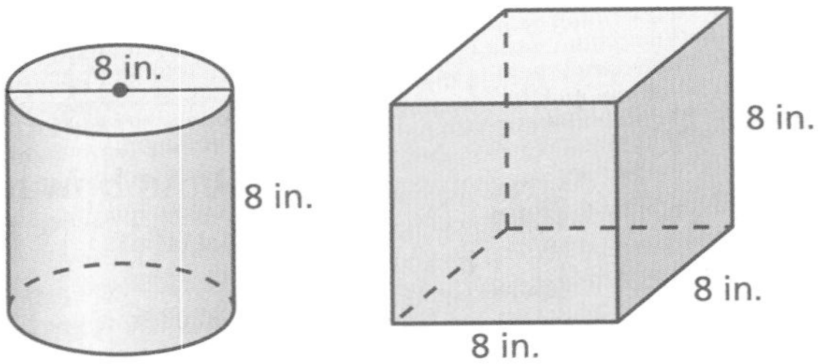

Practice and Problem Solving

Find the volume of the cylinder. Round your answer to the nearest tenth.

(1) 3.
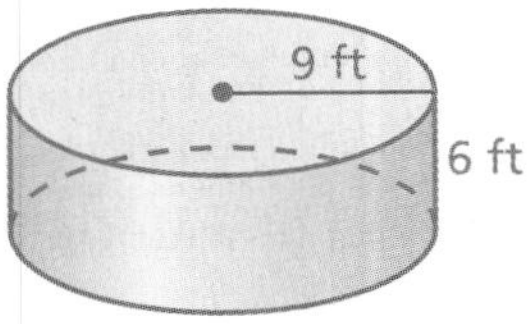

4.
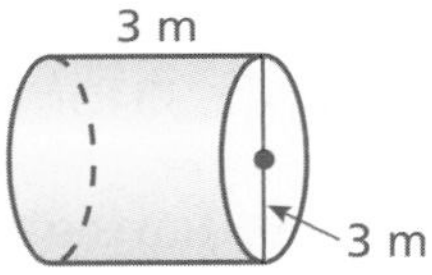

5.
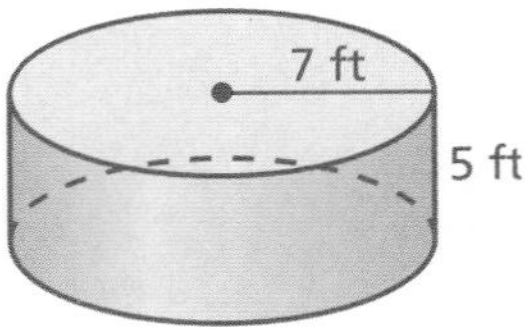

6.
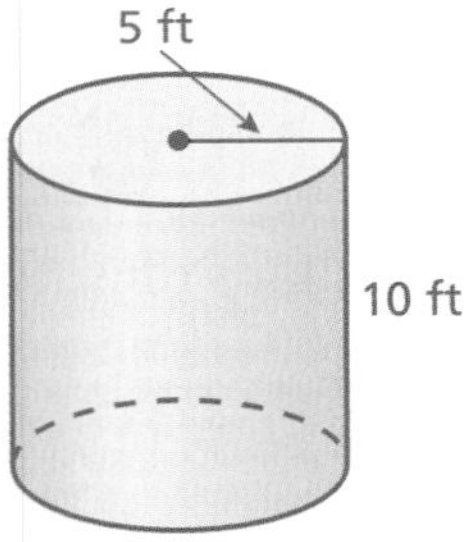

7.
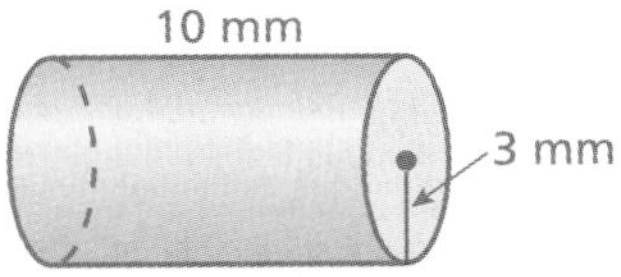

8.
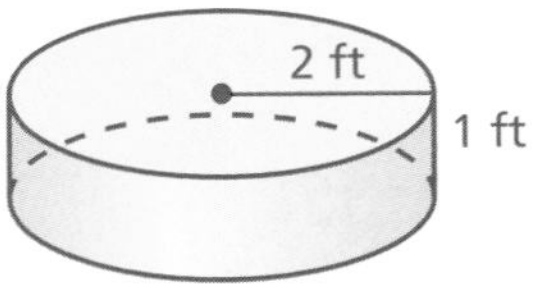

9.
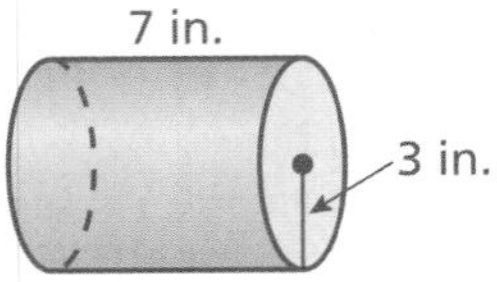

10.
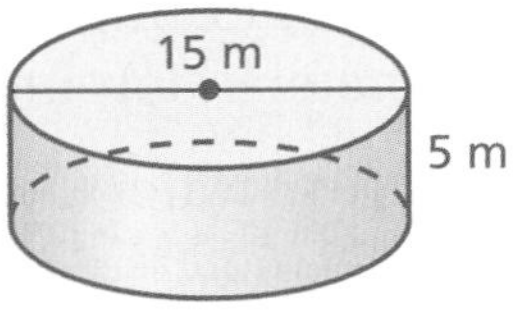

11.
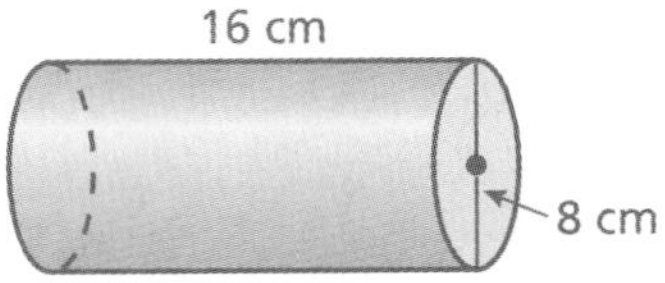

(4) 12. **SWIMMING POOL** A cylindrical swimming pool has a diameter of 16 feet and a height of 4 feet. About how many gallons of water can the pool contain? Round your answer to the nearest whole number. ($1 \text{ ft}^3 \approx 7.5$ gal)

Find the height of the cylinder. Round your answer to the nearest whole number.

② **13.** Volume = 250 ft^3

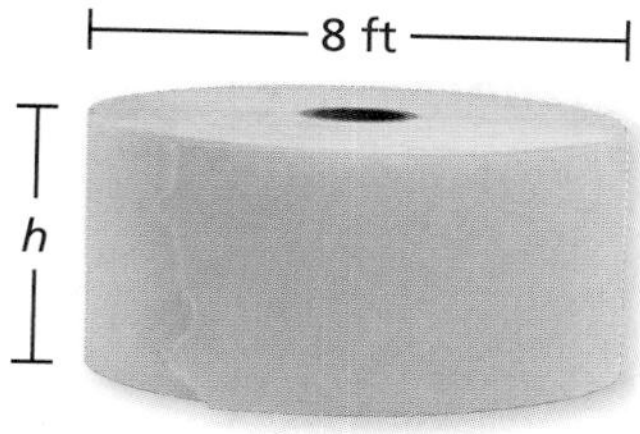

14. Volume = $32{,}000 \text{ in.}^3$

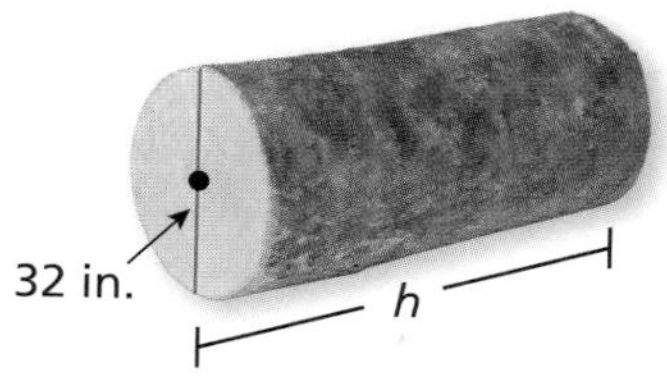

15. Volume = $600{,}000 \text{ cm}^3$

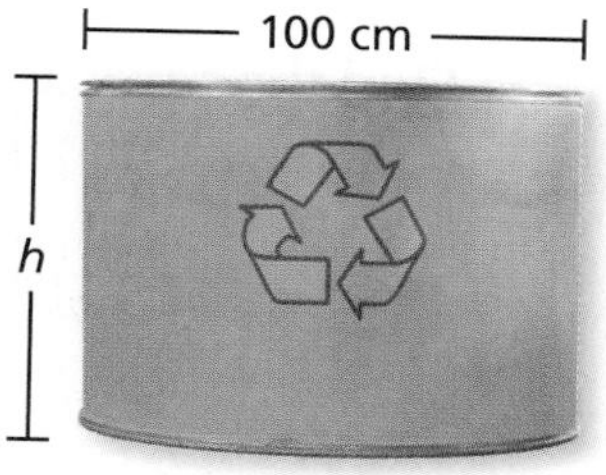

16. CRITICAL THINKING How does the volume of a cylinder change when its diameter is halved? Explain.

Round Hay Bale

17. HAY BALES A traditional "square" bale of hay is actually in the shape of a rectangular prism. Its dimensions are 2 feet by 2 feet by 4 feet. How many "square" bales contain the same amount of hay as one large "round" bale?

18. ROAD ROLLER A tank on the road roller is filled with water to make the roller heavy. The tank is a cylinder that has a height of 6 feet and a radius of 2 feet. One cubic foot of water weighs 62.5 pounds. Find the weight of the water in the tank.

19. VOLUME A cylinder has a surface area of 1850 square meters and a radius of 9 meters. Estimate the volume of the cylinder to the nearest whole number.

20. Critical Thinking Water flows at 2 feet per second through a pipe with a diameter of 8 inches. A cylindrical tank with a diameter of 15 feet and a height of 6 feet collects the water.

a. What is the volume, in cubic inches, of water flowing out of the pipe every second?

b. What is the height, in inches, of the water in the tank after 5 minutes?

c. How many minutes will it take to fill 75% of the tank?

Fair Game Review What you learned in previous grades & lessons

Solve the equation. Check your solution. *(Section 3.4)*

21. $y + 15 = 23$ **22.** $b - 7 = -3$ **23.** $1.3 = x - 3.7$ **24.** $-11.2 = a + 12.4$

25. MULTIPLE CHOICE A couch is 6 feet long. A model of the couch has a scale of 1 in. : 2 ft. How long is the model couch? *(Section 6.5)*

Ⓐ 2 in. Ⓑ 3 in. Ⓒ 6 in. Ⓓ 12 in.

8.3–8.4 Quiz

Check It Out
Progress Check
BigIdeasMath.com

Find the surface area of the cylinder. Round your answer to the nearest tenth. *(Section 8.3)*

1.

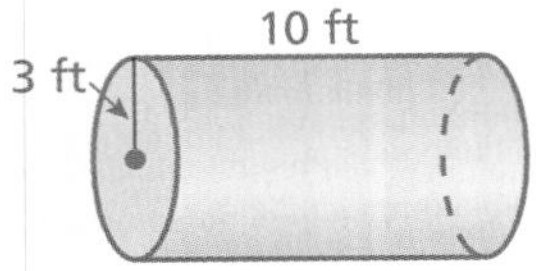

2.

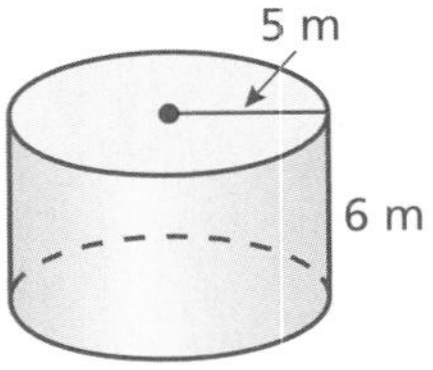

Find the lateral surface area of the cylinder. Round your answer to the nearest tenth. *(Section 8.3)*

3.

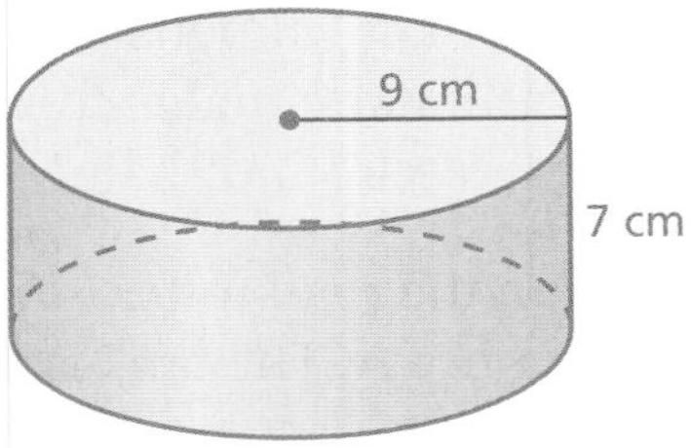

4.

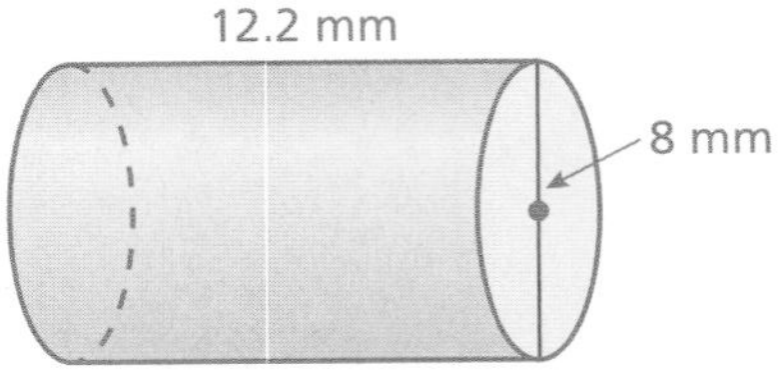

Find the volume of the cylinder. Round your answer to the nearest tenth. *(Section 8.4)*

5.

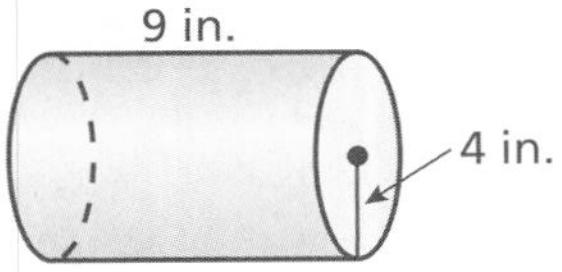

6.

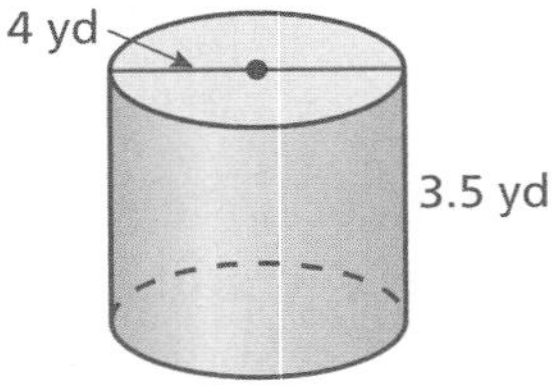

7. MAILING TUBE What is the least amount of material needed to make the mailing tube? *(Section 8.3)*

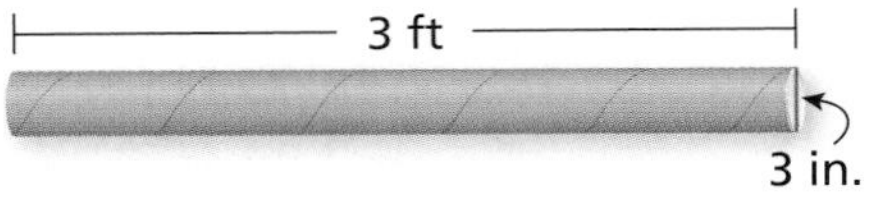

8. TOMATO PASTE How much more paper is used for the label of the large can of tomato paste than for the label of the small can? *(Section 8.3)*

Can A Can B

9. JUICE CAN You buy two cylindrical cans of juice. Each can holds the same amount of juice. What is the height of Can B? *(Section 8.4)*

8 Chapter Review

Check It Out
Vocabulary Help
BigIdeasMath.com

Review Key Vocabulary

rectangular prism, *p. 320* volume, *p. 326* cylinder, *p. 332*

Review Examples and Exercises

8.1 Surface Areas of Rectangular Prisms *(pp. 318–323)*

Find the surface area of the prism.

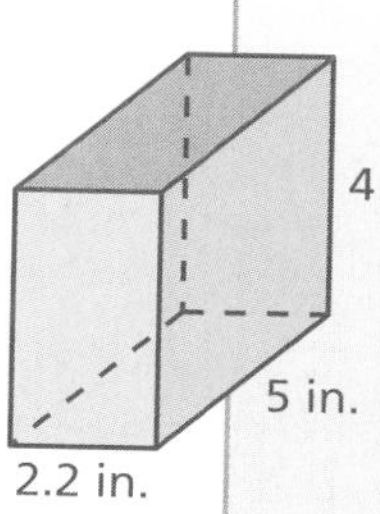

Draw a net.

$$S = 2\ell w + 2\ell h + 2wh$$
$$= 2(5)(2.2) + 2(5)(4) + 2(2.2)(4)$$
$$= 22 + 40 + 17.6$$
$$= 79.6$$

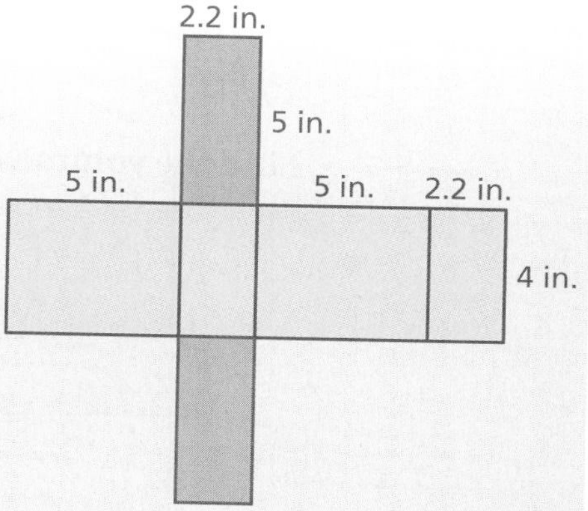

∴ The surface area is 79.6 square inches.

Exercises

Find the surface area of the prism.

1.

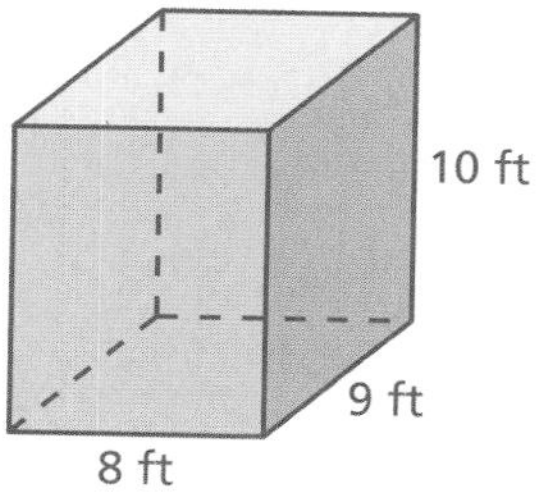

2.

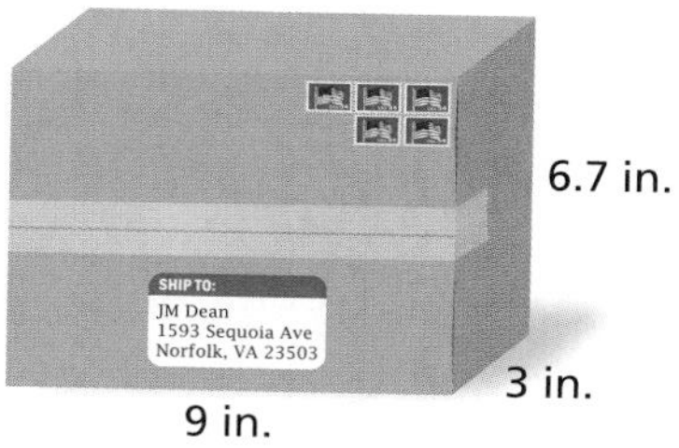

3. **GEOMETRY** Compare the dimensions of the prisms. How many times greater is the surface area of the red prism than the surface area of the blue prism?

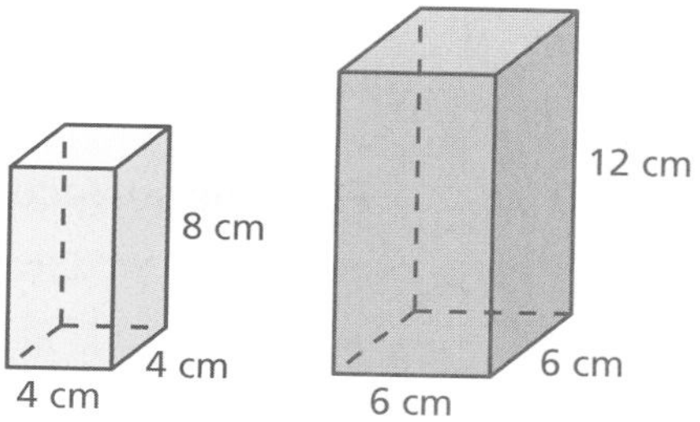

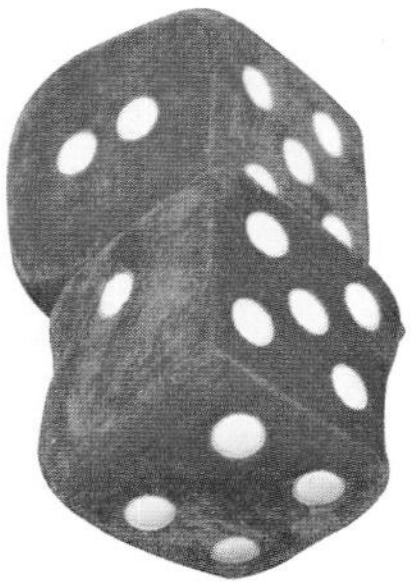

4. **CUBE** A number cube has side lengths of 3 inches. Find the surface area of one cube.

8.2 Volumes of Rectangular Prisms (pp. 324–329)

Find the volume of the prism.

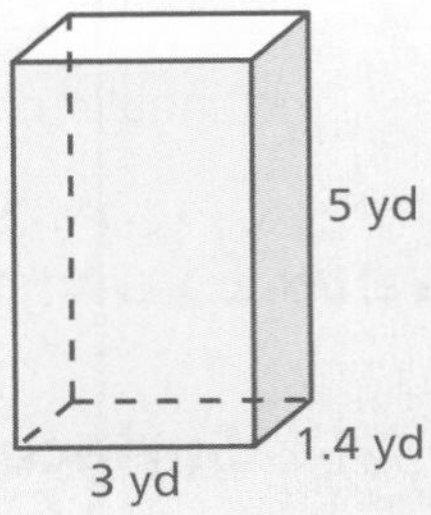

$V = Bh$ — Write formula for volume.

$= 3(1.4) \cdot 5$ — Substitute.

$= 4.2 \cdot 5$ — Multiply.

$= 21$ — Multiply.

∴ The volume is 21 cubic yards.

Exercises

Find the volume of the prism.

5.

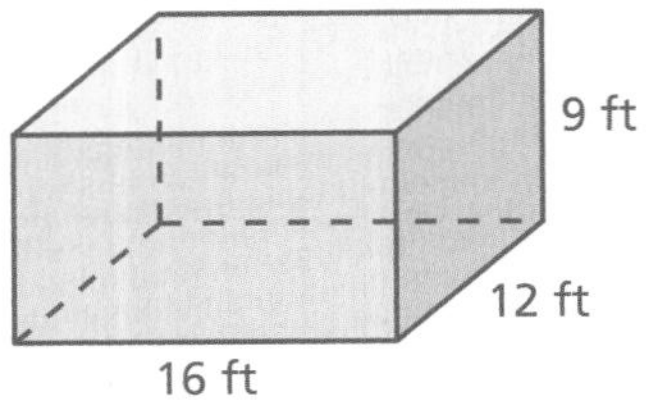

6.

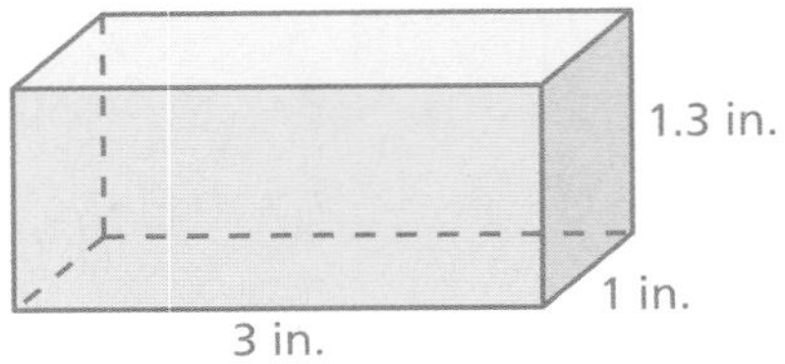

7. GEOMETRY Does doubling the width of a prism double its volume? double its surface area? Explain.

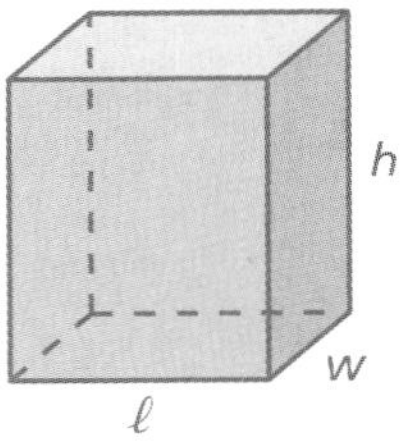

8. GEOMETRY How does tripling all three dimensions of the prism affect its volume?

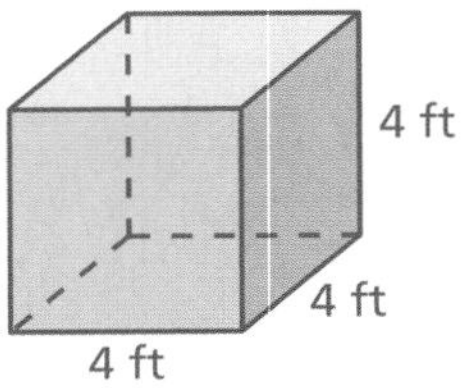

8.3 Surface Areas of Cylinders (pp. 332–337)

Find the surface area of the cylinder. Round your answer to the nearest tenth.

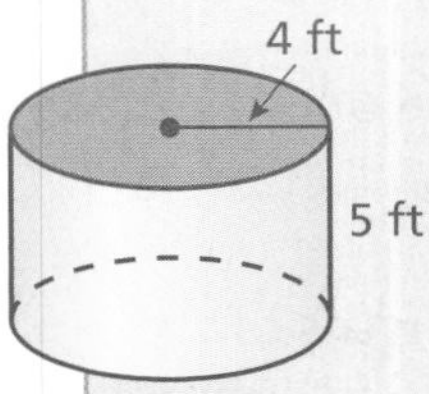

Draw a net.

$S = 2\pi r^2 + 2\pi rh$

$= 2\pi(4)^2 + 2\pi(4)(5)$

$= 32\pi + 40\pi$

$= 72\pi \approx 226.1$

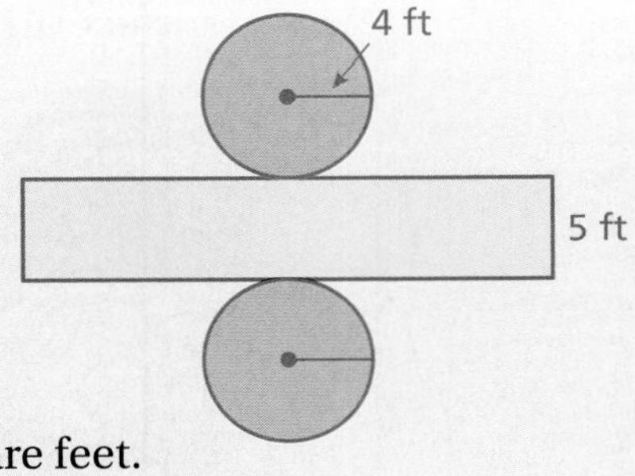

∴ The surface area is about 226.1 square feet.

Exercises

Find the surface area of the cylinder. Round your answer to the nearest tenth.

9.

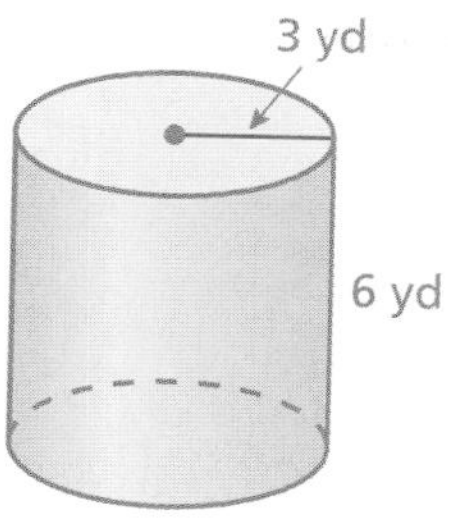

10.

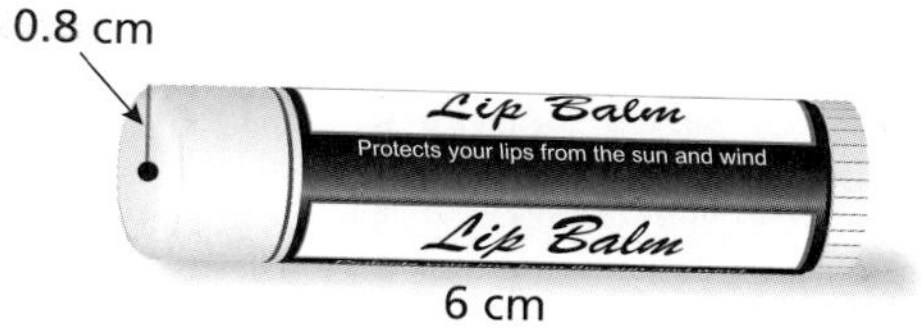

11. ORANGES Find the lateral surface area of the can of mandarin oranges.

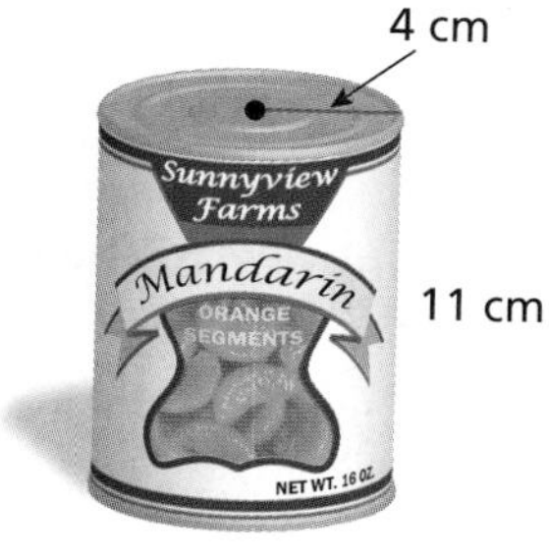

8.4 Volumes of Cylinders (pp. 338–343)

Find the volume of the cylinder. Round your answer to the nearest tenth.

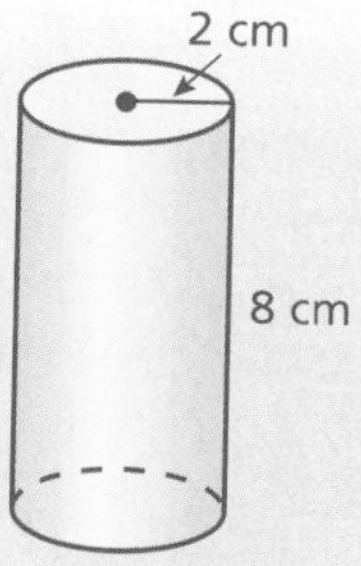

$V = Bh$	Write formula for volume.
$= \pi(2)^2(8)$	Substitute.
$= 32\pi \approx 100.5$	Simplify.

∴ The volume is about 100.5 cubic centimeters.

Exercises

Find the volume of the cylinder. Round your answer to the nearest tenth.

12.

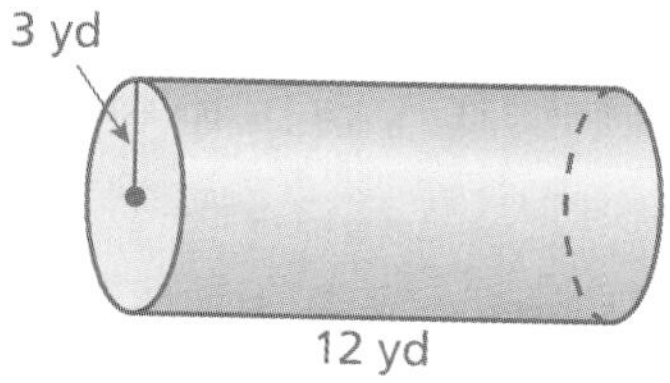

13.

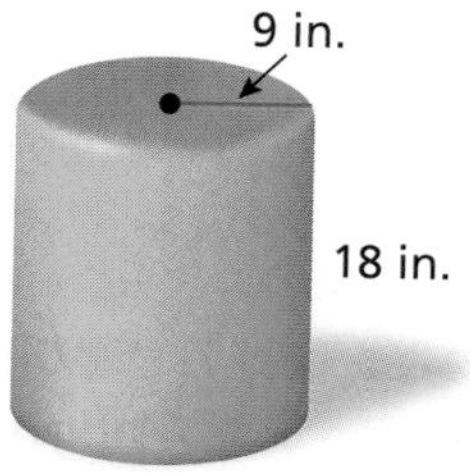

Find the height of the cylinder. Round your answer to the nearest whole number.

14. Volume = 25 in.^{3}

15. Volume = 7599 m^3

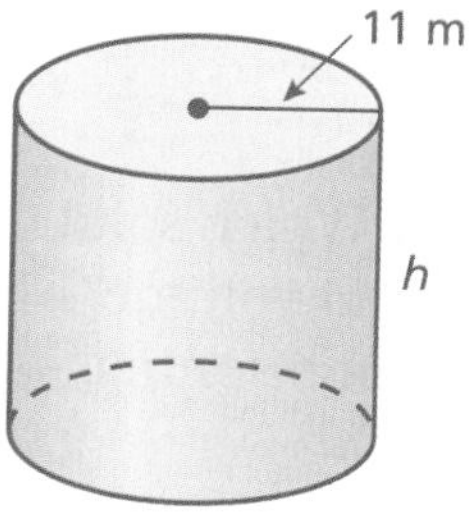

8 Chapter Test

Check It Out
Test Practice
BigIdeasMath.com

Find the surface area and volume of the prism.

1.

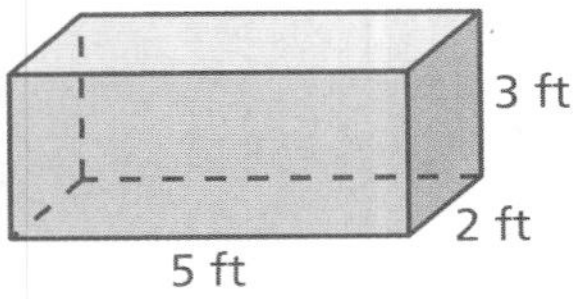

2.

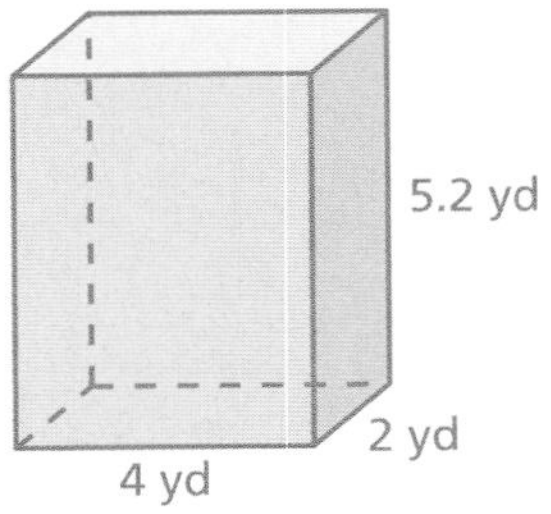

Find the surface area and volume of the cylinder. Round your answer to the nearest tenth.

3.

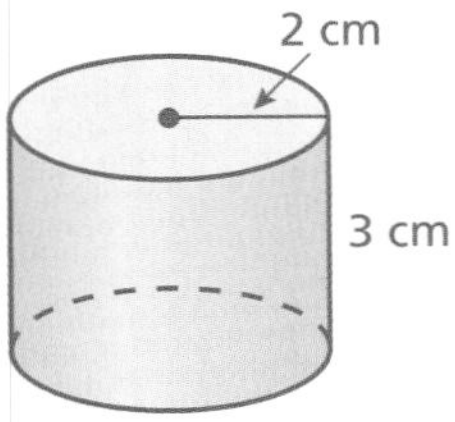

4.

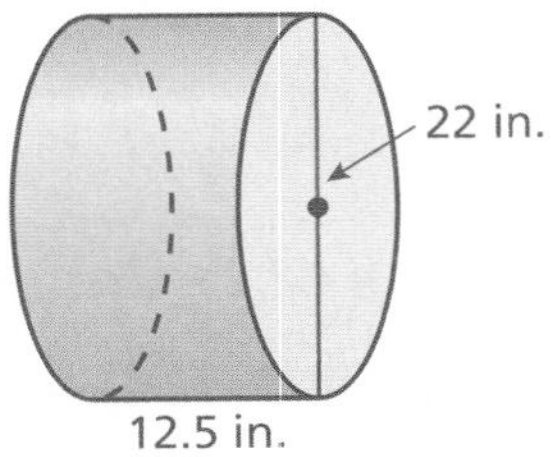

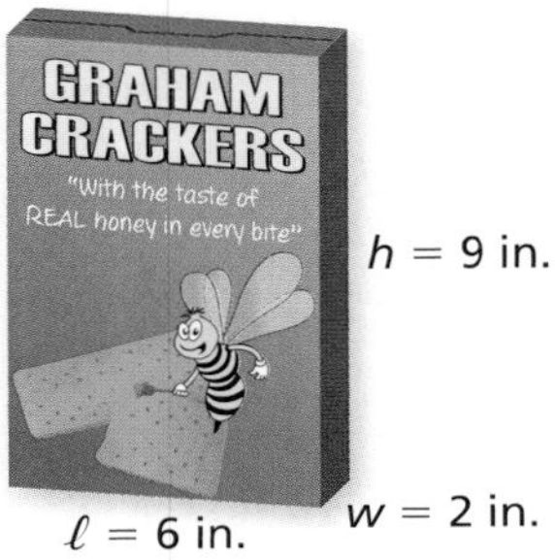

GRAHAM
CRACKERS
"With the taste of REAL honey in every bite"
h = 9 in.
w = 2 in.
ℓ = 6 in.

5. GRAHAM CRACKERS A manufacturer wants to double the volume of the graham cracker box. The manufacturer will either double the height or double the width.

a. Which option uses the least amount of cardboard?

b. What is the volume of the new graham cracker box?

6. MILK Glass A has a diameter of 3.5 inches and a height of 4 inches. Glass B has a radius of 1.5 inches and a height of 5 inches. Which glass can hold more milk?

7. SOUP The label on the can of soup covers about 354.2 square centimeters. What is the height of the can? Round your answer to the nearest whole number.

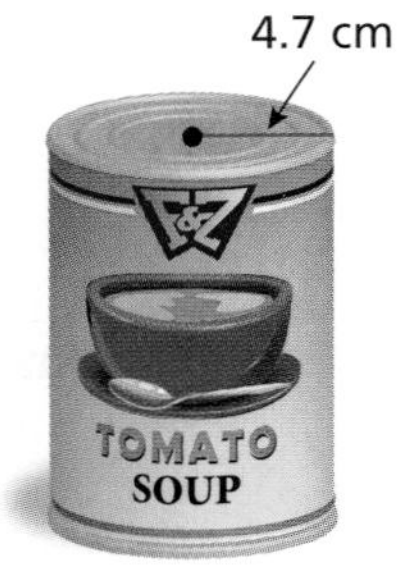

8. REASONING Which solid has the greater volume? Explain your reasoning.

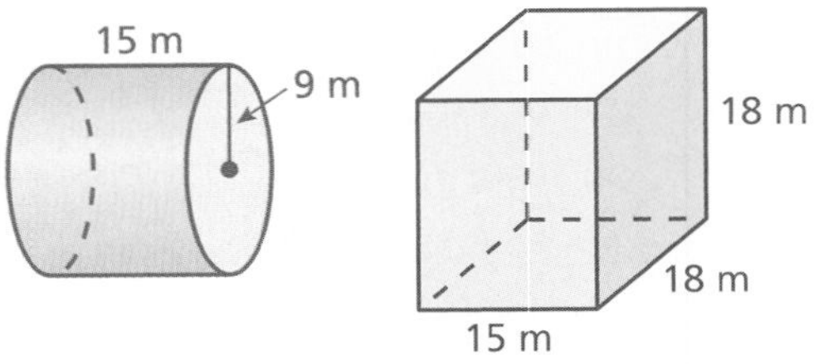

1. A cylinder and its dimensions are shown below.

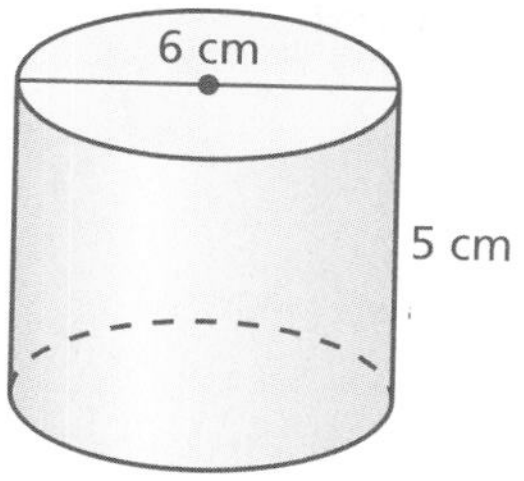

What is the volume of the cylinder? (Use 3.14 for π.)

A. 47.1 cm^3 **C.** 141.3 cm^3

B. 94.2 cm^3 **D.** 565.2 cm^3

Test-Taking Strategy
Answer Easy Questions First
Find the surface area.
(A) 10 ft (C) 10 ft²
(B) 10 ft³ (D) 2 ft³
2ft
1ft
1ft
Neat! Didn't even use a formula.
"Scan the test and answer the easy questions first. You know area is measured in square units."

2. A dance hall charges an initial fee of $150 plus an hourly rate to rent the hall. It costs $477.50 to rent the dance hall for 5 hours. How much is the hourly rate in dollars?

3. Raj was solving the proportion in the box below.

$$\frac{3}{8} = \frac{x-3}{24}$$

$$3 \cdot 24 = (x-3) \cdot 8$$

$$72 = x - 24$$

$$96 = x$$

What should Raj do to correct the error that he made?

F. Set the product of the numerators equal to the product of the denominators.

G. Distribute 8 to get $8x - 24$.

H. Add 3 to each side to get $\frac{3}{8} + 3 = \frac{x}{24}$.

I. Divide both sides by 24 to get $\frac{3}{8} \div 24 = x - 3$.

4. List the ordered pairs shown in the mapping diagram below.

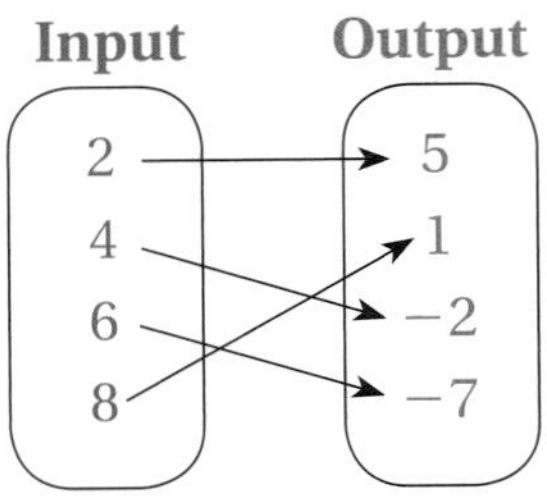

A. (2, 5), (4, −2), (6, −7), (8, 1)

B. (2, −7), (4, −2), (6, 1), (8, 5)

C. (2, 5), (4, 1), (6, −2), (8, −7)

D. (5, 2), (−2, 4), (−7, 6), (1, 8)

5. What are the next three numbers in the arithmetic sequence?

27, 23, 19, 15, . . .

F. 13, 11, 9

G. 19, 23, 27

H. 12, 9, 6

I. 11, 7, 3

6. How much material is needed to make the popcorn container?

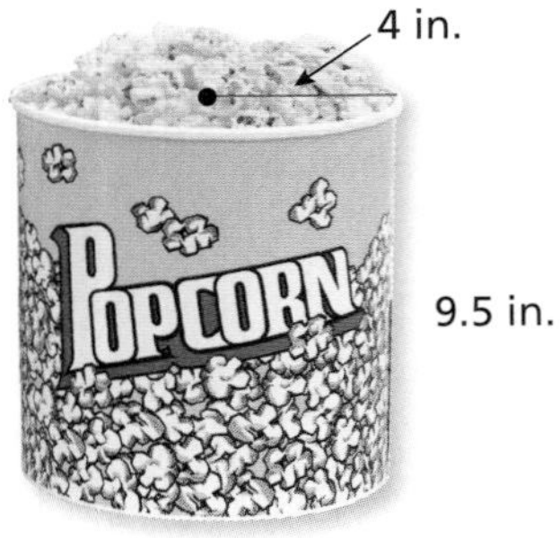

A. 76π in.^{2}

B. 84π in.^{2}

C. 92π in.^{2}

D. 108π in.^{2}

7. To make 10 servings of soup you need 4 cups of broth. You want to know how many servings can be made with 8 pints of broth. Which proportion should you use?

F. $\frac{10}{4} = \frac{x}{8}$

G. $\frac{4}{10} = \frac{x}{16}$

H. $\frac{10}{4} = \frac{8}{x}$

I. $\frac{10}{4} = \frac{x}{16}$

8. A rectangular prism and its dimensions are shown below.

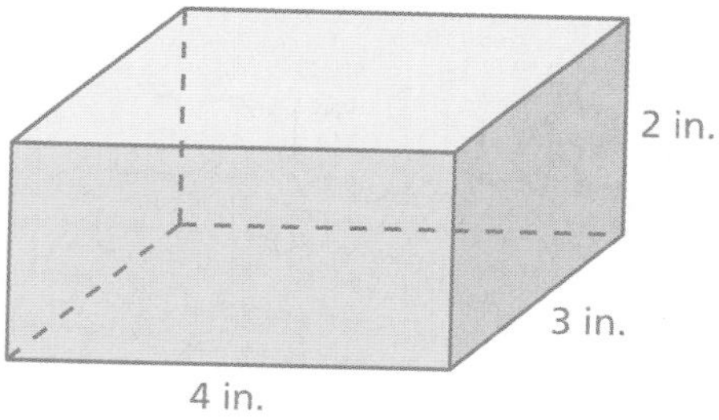

What is the volume of a rectangular prism whose dimensions are three times greater?

9. Which graph represents a linear function?

A.

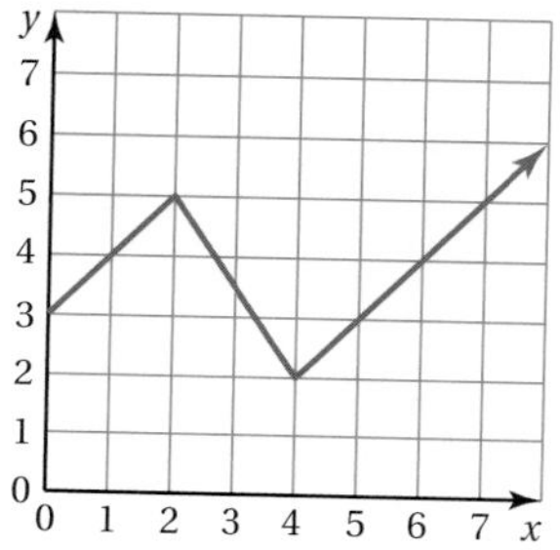

B.

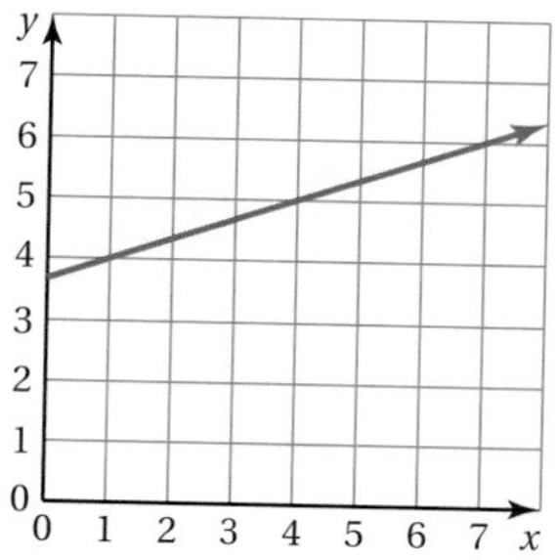

C.

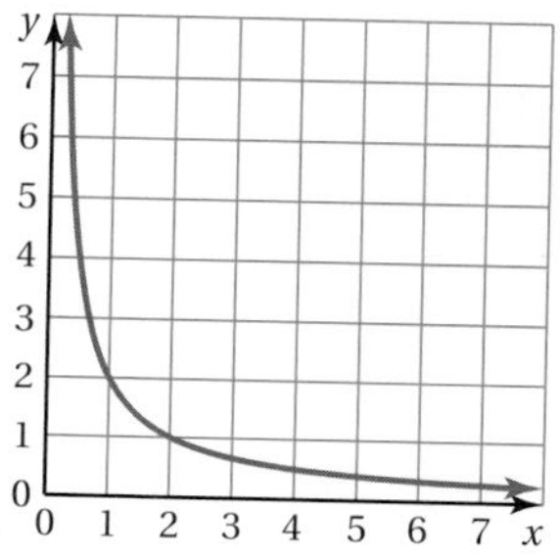

D.

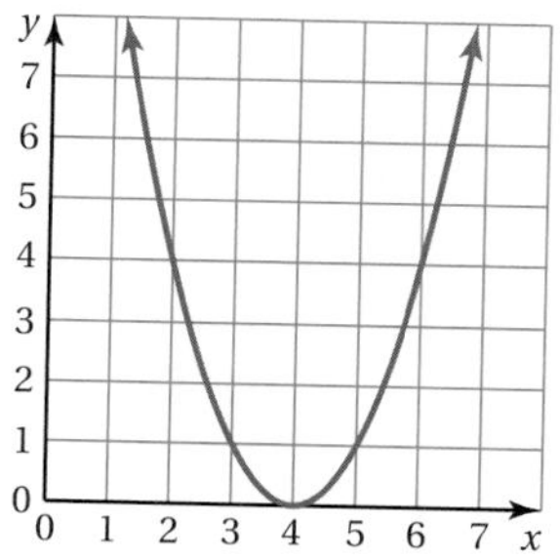

10. The table below shows the cost of buying matinee movie tickets.

Think Solve Explain

Matinee Tickets, *x*	2	3	4	5
Cost, *y*	$9	$13.50	$18	$22.50

Part A Graph the data.

Part B Find and interpret the slope of the line through the points.

Part C How much does it cost to buy 8 matinee movie tickets?

9 Exponents and Scientific Notation

"Here's how it goes, Descartes."
(+)(+)=+
(+)(-)=-
"The friends of my friends are my friends. The friends of my enemies are my enemies."
(-)(+)=-
(-)(-)=+
I hope that hyenas don't have many friends.
"The enemies of my friends are my enemies. The enemies of my enemies are my friends."

1 + 100 + 100 2
FLEAS
"If one flea had 100 babies, and each baby grew up and had 100 babies, ..."
+ 100 3
Help! I can't see the expiration date on my flea collar.
"... and each of those babies grew up and had 100 babies, you would have 1,010,101 fleas."

What You Learned Before

"It's called the Power of Negative One, Descartes!"

Adding and Subtracting Decimals

Example 1 **Find 2.65 + 5.012.**

$$\begin{array}{r} 2.650 \\ +\ 5.012 \\ \hline 7.662 \end{array}$$

Example 2 **Find 3.7 − 0.48.**

$$\begin{array}{r} \scriptstyle 6\ 10 \\ 3.\not{7}\ \not{0} \\ -\ 0.4\ 8 \\ \hline 3.2\ 2 \end{array}$$

Try It Yourself

Find the sum or difference.

1. 2.73 + 1.007 **2.** 3.4 − 1.27 **3.** 0.35 + 0.749 **4.** 1.019 + 0.09

5. 6.03 − 1.008 **6.** 4.21 − 0.007 **7.** 0.228 + 1.205 **8.** 3.003 − 1.9

Multiplying and Dividing Decimals

Example 3 **Find 2.1 • 0.35.**

$$\begin{array}{rl} 2.1 & \leftarrow \text{ 1 decimal place} \\ \times\ 0.3\ 5 & \leftarrow \text{ + 2 decimal places} \\ \hline 1\ 0\ 5 & \\ 6\ 3\ \ & \\ \hline 0.7\ 3\ 5 & \leftarrow \text{ 3 decimal places} \end{array}$$

Example 4 **Find 1.08 ÷ 0.9.**

$0.9\overline{)1.08}$ Multiply each number by 10.

$$\begin{array}{r} 1.2 \\ 9\overline{)10.8} \\ -\ 9\ \ \\ \hline 1\ 8 \\ -\ 1\ 8 \\ \hline 0 \end{array}$$

Place the decimal point above the decimal point in the dividend 10.8.

Try It Yourself

Find the product or quotient.

9. 1.75 • 0.2 **10.** 1.4 • 0.6 **11.** $\begin{array}{r} 7.03 \\ \times\ 4.3 \\ \hline \end{array}$ **12.** $\begin{array}{r} 0.894 \\ \times\ 0.2 \\ \hline \end{array}$

13. 5.40 ÷ 0.09 **14.** 4.17 ÷ 0.3 **15.** $0.15\overline{)3.6}$ **16.** $0.004\overline{)7.2}$

9.1 Exponents

STANDARDS OF LEARNING
7.1

Essential Question How can you use exponents to write numbers?

The expression 3^5 is called a **power**. The **base** is 3. The **exponent** is 5.

base → 3^5 ← exponent

1 ACTIVITY: Using Exponent Notation

Work with a partner.

a. Copy and complete the table.

Power	Repeated Multiplication Form	Value
$(-3)^1$	-3	-3
$(-3)^2$	$(-3) \cdot (-3)$	9
$(-3)^3$		
$(-3)^4$		
$(-3)^5$		
$(-3)^6$		
$(-3)^7$		

b. Describe what is meant by the expression $(-3)^n$. How can you find the value of $(-3)^n$?

2 ACTIVITY: Using Exponent Notation

Work with a partner.

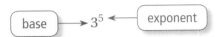

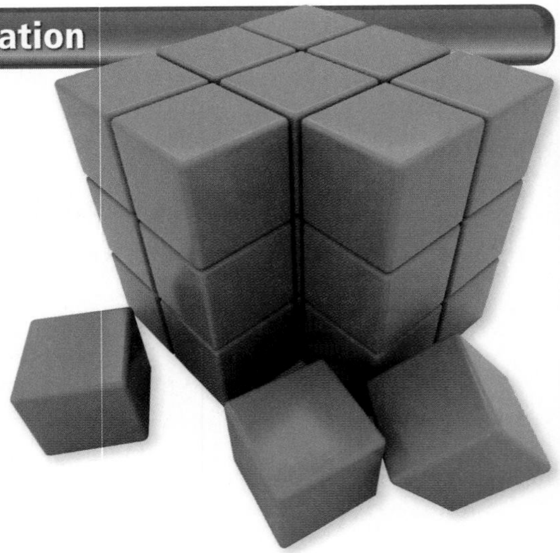

a. The cube at the right has \$3 in each of its small cubes. Write a single power that represents the total amount of money in the large cube.

b. Evaluate the power to find the total amount of money in the large cube.

3 ACTIVITY: Writing Powers as Whole Numbers

Work with a partner. Write each distance as a whole number. Which numbers do you know how to write in words? For instance, in words, 10^3 is equal to *one thousand*.

a. 10^{26} meters: Diameter of observable universe

b. 10^{21} meters: Diameter of Milky Way Galaxy

c. 10^{16} meters: Diameter of Solar System

d. 10^{7} meters: Diameter of Earth

e. 10^{6} meters: Length of Lake Erie shoreline

f. 10^{5} meters: Width of Lake Erie

4 ACTIVITY: Writing a Power

Work with a partner. Write the number of kits, cats, sacks, and wives as a power.

As I was going to St. Ives
I met a man with seven wives
And every wife had seven sacks
And every sack had seven cats
And every cat had seven kits
Kits, cats, sacks, wives
How many were going to St. Ives?

Nursery Rhyme, 1730

What Is Your Answer?

5. **IN YOUR OWN WORDS** How can you use exponents to write numbers? Give some examples of how exponents are used in real life.

Practice Use what you learned about exponents to complete Exercises 3–5 on page 358.

9.1 Lesson

Check It Out
Lesson Tutorials
BigIdeasMath.com

Key Vocabulary
power, *p. 356*
base, *p. 356*
exponent, *p. 356*

A **power** is a product of repeated factors. The **base** of a power is the common factor. The **exponent** of a power indicates the number of times the base is used as a factor.

base → $\frac{1}{2}$; exponent → 5

$$\left(\frac{1}{2}\right)^5 = \frac{1}{2} \cdot \frac{1}{2} \cdot \frac{1}{2} \cdot \frac{1}{2} \cdot \frac{1}{2}$$

power — $\frac{1}{2}$ is used as a factor 5 times.

EXAMPLE 1 Writing Expressions Using Exponents

Write each product using exponents.

a. $(-7) \cdot (-7) \cdot (-7)$

Because -7 is used as a factor 3 times, its exponent is 3.

So, $(-7) \cdot (-7) \cdot (-7) = (-7)^3$.

Study Tip
Use parentheses to write powers with negative bases.

b. $\pi \cdot \pi \cdot r \cdot r \cdot r$

Because π is used as a factor 2 times, its exponent is 2. Because r is used as a factor 3 times, its exponent is 3.

So, $\pi \cdot \pi \cdot r \cdot r \cdot r = \pi^2 r^3$.

On Your Own

Now You're Ready
Exercises 3–10

Write the product using exponents.

1. $\frac{1}{4} \cdot \frac{1}{4} \cdot \frac{1}{4} \cdot \frac{1}{4} \cdot \frac{1}{4}$

2. $0.3 \cdot 0.3 \cdot 0.3 \cdot 0.3 \cdot x \cdot x$

EXAMPLE 2 Evaluating Expressions

Evaluate each expression.

a. $(-2)^4$

$(-2)^4 = (-2) \cdot (-2) \cdot (-2) \cdot (-2)$ — Write as repeated multiplication.

$= 16$ — Simplify.

The base is -2.

b. -2^4

$-2^4 = -(2 \cdot 2 \cdot 2 \cdot 2)$ — Write as repeated multiplication.

$= -16$ — Simplify.

The base is 2.

Multi-Language Glossary at BigIdeasMath.com.

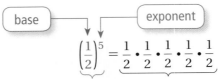

EXAMPLE 3 Using Order of Operations

a. Evaluate $3^3 - 8^2 \div 2$.

$3^3 - 8^2 \div 2 = 27 - 64 \div 2$	Evaluate the powers.
$= 27 - 32$	Divide.
$= -5$	Subtract.

Study Tip

When evaluating expressions, treat absolute value symbols as grouping symbols.

b. Evaluate $\frac{1}{2}\left| -1^5 - 7 \right|$.

$\frac{1}{2}\left| -1^5 - 7 \right| = \frac{1}{2}\left| -1 - 7 \right|$	Evaluate the power.
$= \frac{1}{2}\left| -8 \right|$	Subtract.
$= \frac{1}{2}(8) = 4$	Simplify.

On Your Own

Now You're Ready
Exercises 11–16 and 22–27

Evaluate the expression.

3. -5^4
4. $\left(-\frac{1}{6}\right)^3$
5. $\left| -3^3 \div 27 \right|$
6. $4^2 - 2^5 \cdot 0.5$

EXAMPLE 4 Real-Life Application

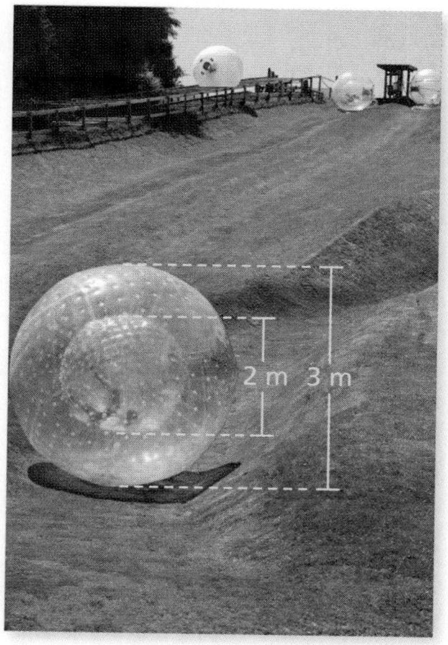

In sphering, a person is secured inside a small, hollow sphere that is surrounded by a larger sphere. The space between the spheres is inflated with air. What is the volume of the inflated space? (The volume V of a sphere is $V = \frac{4}{3}\pi r^3$. Use 3.14 for π.)

Outer sphere		*Inner sphere*
$V = \frac{4}{3}\pi r^3$	Write formula.	$V = \frac{4}{3}\pi r^3$
$= \frac{4}{3}\pi(1.5)^3$	Substitute.	$= \frac{4}{3}\pi(1)^3$
$= \frac{4}{3}\pi(3.375)$	Evaluate the power.	$= \frac{4}{3}\pi(1)$
≈ 14.13	Multiply.	≈ 4.19

So, the volume of the inflated space is about $14.13 - 4.19$, or 9.94 cubic meters.

On Your Own

7. **WHAT IF?** The diameter of the inner sphere is 1.8 meters. What is the volume of the inflated space?

9.1 Exercises

Check It Out
Help with Homework
BigIdeasMath.com

Vocabulary and Concept Check

1. **VOCABULARY** Describe the difference between an exponent and a power. Can the two words be used interchangeably?

2. **REASONING** Which statement is incorrect? Explain.

5^3 The exponent is 3.	5^3 The power is 3.	5^3 The base is 5.	5^3 Five is used as a factor 3 times.

Practice and Problem Solving

Write the product using exponents.

(1) 3. $3 \cdot 3 \cdot 3 \cdot 3$

4. $(-6) \cdot (-6)$

5. $\left(-\frac{1}{2}\right) \cdot \left(-\frac{1}{2}\right) \cdot \left(-\frac{1}{2}\right)$

6. $\frac{1}{3} \cdot \frac{1}{3} \cdot \frac{1}{3}$

7. $\pi \cdot \pi \cdot \pi \cdot x \cdot x \cdot x \cdot x$

8. $(-4) \cdot (-4) \cdot (-4) \cdot y \cdot y$

9. $8 \cdot 8 \cdot 8 \cdot 8 \cdot b \cdot b \cdot b$

10. $(-t) \cdot (-t) \cdot (-t) \cdot (-t) \cdot (-t)$

Evaluate the expression.

(2) 11. 5^2

12. -11^3

13. $(-1)^6$

14. $-\left(\frac{1}{2}\right)^3$

15. $\left(-\frac{1}{2}\right)^3$

16. $-\left(-\frac{1}{2}\right)^3$

17. **NUMBER SENSE** Why are the expressions in Exercises 14 and 15 evaluated differently?

18. **ERROR ANALYSIS** Describe and correct the error in evaluating the expression.

✗ $6^3 = 6 \cdot 3 = 18$

19. **PRIME FACTORIZATION** Write the prime factorization of 675 using exponents.

20. **NUMBER SENSE** Write $-\left(\frac{1}{4} \cdot \frac{1}{4} \cdot \frac{1}{4} \cdot \frac{1}{4}\right)$ using exponents.

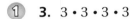

21. **RUSSIAN DOLLS** The largest doll is 12 inches tall. The height of each of the other dolls is $\frac{7}{10}$ the height of the next larger doll. Write an expression for the height of the smallest doll. What is the height of the smallest doll?

Evaluate the expression.

③ **22.** $2 + 7 \cdot (-3)^2$

23. $3^5 - 6^3 \div 4$

24. $(13^2 - 12^2) \div 5$

25. $\frac{1}{2}(4^3 - 6 \cdot 3^2)$

26. $\left|\frac{1}{2}(7 + 5^3)\right|$

27. $\left|\left(-\frac{1}{2}\right)^3 \div \left(\frac{1}{4}\right)^2\right|$

28. MONEY You have a part-time job. One day your boss offers to pay you either $2^h - 1$ or 2^{h-1} dollars for each hour h you work that day. Copy and complete the table. Which option should you choose? Explain.

h	1	2	3	4	5
$2^h - 1$					
2^{h-1}					

29. CARBON-14 DATING Carbon-14 dating is used by scientists to determine the age of a sample.

a. The amount C (in grams) of a 100-gram sample of carbon-14 remaining after t years is represented by the equation $C = 100(0.99988)^t$. Use a calculator to find the amount of carbon-14 remaining after 4 years.

b. What percent of the carbon-14 remains after 4 years?

30. Critical Thinking The frequency (in vibrations per second) of a note on a piano is represented by the equation $F = 440(1.0595)^n$, where n is the number of notes above A-440. Each black or white key represents one note.

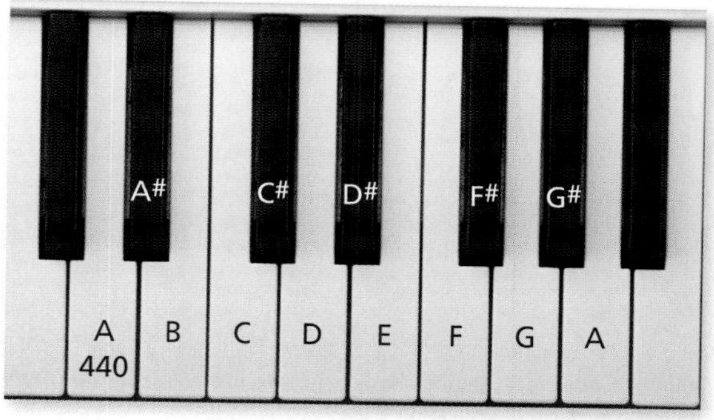

a. How many notes do you take to travel from A-440 to A?

b. What is the frequency of A?

c. Describe the relationship between the number of notes between A-440 and A and the frequency of the notes.

Fair Game Review What you learned in previous grades & lessons

Tell which property is illustrated by the statement. *(Skills Review Handbook)*

31. $8 \cdot x = x \cdot 8$

32. $(2 \cdot 10)x = 2(10 \cdot x)$

33. $3(x \cdot 1) = 3x$

34. MULTIPLE CHOICE The polygons are similar. What is the value of x? *(Section 6.4)*

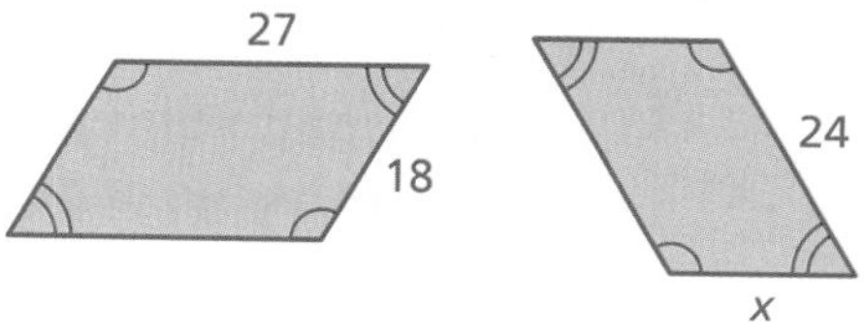

Ⓐ 15

Ⓑ 16

Ⓒ 17

Ⓓ 36

9.2 Zero and Negative Exponents

STANDARDS OF LEARNING
7.1

Essential Question How can you use zero and negative exponents in real life?

Share Your Work at... My.BigIdeasMath.com

1 ACTIVITY: Investigating Powers of 10

In 1977, Charles and Ray Eames made the film *Powers of Ten* to illustrate the relative sizes of things in the Universe. Use the Internet to find and watch the film *Powers of Ten*.

Work with a partner. Some of the numbers given in the film are listed below. Write the exponent for each power of 10.

1 meter $= 10^0$ meter

10 meters $= 10^1$ meters

100 meters $= 10^2$ meters

1000 meters $= 10^3$ meters

10,000 meters $= 10^4$ meters

100,000 meters $= 10^5$ meters

1,000,000 microns $= 10^0$ meter

a. 100,000 microns $= 10^{\square}$ meter

b. 10,000 microns $= 10^{\square}$ meter

c. 1000 microns $= 10^{\square}$ meter

d. 100 microns $= 10^{\square}$ meter

e. 10 microns $= 10^{\square}$ meter

2 ACTIVITY: Writing a Report

Work with a partner. Use the Internet to research Charles and Ray Eames and how they made the film *Powers of Ten*. Then write a report that includes the following information.

- **Explain how the makers of the film knew what the far reaches of space and the inside of an atom looked like.**
- **Are all of the pictures actual photographs? Explain your reasoning.**
- **Describe several careers that involve outer space and atoms.**

3 ACTIVITY: Powers of 10 Matching Game

Work with a partner. Match each picture with its power of 10. Explain your reasoning.

10^5 m 10^2 m 10^0 m 10^{-1} m 10^{-2} m 10^{-5} m

A. 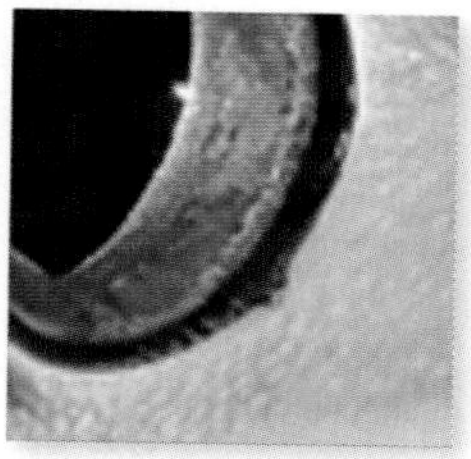

B. 

C. 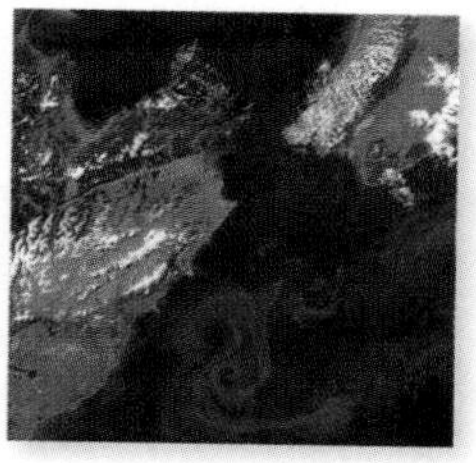

D. 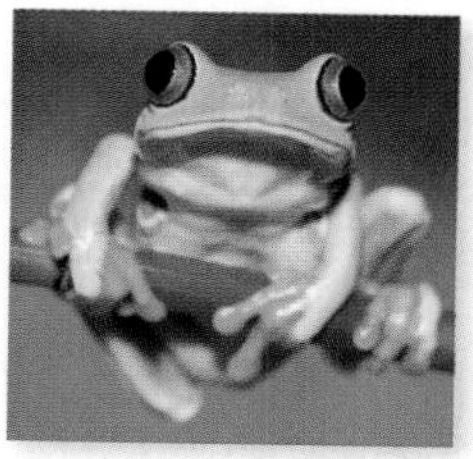

E. 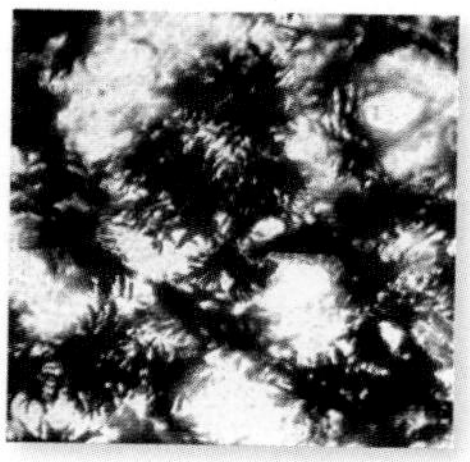

F. 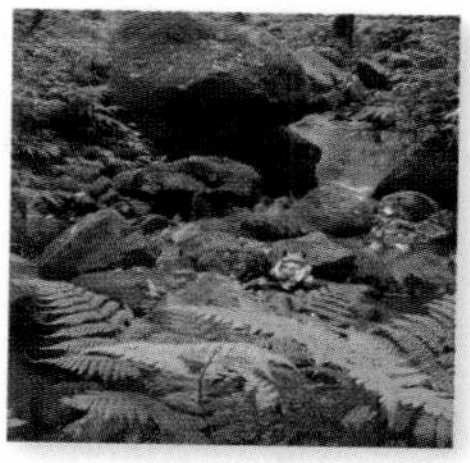

4 ACTIVITY: Making a Powers of 10 Matching Game

Create your own matching game. Use the Internet to find at least 3 pictures that can be represented by different powers of 10. Trade games with a partner and play your partner's matching game.

What Is Your Answer?

5. **IN YOUR OWN WORDS** How can you use zero and negative exponents in real life? Give an example.

Practice Use what you learned about zero and negative exponents to complete Exercises 5–7 on page 364.

9.2 Lesson

Check It Out
Lesson Tutorials
BigIdeasMath.com

Key Ideas

Zero Exponents

Words For any nonzero number a, $a^0 = 1$. The power 0^0 is *undefined.*

Numbers $4^0 = 1$

Algebra $a^0 = 1$, where $a \neq 0$

Negative Exponents

Words For any integer n and any nonzero number a, a^{-n} is the reciprocal of a^n.

Numbers $4^{-2} = \frac{1}{4^2}$

Algebra $a^{-n} = \frac{1}{a^n}$, where $a \neq 0$

EXAMPLE 1 Evaluating Expressions

a. $3^{-4} = \frac{1}{3^4}$ Definition of negative exponent

$= \frac{1}{81}$ Evaluate the power.

b. $\frac{1}{6^{-3}} = 6^{-(-3)}$ Definition of negative exponent

$= 6^3$ Simplify.

$= 216$ Evaluate the power.

Study Tip

Notice that Example 1b uses the rule $\frac{1}{a^n} = a^{-n}$.

c. $5^0 + (-2)^{-4} = 1 + (-2)^{-4}$ Definition of zero exponent

$= 1 + \frac{1}{(-2)^4}$ Definition of negative exponent

$= 1 + \frac{1}{16}$ Evaluate the power.

$= 1\frac{1}{16}$ Simplify.

On Your Own

Now You're Ready
Exercises 8–15

Evaluate the expression.

1. 4^{-2}
2. $(-2)^{-5}$
3. $\frac{1}{(-3)^{-3}}$
4. $5^{-2} + 4^0$

EXAMPLE 2 Simplifying Expressions

a. $-5x^0 = -5(1)$ — Definition of zero exponent

$= -5$ — Multiply.

b. $7y^{-6} = 7\left(\frac{1}{y^6}\right)$ — Definition of negative exponent

$= \frac{7}{y^6}$ — Multiply.

On Your Own

Now You're Ready
Exercises 19–26

Simplify. Write your answer using only positive exponents.

5. $8x^{-2}$
6. $b^0 \cdot b^{-10}$
7. $15n^0$

EXAMPLE 3 Real-Life Application

A drop of water leaks from a faucet every second. How many liters of water leak from the faucet in 1 hour?

Convert 1 hour to seconds.

$$1\ \cancel{h} \times \frac{60\ \cancel{min}}{1\ \cancel{h}} \times \frac{60\ sec}{1\ \cancel{min}} = 3600\ sec$$

Drop of water: 50^{-2} L

Water leaks from the faucet at a rate of 50^{-2} liter per second. Multiply the time by the rate.

$3600 \cdot 50^{-2} = 3600 \cdot \frac{1}{50^2}$ — Definition of negative exponent

$= 3600 \cdot \frac{1}{2500}$ — Evaluate the power.

$= \frac{3600}{2500}$ — Multiply.

$= 1\frac{11}{25}$ — Divide.

$= 1.44$ — Simplify.

So, 1.44 liters of water leak from the faucet in 1 hour.

On Your Own

8. **WHAT IF?** The faucet leaks water at a rate of 5^{-5} liter per second. How many liters of water leak from the faucet in 1 hour?

9.2 Exercises

Vocabulary and Concept Check

1. **VOCABULARY** If a is a nonzero number, does the value of a^0 depend on the value of a? Explain.
2. **WRITING** Explain how to evaluate 10^{-3}.
3. **NUMBER SENSE** Without evaluating, order 5^0, 5^4, and 5^{-5} from least to greatest.
4. **DIFFERENT WORDS, SAME QUESTION** Which is different? Find "both" answers.

Rewrite $\frac{1}{3 \cdot 3 \cdot 3}$ using a negative exponent.	Write the power 3 to the negative third.
Write $\frac{1}{3}$ cubed as a power.	Write $(-3) \cdot (-3) \cdot (-3)$ as a power.

Practice and Problem Solving

Complete the statement using < or >.

5. 10^{12} ☐ 10^{18}
6. 10^{-4} ☐ 10^{-7}
7. 10^{9} ☐ 10^{-9}

Evaluate the expression.

1 8. 6^{-2}
9. 158^0
10. $(-4)^{-3}$
11. $\frac{1}{12^{-2}}$
12. $4 \cdot 2^{-4} + 5$
13. $(-2)^{-8} \cdot (-2)^8$
14. $3^{-3} \cdot 3^{-2}$
15. $\frac{1}{(-7)^{-2}} + 3^0$

16. **ERROR ANALYSIS** Describe and correct the error in evaluating the expression.

$$\begin{aligned} (4)^{-3} &= (-4)(-4)(-4) \\ &= -64 \end{aligned}$$

17. **SAND** The mass of a grain of sand is about 10^{-3} gram. About how many grains of sand are in the bag of sand?

18. **CRITICAL THINKING** How can you write the number 1 as a power with a base of 2? 10?

Simplify. Write your answer using only positive exponents.

19. $6y^{-4}$ **20.** $8^{-2} \cdot a^7$ **21.** 9^2c^{-4} **22.** $\frac{5}{b^{-3}}$

23. $\frac{8}{2x^{-6}}$ **24.** $3d^{-4} \cdot 4d^4$ **25.** $m^{-2} \cdot n^3$ **26.** $\frac{3^{-2} \cdot k^0 \cdot w^0}{w^{-6}}$

METRIC UNITS **In Exercises 27–30, use the table.**

27. How many millimeters are in a decimeter?

28. How many micrometers are in a centimeter?

29. How many nanometers are in a millimeter?

30. How many micrometers are in a meter?

Unit of Length	Length
decimeter	10^{-1} m
centimeter	10^{-2} m
millimeter	10^{-3} m
micrometer	10^{-6} m
nanometer	10^{-9} m

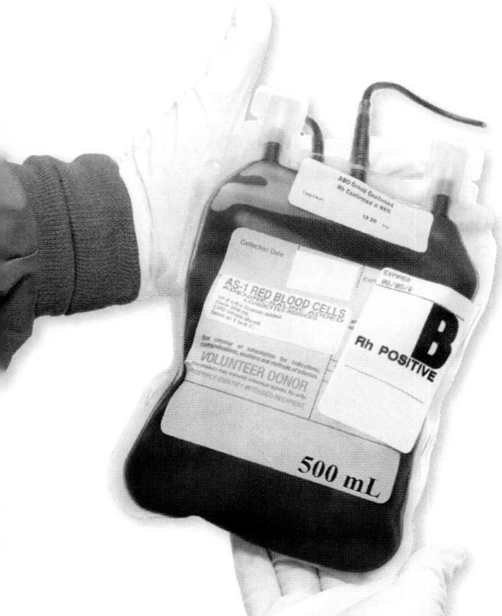

31. **BACTERIA** A species of bacteria is 10 micrometers long. A virus is 10,000 times smaller than the bacteria.

a. Using the table above, find the length of the virus in meters.

b. Is the answer to part (a) *less than, greater than,* or *equal to* one nanometer?

32. **BLOOD DONATION** Every 2 seconds, someone in the United States needs blood. A sample blood donation is shown. ($1 \text{ mm}^3 = 10^{-3} \text{ mL}$)

a. One cubic millimeter of blood contains about 10^4 white blood cells. How many white blood cells are in the donation? Write your answer in words.

b. One cubic millimeter of blood contains about 5×10^6 red blood cells. How many red blood cells are in the donation? Write your answer in words.

c. Compare your answers for parts (a) and (b).

33. **Reasoning** The rule for negative exponents states that $a^{-n} = \frac{1}{a^n}$. Explain why this rule does not apply when $a = 0$.

Fair Game Review What you learned in previous grades & lessons

The vertices of a polygon are given. Draw the polygon and its image after a dilation with the given scale factor. Identify the type of dilation. *(Section 7.4)*

34. $W(1, 4), X(3, 4), Y(3, 1), Z(1, 1); k = 4$ **35.** $Q(-4, 4), R(6, 8), S(6, 4); k = \frac{1}{2}$

36. **MULTIPLE CHOICE** What is the volume of the cylinder? Round your answer to the nearest tenth. *(Section 8.4)*

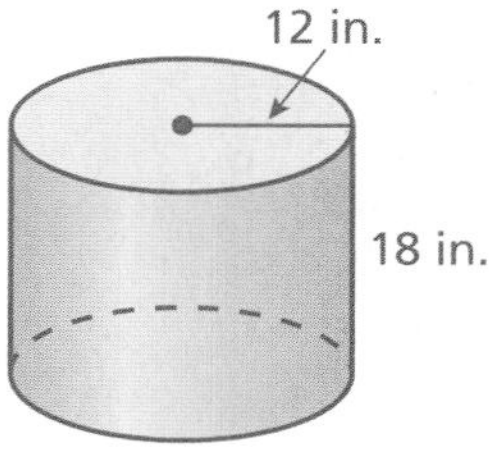

Ⓐ 678.2 in.^{3} Ⓑ 2260.8 in.^{3}
Ⓒ 8138.9 in.^{3} Ⓓ 12,208.3 in.^{3}

9 Study Help

Check It Out
Graphic Organizer
BigIdeasMath.com

You can use a **notetaking organizer** to write notes, vocabulary, and questions about a topic. Here is an example of a notetaking organizer for powers.

Write important vocabulary or formulas in this space. →

A power is a product of repeated factors.

The base of a power is the common factor.

The exponent indicates how many times the base is used as a factor.

base → 5^3 ← exponent; 5^3 = power

Powers

For 5^3, 5 is used as a factor 3 times.

$5^3 = 5 \bullet 5 \bullet 5 = 125$

$(-5)^3 = (-5) \bullet (-5) \bullet (-5) = -125$

$-5^3 = -(5 \bullet 5 \bullet 5) = -125$

← Write your notes about the topic in this space.

How do you evaluate a power when the exponent is zero?

← Write your questions about the topic in this space.

On Your Own

Make a notetaking organizer to help you study these topics.

1. zero exponents
2. negative exponents

After you complete this chapter, make notetaking organizers for the following topics.

3. writing numbers in scientific notation
4. writing numbers in standard form
5. square roots
6. perfect squares

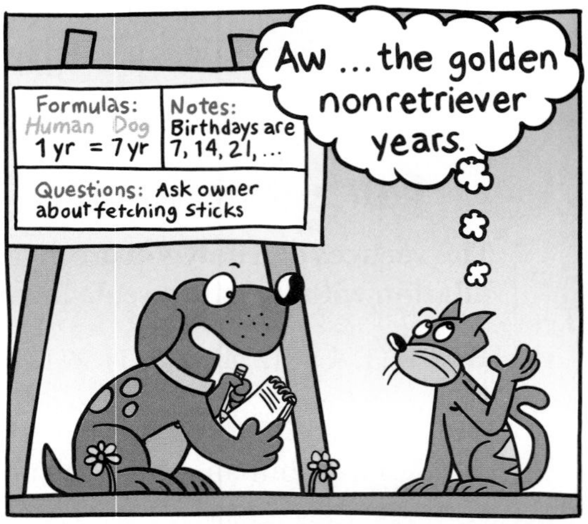

"My notetaking organizer has me thinking about retirement... I won't have to fetch sticks anymore."

Check It Out
Progress Check
BigIdeasMath.com

Write the product using exponents. *(Section 9.1)*

1. $(-5) \cdot (-5) \cdot (-5) \cdot (-5)$

2. $\frac{1}{6} \cdot \frac{1}{6} \cdot \frac{1}{6} \cdot \frac{1}{6} \cdot \frac{1}{6}$

3. $(-x) \cdot (-x) \cdot (-x) \cdot (-x) \cdot (-x) \cdot (-x)$

4. $7 \cdot 7 \cdot m \cdot m \cdot m$

Evaluate the expression. *(Section 9.1 and Section 9.2)*

5. 5^4

6. $(-2)^6$

7. $(-3)^{-3}$

8. $5^0 \cdot 4^{-2}$

Simplify. Write your answer using only positive exponents. *(Section 9.2)*

9. $8d^{-6}$

10. $\frac{3}{x^{-2}}$

11. BALL BOUNCE A ball is dropped from a height of 64 feet. Each time it hits the ground, it bounces $\frac{3}{4}$ of its previous height. Use exponents to write an expression for the height of the third bounce. Find the height of the third bounce. *(Section 9.1)*

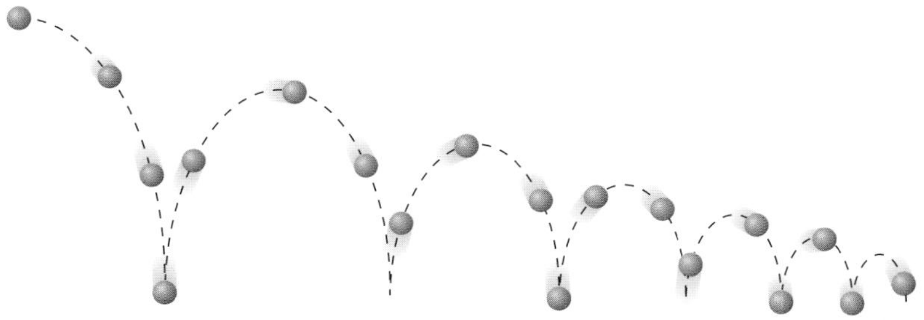

12. SEQUENCE The nth term of a sequence can be found by evaluating $10^n - 1$. Copy and complete the table to find the first four terms of the sequence. *(Section 9.1)*

n	$10^n - 1$
1	
2	
3	
4	

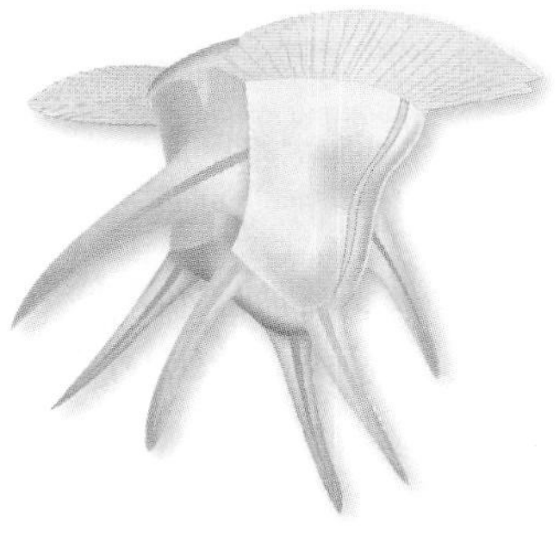

13. ORGANISM A one-celled, aquatic organism called a dinoflagellate is 1000 micrometers long. *(Section 9.2)*

a. One micrometer is 10^{-6} meter. What is the length of the dinoflagellate in meters?

b. Is the length of the dinoflagellate equal to 1 millimeter or 1 kilometer? Explain.

9.3 Reading Scientific Notation

Essential Question How can you read numbers that are written in scientific notation?

STANDARDS OF LEARNING
7.1

1 ACTIVITY: Very Large Numbers

Work with a partner.

- Use a calculator. Experiment with multiplying large numbers until your calculator gives an answer that is *not* in standard form.
- When the calculator at the right was used to multiply 2 billion by 3 billion, it listed the result as

  6.0E+18.

- Multiply 2 billion by 3 billion by hand. Use the result to explain what 6.0E+18 means.
- Check your explanation using products of other large numbers.
- Why didn't the calculator show the answer in standard form?
- Experiment to find the maximum number of digits your calculator displays. For instance, if you multiply 1000 by 1000 and your calculator shows 1,000,000, then it can display 7 digits.

2 ACTIVITY: Very Small Numbers

Work with a partner.

- Use a calculator. Experiment with multiplying very small numbers until your calculator gives an answer that is *not* in standard form.
- When the calculator at the right was used to multiply 2 billionths by 3 billionths, it listed the result as

  6.0E–18.

- Multiply 2 billionths by 3 billionths by hand. Use the result to explain what 6.0E–18 means.
- Check your explanation using products of other very small numbers.

3 ACTIVITY: Reading Scientific Notation

Work with a partner.

Each description gives an example of a number written in scientific notation. Answer the question in the description. Write your answer in standard form.

a. Nearly 1.0×10^5 dust mites can live in 1 square yard of carpet.

How many dust mites can live in 100 square yards of carpet?

b. A micron is about 4.0×10^{-5} inch. The length of a dust mite is 250 microns.

How long is a dust mite in inches?

c. About 1.0×10^{15} bacteria live in a human body.

How many bacteria are living in the humans in your classroom?

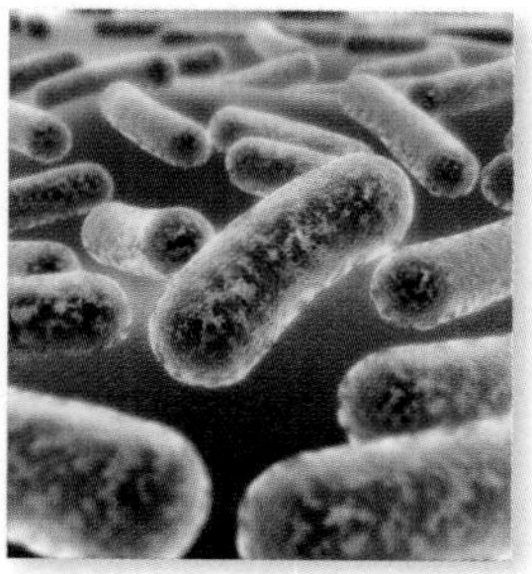

d. A micron is about 4.0×10^{-5} inch. The length of a bacterium is about 0.5 micron.

How many bacteria could lie end-to-end on your finger?

e. Earth has only about 1.5×10^8 kilograms of gold. Earth has a mass of 6.0×10^{24} kilograms.

What percent of Earth's mass is gold?

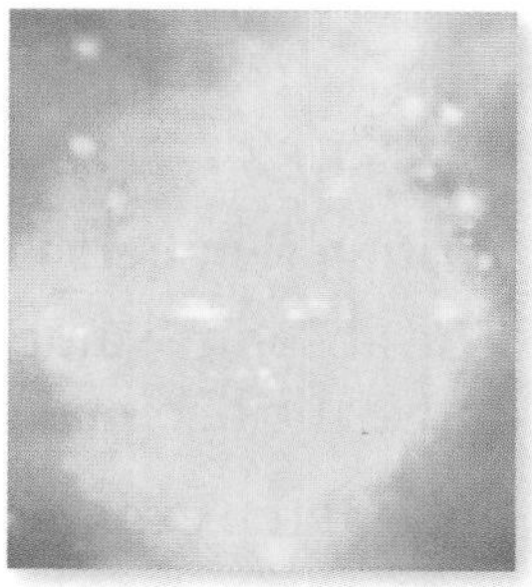

f. A gram is about 0.035 ounce. An atom of gold weighs about 3.3×10^{-22} gram.

How many atoms are in an ounce of gold?

What Is Your Answer?

4. **IN YOUR OWN WORDS** How can you read numbers that are written in scientific notation? Why do you think this type of notation is called "scientific notation"? Why is scientific notation important?

Practice Use what you learned about reading scientific notation to complete Exercises 3–5 on page 372.

9.3 Lesson

Check It Out
Lesson Tutorials
BigIdeasMath.com

Key Vocabulary
scientific notation, *p. 370*

Study Tip
Scientific notation is used to write very small and very large numbers.

Key Idea

Scientific Notation

A number is written in **scientific notation** when it is represented as the product of a factor and a power of 10. The factor must be greater than or equal to 1 and less than 10.

The factor is greater than or equal to 1 and less than 10. → 8.3×10^{-7} ← The power of 10 has an integer exponent.

EXAMPLE 1 Identifying Numbers Written in Scientific Notation

Tell whether the number is written in scientific notation. Explain.

a. 5.9×10^{-6}

∴ The factor is greater than or equal to 1 and less than 10. The power of 10 has an integer exponent. So, the number is written in scientific notation.

b. 0.9×10^{8}

∴ The factor is less than 1. So, the number is not written in scientific notation.

Key Idea

Writing Numbers in Standard Form

The absolute value of the exponent indicates how many places to move the decimal point.

- If the exponent is negative, move the decimal point to the left.
- If the exponent is positive, move the decimal point to the right.

EXAMPLE 2 Writing Numbers in Standard Form

a. Write 3.22×10^{-4} in standard form.

$3.22 \times 10^{-4} = 0.000322$ Move decimal point $|-4| = 4$ places to the left.

4

b. Write 7.9×10^{5} in standard form.

$7.9 \times 10^{5} = 790{,}000$ Move decimal point $|5| = 5$ places to the right.

5

Multi-Language Glossary at BigIdeasMath.com.

On Your Own

Now You're Ready
Exercises 6–23

1. Is 12×10^4 written in scientific notation? Explain.

Write the number in standard form.

2. 6×10^7
3. 9.9×10^{-5}
4. 1.285×10^4

EXAMPLE 3 Comparing Numbers in Scientific Notation

An object with a lesser density than water will float. An object with a greater density than water will sink. Use each given density (in kilograms per cubic meter) to explain what happens when you place a brick and an apple in water.

Water: 1.0×10^3 **Brick:** 1.84×10^3 **Apple:** 6.41×10^2

Write each density in standard form.

Water	Brick	Apple
$1.0 \times 10^3 = 1000$	$1.84 \times 10^3 = 1840$	$6.41 \times 10^2 = 641$

∴ The apple is less dense than water, so it will float. The brick is denser than water, so it will sink.

EXAMPLE 4 Real-Life Application

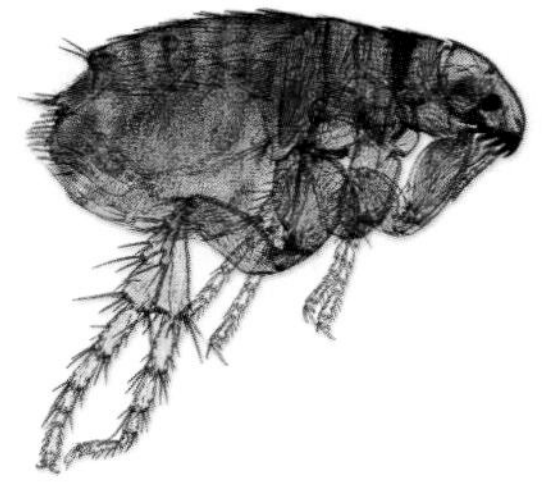

A female flea consumes about 1.4×10^{-5} liter of blood per day.

A dog has 100 female fleas. How many milliliters of blood do the fleas consume per day?

$$1.4 \times 10^{-5} \cdot 100 = 0.000014 \cdot 100 \quad \text{Write in standard form.}$$

$$= 0.0014 \quad \text{Multiply.}$$

∴ The fleas consume about 0.0014 liter, or 1.4 milliliters of blood per day.

On Your Own

Now You're Ready
Exercise 27

5. **WHAT IF?** In Example 3, the density of lead is 1.14×10^4 kilograms per cubic meter. What happens when lead is placed in water?

6. **WHAT IF?** In Example 4, a dog has 75 female fleas. How many milliliters of blood do the fleas consume per day?

9.3 Exercises

Check It Out
Help with Homework
BigIdeasMath.com

Vocabulary and Concept Check

1. **WRITING** Describe the difference between scientific notation and standard form.
2. **WHICH ONE DOESN'T BELONG?** Which number does *not* belong with the other three? Explain.

2.8×10^{15}	4.3×10^{-30}	1.05×10^{28}	10×9.2^{-13}

Practice and Problem Solving

Write your answer in standard form.

3. A micrometer is 1.0×10^{-6} meter. How long is 150 micrometers in meters?
4. An acre is about 4.05×10^{7} square centimeters. How many square centimeters are in 4 acres?
5. A cubic millimeter is about 6.1×10^{-5} cubic inches. How many cubic millimeters are in 1.22 cubic inches?

Tell whether the number is written in scientific notation. Explain.

(1)
6. 1.8×10^{9}
7. 3.45×10^{14}
8. 0.26×10^{-25}
9. 10.5×10^{12}
10. 46×10^{-17}
11. 5×10^{-19}
12. 7.814×10^{-36}
13. 0.999×10^{42}
14. 6.022×10^{23}

Write the number in standard form.

(2)
15. 7×10^{7}
16. 8×10^{-3}
17. 5×10^{2}
18. 2.7×10^{-4}
19. 4.4×10^{-5}
20. 2.1×10^{3}
21. 1.66×10^{9}
22. 3.85×10^{-8}
23. 9.725×10^{6}

24. **ERROR ANALYSIS** Describe and correct the error in writing the number in standard form.

✗ $4.1 \times 10^{-6} = 4100000$

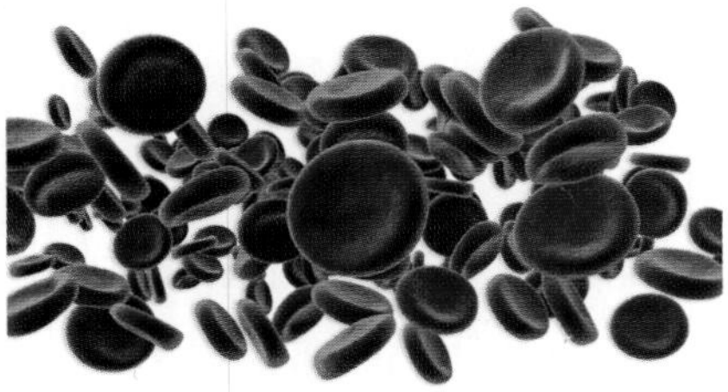
2.7×10^{8} platelets per milliliter

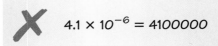

25. **PLATELETS** Platelets are cell-like particles in the blood that help form blood clots.
    a. How many platelets are in 3 milliliters of blood? Write your answer in standard form.
    b. An adult body contains about 5 liters of blood. How many platelets are in an adult body?

26. REASONING A googol is 1.0×10^{100}. How many zeros are in a googol?

3 **27. STARS** The table shows the surface temperatures of five stars.

a. Which star has the highest surface temperature?

b. Which star has the lowest surface temperature?

Star	Betelgeuse	Bellatrix	Sun	Aldebaran	Rigel
Surface Temperature (°F)	6.2×10^3	3.8×10^4	1.1×10^4	7.2×10^3	2.2×10^4

28. CORAL REEF The area of the Florida Keys National Marine Sanctuary is about 9.6×10^3 square kilometers. The area of the Florida Reef Tract is about 16.2% of the area of the sanctuary. What is the area of the Florida Reef Tract in square kilometers?

29. REASONING A gigameter is 1.0×10^6 kilometers. How many square kilometers are in 5 square gigameters?

30. WATER There are about 1.4×10^9 cubic kilometers of water on Earth. About 2.5% of the water is fresh water. How much fresh water is on Earth?

31. Critical Thinking The table shows the speed of light through five media.

a. In which medium does light travel the fastest?

b. In which medium does light travel the slowest?

Medium	Speed
Air	6.7×10^8 mi/h
Glass	6.6×10^8 ft/sec
Ice	2.3×10^5 km/sec
Vacuum	3.0×10^8 m/sec
Water	2.3×10^{10} cm/sec

Fair Game Review What you learned in previous grades & lessons

Write the product using exponents. *(Section 9.1)*

32. $4 \cdot 4 \cdot 4 \cdot 4 \cdot 4$

33. $3 \cdot 3 \cdot 3 \cdot y \cdot y \cdot y$

34. $(-2) \cdot (-2) \cdot (-2)$

35. MULTIPLE CHOICE What is the volume of the prism? *(Section 8.2)*

Ⓐ 13 ft^3 Ⓑ 13.5 ft^3

Ⓒ 65 ft^3 Ⓓ 111 ft^3

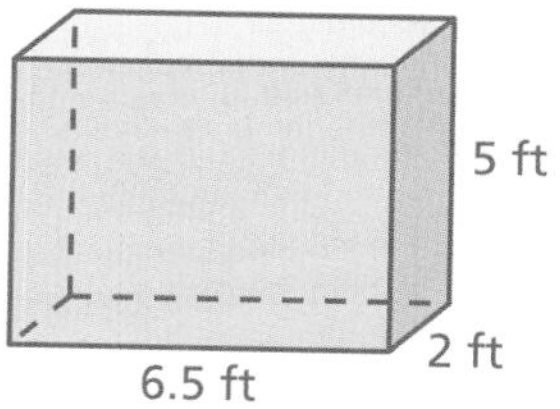

9.4 Writing Scientific Notation

STANDARDS OF LEARNING
7.1

Essential Question How can you write a number in scientific notation?

1 ACTIVITY: Finding pH Levels

Work with a partner. In chemistry, pH is a measure of the activity of dissolved hydrogen ions (H^+). Liquids with low pH values are called acids. Liquids with high pH values are called bases.

Find the pH of each liquid. Is the liquid a base, neutral, or an acid?

a. Lime juice:
$[H^+] = 0.01$

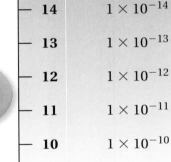

b. Egg:
$[H^+] = 0.00000001$

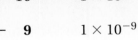

c. Distilled water:
$[H^+] = 0.0000001$

d. Ammonia water:
$[H^+] = 0.00000000001$

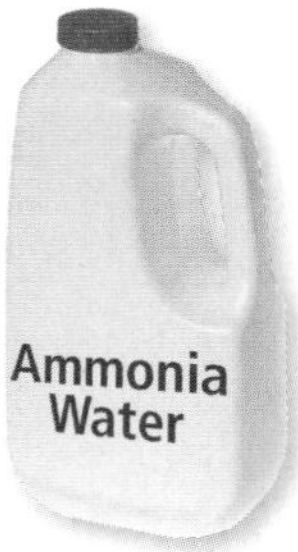

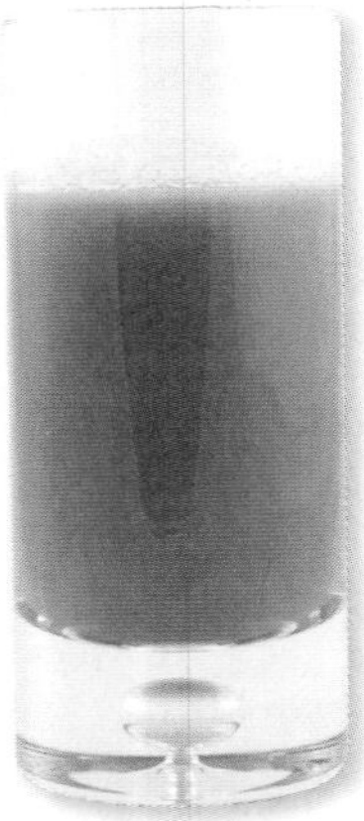

e. Tomato juice:
$[H^+] = 0.0001$

f. Hydrochloric acid:
$[H^+] = 1$

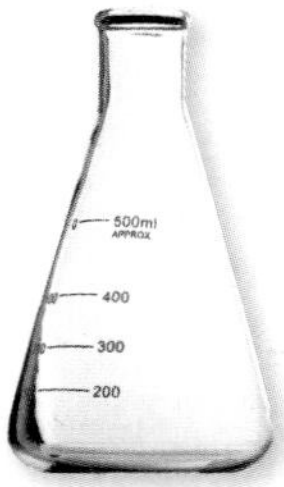

pH	$[H^+]$	
14	1×10^{-14}	Bases
13	1×10^{-13}	
12	1×10^{-12}	
11	1×10^{-11}	
10	1×10^{-10}	
9	1×10^{-9}	
8	1×10^{-8}	
7	1×10^{-7}	**Neutral**
6	1×10^{-6}	Acids
5	1×10^{-5}	
4	1×10^{-4}	
3	1×10^{-3}	
2	1×10^{-2}	
1	1×10^{-1}	
0	1×10^{0}	

2 ACTIVITY: Writing Scientific Notation

Work with a partner. Match each planet with its description. Then write each of the following in scientific notation.

- **Distance from the Sun (in miles)**
- **Distance from the Sun (in feet)**
- **Mass (in kilograms)**

a. Distance: 1,800,000,000 miles
Mass: 87,000,000,000,000,000,000,000,000 kg

b. Distance: 67,000,000 miles
Mass: 4,900,000,000,000,000,000,000,000 kg

c. Distance: 890,000,000 miles
Mass: 570,000,000,000,000,000,000,000,000 kg

d. Distance: 93,000,000 miles
Mass: 6,000,000,000,000,000,000,000,000 kg

e. Distance: 140,000,000 miles
Mass: 640,000,000,000,000,000,000,000 kg

f. Distance: 2,800,000,000 miles
Mass: 100,000,000,000,000,000,000,000,000 kg

g. Distance: 480,000,000 miles
Mass: 1,900,000,000,000,000,000,000,000,000 kg

h. Distance: 36,000,000 miles
Mass: 330,000,000,000,000,000,000,000 kg

3 ACTIVITY: Making a Scale Drawing

Work with a partner. The illustration in Activity 2 is not drawn to scale. Make a scale drawing of the distances in our solar system.

- **Cut a sheet of paper into three strips of equal width. Tape the strips together.**
- **Draw a long number line. Label the number line in hundreds of millions of miles.**
- **Locate each planet's position on the number line.**

What Is Your Answer?

4. IN YOUR OWN WORDS How can you write a number in scientific notation?

Practice Use what you learned about writing scientific notation to complete Exercises 3–5 on page 378.

9.4 Lesson

Check It Out
Lesson Tutorials
BigIdeasMath.com

Key Idea

Writing Numbers in Scientific Notation

Step 1: Move the decimal point to the right of the leading nonzero digit.

Step 2: Count the number of places you moved the decimal point. This indicates the exponent of the power of 10, as shown below.

Number greater than or equal to 10

Use a positive exponent when you move the decimal point to the left.

$$8600 = 8.6 \times 10^3$$

3

Number between 0 and 1

Use a negative exponent when you move the decimal point to the right.

$$0.0024 = 2.4 \times 10^{-3}$$

3

Study Tip

When you write a number greater than or equal to 1 and less than 10 in scientific notation, use zero as the exponent.

$6 = 6 \times 10^0$

EXAMPLE 1 Writing Large Numbers in Scientific Notation

Google purchased YouTube for $1,650,000,000. Write this number in scientific notation.

Move the decimal point 9 places to the left.

$$1{,}650{,}000{,}000 = 1.65 \times 10^9$$

9

The number is greater than 10. So, the exponent is positive.

EXAMPLE 2 Writing Small Numbers in Scientific Notation

The 2004 Indonesian earthquake slowed the rotation of Earth, making the length of a day 0.00000268 second shorter. Write this number in scientific notation.

Move the decimal point 6 places to the right.

$$0.00000268 = 2.68 \times 10^{-6}$$

6

The number is between 0 and 1. So, the exponent is negative.

On Your Own

Now You're Ready
Exercises 3–11

Write the number in scientific notation.

1. 50,000
2. 25,000,000
3. 683
4. 0.005
5. 0.00000033
6. 0.000506

EXAMPLE 3 Standardized Test Practice

An album receives an award when it sells 10,000,000 copies.

An album has sold 8,780,000 copies. How many more copies does it need to sell to receive the award?

Ⓐ 1.22×10^{-7} Ⓑ 1.22×10^{-6}

Ⓒ 1.22×10^{6} Ⓓ 1.22×10^{7}

Use a model to solve the problem.

Remaining sales needed for award = Sales required for award − Current sales total

$= 10{,}000{,}000 - 8{,}780{,}000$

$= 1{,}220{,}000$

$= 1.22 \times 10^6$

The album must sell 1.22×10^6 more copies to receive the award. The correct answer is Ⓒ.

EXAMPLE 4 Real-Life Application

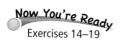

The table shows when the last three geologic eras began. Order the eras from earliest to most recent.

Era	Began
Paleozoic Era	5.42×10^8 years ago
Cenozoic Era	6.55×10^7 years ago
Mesozoic Era	2.51×10^8 years ago

Step 1 Compare the powers of 10.

Because $10^7 < 10^8$,

$6.55 \times 10^7 < 5.42 \times 10^8$ and

$6.55 \times 10^7 < 2.51 \times 10^8$.

Step 2 Compare the factors when the powers of 10 are the same.

Because $2.51 < 5.42$,

$2.51 \times 10^8 < 5.42 \times 10^8$.

From greatest to least, the order is 5.42×10^8, 2.51×10^8, and 6.55×10^7.

So, the eras in order from earliest to most recent are the Paleozoic Era, Mesozoic Era, and Cenozoic Era.

Common Error

To use the method in Example 4, the numbers must be written in scientific notation.

On Your Own

Now You're Ready
Exercises 14–19

7. **WHAT IF?** In Example 3, an album has sold 955,000 copies. How many more copies does it need to sell to receive the award? Write your answer in scientific notation.

8. The Tyrannosaurus Rex lived 7.0×10^7 years ago. Consider the eras given in Example 4. During which era did the Tyrannosaurus Rex live?

9.4 Exercises

Check It Out
Help with Homework
BigIdeasMath.com

Vocabulary and Concept Check

1. **REASONING** How do you know whether a number written in standard form will have a positive or negative exponent when written in scientific notation?

2. **WRITING** When is it appropriate to use scientific notation instead of standard form?

Practice and Problem Solving

Write the number in scientific notation.

(1) (2)

3. 0.0021
4. 5,430,000
5. 321,000,000
6. 0.00000625
7. 0.00004
8. 10,700,000
9. 45,600,000,000
10. 0.000000000009256
11. 840,000

ERROR ANALYSIS Describe and correct the error in writing the number in scientific notation.

12. ✗ 0.000036 → 5 → 3.6×10^5

13. ✗ 72,500,000 → 6 → 72.5×10^6

Order the numbers from least to greatest.

(4)

14. $1.2 \times 10^8,\ 1.19 \times 10^8,\ 1.12 \times 10^8$
15. $6.8 \times 10^{-5},\ 6.09 \times 10^{-5},\ 6.78 \times 10^{-5}$
16. $5.76 \times 10^{12},\ 9.66 \times 10^{11},\ 5.7 \times 10^{10}$
17. $4.8 \times 10^{-6},\ 4.8 \times 10^{-5},\ 4.8 \times 10^{-8}$
18. $9.9 \times 10^{-15},\ 1.01 \times 10^{-14},\ 7.6 \times 10^{-15}$
19. $5.78 \times 10^{23},\ 6.88 \times 10^{-23},\ 5.82 \times 10^{23}$

20. **HAIR** What is the diameter of a human hair in scientific notation?

Diameter: 0.000099 meter

21. **EARTH** What is the circumference of Earth in scientific notation?

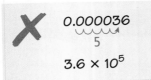

Circumference at the equator: about 40,100,000 meters

22. **WATERFALLS** During high flow, more than 44,380,000 gallons of water go over Niagara Falls every minute. Write this number in scientific notation.

Order the numbers from least to greatest.

23. $\frac{68{,}500}{10}$, 680, 6.8×10^3

24. $\frac{5}{241}$, 0.02, 2.1×10^{-2}

25. 6.3%, 6.25×10^{-3}, $6\frac{1}{4}$, 0.625

26. 3033.4, 305%, $\frac{10{,}000}{3}$, 3.3×10^2

27. SPACE SHUTTLE The total power of a space shuttle during launch is the sum of the power from its solid rocket boosters and the power from its main engines. The power from the solid rocket boosters is 9.75×10^9 watts. What is the power from the main engines?

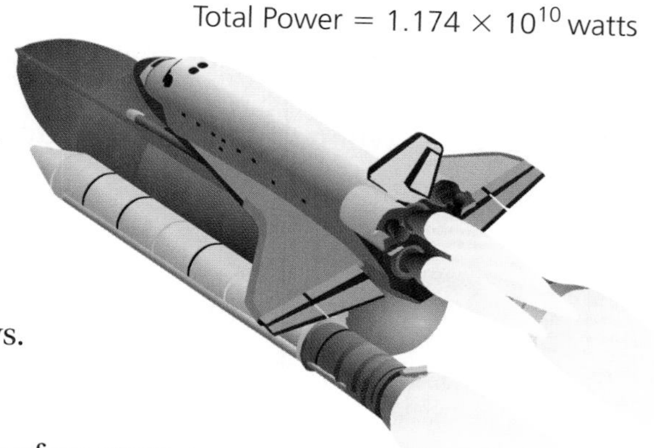

28. NUMBER SENSE Write 670 million in three ways.

Equivalent to 1 Atomic Mass Unit:
8.3×10^{-24} carat
1.66×10^{-21} milligram

29. ATOMIC MASS The mass of an atom or molecule is measured in atomic mass units. Which is greater, a *carat* or a *milligram*? Explain.

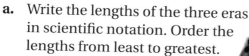

30. Reasoning In Example 4, the Paleozoic Era ended when the Mesozoic Era began. The Mesozoic Era ended when the Cenozoic Era began. The Cenozoic Era is the current era.

a. Write the lengths of the three eras in scientific notation. Order the lengths from least to greatest.

b. Make a timeline to show when the three eras occurred and how long each era lasted.

c. What do you notice about the lengths of the three eras? Use the Internet to determine whether your observation is true for *all* of the geologic eras. Explain your results.

Fair Game Review What you learned in previous grades & lessons

Evaluate the expression. *(Section 9.1)*

31. 4^2

32. $(-5)^2$

33. -7^2

34. What is the surface area of the prism? *(Section 8.1)*

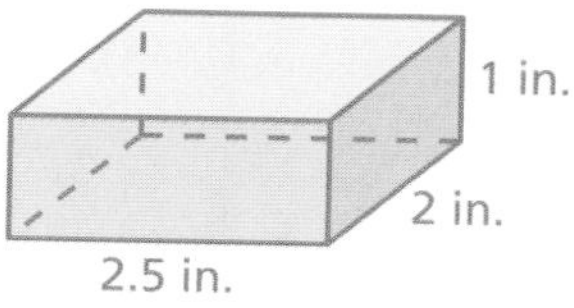

(A) 5 in.^{2}

(B) 5.5 in.^{2}

(C) 10 in.^{2}

(D) 19 in.^{2}

9.5 Square Roots

STANDARDS OF LEARNING
7.1

Essential Question How can you find the side length of a square when you are given the area of the square?

When you multiply a number by itself, you square the number.

Symbol for squaring is the exponent 2. → $4^2 = 4 \cdot 4$

$= 16$ 4 squared is 16.

To "undo" this, take the **square root** of the number.

Symbol for square root is a radical sign. → $\sqrt{16} = \sqrt{4^2} = 4$ The square root of 16 is 4.

1 ACTIVITY: Finding Square Roots

Work with a partner. Use a square root symbol to write the side length of the square. Then find the square root. Check your answer by multiplying.

a. Sample: $s = \sqrt{121} = 11$ ft

Area = 121 ft^2

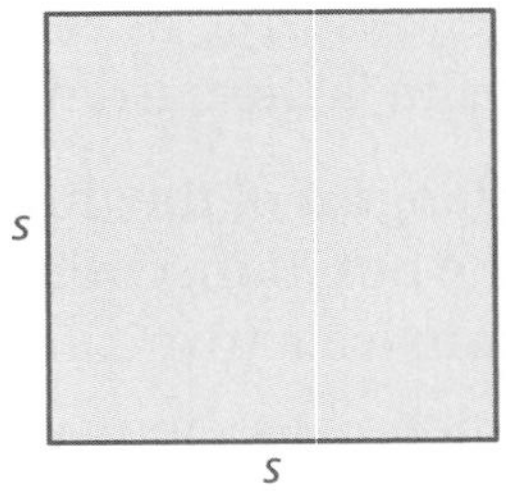

Check

$$\begin{array}{r} 11 \\ \times\ 11 \\ \hline 11 \\ 110 \\ \hline 121 \end{array} \checkmark$$

The side length of the square is 11 feet.

b. Area = 81 yd^2

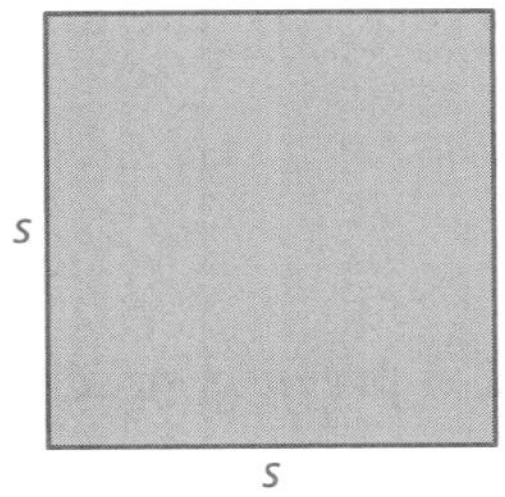

c. Area = 324 cm^2

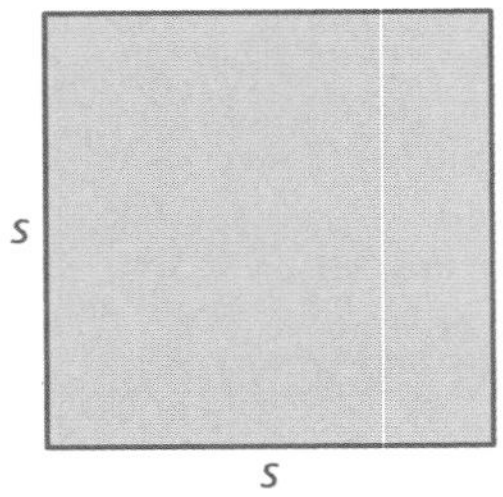

d. Area = 361 mi^2

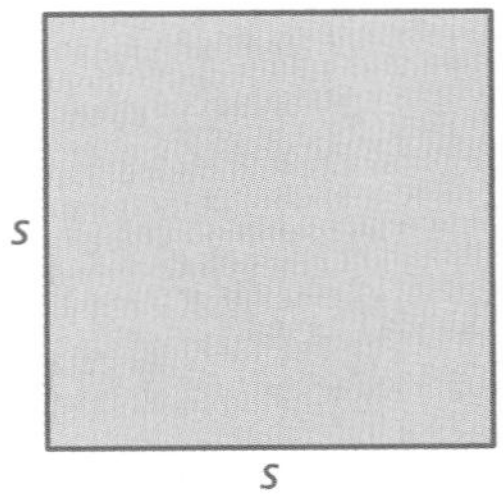

e. Area = 289 in.^{2}

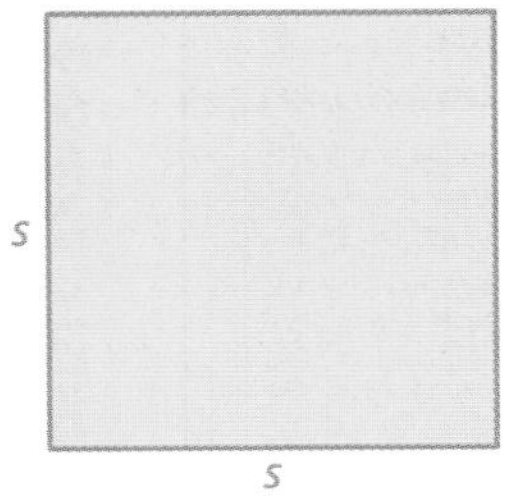

f. Area = 441 in.^{2}

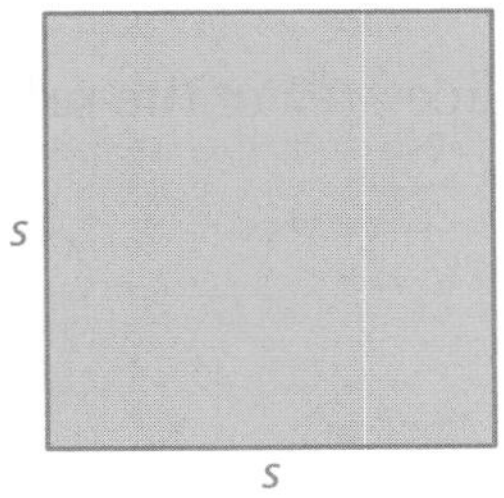

g. Area = 900 ft^2

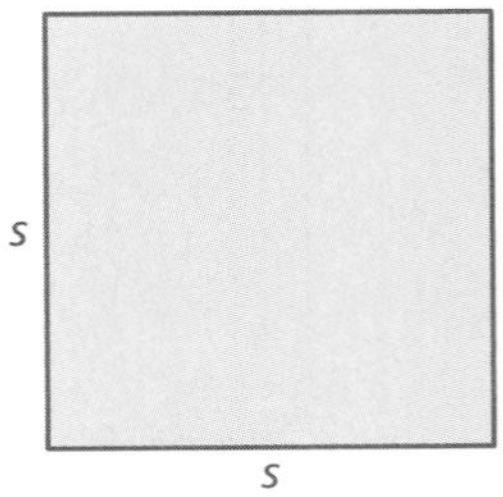

2 ACTIVITY: Describing Patterns

Work with a partner.

a. On a piece of grid paper, draw a sequence of small squares that represent the odd numbers.

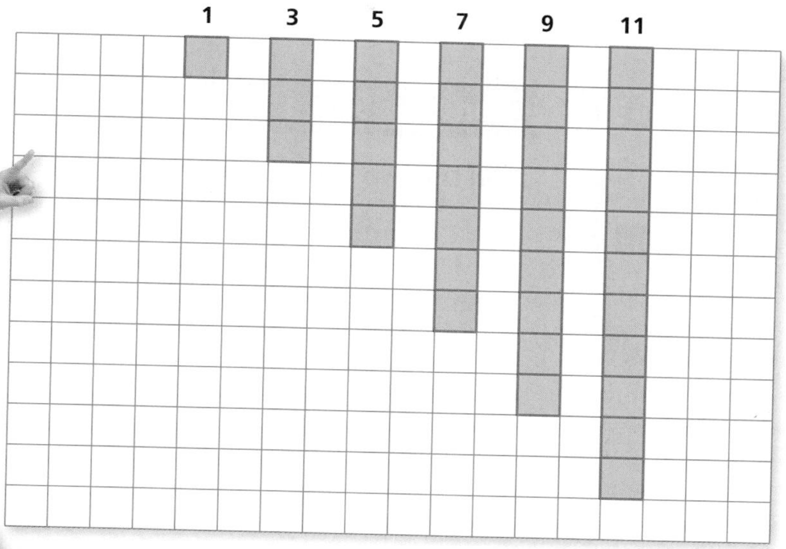

b. Draw a new sequence that represents the sums of terms in the sequence in part (a).

$$1, \qquad 1 + 3, \qquad 1 + 3 + 5, \ldots$$

c. Organize your results in a table.

d. Describe the new sequence.

What Is Your Answer?

3. IN YOUR OWN WORDS How can you find the side length of a square when you are given the area of the square?

4. CONJECTURE Use the results of Activity 2 to write a conjecture about the sum of the first n odd numbers. On a different piece of grid paper, rearrange the small squares to explain *why* your conjecture is true.

Practice Use what you learned about square roots to complete Exercises 5–7 on page 384.

9.5 Lesson

Check It Out
Lesson Tutorials
BigIdeasMath.com

Key Vocabulary
square root, *p. 382*
perfect square, *p. 382*
radical sign, *p. 382*
radicand, *p. 382*

A **square root** of a number is a number that when multiplied by itself, equals the given number. Every positive number has a positive *and* a negative square root. A **perfect square** is a number with integers as its square roots.

EXAMPLE 1 Finding Square Roots of a Perfect Square

Study Tip
Zero has one square root, which is 0.

Find the two square roots of 9.

$3 \cdot 3 = 9$ and $(-3) \cdot (-3) = 9$

So, the square roots of 9 are 3 and -3.

The symbol $\sqrt{}$ is called a **radical sign**. It is used to represent a square root. The number under the radical sign is called the **radicand**.

Positive Square Root $\sqrt{}$	Negative Square Root $-\sqrt{}$	Both Square Roots $\pm\sqrt{}$
$\sqrt{25} = 5$	$-\sqrt{25} = -5$	$\pm\sqrt{25} = \pm 5$

EXAMPLE 2 Finding Square Roots

Find each square root.

a. $\sqrt{81}$

$\sqrt{81}$ represents the *positive* square root.

Because $9^2 = 81$, $\sqrt{81} = \sqrt{9^2} = 9$.

b. $-\sqrt{16}$

$-\sqrt{16}$ represents the *negative* square root.

Because $4^2 = 16$, $-\sqrt{16} = -\sqrt{4^2} = -4$.

On Your Own

Now You're Ready
Exercises 8–17

Find the two square roots of the number.

1. 4 **2.** 49 **3.** 100

Find the square root(s).

4. $-\sqrt{36}$ **5.** $\pm\sqrt{1}$ **6.** $\sqrt{169}$

Multi-Language Glossary at BigIdeasMath.com.

EXAMPLE 3 Evaluating Expressions Involving Square Roots

Evaluate each expression.

a. $2\sqrt{64} - 18$

$2\sqrt{64} - 18 = 2(8) - 18$	Evaluate the square root.
$= 16 - 18$	Multiply.
$= -2$	Subtract.

b. $3\sqrt{49} + 5\sqrt{16}$

$3\sqrt{49} + 5\sqrt{16} = 3(7) + 5(4)$	Evaluate the square roots.
$= 21 + 20$	Multiply.
$= 41$	Add.

Squaring a positive number and finding a square root are inverse operations. Use this relationship to solve equations involving squares.

EXAMPLE 4 Real-Life Application

In a game of marbles, players try to knock other marbles out of a circle that has an area of 78.5 square feet. What is the diameter of the circle? Use 3.14 for π.

$A = \pi r^2$	Write the formula for the area of a circle.
$78.5 \approx 3.14r^2$	Substitute 78.5 for A and 3.14 for π.
$25 = r^2$	Divide each side by 3.14.
$\sqrt{25} = \sqrt{r^2}$	Distance is positive, so take the positive square root of each side.
$5 = r$	Simplify.

The diameter of a circle is two times the radius r. So, the diameter is $2 \cdot 5 = 10$ feet.

On Your Own

Now You're Ready
Exercises 19–24 and 27–29

Evaluate the expression.

7. $15 - \sqrt{121}$ **8.** $8\sqrt{100} + 6$ **9.** $7\sqrt{4} - 2\sqrt{121}$

10. The area of a circle is 706.5 square meters. Find the radius of the circle. Use 3.14 for π.

9.5 Exercises

Check It Out
Help with Homework
BigIdeasMath.com

Vocabulary and Concept Check

1. **VOCABULARY** Explain why 50 is *not* a perfect square.
2. **VOCABULARY** What is the radicand in the expression $2\sqrt{36} + 4$? Explain.
3. **NUMBER SENSE** What is the inverse operation of squaring a positive number? Give an example.
4. **WHICH ONE DOESN'T BELONG?** Which expression does *not* belong with the other three? Explain your reasoning.

$\sqrt{49}$	$\pm\sqrt{169}$	$-\sqrt{121}$	$\sqrt{109}$

Practice and Problem Solving

Find the side length of the square. Check your answer by multiplying.

5. Area = 4 ft^2

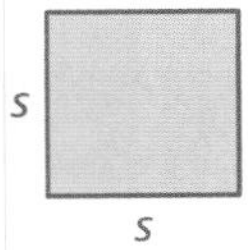

6. Area = 9 cm^2

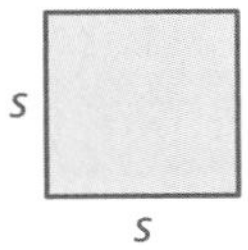

7. Area = 49 $in.^2$

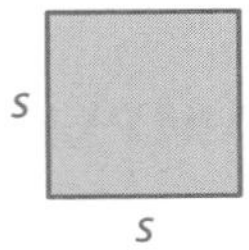

Find the two square roots of the number.

1 8. 16 9. 121 10. 64 11. 169

Find the square root(s).

2 12. $\sqrt{9}$ 13. $-\sqrt{144}$ 14. $-\sqrt{25}$

15. $\sqrt{49}$ 16. $\pm\sqrt{36}$ 17. $\pm\sqrt{196}$

18. **ERROR ANALYSIS** Describe and correct the error in finding the square roots.

✗ $\pm\sqrt{81} = 9$

Evaluate the expression.

3 19. $29 - \sqrt{225}$ 20. $3\sqrt{64} + 12$ 21. $54 - 4\sqrt{16}$

22. $6\sqrt{9} + 4\sqrt{36}$ 23. $7\sqrt{25} - 2\sqrt{256}$ 24. $27\sqrt{4} + \sqrt{400}$

25. **SCHOOL SPIRIT** A school spirit sign is shaped like a square. The area of the sign is 36 square feet. What is the length of one side of the sign?

26. **NUMBER SENSE** Can a square root of a perfect square be another perfect square? Give an example to support your reasoning.

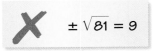

Find the radius of the circle. Use 3.14 for π.

4 **27.** Area = 12.56 cm^2

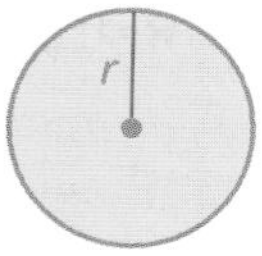

28. Area = 50.24 ft^2

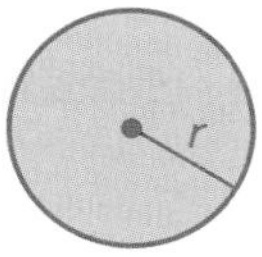

29. Area = 200.96 $in.^2$

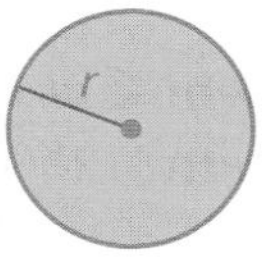

30. PERIMETER Write and evaluate an expression to find the perimeter of the GPS.

31. REASONING What is the square root of c when $a^2 = c$? Explain.

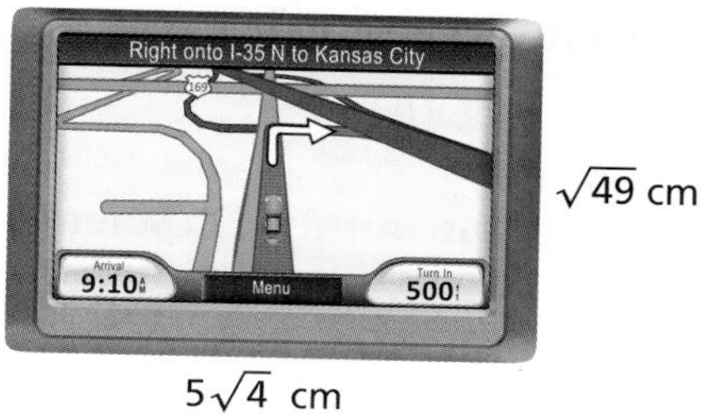

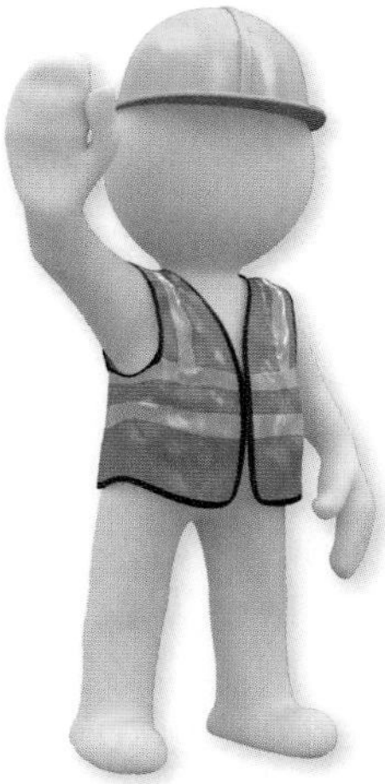

32. VOLUME The height of a cylinder is 4 centimeters. The volume of the cylinder is 314 cubic centimeters. What is the radius of the cylinder? Use 3.14 for π.

33. FALLING OBJECT The distance d (in feet) a hammer falls at a construction site is represented by $d = 16t^2$, where t is the time (in seconds) it takes for the hammer to hit the ground. How long does it take the hammer to hit the ground when it falls from a height of 400 feet?

34. Critical Thinking The perimeter of Square $ABCD$ is 12. The ratio of the areas of Square $ABCD$ and Square $EFGH$ is $\frac{9}{36}$. What is the perimeter of Square $EFGH$?

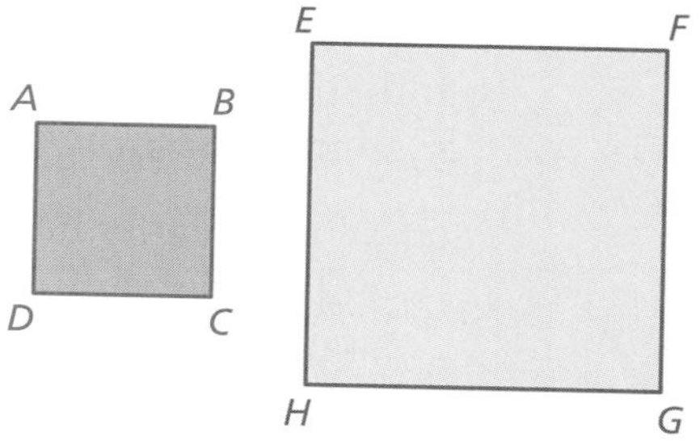

Fair Game Review What you learned in previous grades & lessons

Tell whether the two figures are similar. Explain your reasoning. *(Section 6.2)*

35.

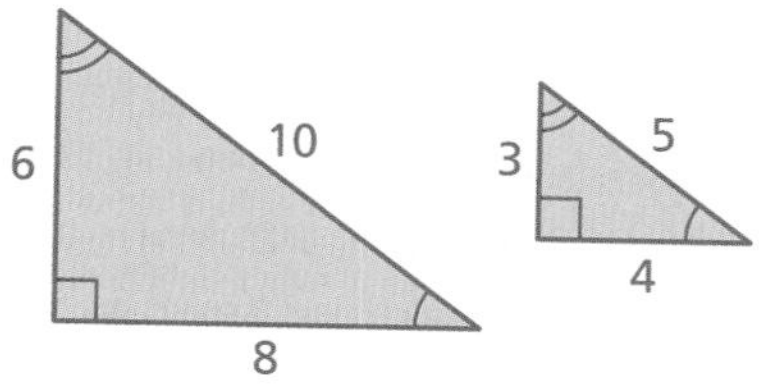

36.

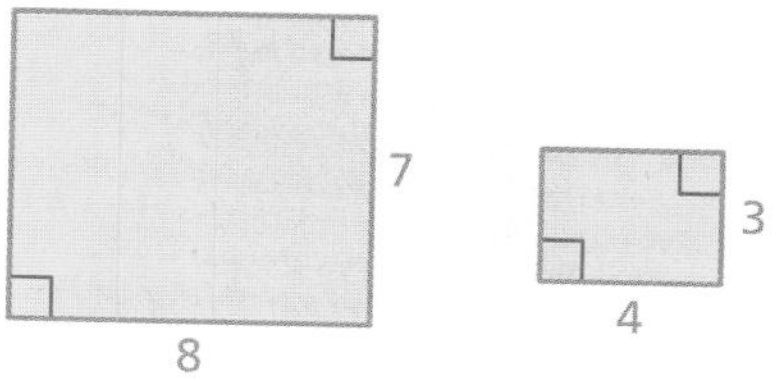

37. MULTIPLE CHOICE What is the surface area of the cylinder? Round your answer to the nearest tenth. *(Section 8.3)*

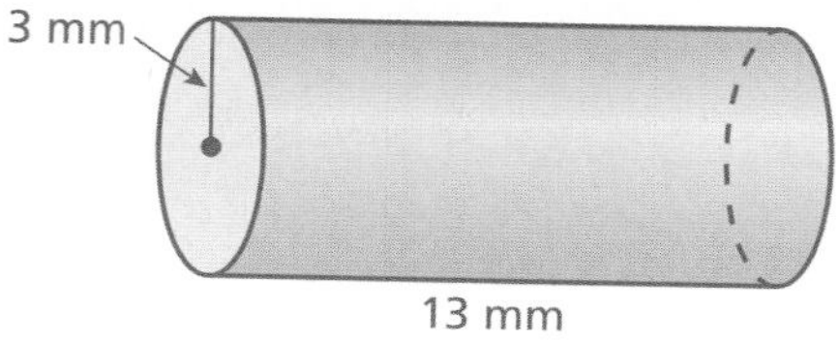

Ⓐ 134.5 mm^2 Ⓑ 263.8 mm^2

Ⓒ 301.4 mm^2 Ⓓ 367.4 mm^2

9.3–9.5 Quiz

Check It Out
Progress Check
BigIdeasMath.com

Tell whether the number is written in scientific notation. Explain. *(Section 9.3)*

1. 23×10^9

2. 0.6×10^{-7}

Write the number in standard form. *(Section 9.3)*

3. 8×10^6

4. 1.6×10^{-2}

Write the number in scientific notation. *(Section 9.4)*

5. 0.00524

6. 892,000,000

Order the numbers from least to greatest. *(Section 9.4)*

7. $5.3 \times 10^{-5}, 5.24 \times 10^{-4}, 5.2 \times 10^{-4}$

8. $7.35 \times 10^{18}, 2.4 \times 10^{19}, 6.02 \times 10^{18}$

Find the square root(s). *(Section 9.5)*

9. $\sqrt{16}$

10. $-\sqrt{121}$

11. $\pm\sqrt{9}$

Evaluate the expression. *(Section 9.5)*

12. $6\sqrt{25} - 12$

13. $4 - \sqrt{64}$

14. $2\sqrt{36} + 5\sqrt{16}$

15. PLANETS The table shows the equatorial radii of the eight planets in our solar system. *(Section 9.3 and Section 9.4)*

a. Which planet has the second smallest equatorial radius? Write its radius in standard form.

b. Which planet has the second greatest equatorial radius? Write its radius in standard form.

Planet	Equatorial Radius (km)
Mercury	2.44×10^3
Venus	6.05×10^3
Earth	6.38×10^3
Mars	3.4×10^3
Jupiter	7.15×10^4
Saturn	6.03×10^4
Uranus	2.56×10^4
Neptune	2.48×10^4

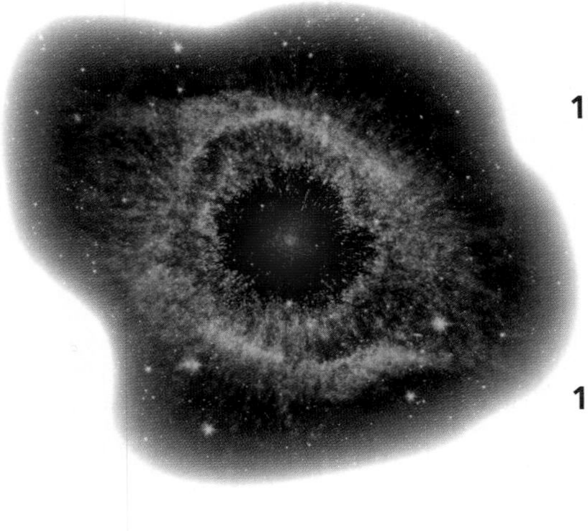

16. OORT CLOUD The Oort Cloud is a spherical cloud that surrounds our solar system. It extends about 18,000,000,000,000 miles from the Sun. Write this number in scientific notation. *(Section 9.4)*

17. VOLUME The volume of the cylinder is 7630.2 cubic centimeters. What is the radius of the cylinder? Use 3.14 for π. *(Section 9.5)*

9 Chapter Review

Check It Out
Vocabulary Help
BigIdeasMath .com

Review Key Vocabulary

power, *p. 356*
base, *p. 356*
exponent, *p. 356*
scientific notation, *p. 370*
square root, *p. 382*
perfect square, *p. 382*
radical sign, *p. 382*
radicand, *p. 382*

Review Examples and Exercises

9.1 Exponents *(pp. 354–359)*

Write $(-4) \cdot (-4) \cdot (-4) \cdot y \cdot y$ using exponents.

Because -4 is used as a factor 3 times, its exponent is 3. Because y is used as a factor 2 times, its exponent is 2.

So, $(-4) \cdot (-4) \cdot (-4) \cdot y \cdot y = (-4)^3 y^2$.

Exercises

Write the product using exponents.

1. $(-9) \cdot (-9) \cdot (-9) \cdot (-9) \cdot (-9)$
2. $2 \cdot 2 \cdot 2 \cdot n \cdot n$

Evaluate the expression.

3. 6^3
4. $-\left(\frac{1}{2}\right)^4$
5. $(-3)^5$
6. $(61 - 4^2) \div 5$
7. $8\left|2^3 - 7^2\right|$
8. $\left|\frac{1}{2}(16 - 6^3)\right|$

9.2 Zero and Negative Exponents *(pp. 360–365)*

a. $10^{-3} = \frac{1}{10^3}$ — Definition of negative exponent

$= \frac{1}{1000}$ — Evaluate the power.

b. $(11^0 + 7) \cdot 6^{-2} = (1 + 7) \cdot 6^{-2}$ — Evaluate power inside parentheses.

$= 8 \cdot 6^{-2}$ — Add.

$= 8 \cdot \frac{1}{6^2}$ — Definition of negative exponent

$= 8 \cdot \frac{1}{36}$ — Evaluate the power.

$= \frac{8}{36}$ — Multiply.

$= \frac{2}{9}$ — Simplify.

Exercises

Evaluate the expression.

9. 2^{-4} **10.** 95^0 **11.** $22 - 4^0 + \frac{1}{3^{-3}}$

Simplify. Write your answer using only positive exponents.

12. $32z^{-5}$ **13.** $3x^{-7} \cdot y^2$ **14.** $\frac{4^{-2} \cdot w^0 \cdot k}{k^{-3}}$

9.3 Reading Scientific Notation *(pp. 368–373)*

a. Write 5.9×10^4 in standard form.

$5.9 \times 10^4 = 59{,}000$ (4) Move decimal point 4 places to the right.

b. Write 7.31×10^{-6} in standard form.

$7.31 \times 10^{-6} = 0.00000731$ (6) Move decimal point 6 places to the left.

Exercises

Tell whether the number is written in scientific notation. Explain.

15. 0.9×10^9 **16.** 3.04×10^{-11} **17.** 15×10^{26}

Write the number in standard form.

18. 2×10^7 **19.** 4.8×10^{-3} **20.** 6.25×10^5

9.4 Writing Scientific Notation *(pp. 374–379)*

a. In 2010, the population of the United States was about 309,000,000. Write this number in scientific notation.

Move the decimal point 8 places to the left.

$309{,}000{,}000 = 3.09 \times 10^8$ (8)

The number is greater than 10. So, the exponent is positive.

b. The cornea of an eye is 0.00056 meter thick. Write this number in scientific notation.

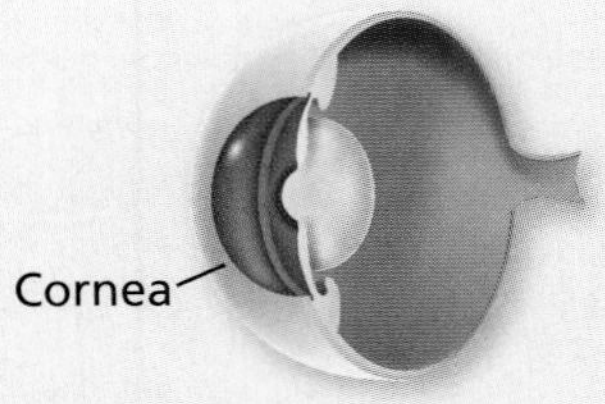

Move the decimal point 4 places to the right.

$0.00056 = 5.6 \times 10^{-4}$ (4)

The number is between 0 and 1. So, the exponent is negative.

Exercises

Write the number in scientific notation.

21. 0.00036 **22.** 800,000 **23.** 79,200,000

Order the numbers from least to greatest.

24. $8.56 \times 10^{32}, 8.01 \times 10^{32}, 8.07 \times 10^{32}$

25. $6.3 \times 10^{-11}, 7.9 \times 10^{-9}, 7.4 \times 10^{-13}$

26. DISTANCE The distance between the North Pole and the South Pole is about 65,600,000 feet. Write this number in scientific notation.

9.5 Square Roots (pp. 380–385)

Find the square root(s).

a. $\sqrt{144}$

$\sqrt{144}$ represents the *positive* square root.

Because $12^2 = 144$, $\sqrt{144} = \sqrt{12^2} = 12$.

b. $-\sqrt{64}$

$-\sqrt{64}$ represents the *negative* square root.

Because $8^2 = 64$, $-\sqrt{64} = -\sqrt{8^2} = -8$.

c. $\pm\sqrt{225}$

$\pm\sqrt{225}$ represents both the *positive and negative* square roots.

Because $15^2 = 225$, $\pm\sqrt{225} = \pm\sqrt{15^2} = 15$ and -15.

Exercises

Find the two square roots of the number.

27. 25 **28.** 36 **29.** 196

Find the square root(s).

30. $\pm\sqrt{100}$ **31.** $\sqrt{289}$ **32.** $-\sqrt{625}$

Evaluate the expression.

33. $28 - 9\sqrt{4}$ **34.** $8\sqrt{16} - 2\sqrt{49}$ **35.** $6\sqrt{25} + 17$

Check It Out
Test Practice
BigIdeasMath.com

Write the product using exponents.

1. $(-15) \cdot (-15) \cdot (-15)$

2. $\left(\frac{1}{12}\right) \cdot \left(\frac{1}{12}\right) \cdot \left(\frac{1}{12}\right) \cdot \left(\frac{1}{12}\right) \cdot \left(\frac{1}{12}\right)$

Evaluate the expression.

3. -2^3

4. $10 + 3^3 \div 9$

5. $5^{-2} \cdot 5^2$

6. $\left|(-3)^{-3}\right| + (3 - 4^0)$

Simplify. Write your answer using only positive exponents.

7. $\dfrac{8}{m^{-4}}$

8. $2^{-3} \cdot b^2 \cdot w^0$

Write the number in standard form.

9. 3×10^7

10. 9.05×10^{-3}

Find the square root(s).

11. $-\sqrt{9}$

12. $\pm\sqrt{121}$

Evaluate the expression.

13. $12 + 8\sqrt{16}$

14. $6\sqrt{64} - 8\sqrt{36}$

15. **HAMSTER** A hamster toy is in the shape of a sphere. The volume V of a sphere is represented by $V = \frac{4}{3}\pi r^3$, where r is the radius of the sphere. What is the volume of the toy? Round your answer to the nearest cubic centimeter.

16. **TASTE BUDS** There are about 10,000 taste buds on a human tongue. Write this number in scientific notation.

17. **CILIA** Hair-like projections called cilia line the respiratory system to help remove inhaled dust and potentially harmful microbes. What is the length of 10 cilia in meters? (1 micrometer $= 10^{-6}$ meter)

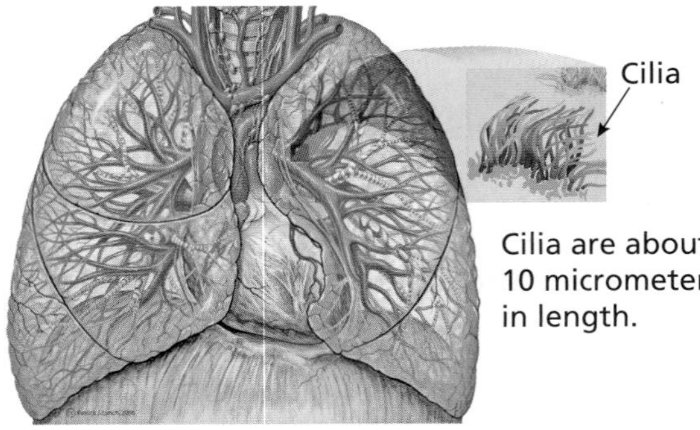

Cilia are about 10 micrometers in length.

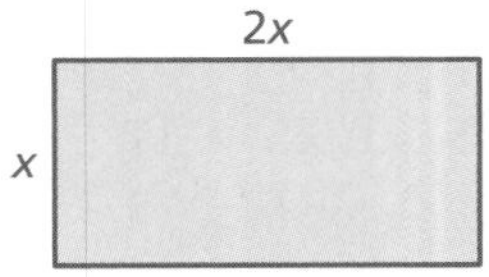

18. **GEOMETRY** The area of the rectangle is 242 square feet. Find the length and width of the rectangle.

Test-Taking Strategy
Use Intelligent Guessing

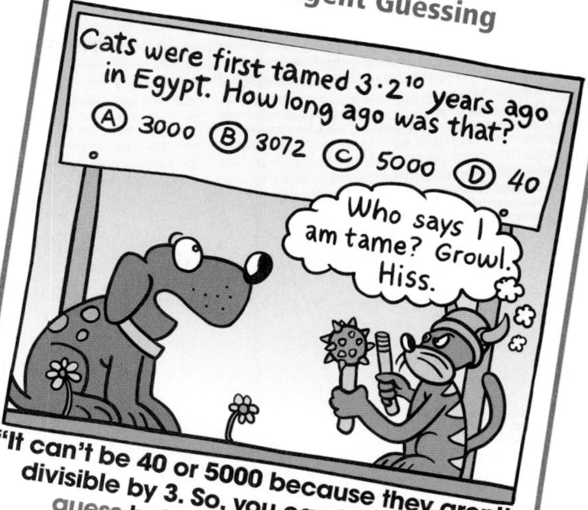

Cats were first tamed $3 \cdot 2^{10}$ years ago in Egypt. How long ago was that?
(A) 3000 (B) 3072 (C) 5000 (D) 40
Who says I am tame? Growl. Hiss.

"It can't be 40 or 5000 because they aren't divisible by 3. So, you can intelligently guess between 3000 and 3072."

1. The length of a paramecium is about 0.0005 meter. What is this length in scientific notation?

   A. 5.0×10^{-4} m
   B. 5.0×10^{-3} m
   C. 5.0×10^{3} m
   D. 5.0×10^{4} m

2. Callie was evaluating the expression in the box below.

$$\begin{aligned} 12 - 6^2 \div (-3) &= 12 - 6 \cdot 6 \div (-3) \\ &= 12 - 36 \div (-3) \\ &= 12 - 12 \\ &= 0 \end{aligned}$$

   What should Callie do to correct the error that she made?

   F. Divide 6 by -3.
   G. First subtract 6 from 12.
   H. Divide 36 by -3 to get -12 instead of 12.
   I. Subtract 36 from 12 before dividing by -3.

3. What is the slope of the line through the points $(0, 0)$ and $(-4, 8)$?

4. What is the value of x?

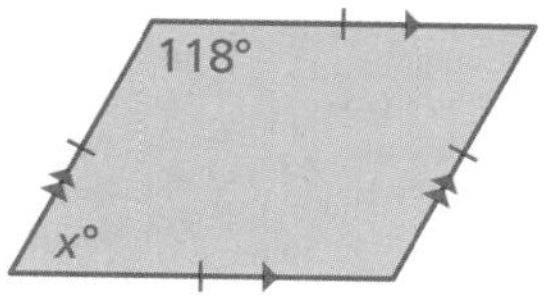

   A. 62
   B. 72
   C. 118
   D. 124

5. Which is *not* a solution of the inequality shown below?

$$-4.2 < \frac{y}{5}$$

F. -21

G. -10

H. 0

I. 21

6. Lexi wants to draw a square with an area of 81 square feet to play a game with her classmates. What length s should she make each side?

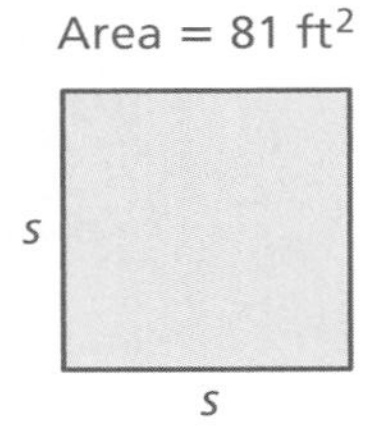

A. 6 ft

B. 9 ft

C. $20\frac{1}{4}$ ft

D. $40\frac{1}{2}$ ft

7. Which function rule is shown by the table?

Input, x	-1	1	3	5
Output, y	-3	3	9	15

F. $y = 3x$

G. $y = -3x$

H. $y = x + 6$

I. $y = 2x + 6$

8. Find $(-5)^{-4}$.

9. Mercury *transits* directly between Earth and the Sun a few times each century. How far was Mercury from Earth during the 2006 transit shown below?

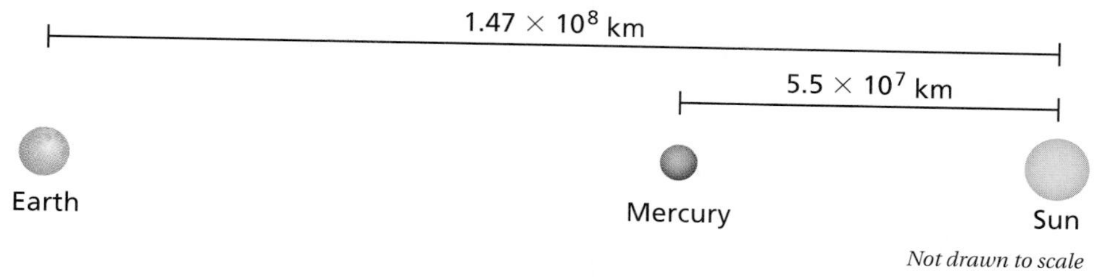

A. 9.2×10^6 km

B. 9.2×10^7 km

C. 1.415×10^7 km

D. 2.02×10^7 km

10. Which transformation is shown below?

F. translation

G. reflection

H. rotation

I. dilation

11. The shipping box is redesigned to double its volume. Only one of its dimensions is changed.

Think Solve Explain

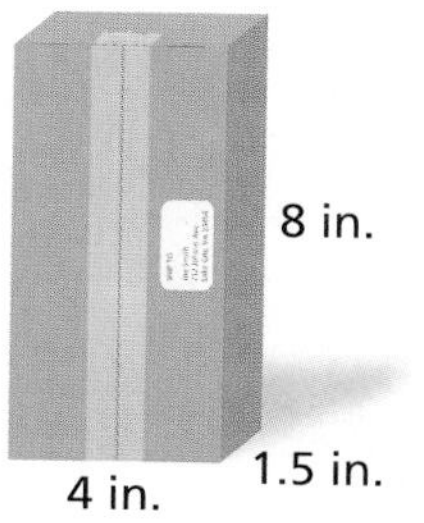

What is the least amount of cardboard needed to make the new box? Show your work and explain your reasoning.

10 Data Analysis and Probability

"I'm named after Sir Isaac Newton, the inventor of calculus."

"And you're named after René Descartes, the inventor of the Cartesian plane."

"Did you know that the letters of 'SIR ISAAC NEWTON' can be rearranged to form 'IS NOW CARTESIAN'?"

"Let's spin to see who has to lick the lunch bowls clean."

"If it's orange, you lick the bowls. If it's brown, I lick the bowls."

What You Learned Before

"If I go as a hyena to the costume party, what is the probability that Fluffy will recognize me?"

Analyzing Bar Graphs

Example 1 **The bar graph shows the favorite colors of the students in a class. How many students said their favorite color is blue?**

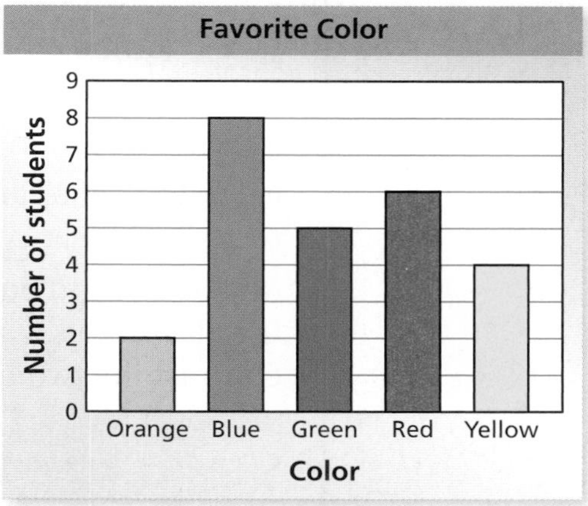

The height of the bar labeled "Blue" is 8.

So, 8 students said their favorite color is blue.

Try It Yourself

1. What color was chosen the least?
2. How many students said green or red is their favorite color?
3. How many students did *not* choose yellow as their favorite color?
4. How many students are in the class?

Writing Ratios

Example 2 **There are 32 football players and 16 cheerleaders at your school. Write the ratio of cheerleaders to football players.**

cheerleaders → football players → $\frac{16}{32} = \frac{1}{2}$ Write in simplest form.

The ratio of cheerleaders to football players is $\frac{1}{2}$.

Try It Yourself

Write the ratio in simplest form.

5. black beads to blue beads
6. green beads to blue beads
7. blue beads : red beads
8. red beads : green beads
9. green beads : total number of beads

10.1 Histograms

STANDARDS OF LEARNING
7.11

Essential Question How can you use tables and graphs to help organize data?

1 ACTIVITY: Conducting an Experiment

Work with a partner.

a. Roll a number cube 20 times. Record your results in a tally chart.

b. Make a bar graph of the totals.

c. Go to the board and enter your totals in the class tally chart.

d. Make a second bar graph showing the class totals. Compare and contrast the two bar graphs.

	Tally Chart
1	
2	
3	
4	
5	
6	

Key: | = 1   ~~||||~~ = 5

2 ACTIVITY: Organizing Data

Work with a partner. You are judging a paper airplane contest. Each contestant flies his or her paper airplane 20 times. Make a tally chart and a graph of the distances.

Sample:

20.5 ft, 24.5 ft, 18.5 ft, 19.5 ft, 21.0 ft, 14.0 ft, 12.5 ft, 20.5 ft, 17.5 ft, 24.5 ft, 19.5 ft, 17.0 ft, 18.5 ft, 12.0 ft, 21.5 ft, 23.0 ft, 13.5 ft, 19.0 ft, 22.5 ft, 19.0 ft

Tally Chart

Interval	Tally	Total
10.0–12.9	II	2
13.0–15.9	II	2
16.0–18.9	IIII	4
19.0–21.9	~~IIII~~ III	8
22.0–24.9	IIII	4

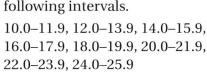

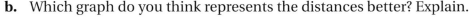

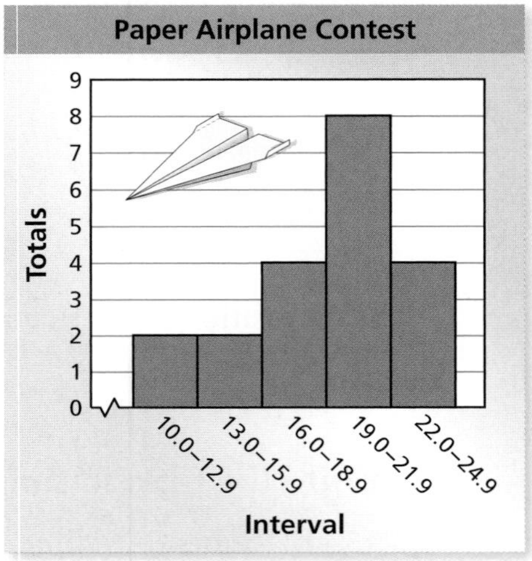

a. Make a different tally chart and graph of the distances using the following intervals.

10.0–11.9, 12.0–13.9, 14.0–15.9, 16.0–17.9, 18.0–19.9, 20.0–21.9, 22.0–23.9, 24.0–25.9

b. Which graph do you think represents the distances better? Explain.

3 ACTIVITY: Developing an Experiment

Work with a partner.

a. Design and make a paper airplane from a single sheet of $8\frac{1}{2}$-by-11-inch paper.

Sample:

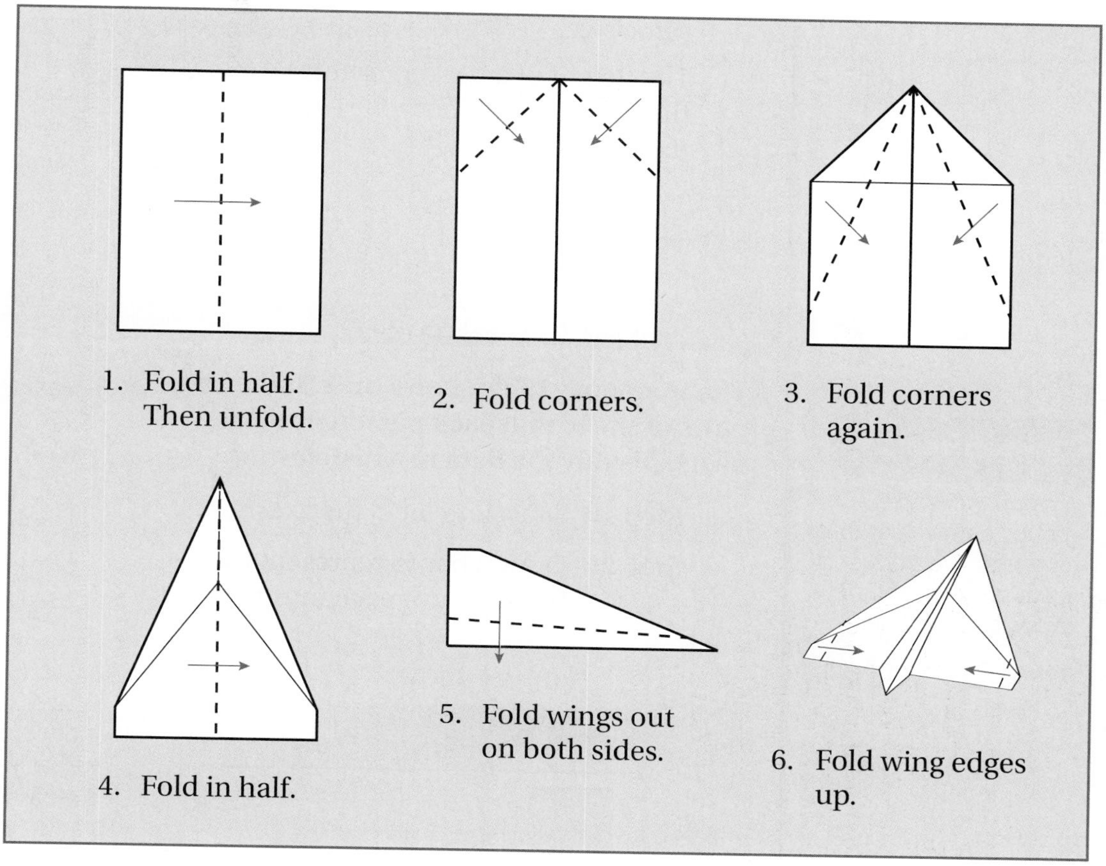
1. Fold in half. Then unfold.
2. Fold corners.
3. Fold corners again.
4. Fold in half.
5. Fold wings out on both sides.
6. Fold wing edges up.

b. Fly the airplane 20 times. Keep track of the distance flown each time.

c. Organize your results in a tally chart and a graph. What is the mean distance flown by the airplane?

What Is Your Answer?

4. IN YOUR OWN WORDS How can you use tables and graphs to help organize data? Give examples of careers in which the organization of data is important.

Practice Use what you learned about organizing data to complete Exercises 4 and 5 on page 401.

10.1 Lesson

Check It Out
Lesson Tutorials
BigIdeasMath.com

Key Vocabulary
histogram, *p. 398*

Key Idea

Histograms

A **histogram** is a bar graph that shows the frequency of data values in intervals of the same size.

The height of a bar represents the frequency of the values in the interval.

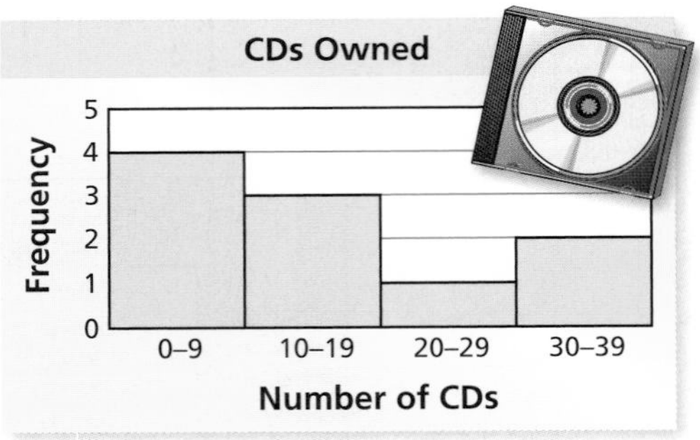

EXAMPLE 1 Making a Histogram

Remember

A *frequency table* groups data values into intervals. The *frequency* is the number of data values in an interval.

The frequency table shows the number of pairs of shoes that each person in a class owns. Display the data in a histogram.

Pairs of Shoes	Frequency
1–3	11
4–6	4
7–9	0
10–12	3
13–15	6

Step 1: Draw and label the axes.

Step 2: Draw a bar to represent the frequency of each interval.

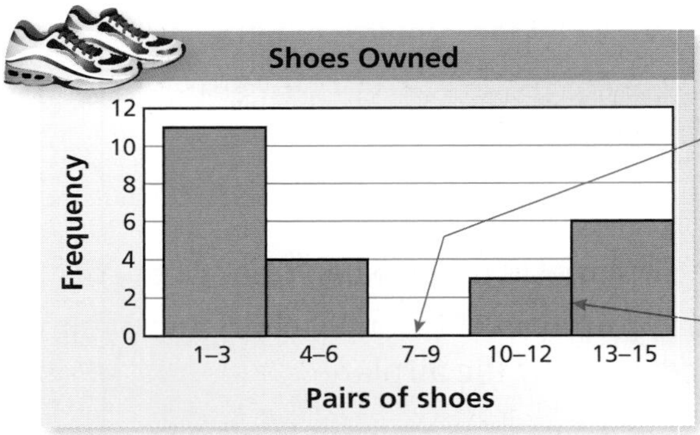

Include any interval with a frequency of 0. The bar height is 0.

There is no space between the bars of a histogram.

On Your Own

Now You're Ready
Exercises 6–8

1. The frequency table shows the ages of people riding a roller coaster. Display the data in a histogram.

Age	10–19	20–29	30–39	40–49	50–59
Frequency	16	11	5	2	4

Multi-Language Glossary at BigIdeasMath.com.

EXAMPLE 2 Using a Histogram

The histogram shows the winning speeds at the Daytona 500. (a) Which interval contains the most data values? (b) How many of the winning speeds are less than 140 miles per hour? (c) How many of the winning speeds are at least 160 miles per hour?

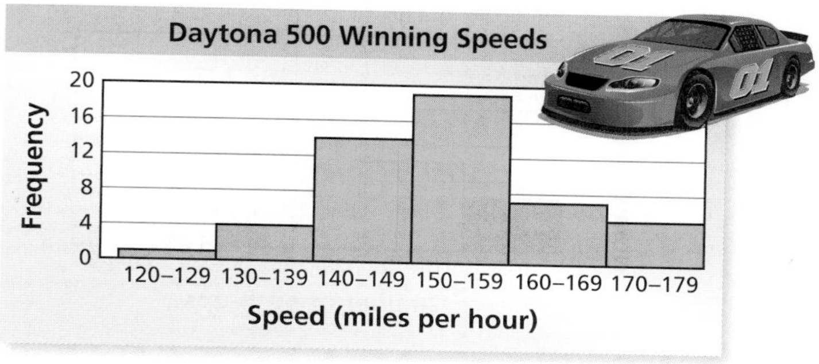

a. The interval with the tallest bar contains the most data values.

So, the 150–159 miles per hour interval contains the most data values.

b. One winning speed is in the 120–129 miles per hour interval and four winning speeds are in the 130–139 miles per hour interval.

So, $1 + 4 = 5$ winning speeds are less than 140 miles per hour.

c. Seven winning speeds are in the 160–169 miles per hour interval and five winning speeds are in the 170–179 miles per hour interval.

So, $7 + 5 = 12$ winning speeds are at least 160 miles per hour.

On Your Own

Now You're Ready
Exercises 10–13

2. The histogram shows the number of hours that students in a class slept last night.

   a. How many students slept at least 8 hours?

   b. How many students slept less than 12 hours?

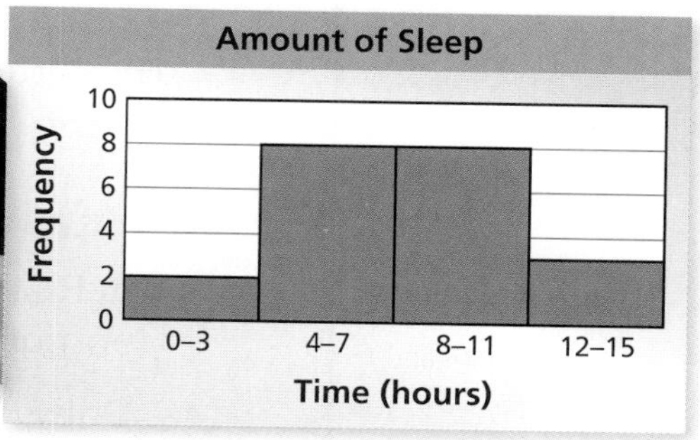

EXAMPLE 3 Comparing Data Displays

The data displays show how many push-ups students in a class completed for a physical fitness test. Which data display can you use to find how many students are in the class? Explain.

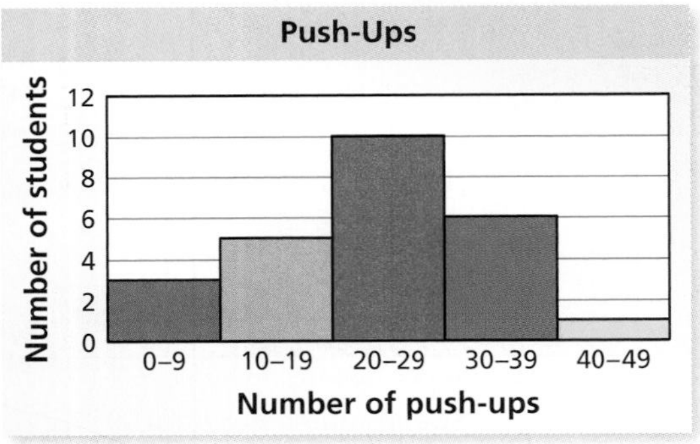

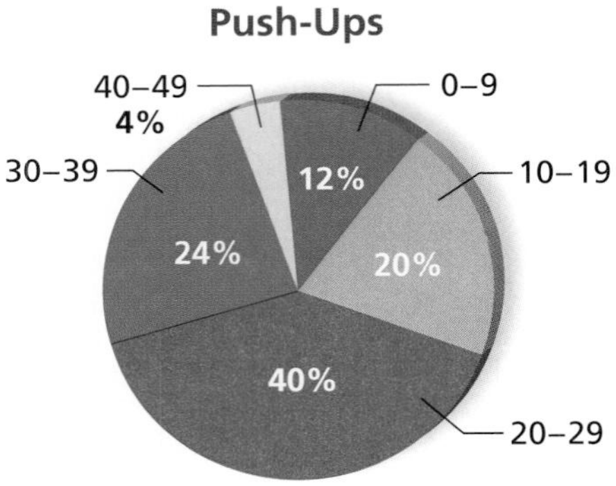

You can use the histogram because it shows the number of students in each interval. The sum of these values represents the number of students in the class. You cannot use the circle graph because it does not show the number of students in each interval.

EXAMPLE 4 Standardized Test Practice

Which statement *cannot* be made using the data displays in Example 3?

(A) Twelve percent of the class completed less than 10 push-ups.

(B) Five students completed at least 10 and at most 19 push-ups.

(C) At least one student completed more than 39 push-ups.

(D) Twenty-nine percent of the class completed 30 or more push-ups.

The circle graph shows that 12% completed 0–9 push-ups. So, Statement A can be made.

In the histogram, the bar height for the 10–19 interval is 5 and the bar height for the 40–49 interval is 1. So, Statements B and C can be made.

The circle graph shows that 24% completed 30–39 push-ups and 4% completed 40–49 push-ups. So, 24% + 4% = 28% completed 30 or more push-ups. Statement D cannot be made.

The correct answer is (D).

On Your Own

Now You're Ready
Exercises 14 and 15

3. In Example 3, which data display should you use to describe the portion of the entire class that completed 30–39 push-ups?

4. Draw two more conclusions from the data displays in Example 3.

10.1 Exercises

Check It Out
Help with Homework
BigIdeasMath.com

Vocabulary and Concept Check

1. **VOCABULARY** Which graph is a histogram? Explain your reasoning.

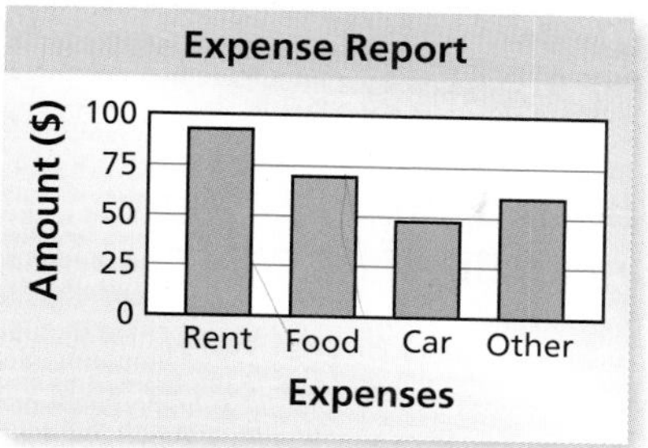

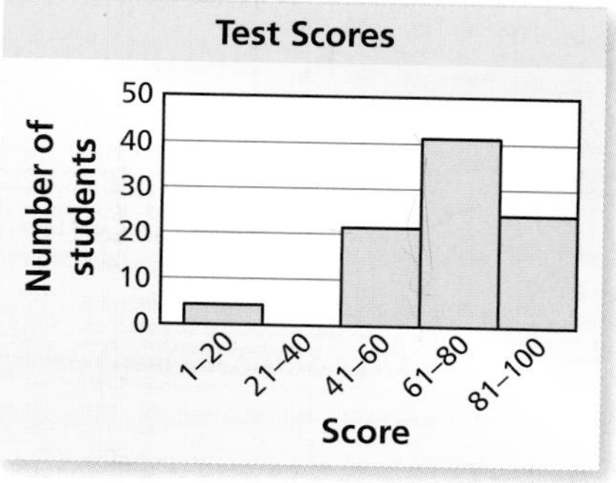

2. **REASONING** Describe the outliers in the histogram.

3. **CRITICAL THINKING** How can you tell when an interval of a histogram has a frequency of zero?

Practice and Problem Solving

Make a tally chart and a bar graph of the data.

4.

Number of Pets			
1	1	1	4
3	1	2	2
2	4	1	2

5.

Points Scored				
7	10	9	8	9
9	7	6	9	9
8	8	9	9	10

Display the data in a histogram.

1 6.

States Visited	
States	**Frequency**
1–5	12
6–10	14
11–15	6
16–20	3

7.

Chess Team	
Wins	**Frequency**
10–13	3
14–17	4
18–21	4
22–25	2

8.

Movies Watched	
Movies	**Frequency**
0–1	5
2–3	11
4–5	8
6–7	1

9. **ERROR ANALYSIS** Describe and correct the error made in displaying the data in a histogram.

Confirmed Flu Cases per School	
Cases	**Frequency**
0–2	3
3–5	7
6–8	9
9–11	12

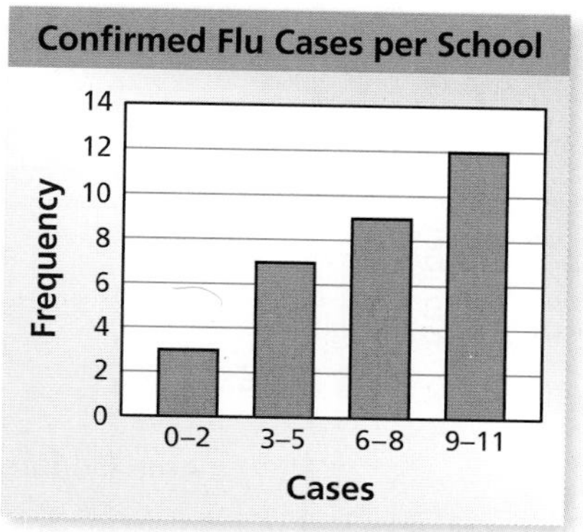

2 **10. MAGAZINES** The histogram shows the number of magazines read last month by students in a class.

a. Which interval contains the fewest data values?

b. How many students are in the class?

c. What percent of the students read less than six magazines?

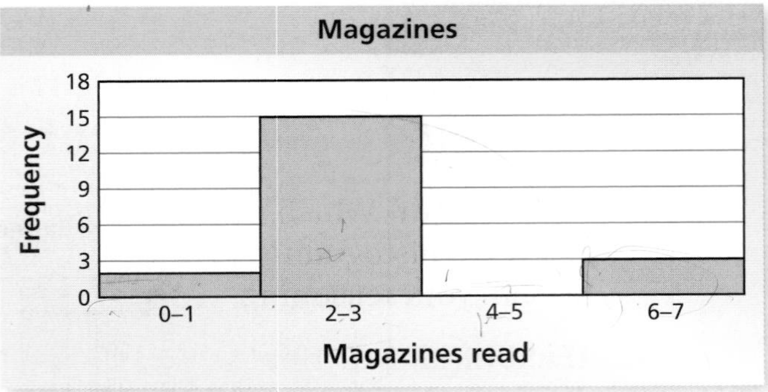

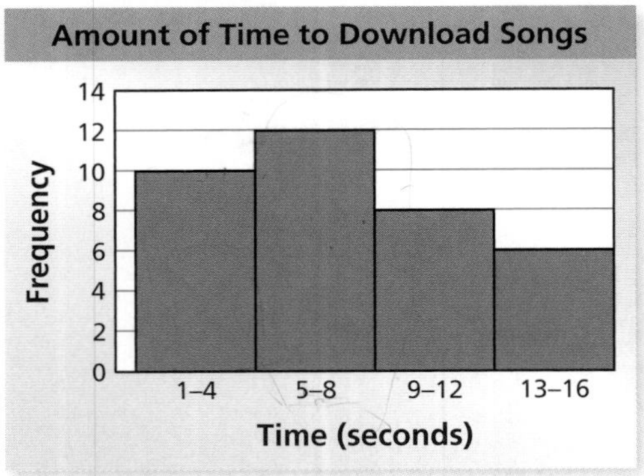

11. ERROR ANALYSIS Describe and correct the error made in reading the histogram.

"12% of the songs took 5–8 seconds to download."

12. VOTING The histogram shows the percent of the voting age population that voted in a recent presidential election. Explain whether each statement is supported by the graph.

a. Only 40% of one state voted.

b. Most states had between 50% and 64.9% that voted.

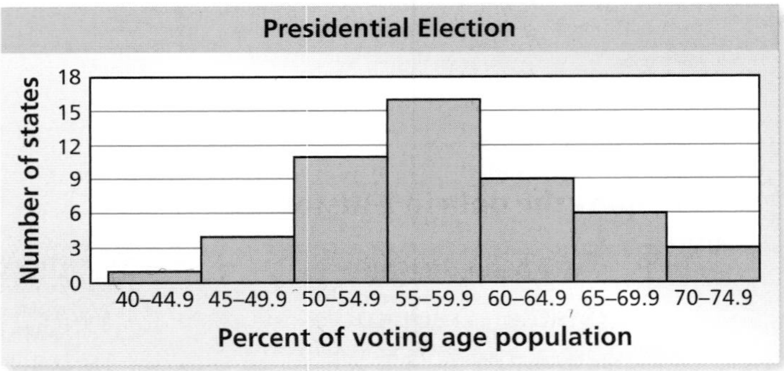

13. AREA The histograms show the areas of counties in Pennsylvania and Indiana. Which state do you think has the greater area? Explain.

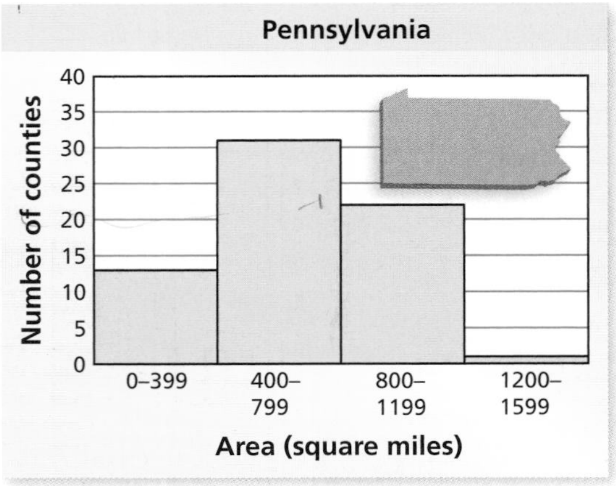

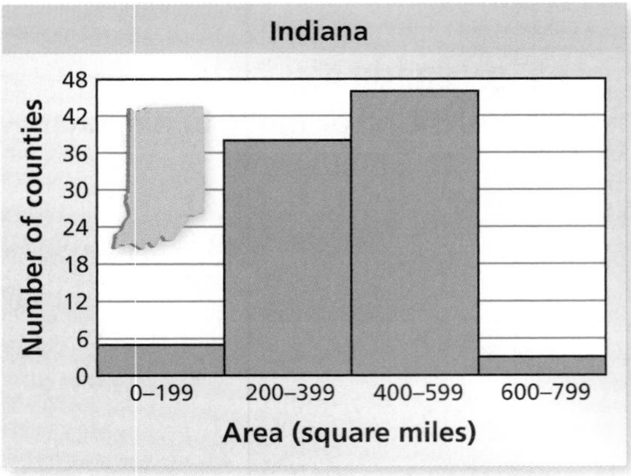

③ **14. GARBAGE** The data displays show how many pounds of garbage apartment residents produced in one week. Which data display can you use to find how many residents produced more than 25 pounds of garbage? Explain.

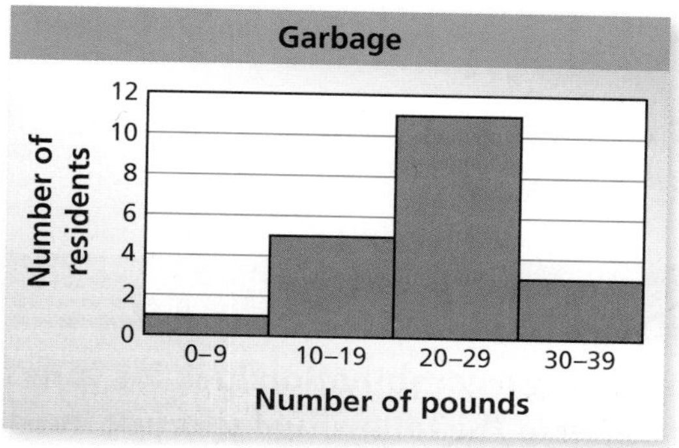

Garbage

Stem	Leaf
0	9
1	0 5 8 8 9
2	1 2 5 5 6 7 7 7 9 9 9
3	2 3 3

Key: 1|5 = 15 pounds

15. REASONING Determine whether each statement can be made using the data displays in Exercise 14. Explain your reasoning.

a. One resident produced 10 pounds of garbage.

b. Twelve residents produced between 20 and 29 pounds of garbage.

16. NUMBER SENSE Can you find the mean, median, mode, and range of the data in Exercise 7? If so, find them. If not, explain why.

17. Critical Thinking The table shows the weights of guide dogs enrolled in a training program.

a. Make a histogram of the data starting with the interval 51–55.

b. Make another histogram of the data using a different sized interval.

c. Compare and contrast the two histograms.

Weight (lb)					
81	88	57	82	70	85
71	51	82	77	79	77
83	80	54	80	81	73
59	84	75	76	68	78
83	78	55	67	85	79

Fair Game Review What you learned in previous grades & lessons

Find the percent of the number. *(Skills Review Handbook)*

18. 25% of 180 **19.** 30% of 90 **20.** 16% of 140 **21.** 64% of 80

22. MULTIPLE CHOICE Two rectangles are similar. The smaller rectangle has a length of 8 feet. The larger rectangle has a length of 14 feet. What is the ratio of the area of the smaller rectangle to the area of the larger rectangle? *(Section 6.4)*

Ⓐ 7 : 4 Ⓑ 4 : 7 Ⓒ 9 : 16 Ⓓ 16 : 49

10.2 Fundamental Counting Principle

STANDARDS OF LEARNING
7.10

Essential Question How can you count the number of different combinations that are possible on a lock?

1 ACTIVITY: Comparing Combination Locks

Work with a partner. You are buying a combination lock for your locker. You have three choices. For which of the following locks is it most difficult to guess the combination? Explain your reasoning.

a. This lock has 3 wheels, each numbered from 0 to 9.

b. This lock is numbered from 0 to 39. Each combination uses three numbers in a right, left, right pattern.

c. This lock has 4 wheels:
Wheel 1: 0–9
Wheel 2: A–J
Wheel 3: K–T
Wheel 4: 0–9

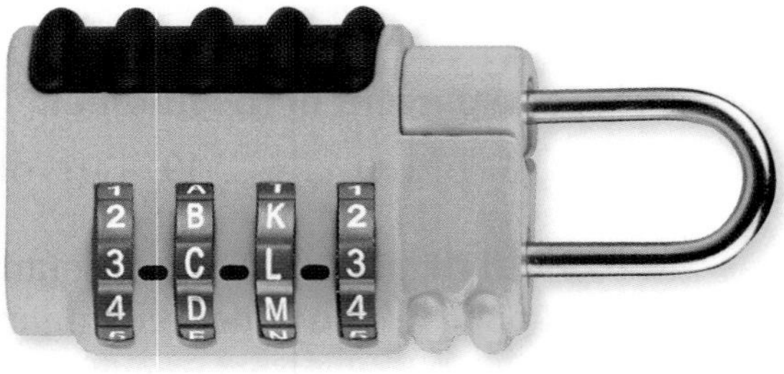

2 ACTIVITY: Comparing Password Security

Work with a partner. Which password requirement is most secure? Explain your reasoning. Include the number of different passwords that are possible for each requirement.

a. The password must have four digits.

b. The password must have five digits.

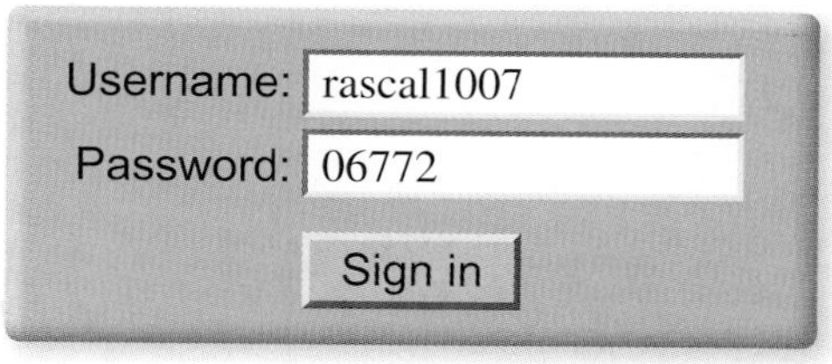

c. The password must have six letters.

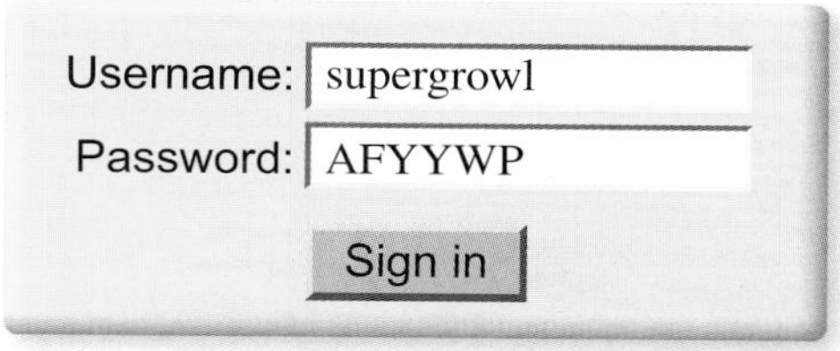

d. The password must have eight digits or letters.

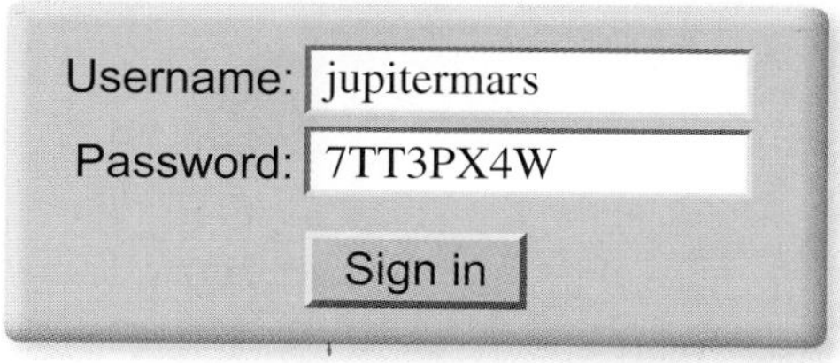

What Is Your Answer?

3. **IN YOUR OWN WORDS** How can you count the number of different combinations that are possible on a lock?

4. **SECURITY** A hacker uses a software program to guess the passwords in Activity 2. The program checks 600 passwords per minute. How long will it take the program to guess each of the four types of passwords?

Practice Use what you learned about lock combinations to complete Exercise 3 on page 408.

10.2 Lesson

Check It Out
Lesson Tutorials
BigIdeasMath.com

Key Vocabulary
outcomes, *p. 406*
event, *p. 406*
Fundamental Counting Principle, *p. 406*

The possible results of an action are called **outcomes**. A collection of one or more outcomes is an **event**.

One way to find the total number of possible outcomes is to use a tree diagram.

EXAMPLE 1 Using a Tree Diagram

Crust

- Thin Crust
- Stuffed Crust

Style

- Hawaiian
- Mexican
- Pepperoni
- Veggie

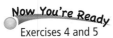

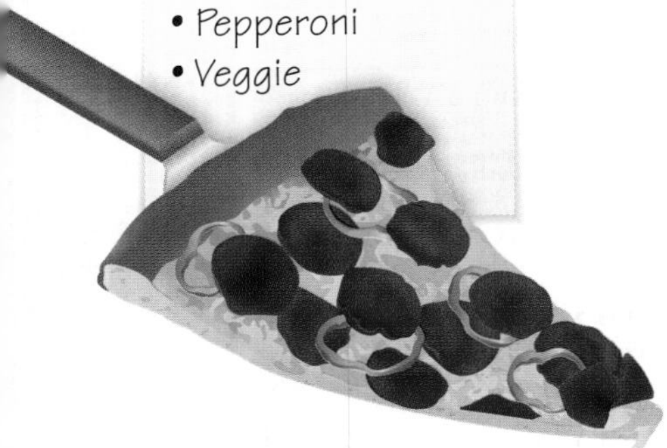

The choices at a pizza shop are shown. How many different pizzas are possible?

Use a tree diagram to find the total number of possible outcomes.

Crust	*Style*	***Outcome***
Thin	Hawaiian	Thin Crust Hawaiian
	Mexican	Thin Crust Mexican
	Pepperoni	Thin Crust Pepperoni
	Veggie	Thin Crust Veggie
Stuffed	Hawaiian	Stuffed Crust Hawaiian
	Mexican	Stuffed Crust Mexican
	Pepperoni	Stuffed Crust Pepperoni
	Veggie	Stuffed Crust Veggie

∴ There are 8 different outcomes. So, there are 8 different pizzas possible.

On Your Own

Now You're Ready
Exercises 4 and 5

1. **WHAT IF?** The pizza shop adds a “deep dish” crust. How many pizzas are possible?

Another way to find the total number of possible outcomes is to use the **Fundamental Counting Principle.**

Key Idea

Fundamental Counting Principle

An event M has m possible outcomes and event N has n possible outcomes. The total number of outcomes of event M followed by event N is $m \times n$.

Study Tip

The Fundamental Counting Principle can be extended to more than two events.

EXAMPLE 2 Finding the Total Number of Possible Outcomes

Find the total number of possible outcomes of rolling a number cube and flipping a coin.

Method 1: Use a tree diagram. Let H = Heads and T = Tails.

Number cube roll	1	2	3	4	5	6
Coin flip	H T	H T	H T	H T	H T	H T
Outcome	1H 1T	2H 2T	3H 3T	4H 4T	5H 5T	6H 6T

There are 12 possible outcomes.

Method 2: Use the Fundamental Counting Principle.
Identify the number of possible outcomes of each event.

Event 1: Rolling a number cube has 6 possible outcomes.

Event 2: Flipping a coin has 2 possible outcomes.

number cube, coin, total

$$6 \times 2 = 12$$

There are 12 possible outcomes.

EXAMPLE 3 Finding the Total Number of Possible Outcomes

How many different outfits can you make from the T-shirts, jeans, and shoes in the closet?

Use the Fundamental Counting Principle. Identify the number of possible outcomes for each event.

Event 1: Choosing a T-shirt has 7 possible outcomes.

Event 2: Choosing jeans has 4 possible outcomes.

Event 3: Choosing shoes has 3 possible outcomes.

$7 \times 4 \times 3 = 84$ Fundamental Counting Principle

You can make 84 different outfits.

On Your Own

Now You're Ready
Exercises 6–9

2. Find the total number of possible outcomes of spinning the spinner and choosing a number from 1 to 5.

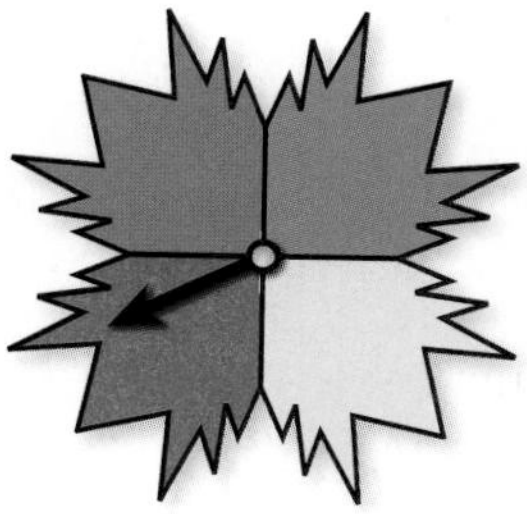

3. How many different outfits can you make from 4 T-shirts, 5 pairs of jeans, and 5 pairs of shoes?

10.2 Exercises

Check It Out
Help with Homework
BigIdeasMath.com

Vocabulary and Concept Check

1. **VOCABULARY** Explain how to use the Fundamental Counting Principle.

2. **WRITING** Describe two ways to find the total number of possible outcomes of spinning the spinner and rolling the number cube.

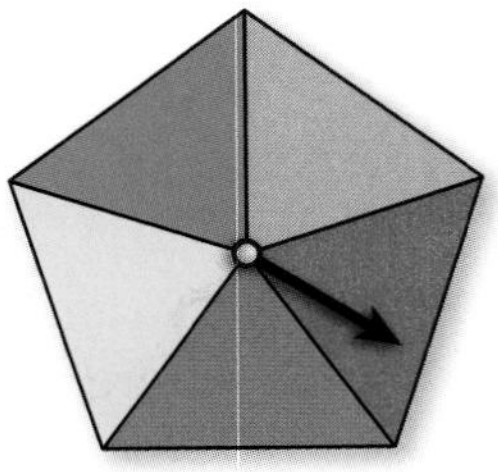

Practice and Problem Solving

3. The lock is numbered from 0 to 49. Each combination uses three numbers in a right, left, right pattern. Find the total number of possible combinations for the lock.

Use a tree diagram to find the total number of possible outcomes.

1 4.

Birthday Party	
Event	Miniature golf, Laser tag, Roller skating
Time	1 P.M.–3 P.M., 6 P.M.–8 P.M.

5.

New School Mascot	
Type	Lion, Bear, Hawk, Dragon
Style	Realistic, Cartoon

Use the Fundamental Counting Principle to find the total number of possible outcomes.

2 3 6.

Beverage	
Size	Small, Medium, Large
Flavor	Root beer, Cola, Diet cola, Iced tea, Lemonade, Water, Coffee

7.

MP3 Player	
Memory	2 GB, 4 GB, 8 GB, 16 GB
Color	Silver, Green, Blue, Pink, Black

8.

Clown	
Suit	Dots, Stripes, Checkerboard
Wig	One color, Multicolor
Talent	Balloon animals, Juggling, Unicycle, Magic

9.

Meal	
Appetizer	Nachos, Soup, Spinach dip, Salad, Fruit
Entrée	Chicken, Beef, Spaghetti, Fish
Dessert	Cake, Cookies, Ice cream

10. **NOTE CARDS** A store sells three types of note cards. There are three sizes of each type. Show two ways to find the total number of note cards the store sells.

11. **ERROR ANALYSIS** Describe and correct the error in using the Fundamental Counting Principle to find the total number of ways a true-false quiz with five questions can be answered.

> ✗ $2 + 2 + 2 + 2 + 2 = 10$
>
> The quiz can be answered 10 different ways.

12. **MARBLES** You randomly choose one of the marbles. Without replacing the first marble, you choose a second marble.
    - **a.** Use a tree diagram to find the total number of possible outcomes.
    - **b.** Can you use the Fundamental Counting Principle to find the total number of possible outcomes? Explain.

13. **TRAINS** Your model train has one engine and eight train cars. Use the Fundamental Counting Principle to find the total number of ways you can arrange the train. (The engine must be first.)

14. Critical Thinking You have been assigned a 9-digit identification number.
    - **a.** Why should you use the Fundamental Counting Principle instead of a tree diagram to find the total number of possible identification numbers?
    - **b.** How many identification numbers are possible?
    - **c.** **RESEARCH** Use the Internet to find out why the possible number of Social Security numbers is not the same as your answer to part (b).

Fair Game Review What you learned in previous grades & lessons

Write the number in scientific notation. *(Section 9.4)*

15. 2,560,000,000 **16.** 0.00000976 **17.** 320,400,000,000

18. **MULTIPLE CHOICE** The two triangles are similar. What is the value of x? *(Section 6.4)*

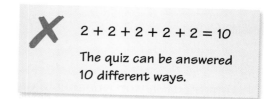

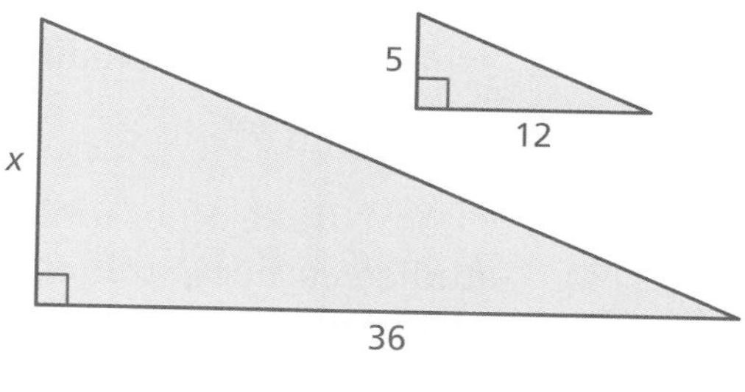

Ⓐ 1.7 Ⓑ 15

Ⓒ 20 Ⓓ 86.4

10 Study Help

Check It Out
Graphic Organizer
BigIdeasMath.com

You can use a **word magnet** to organize information associated with a vocabulary word. Here is an example of a word magnet for histogram.

Histogram

A histogram is a bar graph that shows the frequency of data values in intervals of the same size.

The height of a bar represents the frequency of the values in the interval.

You can make a histogram from a frequency table. A frequency table groups data values into intervals. The frequency is the number of data values in an interval.

The histogram shows the number of times students were absent from homeroom this year.

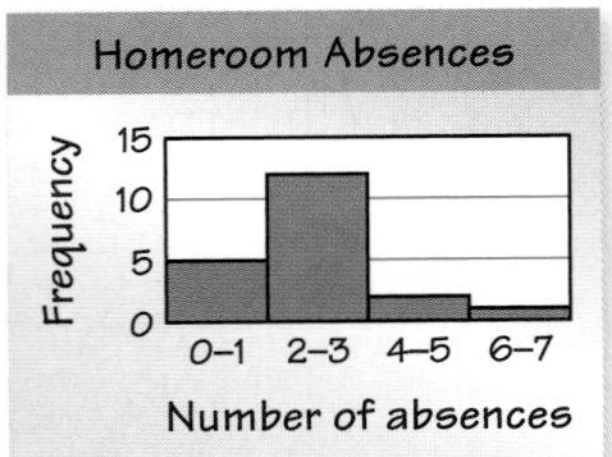

On Your Own

Make a word magnet to help you study this topic.

1. Fundamental Counting Principle

After you complete this chapter, make word magnets for the following topics.

2. theoretical probability
3. experimental probability
4. compound event
5. Choose three other topics that you studied earlier in this course. Make a word magnet for each topic.

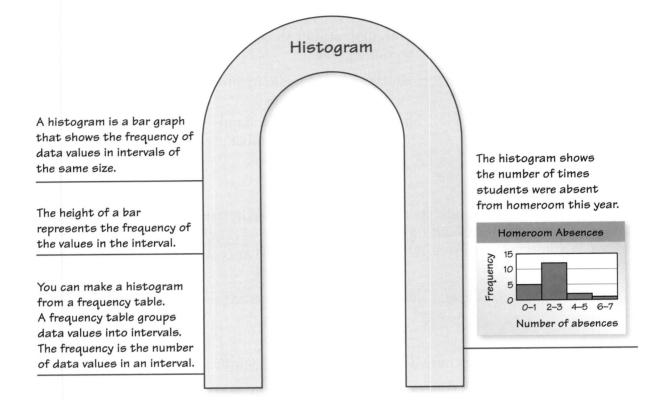

"How do you like the word magnet I made for 'Beagle'?"

10.1–10.2 Quiz

Check It Out
Progress Check
BigIdeasMath.com

Display the data in a histogram. *(Section 10.1)*

1.

Soccer Team Goals	
Goals per Game	Frequency
0–1	5
2–3	4
4–5	0
6–7	1

2.

Minutes Practiced	
Minutes	Frequency
0–19	8
20–39	10
40–59	11
60–79	2

3.

Poems Written for Class	
Poems	Frequency
0–4	6
5–9	16
10–14	4
15–19	2
20–24	2

Use the Fundamental Counting Principle to find the total number of possible outcomes. *(Section 10.2)*

4.

Calculator	
Type	Basic display, Scientific, Graphing, Financial
Color	Black, White, Silver

5.

Vacation	
Destination	Florida, Italy, Mexico, England
Length	One week, Two weeks

6. REBOUNDS The histogram shows the number of rebounds per game for a middle school basketball player this season. *(Section 10.1)*

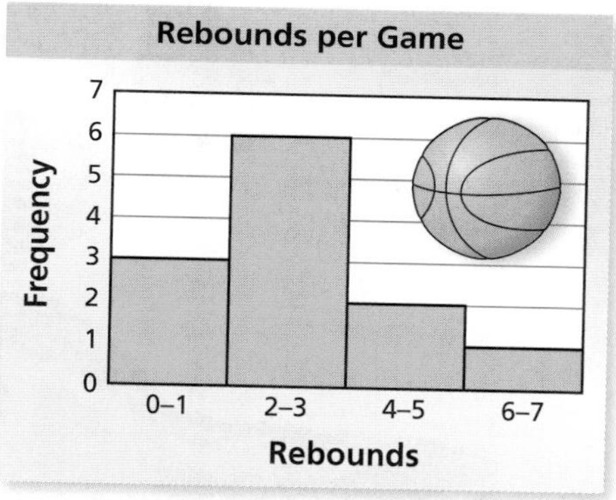

a. Which interval contains the most data values?

b. How many games were played by the player this season?

c. In what percent of the games did the player have 4 or more rebounds?

7. REASONING Determine whether each statement can be made using the histogram in Exercise 6. Explain your reasoning. *(Section 10.1)*

a. The player had 0 rebounds in at least 1 game.

b. The player had at most 3 rebounds in 9 games.

8. OUTCOMES Use the Fundamental Counting Principle to find the total number of possible outcomes when flipping the coin, rolling the number cube, and spinning the spinner. *(Section 10.2)*

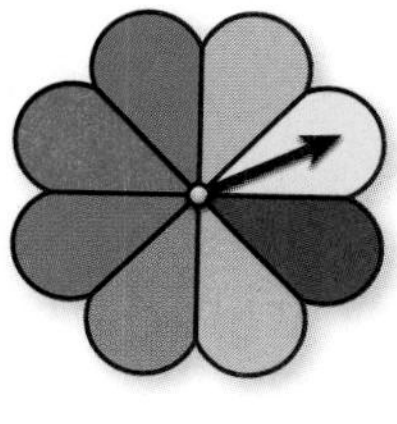

10.3 Theoretical Probability

STANDARDS OF LEARNING
7.9

Share Your Work at... My.BigIdeasMath.com

Essential Question How can you find a theoretical probability?

1 ACTIVITY: Black and White Spinner Game

Work with a partner. You work for a game company. You need to create a game that uses the spinner below.

a. Write rules for a game that uses the spinner. Then play it.

b. After playing the game, do you want to revise the rules? Explain.

c. Each pie-shaped section of the spinner is the same size. What is the measure of the central angle of each section?

d. What is the probability that the spinner will land on 1? Explain.

2 ACTIVITY: Changing the Spinner

Work with a partner. For each spinner, find the probability of landing on each number. Do your rules from Activity 1 make sense for these spinners? Explain.

a.

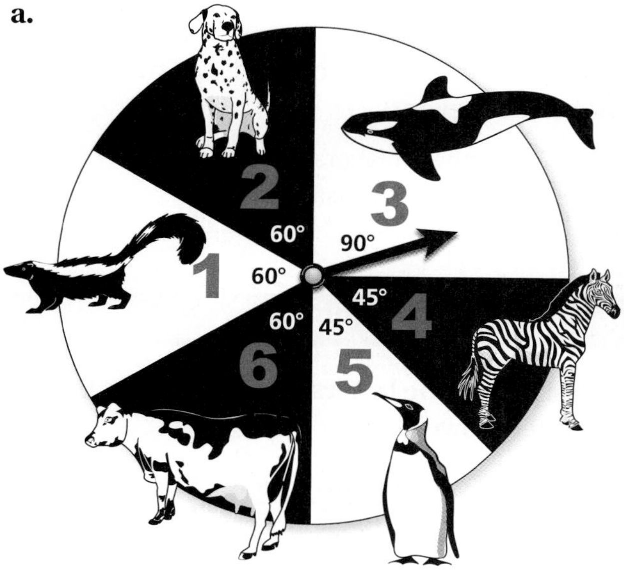

b.

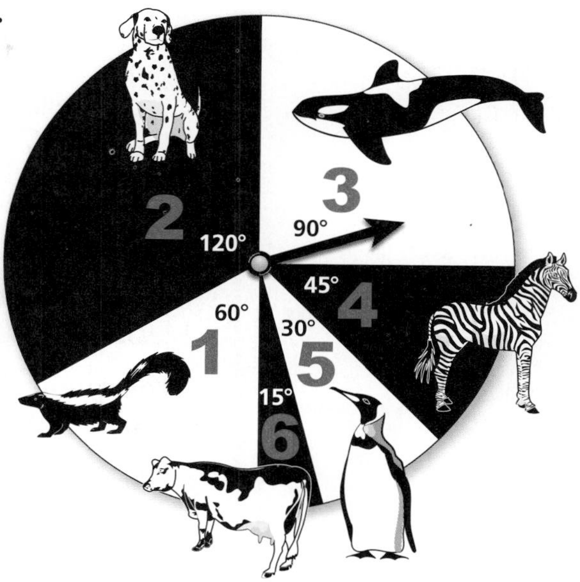

3 ACTIVITY: Is This Game Fair?

Work with a partner. Apply the following rules to each spinner in Activities 1 and 2. Is the game fair? If not, who has the better chance of winning?

- **Take turns spinning the spinner.**
- **If the spinner lands on an odd number, Player 1 wins.**
- **If the spinner lands on an even number, Player 2 wins.**

What Is Your Answer?

4. **IN YOUR OWN WORDS** How can you find a theoretical probability?
5. Find and describe a career in which probability is used. Explain why probability is used in that career.
6. Two people play the following game.

   Each player has 6 cards numbered 1, 2, 3, 4, 5, and 6. At the same time, each player holds up one card. If the product of the two numbers is odd, Player 1 wins. If the product is even, Player 2 wins. Continue until both players are out of cards. Which player is more likely to win? Why?

Practice Use what you learned about theoretical probability to complete Exercises 4–7 on page 416.

10.3 Lesson

Check It Out
Lesson Tutorials
BigIdeasMath.com

Key Vocabulary

theoretical probability, *p. 414*

fair experiment, *p. 415*

Key Idea

Theoretical Probability

When all possible outcomes are equally likely, the **theoretical probability** of an event is the ratio of the number of favorable outcomes to the number of possible outcomes. The probability of an event is written as $P(\text{event})$.

$$P(\text{event}) = \frac{\text{number of favorable outcomes}}{\text{number of possible outcomes}}$$

EXAMPLE 1 Finding a Theoretical Probability

You randomly choose one of the letters shown. What is the theoretical probability of choosing a vowel?

$$P(\text{event}) = \frac{\text{number of favorable outcomes}}{\text{number of possible outcomes}}$$

$$P(\text{vowel}) = \frac{3}{7}$$

There are 3 vowels.

There is a total of 7 letters.

∴ The probability of choosing a vowel is $\frac{3}{7}$ or about 43%.

EXAMPLE 2 Using a Theoretical Probability

The theoretical probability that you randomly choose a green marble from a bag is $\frac{3}{8}$. There are 40 marbles in the bag. How many are green?

$$P(\text{green}) = \frac{\text{number of green marbles}}{\text{total number of marbles}}$$

$\frac{3}{8} = \frac{n}{40}$ Substitute. Let n be the number of green marbles.

$15 = n$ Multiply each side by 40.

∴ There are 15 green marbles in the bag.

On Your Own

Now You're Ready
Exercises 4–11

1. In Example 1, what is the theoretical probability of choosing an X?
2. The theoretical probability that you spin an odd number on a spinner is 0.6. The spinner has 10 sections. How many sections have odd numbers?

An experiment is **fair** if all of its possible outcomes are equally likely.

Study Tip

A game is fair if every player has the same probability of winning.

The spinner is equally likely to land on 1 or 2. The spinner is fair.

The spinner is more likely to land on 1 than on either 2 or 3. The spinner is *not* fair.

EXAMPLE 3 Making a Prediction

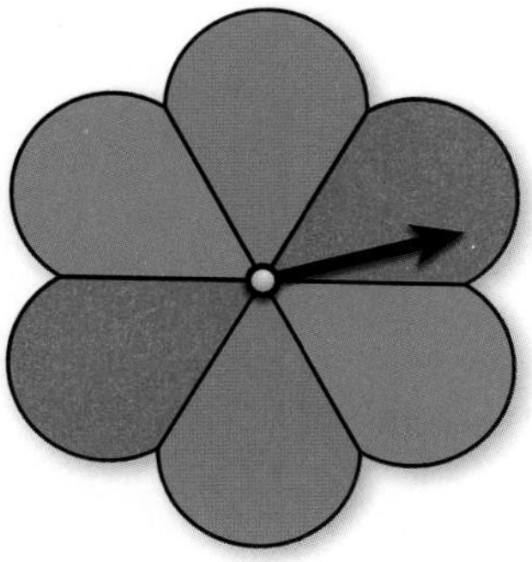

Scoring Rules:

- You get one point when the spinner lands on blue or green.
- Your friend gets one point when the spinner lands on red.
- The first person to get 10 points wins.

You and your friend play the game. (a) Is the spinner fair? (b) Is the game fair? (c) Predict the number of turns it will take you to win.

a. Yes, the spinner is fair because it is equally likely to land on red, blue, or green.

b. Find and compare the theoretical probabilities of the events.

You: $P(\text{blue or green}) = \dfrac{\text{number of blue or green sections}}{\text{total number of sections}}$

$$= \frac{4}{6} = \frac{2}{3}$$

Your friend: $P(\text{red}) = \dfrac{\text{number of red sections}}{\text{total number of sections}}$

$$= \frac{2}{6} = \frac{1}{3}$$

It is more likely that the spinner will land on blue or green than on red. Because your probability is greater, the game is *not* fair.

c. Write and solve an equation using P(blue or green) found in part (b). Let x be the number of turns it will take you to win.

$\frac{2}{3}x = 10$ Write equation.

$x = 15$ Multiply each side by $\frac{3}{2}$.

So, you can predict that it will take you 15 turns to win.

On Your Own

Now You're Ready
Exercises 12–14

3. WHAT IF? In Example 3, you get one point when the spinner lands on blue or green. Your friend gets one point when the spinner lands on red or blue. The first person to get 5 points wins. Is the game fair? Explain.

10.3 Exercises

Check It Out
Help with Homework
BigIdeasMath.com

Vocabulary and Concept Check

1. **VOCABULARY** An event has a theoretical probability of 0.5. What does this mean?

2. **OPEN-ENDED** Describe an event that has a theoretical probability of $\frac{1}{4}$.

3. **WHICH ONE DOESN'T BELONG?** Which spinner does *not* belong with the other three? Explain your reasoning.

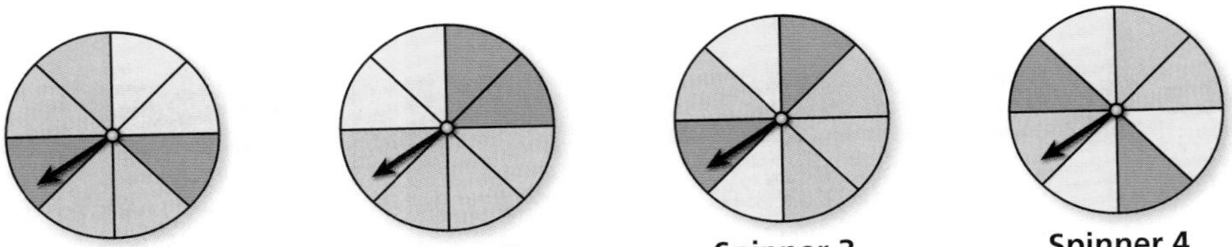

Practice and Problem Solving

Use the spinner to determine the theoretical probability of the event.

(1) 4. Spinning red

5. Spinning a 1

6. Spinning an odd number

7. Spinning a multiple of 2

8. Spinning a number less than 7

9. Spinning a 7

10. **LETTERS** Each letter of the alphabet is printed on an index card. What is the theoretical probability of randomly choosing any letter except Z?

(2) 11. **GAME SHOW** On a game show, a contestant randomly chooses a chip from a bag that contains numbers and strikes. The theoretical probability of choosing a strike is $\frac{3}{10}$. There are 30 chips in the bag. How many are strikes?

A number cube is rolled. Determine if the game is fair. If it is *not* fair, who has the greater probability of winning?

(3) 12. You win if the number is odd. Your friend wins if the number is even.

13. You win if the number is less than 3. If it is not less than 3, your friend wins.

14. **SCORING POINTS** You get one point if a 1 or a 2 is rolled on the number cube. Your friend gets one point if a 5 or a 6 is rolled. The first person to 5 points wins.

   a. Is the number cube fair? Is the game fair? Explain.

   b. Predict the number of turns it will take you to win.

15. **HISTORY** You write a report about your favorite president. Your friend writes a report on a randomly chosen president. What is the theoretical probability that you write reports on the same president?

16. **BIRTHDAYS** The bar graph shows the birthday months of all 200 employees at a local business.

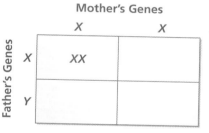

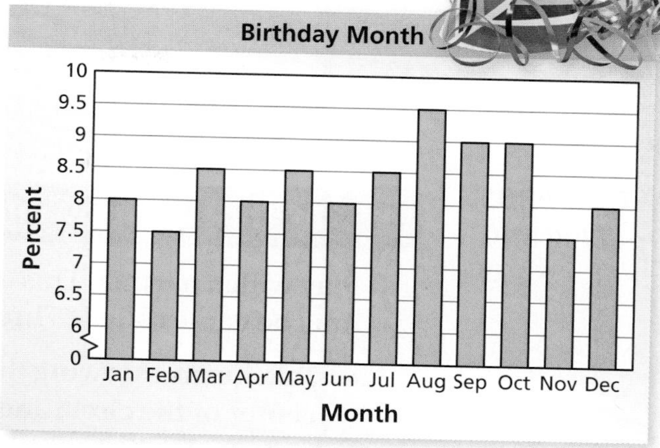

a. What is the theoretical probability of randomly choosing a person at the business who was born in a month with an R in its name?

b. What is the theoretical probability of randomly choosing a person at the business who has a birthday in the first half of the year?

17. **SCHEDULING** There are 16 females and 20 males in a class.

a. What is the theoretical probability of randomly choosing a female from the class?

b. One week later, there are 45 students in the class. The theoretical probability of randomly selecting a female is the same as last week. How many males joined the class?

Mother's Genes

Father's Genes	*X*	*X*
X	*XX*	
Y		

A Punnett square is a grid used to show possible gene combinations for the offspring of two parents. In the Punnett square shown, a boy is represented by *XY*. A girl is represented by *XX*.

18. Complete the Punnett square.

19. Explain why the probability of two parents having a boy or having a girl is equally likely.

20. Critical Thinking Two parents each have the gene combination *Cs*. The gene *C* is for curly hair. The gene *s* is for straight hair.

a. Make a Punnett square for the two parents. If all outcomes are equally likely, what is the probability of a child having the gene combination *CC*?

b. Any gene combination that includes a *C* results in curly hair. If all outcomes are equally likely, what is the probability of a child having curly hair?

Fair Game Review What you learned in previous grades & lessons

Multiply. *(Section 3.3)*

21. $\frac{1}{2} \times \frac{1}{2}$

22. $-\frac{1}{6} \times \frac{2}{3}$

23. $-\frac{3}{5} \times \frac{7}{8}$

24. $\frac{4}{5} \times \frac{1}{36}$

25. **MULTIPLE CHOICE** What is the mean of the numbers 11, 6, 12, 22, 7, 8, and 4? *(Skills Review Handbook)*

Ⓐ 4 Ⓑ 8 Ⓒ 10 Ⓓ 70

10.4 Experimental Probability

STANDARDS OF LEARNING
7.9

Essential Question What is meant by experimental probability?

1 ACTIVITY: Throwing Sticks

Play with a partner. This game is based on an Apache game called "Throw Sticks."

- **Take turns throwing three sticks into the center of the circle and moving around the circle according to the chart.**
- **If your opponent lands on or passes your playing piece, you must start over.**
- **The first player to pass his or her starting point wins.**

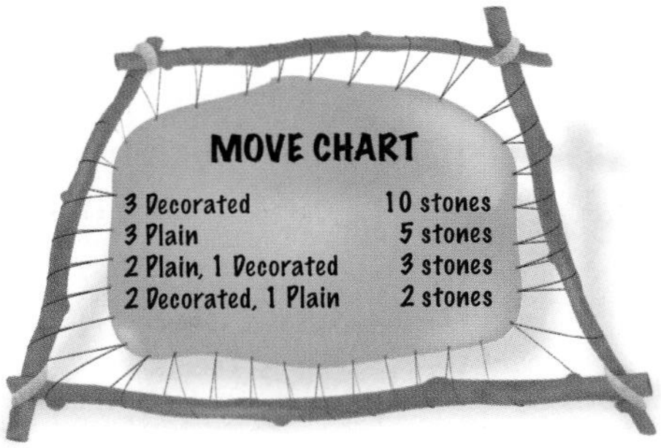

MOVE CHART
3 Decorated 10 stones
3 Plain 5 stones
2 Plain, 1 Decorated 3 stones
2 Decorated, 1 Plain 2 stones

Each stick has one plain side and one decorated side.

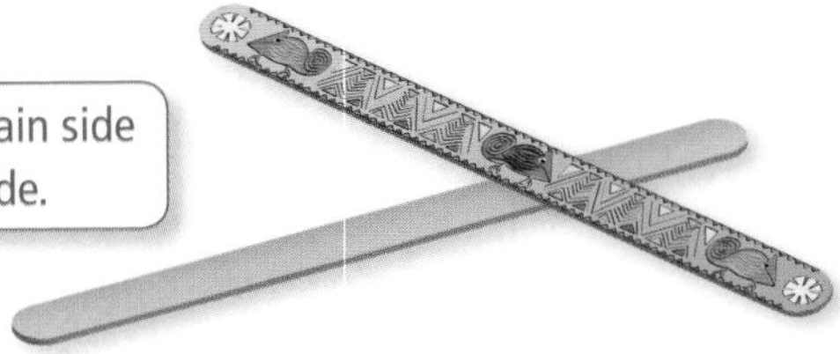

The game board has 40 stones arranged in a circle. The stones are placed in groups of 10.

Players start on opposite sides of the circle.

Player 1 Starting Point

Player 2 Starting Point

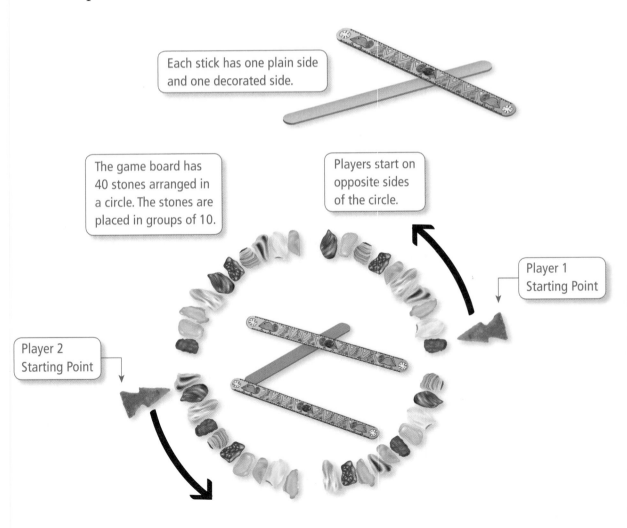

2 ACTIVITY: Conducting an Experiment

Work with a partner. Throw the 3 sticks 32 times. Tally the results using the outcomes listed below. Organize the results in a bar graph. Use the bar graph to estimate the probability of each outcome. These are called experimental probabilities.

a. PPP

b. DPP

c. DDP

d. DDD

3 ACTIVITY: Analyzing the Possibilities

Work with a partner. A tree diagram helps you see different ways that the same outcome can occur.

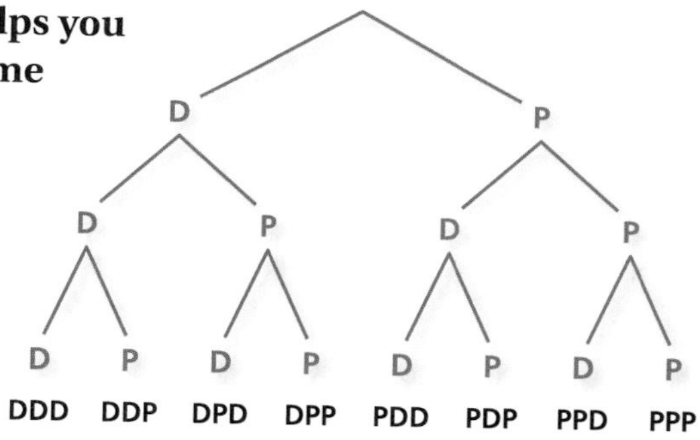

a. Find the number of ways that each outcome can occur.

- Three Ps
- One D and two Ps
- Two Ds and one P
- Three Ds

b. Find the theoretical probability of each outcome.

c. Compare and contrast your experimental and theoretical probabilities.

What Is Your Answer?

4. **IN YOUR OWN WORDS** What is meant by experimental probability?

5. Give a real-life example of experimental probability.

Practice Use what you learned about experimental probability to complete Exercises 3–6 on page 422.

10.4 Lesson

Check It Out
Lesson Tutorials
BigIdeasMath.com

Key Vocabulary
experimental probability, *p. 420*

Key Idea

Experimental Probability

Probability that is based on repeated trials of an experiment is called **experimental probability**.

$$P(\text{event}) = \frac{\text{number of times the event occurs}}{\text{total number of trials}}$$

EXAMPLE 1 Standardized Test Practice

Thirteen out of 20 emails in your inbox are junk emails. What is the experimental probability that your next email is junk?

Ⓐ 35% Ⓑ 45% Ⓒ 55% Ⓓ 65%

$$P(\text{event}) = \frac{\text{number of times the event occurs}}{\text{total number of trials}}$$

$$P(\text{junk}) = \frac{13}{20}$$

You have 13 emails that are junk.

You have a total of 20 emails.

∴ The probability is $\frac{13}{20}$, 0.65, or 65%. The correct answer is Ⓓ.

EXAMPLE 2 Making a Prediction

"April showers bring May flowers." Old Proverb, 1557

It rains 2 out of the last 12 days in March. If this trend continues, how many rainy days would you expect in April?

Find the experimental probability of a rainy day.

$$P(\text{event}) = \frac{\text{number of times the event occurs}}{\text{total number of trials}}$$

$$P(\text{rain}) = \frac{2}{12} = \frac{1}{6}$$

It rains 2 days.

There is a total of 12 days.

To make a prediction, multiply the probability of a rainy day by the number of days in April.

$$\frac{1}{6} \cdot 30 = 5$$

∴ You can predict that there will be 5 rainy days in April.

On Your Own

Now You're Ready
Exercises 7–16

1. In Example 1, what is the experimental probability that your next email is *not* junk?

2. At a clothing company, an inspector finds 5 defective pairs in a shipment of 200 jeans.

   a. What is the experimental probability of a pair of jeans being defective?

   b. About how many would you expect to be defective in a shipment of 5000 pairs of jeans?

EXAMPLE 3 Comparing Experimental and Theoretical Probabilities

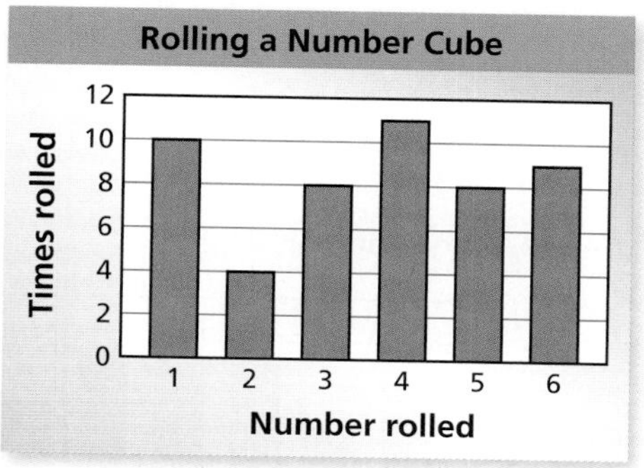

The bar graph shows the results of rolling a number cube 50 times. What is the experimental probability of rolling an odd number? How does this compare with the theoretical probability of rolling an odd number?

Find the experimental probability of rolling a 1, 3, or 5.

The bar graph shows 10 ones, 8 threes, and 8 fives. So, an odd number was rolled $10 + 8 + 8 = 26$ times in a total of 50 rolls.

Experimental Probability

$$P(\text{event}) = \frac{\text{number of times the event occurs}}{\text{total number of trials}}$$

$$P(\text{odd}) = \frac{26}{50}$$

An odd number was rolled 26 times.

There was a total of 50 rolls.

$$= \frac{13}{25}$$

Theoretical Probability

$$P(\text{event}) = \frac{\text{number of favorable outcomes}}{\text{number of possible outcomes}}$$

$$P(\text{odd}) = \frac{3}{6}$$

There are 3 odd numbers.

There is a total of 6 numbers.

$$= \frac{1}{2}$$

∴ The experimental probability is $\frac{13}{25} = 0.52 = 52\%$. The theoretical probability is $\frac{1}{2} = 0.5 = 50\%$. The experimental and theoretical probabilities are similar.

On Your Own

Now You're Ready
Exercise 18

3. In Example 3, what is the experimental probability of rolling a number greater than 1? How does this compare with the theoretical probability of rolling a number greater than 1?

10.4 Exercises

Check It Out
Help with Homework
BigIdeasMath.com

Vocabulary and Concept Check

1. **VOCABULARY** Describe how to find the experimental probability of an event.
2. **REASONING** You flip a coin 10 times and find the experimental probability of flipping tails to be 0.7. Does this seem reasonable? Explain.

Practice and Problem Solving

You have three sticks. Each stick has one red side and one blue side. You throw the sticks 10 times and record the results. Use the table to find the experimental probability of the event.

Outcome	Frequency
3 red	4
3 blue	0
2 blue, 1 red	2
2 red, 1 blue	4

3. Tossing 3 red
4. Tossing 2 blue, 1 red
5. Tossing 2 red, 1 blue
6. *Not* tossing all red

Use the bar graph to find the experimental probability of the event.

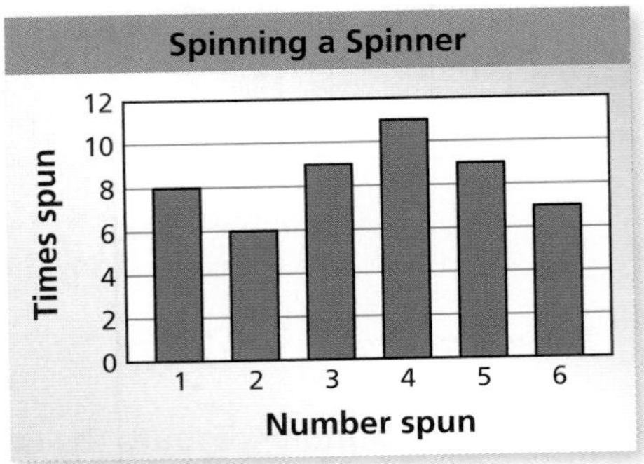

1 7. Spinning a 6
8. Spinning an even number
9. *Not* spinning a 1
10. Spinning a number less than 3
11. Spinning a 1 or a 3
12. Spinning a 7

13. **ERROR ANALYSIS** Describe and correct the error in finding $P(4)$ using the bar graph.

✗ $P(4) = \dfrac{\text{number of favorable outcomes}}{\text{number of possible outcomes}} = \dfrac{1}{6}$

14. **EGGS** You check 20 cartons of eggs. Three of the cartons have at least one cracked egg. What is the experimental probability that a carton of eggs has at least one cracked egg?

2 15. **BOARD GAME** There are 105 lettered tiles in a board game. You choose the tiles shown. How many of the 105 tiles would you expect to be vowels?

16. **CARDS** You have a package of 20 assorted thank-you cards. You pick the four cards shown. How many of the 20 cards would you expect to have flowers on them?

17. **MUSIC** During a 24-hour period, the ratio of pop songs played to rap songs played on a radio station is 60 : 75.

   a. What is the experimental probability that the next song played is rap?

   b. Out of the next 90 songs, how many would you expect to be pop?

3 18. **FLIPPING A COIN** You flip a coin 20 times. You flip heads 12 times. Compare your experimental probability of flipping heads with the theoretical probability of flipping heads.

You roll a pair of number cubes 60 times. You record your results in the bar graph shown.

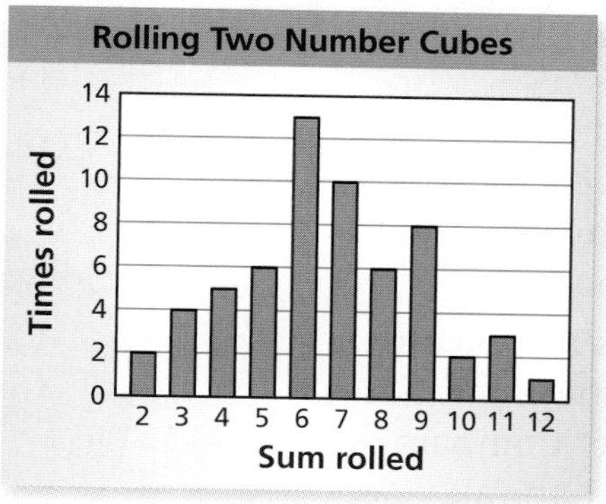

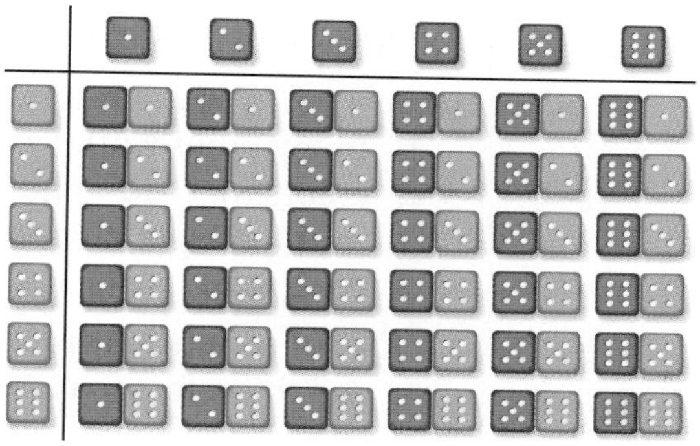

19. Use the bar graph to find the experimental probability of rolling each sum. Which sum is most likely?

20. Use the table to find the theoretical probability of rolling each sum. Which sum is most likely?

21. Compare the probabilities you found in Exercises 19 and 20.

22. **REASONING** Consider the results of Exercises 19 and 20.

   a. Which sum would you expect to be most likely after 500 trials? 1000 trials? 10,000 trials?

   b. Explain how experimental probability is related to theoretical probability as the number of trials increases.

23. Critical Thinking You roll two number cubes. Describe and perform an experiment to find the probability that the product of the two numbers rolled is at least 12. How many times did you roll the number cubes?

Fair Game Review What you learned in previous grades & lessons

Solve the equation. *(Section 3.5)*

24. $5x = 100$

25. $75 = 15x$

26. $2x = -26$

27. $-4x = -96$

28. **MULTIPLE CHOICE** What is the least common multiple of 4, 5, and 25? *(Skills Review Handbook)*

Ⓐ 20 Ⓑ 25 Ⓒ 60 Ⓓ 100

10.5 Compound Events

STANDARDS OF LEARNING
7.10

Essential Question How can you find the probability that a compound event will occur?

1 ACTIVITY: Performing an Experiment

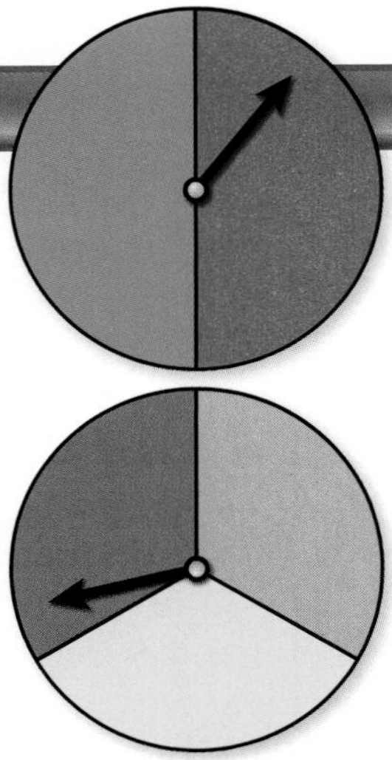

Work with a partner.

a. If you spin each spinner 48 times, how many times would you expect the two colors that the spinners land on to make purple? Explain your reasoning.

b. Spin each spinner 48 times and record your results in the table. How many times did the colors make purple? Compare the result with your answer to part (a).

Trial	Purple	Not Purple
1		
2		
3		
4		
5		
6		
7		
8		
9		
10		
11		
12		
13		
14		
15		
16		

Trial	Purple	Not Purple
17		
18		
19		
20		
21		
22		
23		
24		
25		
26		
27		
28		
29		
30		
31		
32		

Trial	Purple	Not Purple
33		
34		
35		
36		
37		
38		
39		
40		
41		
42		
43		
44		
45		
46		
47		
48		

2 ACTIVITY: Performing an Experiment

Work with a partner.

a. You randomly draw one piece of paper from the bag, keep it, and then draw a second piece of paper. If you do this 48 times, how many times would you expect the two colors chosen to make purple? Explain.

b. Perform the experiment 48 times. How many times did the colors make purple? Compare the result with your answer to part (a).

Trial	Purple	Not Purple
1		
2		
3		
4		
5		
6		
7		
8		
9		
10		
11		
12		
13		
14		
15		
16		

Trial	Purple	Not Purple
17		
18		
19		
20		
21		
22		
23		
24		
25		
26		
27		
28		
29		
30		
31		
32		

Trial	Purple	Not Purple
33		
34		
35		
36		
37		
38		
39		
40		
41		
42		
43		
44		
45		
46		
47		
48		

What Is Your Answer?

3. IN YOUR OWN WORDS How can you find the probability that a compound event will occur?

4. Compare the experiments and their results in Activities 1 and 2.

Practice Use what you learned about the probabilities of compound events to complete Exercises 5 and 9 on page 429.

10.5 Lesson

Check It Out
Lesson Tutorials
BigIdeasMath.com

Key Vocabulary
compound event, *p. 426*
independent events, *p. 426*
dependent events, *p. 427*

A **compound event** consists of two or more events. Two events are **independent events** if the occurrence of one event *does not* affect the likelihood that the other event will occur.

Key Idea

Probability of Independent Events

Words The probability of two independent events A and B is the probability of A times the probability of B.

Symbols $P(A \text{ and } B) = P(A) \cdot P(B)$

- $P(A \text{ and } B)$: probability of both events
- $P(A)$: probability of first event
- $P(B)$: probability of second event

EXAMPLE 1 Finding the Probability of Independent Events

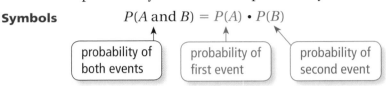

You spin the spinner and flip the coin. What is the probability that you spin a prime number and flip tails?

The outcome of spinning the spinner does not affect the outcome of flipping the coin. So, the events are independent.

$P(\text{prime}) = \dfrac{3}{5}$ — There are 3 prime numbers (2, 3, and 5). There is a total of 5 numbers.

$P(\text{tails}) = \dfrac{1}{2}$ — There is 1 tails side. There is a total of 2 sides.

Use the formula for the probability of independent events.

$$P(A \text{ and } B) = P(A) \cdot P(B)$$

$$P(\text{prime and tails}) = P(\text{prime}) \cdot P(\text{tails})$$

$$= \frac{3}{5} \cdot \frac{1}{2} \qquad \text{Substitute.}$$

$$= \frac{3}{10} \qquad \text{Simplify.}$$

∴ The probability that you spin a prime number and flip tails is $\dfrac{3}{10}$, or 30%.

On Your Own

Now You're Ready
Exercises 5–8

1. What is the probability of spinning a multiple of 2 and flipping heads?

Multi-Language Glossary at BigIdeasMath.com.

Two events are **dependent events** if the occurrence of one event *does* affect the likelihood that the other event will occur.

Key Idea

Probability of Dependent Events

Words The probability of two dependent events A and B is the probability of A times the probability of B after A occurs.

Symbols $P(A \text{ and } B) = P(A) \cdot P(B \text{ after } A)$

- $P(A \text{ and } B)$: probability of both events
- $P(A)$: probability of first event
- $P(B \text{ after } A)$: probability of second event after first event occurs

EXAMPLE 2 Finding the Probability of Dependent Events

You randomly choose a flower from the vase to take home. Your friend randomly chooses another flower from the vase to take home. What is the probability that you choose a purple flower and your friend chooses a yellow flower?

Purple: 7
Yellow: 9
Pink: 12

Choosing a flower changes the number of flowers left in the vase. So, the events are dependent.

$$P(\text{first is purple}) = \frac{7}{28} = \frac{1}{4}$$

There are 7 purple flowers. There is a total of 28 flowers.

$$P(\text{second is yellow}) = \frac{9}{27} = \frac{1}{3}$$

There are 9 yellow flowers. There is a total of 27 flowers left.

Use the formula for the probability of dependent events.

$$P(A \text{ and } B) = P(A) \cdot P(B \text{ after } A)$$

$$P(\text{purple and yellow}) = P(\text{purple}) \cdot P(\text{yellow after purple})$$

$$= \frac{1}{4} \cdot \frac{1}{3} \qquad \text{Substitute.}$$

$$= \frac{1}{12} \qquad \text{Simplify.}$$

∴ The probability of choosing a purple flower and then a yellow flower is $\frac{1}{12}$, or about 8%.

On Your Own

Now You're Ready
Exercises 9–12

2. **WHAT IF?** What is the probability that both flowers are purple?

EXAMPLE 3 Finding the Probability of a Compound Event

A student randomly guesses the answer for each of the multiple choice questions. What is the probability that both questions are answered correctly?

1. In what year did Virginia become a state?

A. 1492 **B.** 1776 **C.** 1788 **D.** 1795 **E.** 2000

2. Which amendment to the Constitution grants citizenship to all persons born in the United States and guarantees them equal protection under the law?

A. 1st **B.** 5th **C.** 12th **D.** 13th **E.** 14th

Choosing the answer for one question does not affect the choice for the second question. So, the events are independent.

Method 1: Use the formula for the probability of independent events.

$$P(\#1 \text{ correct and } \#2 \text{ correct}) = P(\#1 \text{ correct}) \cdot P(\#2 \text{ correct})$$

$$= \frac{1}{5} \cdot \frac{1}{5} \quad \text{Substitute.}$$

$$= \frac{1}{25} \quad \text{Simplify.}$$

∴ The probability of answering both questions correctly is $\frac{1}{25}$, or 4%.

Method 2: Use the Fundamental Counting Principle.

There are 5 choices for each question, so there are $5 \cdot 5 = 25$ possible outcomes. There is only 1 way to answer both questions correctly.

$$P(\#1 \text{ correct and } \#2 \text{ correct}) = \frac{1}{25}$$

∴ The probability of answering both questions correctly is $\frac{1}{25}$, or 4%.

On Your Own

3. The student knows that Choice A is incorrect for both questions. What is the probability that both questions are answered correctly? Compare this probability with the probability in Example 3. What do you notice?

10.5 Exercises

Check It Out
Help with Homework
BigIdeasMath.com

Vocabulary and Concept Check

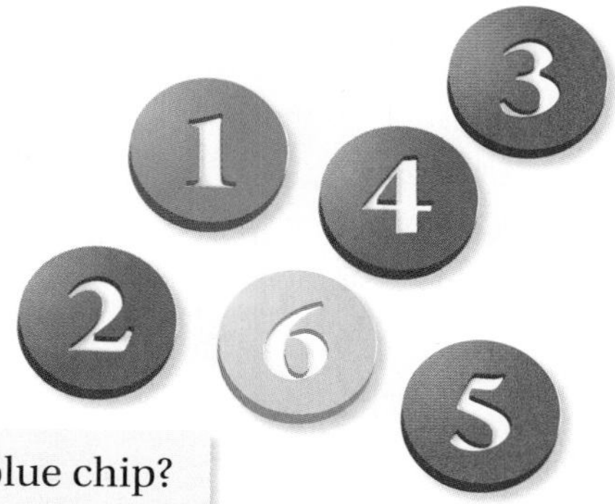

1. **DIFFERENT WORDS, SAME QUESTION** You randomly choose one of the chips. Without replacing the first chip, you choose a second chip. Which question is different? Find "both" answers.

   What is the probability of choosing a 1 and then a blue chip?

   What is the probability of choosing a 1 and then an even number?

   What is the probability of choosing a green chip and then a chip that is *not* red?

   What is the probability of choosing a number less than 2 and then an even number?

2. **WRITING** When is one event dependent on another event?
3. **OPEN-ENDED** Give a real-life example of compound events.
4. **WRITING** How do you find the probability of two events A and B when A and B are independent? dependent?

Practice and Problem Solving

You spin the spinner and flip a coin. Find the probability of the events.

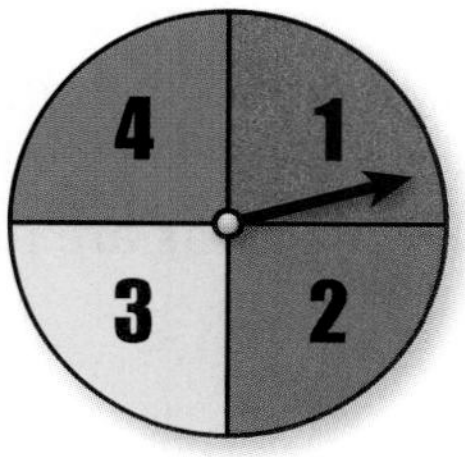

1 5. Spinning a 3 and flipping heads
6. Spinning an even number and flipping tails
7. Spinning a number greater than 1 and flipping tails
8. *Not* spinning a 2 and flipping heads

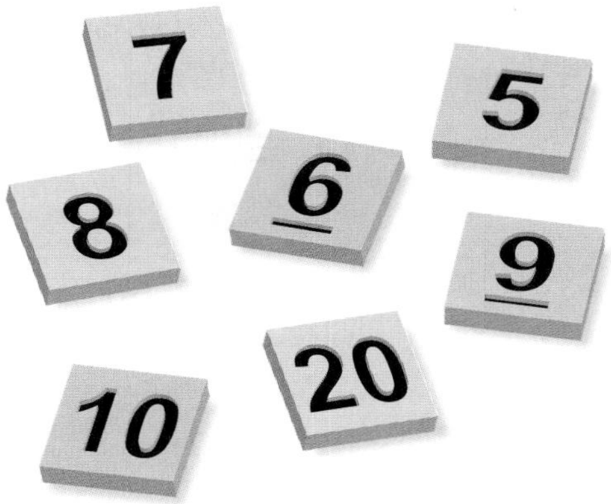

You randomly choose one of the tiles. Without replacing the first tile, you choose a second tile. Find the probability of the events.

2 9. Choosing a 5 and then a 6
10. Choosing an odd number and then a 20
11. Choosing a number less than 7 and then a multiple of 4
12. Choosing two even numbers

13. **ERROR ANALYSIS** Describe and correct the error in finding the probability.

> ✗ You randomly choose one of the marbles. Without replacing the first marble, you choose a second marble. What is the probability of choosing red and then green?
>
> $P(\text{red and green}) = \frac{1}{4} \cdot \frac{1}{4} = \frac{1}{16}$

14. **MAZE** You enter a maze that has one correct path. You come to a fork and randomly take the left path. You come to another fork and randomly take the right path. What is the probability that you are still on the correct path?

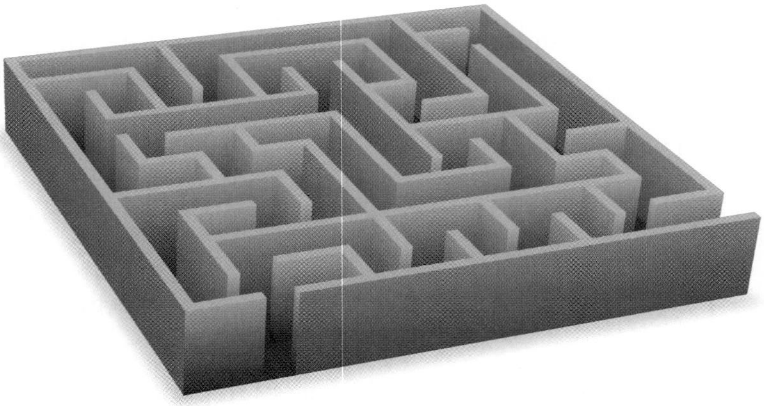

15. **SOCKS** You randomly draw two of the socks from a laundry basket. What is the probability that both socks are blue?

16. **CARNIVAL** At a carnival game, you randomly throw two darts and break two balloons. What is the probability that both of the balloons you break are purple?

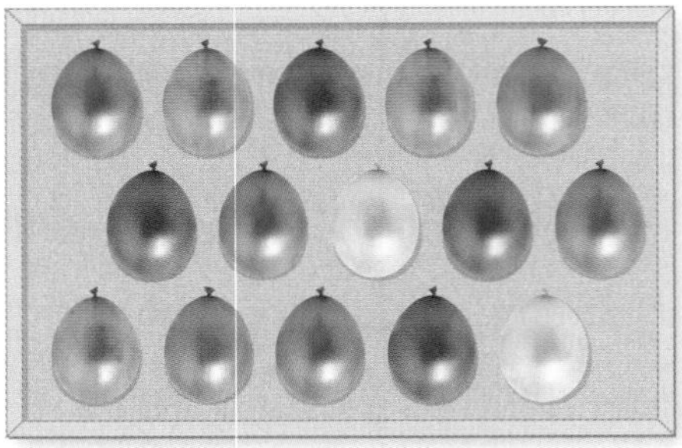

You roll a number cube twice. Find the probability of the events.

17. Rolling two numbers whose sum is 12
18. Rolling a perfect square and then a different perfect square
19. Rolling a number greater than 0 and then a factor of 720
20. Rolling a 5 and then the positive square root of 100
21. The second roll is twice the first roll.
22. The second roll is the square root of the first roll.

23. **LANGUAGES** There are 16 students in your Spanish class. Your teacher randomly chooses one student at a time to take a verbal exam. What is the probability that you are *not* one of the first two students chosen?

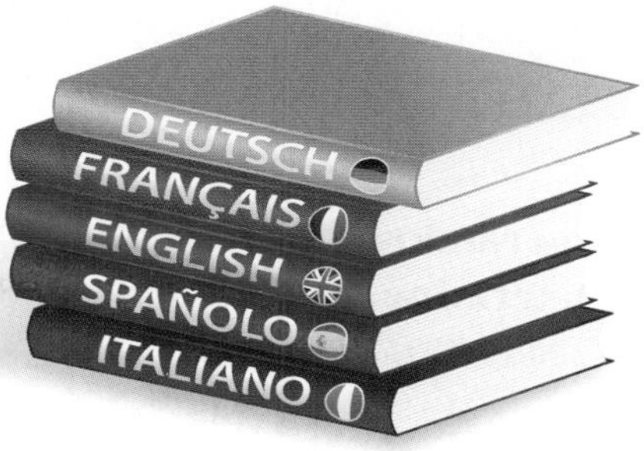

24. **GROUP LEADERS** Your teacher divides your class into two groups and then randomly chooses a leader for each group. The probability that you are chosen to be a leader is $\frac{1}{12}$. The probability that both you and your best friend are chosen is $\frac{1}{132}$.

a. Is your best friend in your group? Explain.

b. What is the probability that your best friend is chosen as a group leader?

c. How many students are in the class?

25. **Number Sense** After ruling out some of the answer choices, you randomly guess the answer for each of the story questions below.

1. Who was the oldest?

A. Ned **B.** Yvonne **C.** Sun Li **D.** Angel **E.** Dusty

2. What city was Stacey from?

A. Raleigh **B.** New York **C.** Roanoke **D.** Dallas **E.** San Diego

a. How can the probability of getting both answers correct be 25%?

b. How can the probability of getting both answers correct be $8\frac{1}{3}\%$?

Fair Game Review What you learned in previous grades & lessons

Evaluate the expression. *(Section 9.2)*

26. 2^{-6}

27. 14^{0}

28. $(-4)^{-1}$

29. $\frac{4}{(-2)^{-4}}$

30. **MULTIPLE CHOICE** What is the radius of the circle? *(Section 9.5)*

Ⓐ 2.5 ft Ⓑ 5 ft

Ⓒ 10 ft Ⓓ 20 ft

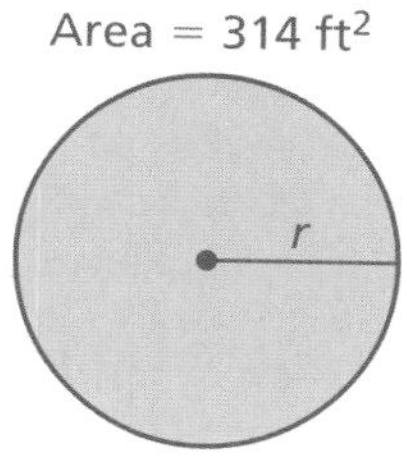

10.3–10.5 Quiz

You randomly choose one push pin from the jar. Determine the theoretical probability of the event.
(Section 10.3)

1. Choosing a yellow pin

2. *Not* choosing a blue pin

3. Choosing a green or red pin

12 Green
6 White
8 Red
4 Blue
10 Yellow

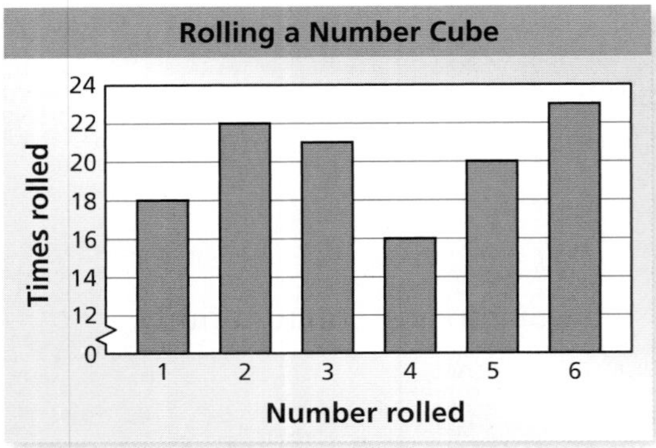

Use the bar graph to find the experimental probability of the event. *(Section 10.4)*

4. Rolling a 4

5. Rolling a multiple of 3

6. Rolling a 2 or a 3

7. Rolling a number less than 7

The spinner is spun. Determine if the game is fair. If it is *not* fair, who has the greater probability of winning? *(Section 10.3)*

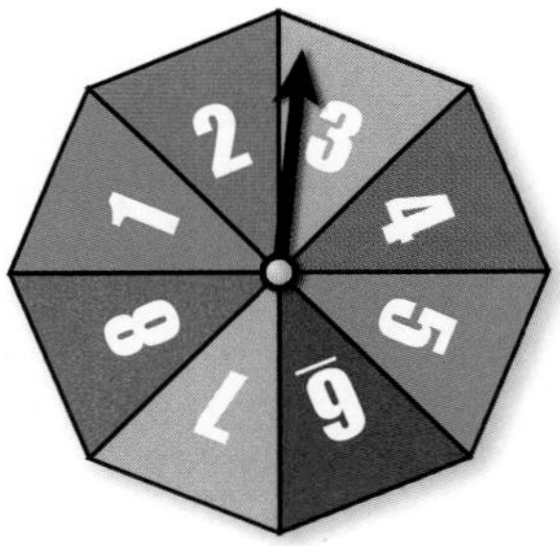

8. You win if the number is even. Your friend wins if the number is odd.

9. You win if the number is less than 4. Your friend wins if the number is 4 or greater.

10. **SPINNER AND COINS** You spin the spinner shown at the right and flip a coin. What is the probability that you spin an even number and flip heads? *(Section 10.5)*

11. **PENS** There are 30 pens in a box. You randomly choose the five pens shown. How many of the 30 pens would you expect to have red ink? *(Section 10.4)*

12. **BLUE PENS** You randomly choose one of the five pens shown. Your friend randomly chooses one of the remaining pens. What is the probability that you and your friend both choose a blue pen? *(Section 10.5)*

13. **APPLES** There are 104 apples in a basket. The probability of randomly choosing a Granny Smith apple from the basket is 25%. How many of the apples are *not* Granny Smith apples? *(Section 10.3)*

Check It Out
Vocabulary Help
BigIdeasMath.com

Review Key Vocabulary

histogram, *p. 398*
outcomes, *p. 406*
event, *p. 406*
Fundamental Counting Principle, *p. 406*
theoretical probability, *p. 414*
fair experiment, *p. 415*
experimental probability, *p. 420*
compound event, *p. 426*
independent events, *p. 426*
dependent events, *p. 427*

Review Examples and Exercises

10.1 Histograms *(pp. 396–403)*

The frequency table shows the number of crafts each member of the Craft Club made for a fundraiser. Display the data in a histogram.

Crafts	Frequency
0–2	10
3–5	8
6–8	5
9–11	0
12–14	2

Step 1: Draw and label the axes.

Step 2: Draw a bar to represent the frequency of each interval.

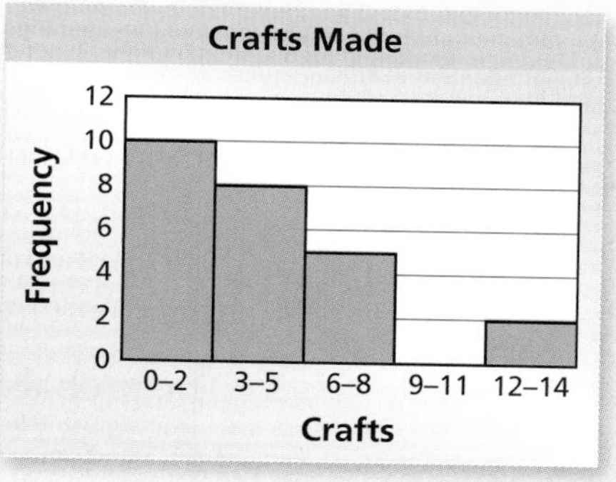

Exercises

Display the data in a histogram.

1.

Heights of Gymnasts	
Heights (in.)	**Frequency**
50–54	1
55–59	8
60–64	5
65–69	2

2.

Minutes Studied	
Minutes	**Frequency**
0–19	5
20–39	9
40–59	12
60–79	3

10.2 Fundamental Counting Principle *(pp. 404–409)*

How many different home theater systems can you make from 6 DVD players, 8 TVs, and 3 brands of speakers?

$6 \times 8 \times 3 = 144$ Fundamental Counting Principle

∴ You can make 144 different home theater systems.

Exercises

3. You have 6 bracelets and 15 necklaces. Find the number of ways you can wear one bracelet and one necklace.

10.3 Theoretical Probability (pp. 412–417)

The theoretical probability that you choose a purple grape from a bag is $\frac{2}{9}$. There are 36 grapes in the bag. How many are purple?

$$P(\text{purple}) = \frac{\text{number of purple grapes}}{\text{total number of grapes}}$$

$\frac{2}{9} = \frac{n}{36}$ — Substitute. Let n be the number of purple grapes.

$8 = n$ — Multiply each side by 36.

∴ There are 8 purple grapes in the bag.

Exercises

4. You get one point when the spinner at the right lands on an odd number. Your friend gets one point when it lands on an even number. The first person to get 5 points wins. Is the game fair? If it is *not* fair, who has the greater probability of winning?

5. The probability that you spin an even number on a spinner is $\frac{2}{3}$. The spinner has 12 sections. How many sections have even numbers?

10.4 Experimental Probability (pp. 418–423)

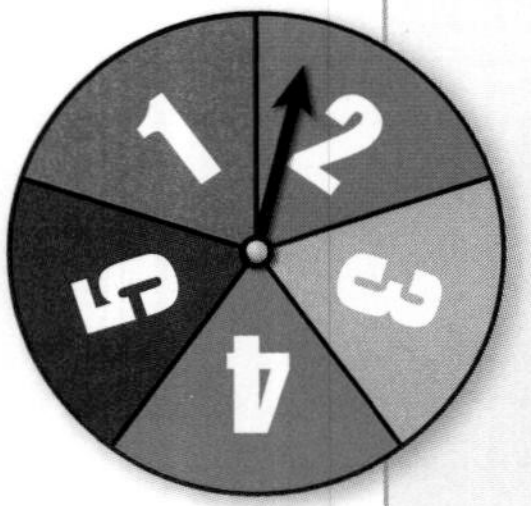

The bar graph shows the results of spinning the spinner 70 times. What is the experimental probability of spinning a 2?

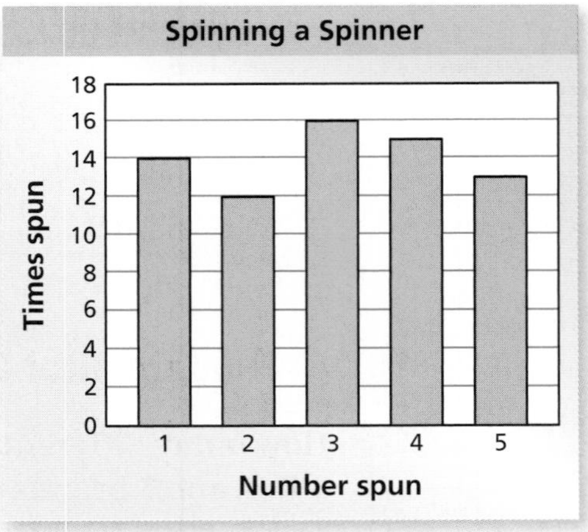

The bar graph shows 14 ones, 12 twos, 16 threes, 15 fours, and 13 fives. So, the spinner landed on two 12 times in 70 spins.

$$P(\text{event}) = \frac{\text{number of times the event occurs}}{\text{total number of trials}}$$

Two was landed on 12 times.

$$P(2) = \frac{12}{70} = \frac{6}{35}$$

There was a total of 70 spins.

∴ The experimental probability is $\frac{6}{35}$, or about 17%.

Exercises

Use the bar graph on page 434 to find the experimental probability of the event.

6. Spinning a 3

7. Spinning an odd number

8. *Not* spinning a 5

9. Spinning a number greater than 3

10.5 Compound Events *(pp. 424–431)*

You randomly choose one of the tiles and flip the coin. What is the probability that you choose a vowel and flip heads?

Choosing one of the tiles does not affect the outcome of flipping the coin. So, the events are independent.

$P(\text{vowel}) = \frac{2}{7}$ ← There are 2 vowels (A and E). ← There is a total of 7 tiles.

$P(\text{tails}) = \frac{1}{2}$ ← There is 1 tails side. ← There is a total of 2 sides.

Use the formula for the probability of independent events.

$P(A \text{ and } B) = P(A) \bullet P(B)$

$= \frac{2}{7} \bullet \frac{1}{2}$ Substitute.

$= \frac{1}{7}$ Simplify.

∴ The probability of choosing a vowel and flipping heads is $\frac{1}{7}$, or about 14%.

Exercises

You randomly choose one of the tiles above and flip the coin. Find the probability of the events.

10. Choosing a blue tile and flipping tails

11. Choosing the letter G and flipping tails

You randomly choose one of the tiles above. Without replacing it, you randomly choose a second tile. Find the probability of the events.

12. Choosing a green tile and then choosing a blue tile

13. Choosing a red tile and then choosing a vowel

10 Chapter Test

Check It Out
Test Practice
BigIdeasMath.com

Use the Fundamental Counting Principle to find the total number of possible outcomes.

1.

	Sunscreen
SPF	10, 15, 30, 45, 50
Type	Lotion, Spray, Gel

2.

	Books
Cover	Paperback, Hard
Type	Mystery, Biography, Cooking

The spinner is spun. Determine if the game is fair. If it is *not* fair, who has the greater probability of winning?

3. You win if the number is odd. Your friend wins if the number is even.

4. You win if the number is less than 5. Your friend wins if the number is greater than 5. If the number is 5, nobody wins.

5. **WATER** The histogram shows the number of glasses of water that the students in a class drink in one day.

   a. Which interval contains the fewest data values?

   b. How many students are in the class?

   c. Health experts recommend drinking at least 8 glasses of water per day. What percent of the students drink the recommended amount?

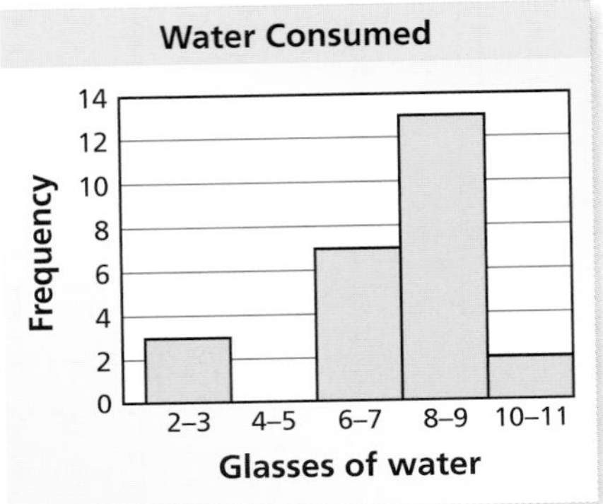

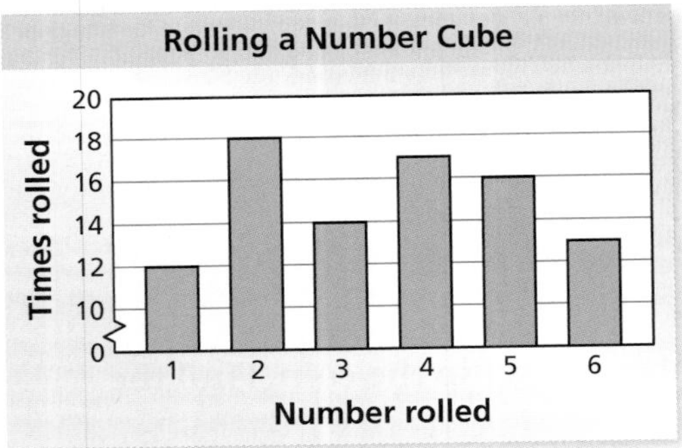

Use the bar graph to find the experimental probability of the event.

6. Rolling a 1 or a 2

7. Rolling an odd number

8. *Not* rolling a 5

9. **MINTS** You have a bag of 60 assorted mints. You randomly choose six mints. Two of the mints you choose are peppermints. How many of the 60 mints would you expect to be peppermints?

10. **BEADS** Thirty percent of the beads in a bag are blue. One bead is randomly chosen and replaced. Then a second bead is chosen. What is the probability that *neither* bead is blue?

1. Which function rule is shown by the table?

Input, x	Output, y
-8	-4
-4	-2
0	0
4	2

A. $y = 2x$ **C.** $y = \frac{x}{2}$

B. $y = -\frac{x}{2}$ **D.** $y = -2x$

Test-Taking Strategy
Read Question Before Answering

Of 2048 cats, how many of them will NOT answer to "Here, kitty kitty"?
(A) 100% (B) 1 (C) 4098/2 (D) 2048
Hey, it means "free food".

"Be sure to read the question before choosing your answer. You may find a word that changes the meaning."

2. What is the value of the expression below when $g = -12$?

$$-6 + g$$

F. 6 **H.** -6

G. 18 **I.** -18

3. Trapezoid *KLMN* is graphed in the coordinate plane shown.

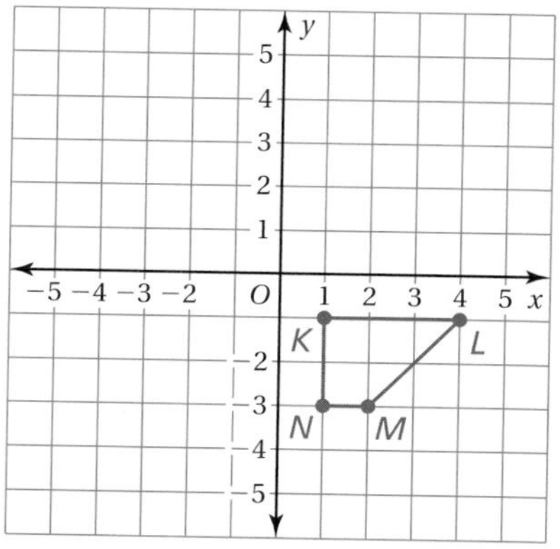

Rotate trapezoid *KLMN* 90° clockwise about the origin. What are the coordinates of point M', the image of point M after the rotation?

A. $(-3, -2)$ **C.** $(-2, 3)$

B. $(-2, -3)$ **D.** $(3, 2)$

4. Each student in your class voted for his or her favorite day of the week. Their votes are shown below:

- Sunday: 6
- Friday: 8
- Saturday: 10
- Other day: 6

A student from your class is picked at random. What is the probability that this student's favorite day of the week is Sunday?

5. A formula for converting a temperature in degrees Celsius C to a temperature in degrees Fahrenheit F is shown below.

$$F = \frac{9}{5}C + 32$$

When the temperature in degrees Celsius is $-9°$, what is the temperature, to the nearest degree, in degrees Fahrenheit?

F. $-27°$

G. $-16°$

H. $16°$

I. $27°$

6. There are 665 tickets in a drawing. The theoretical probability that one of your tickets is drawn is $\frac{1}{35}$. How many tickets do you have in the drawing?

A. 1

B. 19

C. 35

D. 646

7. A cylinder has a diameter of 10 centimeters and a height of 13 centimeters. Stan was computing its surface area in the box below.

$$\begin{aligned} 2\pi r^2 + 2\pi rh &= 2 \cdot 3.14 \cdot 5^2 + 2 \cdot 3.14 \cdot 5 \cdot 13 \\ &= 31.4^2 + 31.4 \cdot 13 \\ &= 985.96 + 408.2 \\ &= 1394.16 \text{ cm}^2 \end{aligned}$$

What should Stan do to correct the error that he made?

F. Use the formula $\pi r^2 h$.

G. Label the answer with the unit cm^3.

H. Square the 5 before multiplying.

I. Distribute the 3.14 to get $3.14 \cdot 5 + 3.14 \cdot 13$.

8. Which expression is *not* equal to the other three?

A. 6

B. -6

C. $|6|$

D. $|-6|$

9. A spinner is divided into 10 congruent sections, as shown below.

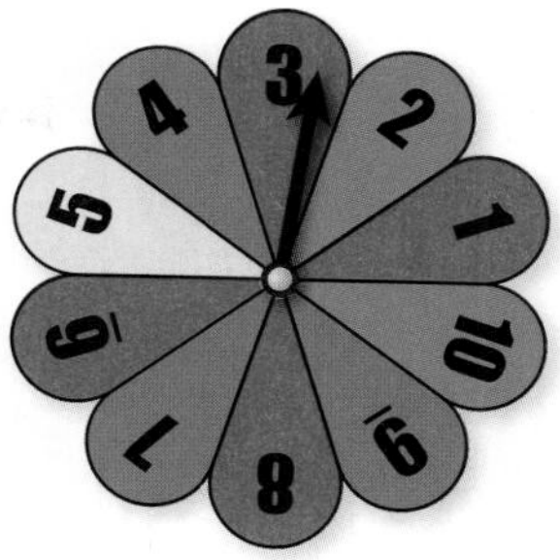

You spin the spinner twice. What is the probability that the arrow will stop on a red section and then on an even number?

10. For a vacation trip, 4 cousins decided to rent a minivan and share the cost equally. They determined that each cousin would have to pay \$300.

The minivan had 2 unused seats, so 2 additional cousins joined them. If the larger group of cousins now share the cost equally, what would each cousin have to pay?

F. \$150

G. \$200

H. \$300

I. \$450

11. Think Solve Explain At the end of the school year, your teacher counted up the number of absences for each student. The results are shown in the histogram below.

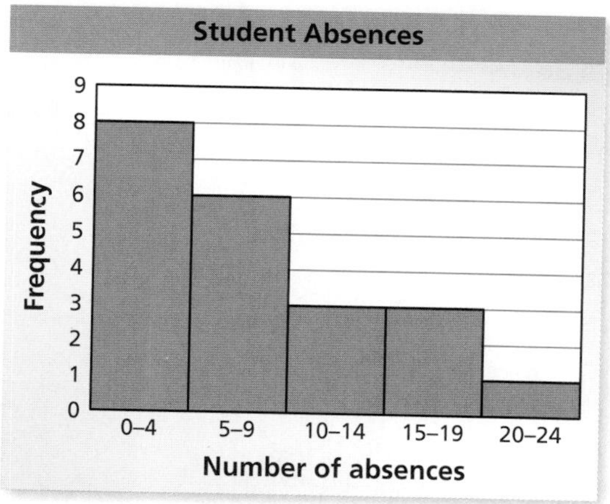

Part A Based on the histogram, how many students had fewer than 10 absences?

Part B What percent of the students had fewer than 10 absences? Show your work.

Appendix A
My Big Ideas Projects

A.1 Literature Project
The Mathematics of Jules Verne

A.2 History Project
Mathematics in Ancient Greece

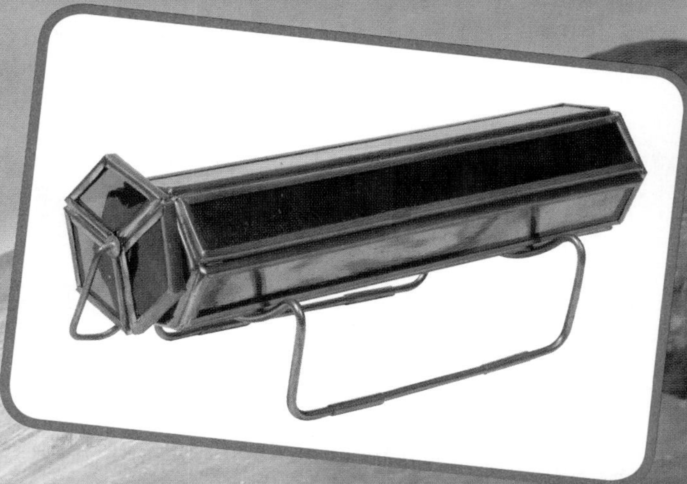

My Big Ideas Projects

A.3 Art Project
Building a Kaleidoscope

A.4 Science Project
Classifying Animals

A.1 Literature Project

STANDARDS OF LEARNING
7.1–16

Share Your Work at... My.BigIdeasMath.com

The Mathematics of Jules Verne

1 Project Overview

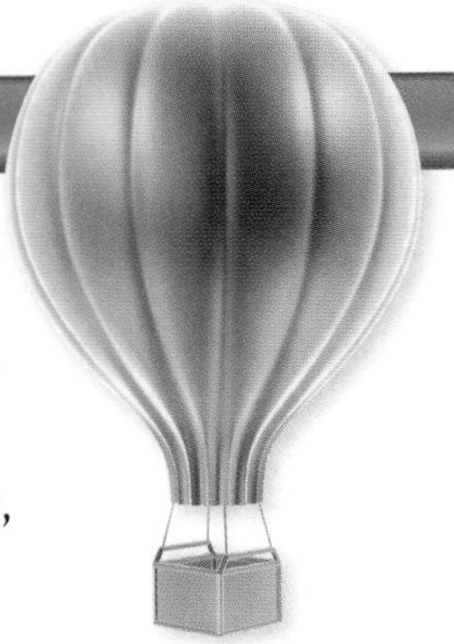

Jules Verne (1828–1905) was a famous French science fiction writer. He wrote about space, air, and underwater travel before aircraft and submarines were commonplace, and before any means of space travel had been devised.

For example, in his 1865 novel *From the Earth to the Moon*, he wrote about three astronauts who were launched from Florida and recovered through a splash landing. The first actual moon landing wasn't until 1969.

Essential Question How does the knowledge of mathematics influence science fiction writing?

Read one of Jules Verne's science fiction novels. Then write a book report about some of the mathematics used in the novel.

Sample: A league is an old measure of distance. It is approximately equal to 4 kilometers. You can convert 20,000 leagues to miles as follows.

$$20{,}000 \text{ leagues} \cdot \frac{4 \text{ km}}{1 \text{ league}} \cdot \frac{1 \text{ mile}}{1.6 \text{ km}} = 50{,}000 \text{ miles}$$

2 Things to Include

- Describe the major events in the plot.
- Write a brief paragraph describing the setting of the story.
- List and identify the main characters. Explain the contribution of each character to the story.
- Explain the major conflict in the story.
- Describe at least four examples of mathematics used in the story.
- Which of Jules Verne's scientific predictions have come true since he wrote the novel?

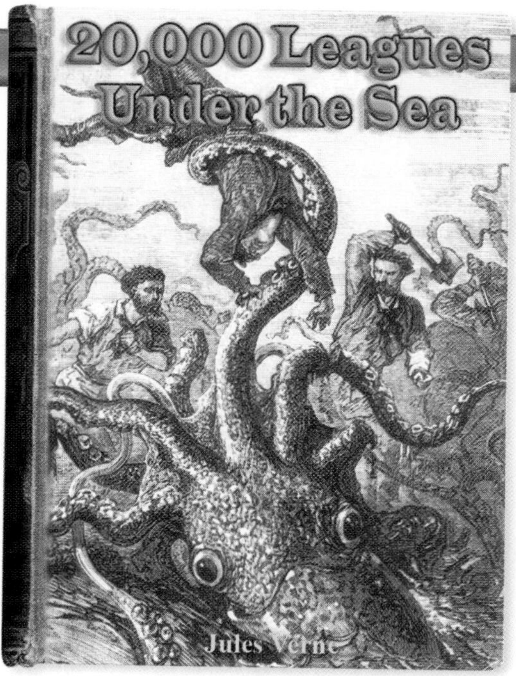

Jules Verne (1828–1905)

3 Things to Remember

- You can download one of Jules Verne's novels at *BigIdeasMath.com.*
- Add your own illustrations to your project.
- Organize your report in a folder, and think of a title for your report.

A.2 History Project

Mathematics in Ancient Greece

STANDARDS OF LEARNING
7.1–16

Share Your Work at... My.BigIdeasMath.com

1 Getting Started

Corinthian Helmet

The ancient Greek period began around 1100 B.C. and lasted until the Roman conquest of Greece in 146 B.C.

The civilization of the ancient Greeks influenced the languages, politics, educational systems, philosophy, science, mathematics, and arts of Western Civilization. It was a primary force in the birth of the Renaissance in Europe between the 14th and 17th centuries.

Essential Question How do you use mathematical knowledge that was originally discovered by the Greeks?

Sample: Ancient Greek symbols for the numbers from 1 through 10 are shown in the table.

<table>
<tr><td>|</td><td>||</td><td>|||</td><td>||||</td><td>Γ</td><td>Γ|</td><td>Γ||</td><td>Γ|||</td><td>Γ||||</td><td>△</td></tr>
<tr><td>1</td><td>2</td><td>3</td><td>4</td><td>5</td><td>6</td><td>7</td><td>8</td><td>9</td><td>10</td></tr>
</table>

These same symbols were used to write the numbers between 11 and 39. Here are some examples.

△Γ||| = 18   △△△Γ = 35   △△|||| = 24

Alexander the Great

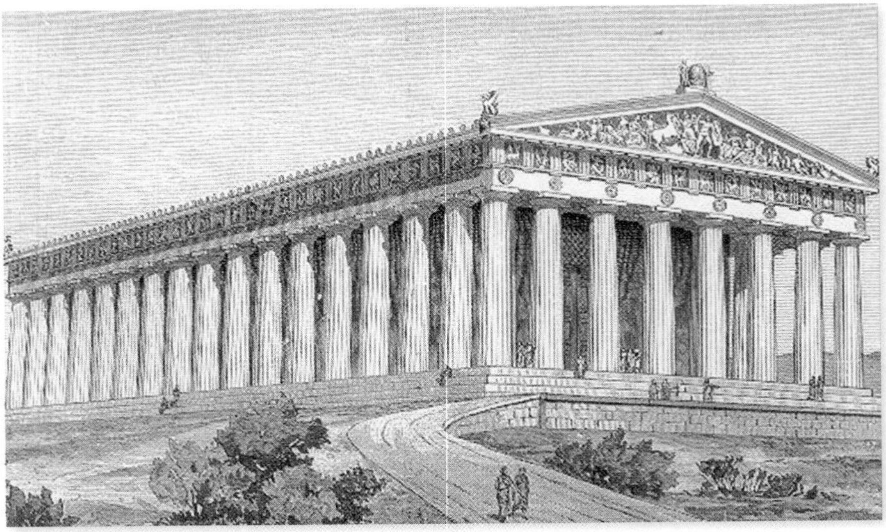
Parthenon

2 Things to Include

- Describe at least one contribution that each of the following people made to mathematics.

  Pythagoras (c. 570 B.C.–c. 490 B.C.)

  Aristotle (c. 384 B.C.–c. 322 B.C.)

  Euclid (c. 300 B.C.)

  Archimedes (c. 287 B.C.–c. 212 B.C.)

  Eratosthenes (c. 276 B.C.–c. 194 B.C.)

- Which of the people listed above was the teacher of Alexander the Great? What subjects did Alexander the Great study when he was in school?
- How did the ancient Greeks represent fractions?
- Describe how the ancient Greeks used mathematics. How does this compare with the ways in which mathematics is used today?

Α α	alpha	Ν ν	nu
Β β	beta	Ξ ξ	xi
Γ γ	gamma	Ο ο	omicron
Δ δ	delta	Π π	pi
Ε ε	epsilon	Ρ ρ	rho
Ζ ζ	zeta	Σ σ	sigma
Η η	eta	Τ τ	tau
Θ θ	theta	Υ υ	upsilon
Ι ι	iota	Φ φ	phi
Κ κ	kappa	Χ χ	chi
Λ λ	lambda	Ψ ψ	psi
Μ μ	mu	Ω ω	omega

3 Things to Remember

- Add your own illustrations to your project.
- Try to include as many different math concepts as possible. Your goal is to include at least one concept from each of the chapters you studied this year.
- Organize your report in a folder, and think of a title for your report.

Greek Pottery

Trireme Greek Warship

A.3 Art Project

STANDARDS OF LEARNING
7.1–16

Share Your Work at... My.BigIdeasMath.com

Building a Kaleidoscope

1 Getting Started

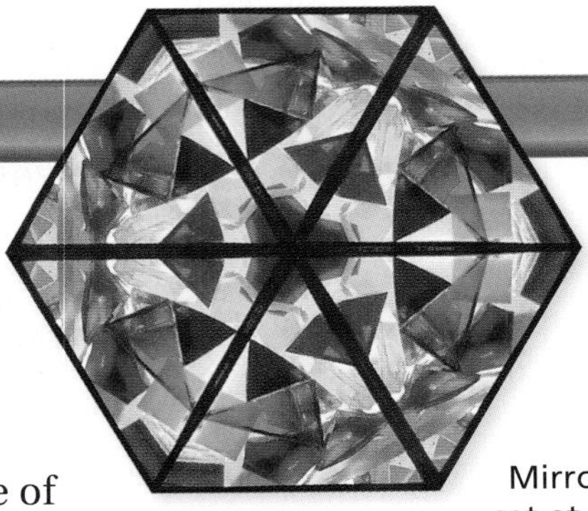

Mirrors set at 60°

A kaleidoscope is a tube of mirrors containing loose colored beads, pebbles, or other small colored objects. You look in one end and light enters the other end, reflecting off the mirrors.

Essential Question How does the knowledge of mathematics help you create a kaleidoscope?

If the angle between the mirrors is 45°, you see 8 duplicate images. If the angle is 60°, you see 6 duplicate images. If the angle is 90°, you see 4 duplicate images. As the tube is rotated, the colored objects tumble, creating various patterns.

Write a report about kaleidoscopes. Discuss the mathematics you need to know in order to build a kaleidoscope.

Sample: A kaleidoscope whose mirrors meet at 60° angles has reflective symmetry and rotational symmetry.

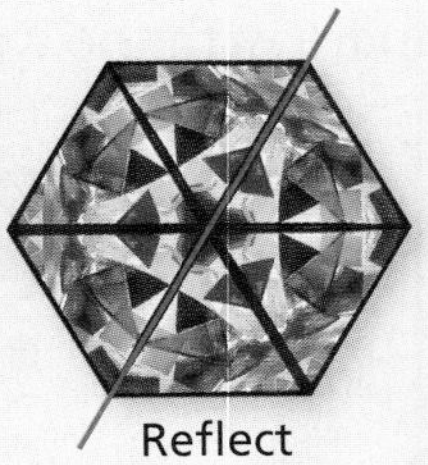

Reflect

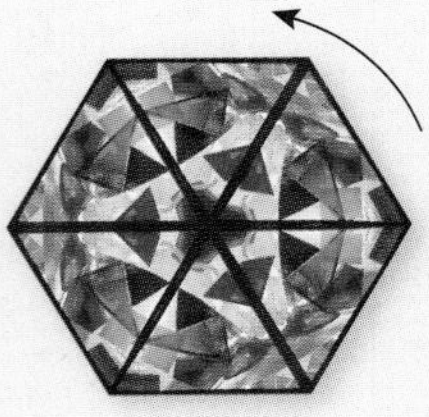

Rotate 120°

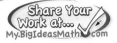

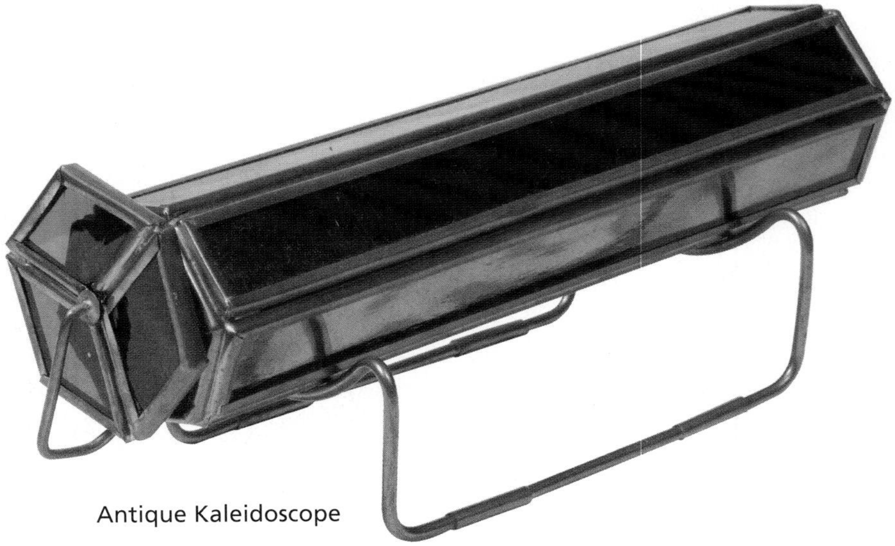

Antique Kaleidoscope

2 Things to Include

- How does the angle at which the mirrors meet affect the number of duplicate images that you see?
- What angles can you use other than 45°, 60°, and 90°? Explain your reasoning.
- Research the history of kaleidoscopes. Can you find examples of kaleidoscopes being used before they were patented by David Brewster in 1816?
- Make your own kaleidoscope.
- Describe the mathematics you used to create your kaleidoscope.

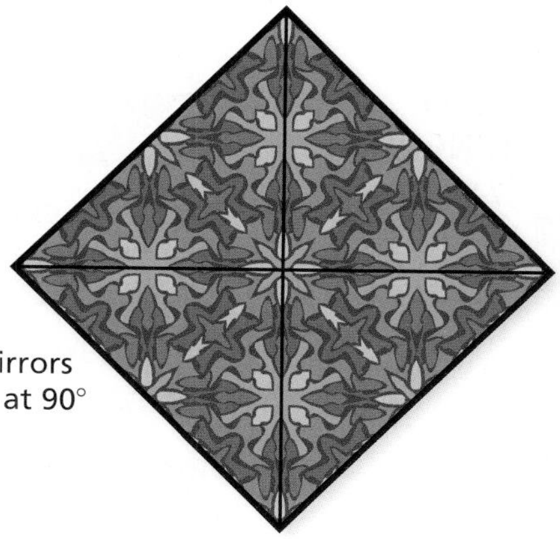

Mirrors set at 90°

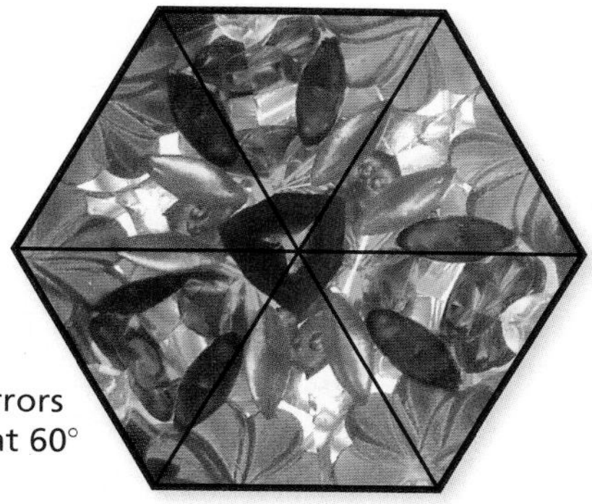

Mirrors set at 60°

3 Things to Think About

- Add your own drawings and pattern creations to your project.
- Organize your report in a folder, and think of a title for your report.

Giant Kaleidoscope, San Diego harbor

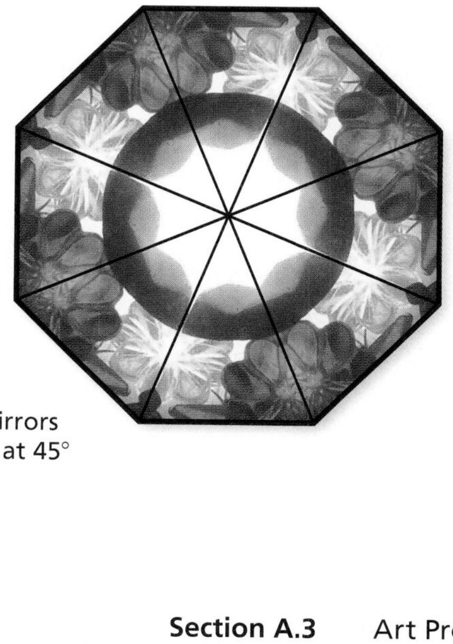

Mirrors set at 45°

A.4 Science Project

STANDARDS OF LEARNING
7.1–16

Share Your Work at... My.BigIdeasMath.com

Classifying Animals

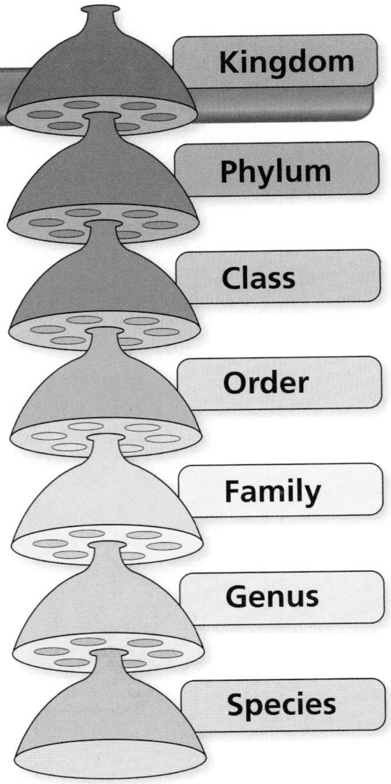

1 Getting Started

Biologists classify animals by placing them in phylums, or groups, with similar characteristics. Latin names, such as Chordata (having a spinal cord) or Arthropoda (having jointed limbs and rigid bodies) are used to describe these groups.

Biological classification is difficult, and scientists are still developing a complete system. There are seven main ranks of life on Earth; kingdom, phylum, class, order, family, genus, and species. However, scientists usually use more than these seven ranks to classify organisms.

Essential Question How does the classification of living organisms help you understand the similarities and differences of animals?

Write a report about how animals are classified. Choose several different animals and list the phylum, class, and order of each animal.

Wasp
Phylum: Arthropoda
Class: Insecta
Order: Hymenoptera
(membranous wing)

Sample: A bat is classified as an animal in the phylum Chordata, class Mammalia, and order Chiroptera. *Chiroptera* is a Greek word meaning "hand-wing."

Bat
Phylum: Chordata
Class: Mammalia
Order: Chiroptera
(hand-wing)

Monkey
Phylum: Chordata
Class: Mammalia
Order: Primate (large brain)

Kangaroo
Phylum: Chordata
Class: Mammalia
Order: Diprotodontia
(two front teeth)

2 Things to Include

- List the different classes of phylum Chordata. Have you seen a member of each class?
- List the different classes of phylum Arthropoda. Have you seen a member of each class?
- Show how you can use graphic organizers to help classify animals. Which types of graphic organizers seem to be most helpful? Explain your reasoning.
- Summarize the number of species in each phlyum in an organized way. Be sure to include fractions, decimals, and percents.

Parrot
Phylum: Chordata
Class: Aves
Order: Psittaciformes
(strong, curved bill)

Frog
Phylum: Chordata
Class: Amphibia
Order: Anura (no tail)

Spider
Phylum: Arthropoda
Class: Arachnida
Order: Araneae (spider)

3 Things to Remember

- Organize your report in a folder, and think of a title for your report.

Cougar
Phylum: Chordata
Class: Mammalia
Order: Carnivora (meat eater)

Lobster
Phylum: Arthropoda
Class: Malacostraca
Order: Decopada (ten footed)

Crocodile
Phylum: Chordata
Class: Reptilia
Order: Crocodilia
(pebble-worm)

Photo Credits

iv Big Ideas Learning, LLC; **viii** *top* ©iStockphoto.com/Angel Rodriguez, ©iStockphoto.com/Ann Marie Kurtz; *bottom* ©iStockphoto.com/daaronj; **ix** *top* ©iStockphoto.com/Jonathan Larsen; *bottom* Peter Close; **x** *top* Chiyacat, Zoom Team; *bottom* Apollofoto; **xi** *top* ©iStockphoto.com/Lisa Thornberg, ©iStockphoto.com/Ann Marie Kurtz; *bottom* Jane Norton; **xii** *top* ©iStockphoto.com/Angel Rodriguez, ©iStockphoto.com/Ann Marie Kurtz; *bottom* Evok20; **xiii** *top* ©iStockphoto.com/Michael Flippo, ©iStockphoto.com/Ann Marie Kurtz; *bottom* ©iStockphoto.com/ronen; **xiv** *top* ©iStockphoto.com/Alistair Cotton; *bottom* Heather Prosch-Jensen; **xv** *top* ©iStockphoto.com/Ron Sumners, ©iStockphoto.com/Ann Marie Kurtz; *bottom* Apollofoto; **xvi** *top* Varina and Jay Patel, ©iStockphoto.com/Ann Marie Kurtz; *bottom* ©iStockphoto.com/Thomas Perkins; **xvii** *top* Cigdem Cooper, ©iStockphoto.com/Andreas Gradin, stephan kerkhofs; *bottom* ©iStockphoto.com/ranplett; **xviii** ©iStockphoto.com/Ekaterina Monakhova; **1** ©iStockphoto.com/Angel Rodriguez, ©iStockphoto.com/Ann Marie Kurtz; **21** ©iStockphoto.com/Ronnie Comeau; **27** Dmitry Melnikov; **34** *Exercise 12* ©iStockphoto.com/Rich Legg; *Exercise 15* CLFProductions; **37** Liem Bahneman; **38** *center* ©iStockphoto.com/susaro; *bottom* Leonard J. DeFrancisci; **42** ©iStockphoto.com/Jonathan Larsen; **47** ©iStockphoto.com/Rich Legg; **48** sonya etchison; **49** Roman Sigaev; **50** *Activity 1a* Anita Patterson Peppers; *Activity 1b* ©iStockphoto.com/Frank Wright; *Activity 1c* Rob Wilson; *Activity 1d* Coprid; *Activity 1e* Jeff R. Clow; **51** *top* ©iStockphoto.com/Josh Hodge; **53** Luc Viatour; **55** Photo courtesy of Chesapeake Bay Bridge-Tunnel.; **57** Allyson Ricketts; **63** ©iStockphoto.com/Brandon Alms; **68** photo25th; **70** ©iStockphoto.com/Vadim Ponomarenko; **72** *top* ©iStockphoto.com/Sandra Layne; *bottom* ©iStockphoto.com/Dawn Nichols; **74** Pete Saloutos; **78** Chiyacat, Zoom Team; **80** ©iStockphoto.com/shantell; **99** ©iStockphoto.com/Franky Sze; **105** ©iStockphoto.com/fotoVoyager; **118** ©iStockphoto.com/bratan007; **122** pandapaw; **123** ©iStockphoto.com/Daniel Van Beek; **129** Alexander Kalina; **130** *top right* ©iStockphoto.com/Martin Firus; *bottom* Robert Pernell; **131** ©iStockphoto.com/Trevor Fisher; **132** Tatiana Popova; **142** ©iStockphoto.com/Lisa Thornberg, ©iStockphoto.com/Ann Marie Kurtz; **149** ©iStockphoto.com/Kevin Panizza; **153** Cedar Point; **154** ©iStockphoto.com/Julie Vader; **155** *top right* ©iStockphoto.com/Sergey Lemeshencko; *center left* ©iStockphoto.com/Robert Rushton; **161** ©iStockphoto.com/Liz McClanahan; **163** ©iStockphoto.com/Mehmet Salih Guler; **164** ©iStockphoto.com/Jacom Stephens; **167** ©iStockphoto.com/DivaNir4a; **168** ©iStockphoto.com/Manuel Angel Diaz Blanco; **169** ©iStockphoto.com/José Luis Gutiérrez; **171** ©iStockphoto.com/Michael Flippo; **180** Rafa Irusta; **181** ©iStockphoto.com/Michael Blackburn; **182** Koksharov Dmitry; **183** Christopher Futcher; **187** *top* Tracy Whiteside; *bottom* ©iStockphoto.com/Paolo Cipriani; **188** ©iStockphoto.com/landbysea; **192** ©iStockphoto.com/Tomasz Zachariasz; **196** ©iStockphoto.com/Angel Rodriguez, ©iStockphoto.com/Ann Marie Kurtz; **216** ©iStockphoto.com/Michael Kemter; **217** ©iStockphoto.com/Vika Valter; **223** NASA; **224** Jean Thompson; **231** *left* Eric Isselée; *right* BORTEL Pavel; **238** ©iStockphoto.com/Michael Flippo, ©iStockphoto.com/Ann Marie Kurtz; **241** ©iStockphoto.com/DIGIcal; **245** Courtesy of Ateliers Jean Nouvel; **246** *top* ©iStockphoto.com/ Viatcheslav Dusaleev; *bottom left* ©iStockphoto.com/Jason Mooy; *bottom center* ©iStockphoto.com/Andres Peiro Palmer; *bottom right* ©iStockphoto.com/Felix Möckel; **255** ©iStockphoto.com/Dennis Guyitt; **266** ©iStockphoto.com/Chris Schmidt; **267** ©iStockphoto.com/Lisa F. Young; **270** ©iStockphoto.com/Dan Moore; **271** ©iStockphoto.com/Aldo Murillo; **272** *center left* Antonio Jorge Nunes, Tom C Amon; *center right* MamaGeek, ©iStockphoto.com/Adam Borkowski; **280** ©iStockphoto.com/Alistair Cotton; **281** *Example 2* ©iStockphoto.com/Denis Pepin; *Exercise 1* ©iStockphoto.com/Carlos Fierro; *Exercise 3* ©iStockphoto.com/Jasmin Awad; *Example 3* ©iStockphoto.com/DNY59; *Exercise 4* ©iStockphoto.com/Rita Jayaraman; *Exercise 5* ©iStockphoto.com/Игорь Безпалов; **286** ©iStockphoto.com/Er Ten Hong; **287** ©iStockphoto.com/Sergey Galushko,©iStockphoto.com/Tryfonov Ievgenii; **288** ©iStockphoto.com/Ingmar Wesemann; **293** *center right* ©iStockphoto.com/asiseeit; *center left* ©iStockphoto.com/Hazlan Abdul Hakim; **301** ©iStockphoto.com/Maksim Shmeljov; **303** Diego Cervo; **316** ©iStockphoto.com/Ron Sumners, ©iStockphoto.com/Ann Marie Kurtz; **319** Dejan Milinkovic; **322** *Exercise 6* ©iStockphoto.com/Ferenc Gerak; *Exercise 8* ©iStockphoto.com/Ljupco; *Exercise 10* ©iStockphoto.com/valzan; **323** Yasonya; **324** Gabriela Trojanowska; **325** Dejan Milinkovic; **328** *Exercise 7* Steyno&Stitch; *Exercise 8* Gil; **331** *Exercise 3* ©iStockphoto.com/Linda & Colin McKie; *Exercise 10* Arkadiusz Komski; **336** ©iStockphoto.com/Tomasz Pietryszek; **337** *center left* Newcastle Drum Centre; *center right* ©iStockphoto.com/scol22; **338** ©iStockphoto.com/Jill Chen; **341** ©iStockphoto.com/camilla wisbauer; **343** *top left and center* ©iStockphoto.com/Achim Prill; *top right* ©iStockphoto.com/Marie-france Bélanger; *center left* ©iStockphoto.com/Matthew Dixon; *center right* ©iStockphoto.com/Nilgun Bostanci; **345** Denise Kappa; **348** ©iStockphoto.com/Daniel Loiselle; **352** ©iStockphoto.com/Varina and Jay Patel, ©iStockphoto.com/Ann Marie Kurtz; **354** ©iStockphoto.com/Franck Boston; **355** *Activity 3a* NASA; *Activity 3b* NASA/JPL-Caltech/R. Hurt (SSC); *Activity 3c* NASA; *Activity 3d* NASA; *Activity 4* Stevyn Colgan; **357** ©iStockphoto.com/Philippa Banks; **358** ©iStockphoto.com/Clotilde Hulin; **359** ©iStockphoto.com/Boris Yankov; **360** 3dimentii; **361** *Activity 3a* Tom C Amon; *Activity 3b* Tom C Amon, HuHu; *Activity 3c* Olga Gabay; *Activity 3d* NASA/MODIS Rapid Response/Jeff Schmaltz; *Activity 3e* Tom C Amon; **363** ©iStockphoto.com/Aliaksandr Autayeu; **364** EugeneF; **365** ©iStockphoto.com/Nancy Louie; **368** ©iStockphoto.com/Kais Tolmats; **369** *center* ©iStockphoto.com/Henrik Jonsson; *top* ©iStockphoto.com/Sebastian Kaulitzki; *bottom* 7artscreensavers.com; **371** *top center* ©iStockphoto.com/Frank Wright; *top left* ©iStockphoto.com/Mark Stay; *bottom left* ©iStockphoto.com/Oliver Sun Kim; *top right* ©iStockphoto.com/Evgeniy Ivanov; **372** ©iStockphoto.com/ChristianAnthony; **373** Microgen; **374** ©iStockphoto.com/camilla wisbauer, ©iStockphoto.com/David Freund, ©iStockphoto.com/Joe Belanger, ©iStockphoto.com/thumb, ©iStockphoto.com/Marie-france Bélanger, ©iStockphoto.com/Susan Trigg; **375** NASA; **376** *center left* Google logo ©Google Inc., 2010. Reprinted with Permission.; *center right* YouTube logo ©Google Inc., 2010. Reprinted with Permission.; **377** *top left* Elaine Barker; *center right* ©iStockphoto.com/breckeni; **378** *bottom left* ©iStockphoto.com/angelhell; *bottom right* ©iStockphoto.com/Jan Rysavy; **379** *top right* BORTEL Pavel; *center right* ©iStockphoto.com/breckeni; **381** JinYoung Lee; **383** ZanyZeus; **385** ©iStockphoto.com/John Kounadeas; **386** NASA, JPL-Caltech, Kate Su (Steward Obs, U. Arizona) et al.; **389** Atlaspix; **390** *center left* Eric Isselée; *bottom right* Patrick J. Lynch, medical illustrator; **394** Cigdem Cooper, ©iStockphoto.com/Andreas Gradin, stephan kerkhofs; **399** *top* ©iStockphoto.com/susaro; *bottom* Arman Zhenikeyev; **400** Tomasz Trojanowski; **403** ©iStockphoto.com/Eric Isselée; **404** *top* John McLaird; *center* Robert Asento; *bottom* Mark Aplet; **408** ©iStockphoto.com/Justin Horrocks; **409** tele52; **414** ©iStockphoto.com/Eric Ferguson; **417** ©iStockphoto.com/Elena Schweitzer; **420** ©iStockphoto.com/Frank Van den Bergh; **427** ©iStockphoto.com/Tim Inman; **428** James Steidl; **430** *top right* Bev Stofko; *center left* Dmitri Mikitenko, vitor costa; *center right* ©iStockphoto.com/Andy Cook; **431** -Albachiaraa-; **432** ©iStockphoto.com/Pawel Kuniec; **A0** ©iStockphoto.com/Björn Kindler; *center left* ©iStockphoto.com/Michael Mattner; *center right* AKaiser; **A1** *top* ©iStockphoto.com/toddmedia; *bottom left* ©iStockphoto.com/Eric Isselée; *bottom right* ©iStockphoto.com/Loretta Hostettler; **A2** sgame; **A3** *top* AKaiser; *bottom left* Photononstop/SuperStock, ©iStockphoto.com/PhantomOfTheOpera; *bottom right* ©iStockphoto.com/Michael Mattner; **A4** ©iStockphoto.com/Duncan Walker; **A5** *top* ©iStockphoto.com/DNY59; *bottom left* ©iStockphoto.com/Boris Katsman; **A6** *top and center* ©iStockphoto.com/Loretta Hostettler; *bottom* ©iStockphoto.com/toddmedia; **A7** *top right* L M Bilder; *center left* ©iStockphoto.com/Clayton Hansen; *bottom left* Bill Whittaker; **A8** *wasp* ©iStockphoto.com/arlindo71; *bat* ©iStockphoto.com/Alexei Zaycev; *monkey and kangaroo* ©iStockphoto.com/Eric Isselée; **A9** *parrot* ©iStockphoto.com/Eric Isselée; *frog* ©iStockphoto.com/BrandonAlms; *spider* ©iStockphoto.com/arlindo71; *cougar* ©iStockphoto.com/Eric Isselée; *lobster* ©iStockphoto.com/Joan Vincent Canto Roig; *crocodile* ©iStockphoto.com/Mehmet Salih Guler

Cartoon illustrations Tyler Stout
Cover image Lechner & Benson Design

Selected Answers

Section 1.1 Integers and Absolute Value
(pages 6 and 7)

1. 9, -1, 15

3. -6; All of the other expressions are equal to 6.

5. 6

7. 10

9. 13

11. 12

13. 8

15. 18

17. 45

19. 125

21. $|-4| < 7$

23. $|-4| > -6$

25. $|5| = |-5|$

27. Because $|-5| = 5$, the statement is incorrect. $|-5| > 4$

29. -8, 5

31. -7, -6, $|5|$, $|-6|$, 8

33. -17, $|-11|$, $|20|$, 21, $|-34|$

35. -4

37. **a.** MATE

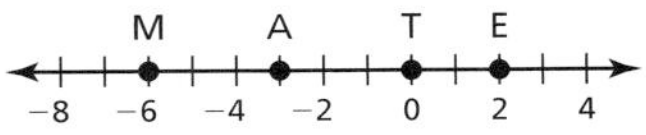

b. TEAM

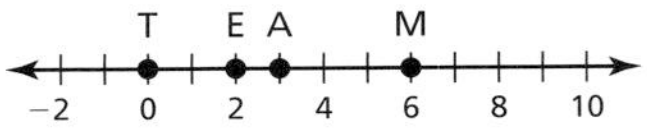

39. $n \geq 0$

41. The number closer to 0 is the greater integer.

43. **a.** Player 3 **b.** Player 2 **c.** Player 1

45. false; The absolute value of zero is zero, which is neither positive nor negative.

47. 144

49. 3170

Section 1.2 Adding Integers
(pages 12 and 13)

1. Change the sign of the integer.

3. positive; 20 has the greater absolute value and is positive.

5. negative; The common sign is a negative sign.

7. false; A positive integer and its absolute value are equal, not opposites.

9. -10

11. 7

13. 0

15. 10

17. -7

19. -11

21. -4

23. -34

25. -10 and -10 are not opposites. $-10 + (-10) = -20$

27. \$48

29. -27

31. 21

33. -85

35. Use the Associative Property to add 13 and -13 first. -8

37. *Sample answer:* Use the Commutative Property to switch the last two terms. -12

39. *Sample answer:* Use the Commutative Property to switch the last two terms. 11

41. $x = -8$

43. $y = 4$

45. $b = 2$

47. $6 + (-3) + 8$

49. Find the number in each row or column that already has two numbers in it before guessing.

51. 8

53. 183

Section 1.3 Subtracting Integers
(pages 18 and 19)

1. You add the integer's opposite.
3. What is 3 less than -2?; -5; 5
5. C
7. B
9. 13
11. -5
13. -10
15. 3
17. 17
19. 1
21. -22
23. -20
25. $-3 - 9$
27. 6
29. 9
31. 7
33. $m = 14$
35. $c = 15$
37. $k = -8$
39. -15 m

41. The range of a data set is the difference between the greatest value and the least value.
43. sometimes; It's positive only if the first integer is greater.
45. never; It's never positive because the first integer is never greater.
47. when a and b both have the same sign, or $a = 0$, or $b = 0$
49. -20
51. 40
53. 1476
55. C

Section 1.4 Multiplying Integers
(pages 26 and 27)

1. **a.** They are the same. **b.** They are different.
3. negative; different signs
5. negative; different signs
7. false; The product of the first two negative integers is positive. The product of the positive result and the third negative integer is negative.
9. -21
11. 12
13. 27
15. 12
17. 0
19. -30
21. 78
23. 121
25. $-240{,}000$
27. 54
29. -105
31. 0
33. -1
35. -36
37. 54
39. The answer should be negative. $-10^2 = -(10 \cdot 10) = -100$
41. -13
43. -7500, 37,500
45. -12
47. **a.**

Month	Price of Skates
June	165 $= \$165$
July	$165 + (-12) = \$153$
August	$165 + 2(-12) = \$141$
September	$165 + 3(-12) = \$129$

b. The price drops \$12 every month.
c. no; yes; In August, you have \$135 but the cost is \$141. In September, you have \$153 and the cost is only \$129.

49. 3
51. 14
53. D

Section 1.5 Dividing Integers
(pages 32 and 33)

1. They have the same sign. They have different signs. The dividend is zero.

3. *Sample answer:* $-4, 2$

5. negative

7. negative

9. -3

11. 3

13. 0

15. -6

17. 7

19. -10

21. undefined

23. 12

25. The quotient should be 0. $0 \div (-5) = 0$

27. 15 pages

29. -8

31. 21

33. 5

35. 4

37. -400 ft/min

39. 5

41. *Sample answer:* $-20, -15, -10, -5, 0$; Started with -10, then pair -15 with -5 and -20 with 0. The sum of the integers must be $5(-10) = -50$.

43.

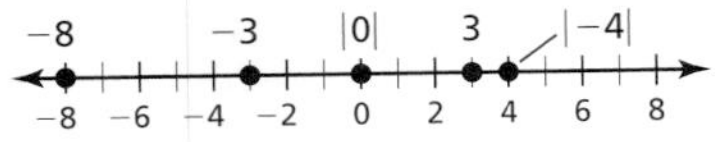

45. B

Section 2.1 Evaluating Algebraic Expressions
(pages 48 and 49)

1. Substitute 3 for x and 9 for y in the expression. Then use the order of operations to evaluate.

$$\begin{aligned} 12(3) - 5(9) &= 36 - 45 \\ &= -9 \end{aligned}$$

3. Evaluate $a + b$ when $a = 12$ and $b = -8$.; 4; -4

5. -35

7. -3

9. 25.2

11. -12

13. -28

15. 2

17. The substitution is incorrect. 6 was substituted instead of -6.

$$\begin{aligned} 12 - a + 4 &= 12 - (-6) + 4 \\ &= 12 + 6 + 4 \\ &= 22 \end{aligned}$$

19. \$10.25

21. -8

23. 53.2

25. 18.98

27. 500 yr

29. Comm. Prop. of Add.; 7

31. Assoc. Prop. of Add.; $1\frac{1}{8}$

33. A

Section 2.2 Writing Expressions and Equations
(pages 54 and 55)

1. When writing an expression, you translate a verbal phrase into a mathematical expression by using numbers, variables, and operations. An equation is a mathematical sentence that equates two expressions.

3. A number x equals 6 minus 10.; $x = 6 - 10$; $x - 6 = 10$

5. \$0.75

7. $-10 - c$

9. $3\left(\frac{y}{4}\right)$

11. $10 - p = 1.2$

13. $\frac{z}{7} = 5$

15. $-5b + 9 = 10$

17. The quotient is not written in the correct order; $\frac{w}{3} = 8$

19. $9\pi = 2\pi x$

21. Check that your equation is correct by using unit analysis to "divide out" the common units in a rate (\$ per hour) problem.

23. **a.** $s + 5.4 = 23$

b. $\frac{1}{4}d = 25$

c. The original bridge-tunnel, which is now the northbound lanes, took 42 months to construct. The current southbound lanes were added later and took 46 months to construct.

25. 17

27. -150

29. D

Section 2.3 Using Formulas to Solve Problems
(pages 62 and 63)

1. *Sample answer:* You substitute value(s) for the variable(s) to find the value of the formula.

3. 48 in.^{2}

5. 108 in.^{2}

7. 30 ft^2

9. **a.** 234 ft^2 **b.**

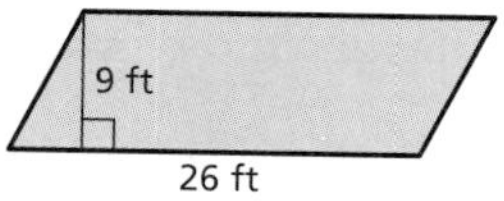

c. 26 ft; The base of the parking space is related to the length of the car.

11. **a.** 192 in.^{3} **b.** almost 13 bowls

13. $4x - 9$

15. $32x - 40 - x^2$

17. 24 karats; If you let $k = 24$, then $P = 100$.

19. Area of black = 252 in.^{2}
Area of yellow = 244 in.^{2}
Area of each blue stripe = 328 in.^{2}

21. $\frac{1}{2}$

23. 1

Section 2.4 Simplifying Algebraic Expressions
(pages 68 and 69)

1. Terms of an expression are separated by addition. The terms in the expression are $3y$, -4, and $-5y$.

3. no; The like terms $3x$ and $2x$ should be combined.

$$\begin{aligned} 3x + 2x - 4 &= (3 + 2)x - 4 \\ &= 5x - 4 \end{aligned}$$

5. Terms: t, 8, $3t$; Like terms: t and $3t$

7. Terms: $2n$, $-n$, -4, $7n$; Like terms: $2n$, $-n$, and $7n$

9. Terms: $1.4y$, 5, -4.2, $-5y^2$, z; Like terms: 5 and -4.2

11. $2x^2$ is not a like term, because x is squared. The like terms are $3x$ and $9x$.

13. $11x + 2$ **15.** $10.7v - 5$ **17.** $y + 5\frac{3}{4}$ **19.** $20p - 4n$ **21.** $10.2x$

23. $15x + 4$ **25.** $7u + t^2$ **27.** $\frac{3}{4}c^2 + c - 6$ **29.** $9 + 3x$

31. *Sample answer:*

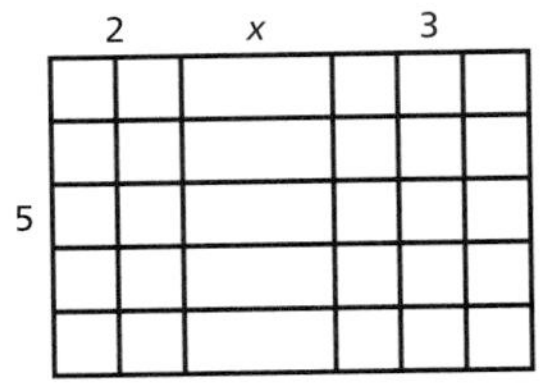

$5x + 25$

33. When you subtract the two red strips, you subtract their intersection twice. So, you need to add it back into the expression once.

35. 0.52 m, 0.545 m, 0.55 m, 0.6 m, 0.65 m

Section 3.1 Rational Numbers
(pages 84 and 85)

1. A number is rational if it can be written as $\frac{a}{b}$, where a and b are integers and $b \neq 0$.

3. rational numbers, integers

5. rational numbers, integers, whole numbers

7. repeating

9. terminating

11. 0.875 **13.** $-0.\overline{7}$ **15.** $1.8\overline{3}$ **17.** $-5.58\overline{3}$

19. The bar should be over both digits to the right of the decimal point. $-\frac{7}{11} = -0.\overline{63}$

21. $\frac{9}{20}$ **23.** $-\frac{39}{125}$ **25.** $-1\frac{16}{25}$ **27.** $-12\frac{81}{200}$

29. $-2.5, -1.1, -\frac{4}{5}, 0.8, \frac{9}{5}$

31. $-\frac{9}{4}, -0.75, -\frac{6}{10}, \frac{5}{3}, 2.1$

33. $-2.4, -2.25, -\frac{11}{5}, \frac{15}{10}, 1.6$

35. spotted turtle

37. $-1.82 < -1.81$

39. $-4\frac{6}{10} > -4.65$

41. $-2\frac{13}{16} < -2\frac{11}{14}$

43. Michelle

45. No; The base of the skating pool is at -10 feet, which is deeper than $-9\frac{5}{6}$ feet.

47. **a.** when a is negative

b. when a and b have the same sign, $a \neq 0 \neq b$

49. $\frac{7}{30}$

51. 21.15

Section 3.2 Adding and Subtracting Rational Numbers
(pages 90 and 91)

1. Because $|-8.46| > |5.31|$, subtract $|5.31|$ from $|-8.46|$ and the sign is negative.

3. What is 3.9 less than -4.8?; -8.7; -0.9

5. $-\frac{5}{14}$

7. $2\frac{3}{10}$

9. -0.9

11. 1.844

13. $1\frac{1}{2}$

15. $\frac{1}{18}$

17. $-18\frac{13}{24}$

19. -2.6

21. 14.963

23. $\frac{3}{8} - \frac{5}{6} = -\frac{11}{24}$

25. $\frac{1}{18}$

27. $-3\frac{9}{10}$

29. No, the cook needs $\frac{1}{12}$ cup more.

31–33. Subtract the least number from the greatest number.

35. $-\frac{n}{4}$

37. $-\frac{b}{24}$

39. 35.88

41. $8\frac{2}{3}$

43. C

Section 3.3 Multiplying and Dividing Rational Numbers
(pages 96 and 97)

1. The same rules for signs of integers are applied to rational numbers.

3. $-\frac{1}{3}$

5. $-\frac{3}{7}$

7. negative

9. positive

11. $-\frac{2}{3}$

13. $-\frac{1}{100}$

15. $2\frac{5}{14}$

17. $3.\overline{63}$

19. -6

21. -2.5875

23. $\frac{1}{3}$

25. $2\frac{1}{2}$

27. $-4\frac{17}{27}$

29. 0.025

31. 47.43

33. -0.064

Section 3.3 Multiplying and Dividing Rational Numbers *(continued) (pages 96 and 97)*

35. The wrong fraction was inverted.

$$-\frac{1}{4} \div \frac{3}{2} = -\frac{1}{4} \times \frac{2}{3}$$
$$= -\frac{2}{12}$$
$$= -\frac{1}{6}$$

37. 8 packages

39. 1.3

41. $-4\frac{14}{15}$

43. $-1\frac{11}{36}$

45. $191\frac{11}{12}$ yd

47. How many spaces are between the boards?

49. **a.** -0.02 in.

b. 0.03 in.; $\dfrac{-0.05 + 0.09 + (-0.04) + (-0.08) + 0.03}{5} = -0.01$

51. -5.4

53. $-8\frac{5}{18}$

Section 3.4 Solving Equations Using Addition or Subtraction *(pages 104 and 105)*

1. Subtraction Property of Equality

3. No, $m = -8$ not -2 in the first equation.

5. $a = 19$

7. $k = -20$

9. $c = 3.6$

11. $q = -\frac{1}{6}$

13. $g = -10$

15. $y = -2.08$

17. $q = -\frac{7}{18}$

19. $w = -1\frac{13}{24}$

21. The 8 should have been subtracted rather than added.

$$\begin{aligned} x + 8 &= 10 \\ \underline{-8} \quad & \underline{-8} \\ x &= 2 \end{aligned}$$

23. $c + 10 = 3$; $c = -7$

25. $p - 6 = -14$; $p = -8$

27. $p + 2.54 = 1.38$; $-\$1.16$ million

29. $x + 8 = 12$; 4 cm

31. $x + 22.7 = 34.6$; 11.9 ft

33. Because your first jump is higher, your second jump went a farther distance than your first jump.

35. $m + 30.3 + 40.8 = 180$; $108.9°$

37. -9

39. 6, -6

41. -56

43. -9

45. B

Section 3.5 Solving Equations Using Multiplication or Division *(pages 110 and 111)*

1. Multiplication is the inverse operation of division, so it can undo division.

3. dividing by 5

5. multiplying by -8

7. $h = 5$

9. $n = -14$

11. $m = -2$

13. $x = -8$

15. $p = -8$

17. $n = 8$

19. $g = -16$

21. $f = 6\frac{3}{4}$

23. They should divide by -4.2.

$$-4.2x = 21$$
$$\frac{-4.2x}{-4.2} = \frac{21}{-4.2}$$
$$x = -5$$

25. $\frac{2}{5}x = \frac{3}{20}$; $x = \frac{3}{8}$

27. $\frac{x}{-1.5} = 21$; $x = -31.5$

29. $\frac{x}{30} = 12\frac{3}{5}$; 378 ft

31 and 33. Sample answers are given.

31. **a.** $-2x = 4.4$ **b.** $\frac{x}{1.1} = -2$

33. **a.** $4x = -5$ **b.** $\frac{x}{5} = -\frac{1}{4}$

35. $-1.26n = -10.08$; 8 days

37. -50 ft

39. $-5, 5$

41. -7

43. 12

45. B

Section 3.6 Solving Two-Step Equations *(pages 116 and 117)*

1. Eliminate the constants on the side with the variable. Then solve for the variable using either division or multiplication.

3. D

5. A

7. $b = -3$

9. $t = -4$

11. $g = 4.22$

13. $p = 3\frac{1}{2}$

15. $h = -3.5$

17. $y = -6.4$

19. Each side should be divided by -3, not 3.

$$-3x + 2 = -7$$
$$-3x = -9$$
$$\frac{-3x}{-3} = \frac{-9}{-3}$$
$$x = 3$$

21. $a = 1\frac{1}{3}$

23. $b = 13\frac{1}{2}$

25. $v = -\frac{1}{30}$

27. $2.5 + 2.25x = 9.25$; 3 games

Section 3.6 Solving Two-Step Equations *(continued)* *(pages 116 and 117)*

29. $v = -5$

31. $d = -12$

33. $m = -9$

35. *Sample answer:* You travel halfway up a ladder. Then you climb down two feet and are 8 feet above the ground. How long is the ladder? $x = 20$

37. the initial fee

39. Find the number of insects remaining and then find the number of insects you caught.

41. decrease the length by 10 cm; $2(25 + x) + 2(12) = 54$

43. $-6\frac{2}{3}$

45. -6.2

Section 3.7 Solving Inequalities Using Addition or Subtraction *(pages 122 and 123)*

1. no; The solution of $r - 5 \le 8$ is $r \le 13$ and the solution of $8 \le r - 5$ is $r \ge 13$.

3. $x < 9$;
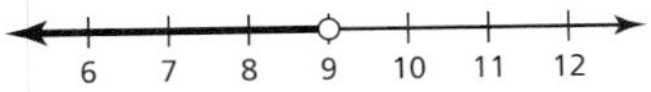

5. $5 \ge y$;

7. $t > 4$;
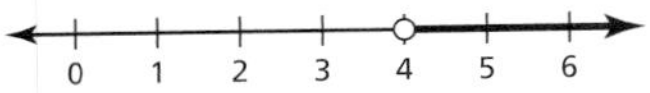

9. $a > -8$;
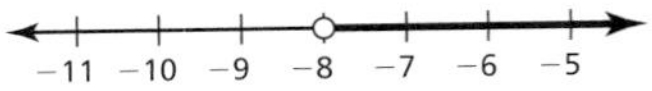

11. $-\frac{3}{5} > d$;
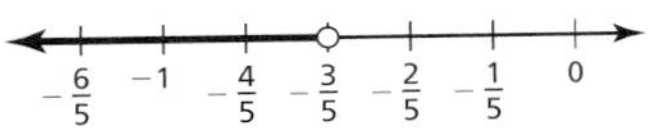

13. $m \le 1$;
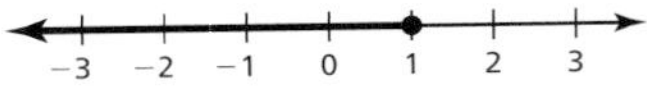

15. $h < -1.5$;
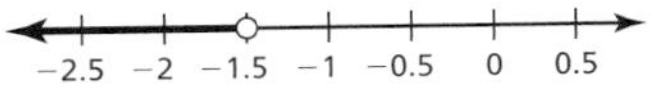

17. $9.5 \ge u$;
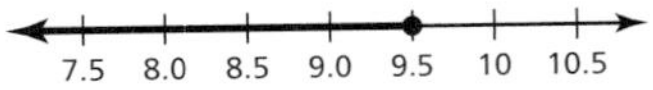

19. **a.** $100 + V \le 700$; $V \le 600$ in.^{3} **b.** $V \le \frac{700}{3}$ in.^{3}

21. $x + 2 > 10$; $x > 8$ m

23. 5

25. **a.** $4500 + x \ge 12{,}000$; $x \ge 7500$ points

b. This changes the number added to x by 60%, so the inequality becomes $7200 + x \ge 12{,}000$. So, you need less points to advance to the next level.

27. 25

29. 10

31. 12

Section 3.8 Solving Inequalities Using Multiplication or Division *(pages 129–131)*

1. Multiply each side of the inequality by 6.

3. *Sample answer:* $\frac{x}{2} \ge 4$, $2x \ge 16$

5. $x \ge -1$

7. $x \le -3$

9. $x \le \frac{3}{2}$

11. $c \le -36$; 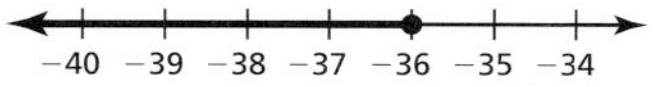

13. $x < -28$; 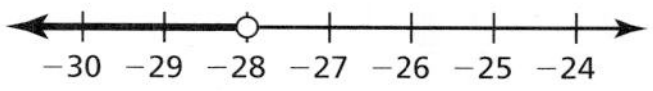

15. $k > 2$; 

17. $y \le -4$; 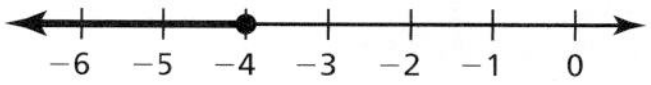

19. The inequality sign should not have been reversed.

$$\frac{x}{2} < -5$$

$$2 \cdot \frac{x}{2} < 2 \cdot (-5)$$

$$x < -10$$

21. $\frac{x}{8} < -2$; $x < -16$

23. $5x > 20$; $x > 4$

25. $0.25x \le 3.65$; $x \le 14.6$; You can make at most 14 copies.

27. $n \ge -5$; 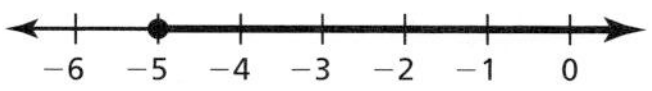

29. $h \le -42$; 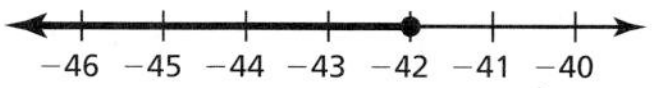

31. $y > \frac{11}{2}$; 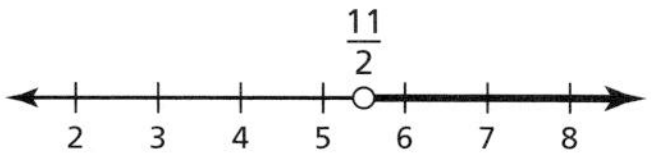

33. $m > -12$; 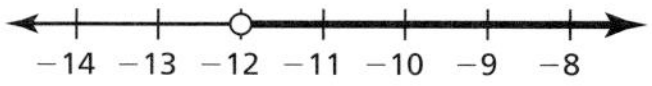

35. $b > 4$; 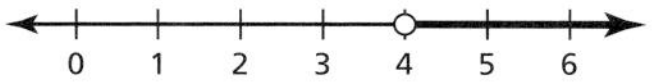

37. no; You need to solve the inequality for x. The solution is $x < 0$. Therefore, numbers greater than 0 are not solutions.

39. $12x \ge 102$; $x \ge 8.5$ cm

41. $\frac{x}{4} < 80$; $x < \$320$

43. *Answer should include, but is not limited to:* The solution should use the correct number of months that the CD has been out. In part (d), an acceptable answer could be never, because the top selling CD could have a higher monthly average.

45. $n \ge -6$ and $n \le -4$;

47. $m < 20$;

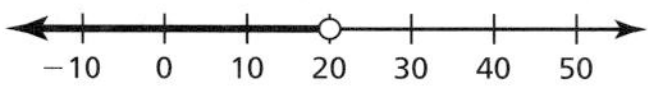

49. $8\frac{1}{4}$

51. 84

Section 4.1 Mapping Diagrams
(pages 148 and 149)

1. the first number; the second number

3. As each input increases by 1, the output increases by 4.

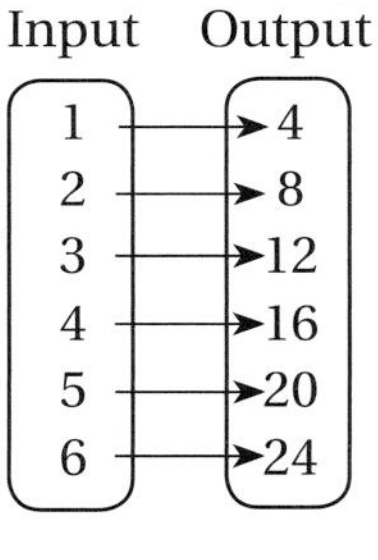

5. As each input increases by 1, the output increases by 5.

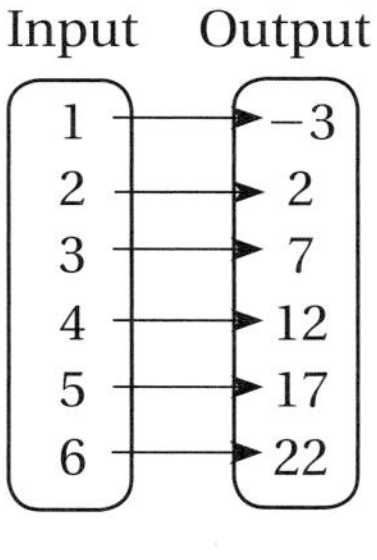

Section 4.1 Mapping Diagrams *(continued)*

(pages 148 and 149)

7. (1, 8), (3, 4), (5, 6), (7, 2)

9. Input Output

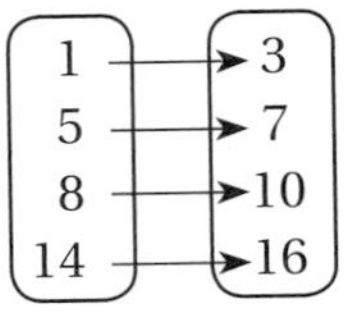

11. Input Output

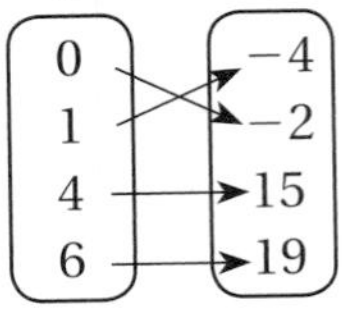

13. The first number of each ordered pair should be an input and the second number should be the output that corresponds to the input.

Input Output

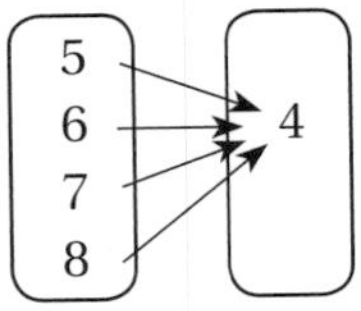

15. Input Output

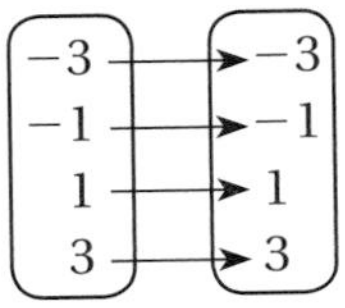

As each input increases by 2, the output increases by 2.

17. Input Output

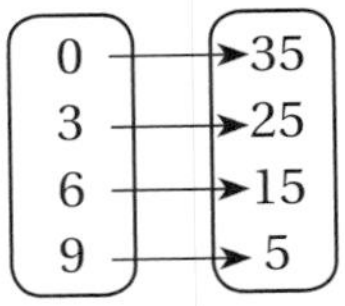

As each input increases by 3, the output decreases by 10.

19. a. Input Output

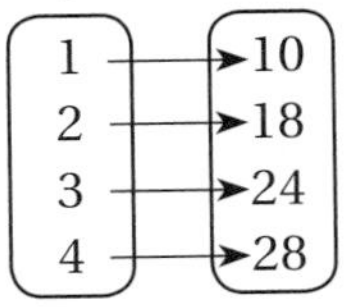

b. The pattern is that for each input increase of 1, the output increases by $2 less than the previous increase. For each additional movie you buy, your cost per movie decreases by $1.

21. $x + 7 = 15; x = 8$

23. C

Section 4.2 Functions as Words and Equations

(pages 154 and 155)

1. input variable: x; output variable: y

3. $y = 4x$

5. $y = x - 5$

7. $y = 6x$

9. $y = x + 11$

11. -35

13. 3.5

15. 3

17. no

19. no

21. yes

23. a. $d = 18s$ **b.** 540 ft

25. -4

27. The profit is equal to the revenue minus the expenses.

29. no; Many rectangles have the same perimeter but different areas.

31.

x	1	2	3
x + 7	8	9	10

33. C

Section 4.3 Input-Output Tables

(pages 160 and 161)

1. Choose the inputs that represent the situation or show the pattern of the function. Pair each input in the table with its resulting output.

3.

Input, *x*	1	2	3	4
Output, *y*	6	7	8	9

5. $y = x + 3$

Input, *x*	−6	−3	0	3
Output, *y*	−3	0	3	6

7. $y = x + 8$

9. $y = \frac{x}{3}$

11. Each output in the table is one-fourth of the input, but the equation would make each output four times each input; $y = \frac{x}{4}$

13.

Input, *x*	−8	−6	−4	−2	8	18
Output, *y*	0	1	2	3	8	13

15. *Sample answer:*

GMT, *x*	6:00	7:00	8:00	9:00	10:00
Eastern Standard Time, *y*	1:00	2:00	3:00	4:00	5:00

$y = x - 5$

17 and 19.

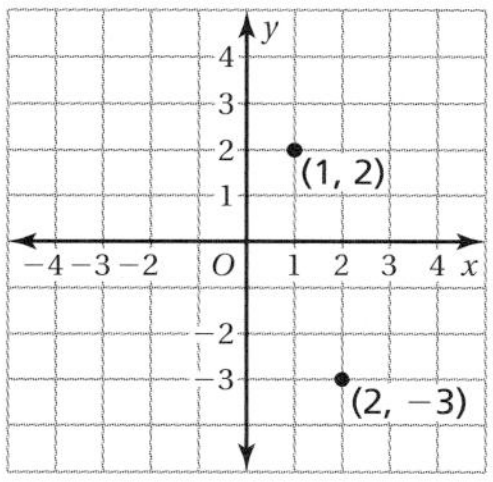

21. B

Section 4.4 Graphs

(pages 168 and 169)

1. Make an input-output table. Plot the ordered pairs. Draw a line through the points.

3. Find points on the graph. Make a mapping diagram or input-output table to show the pattern. Use the pattern to write a function rule.

5.

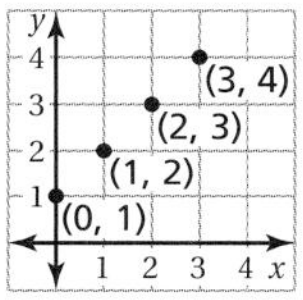

7.

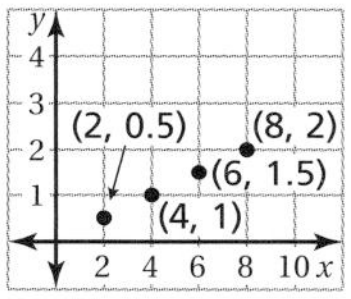

9.

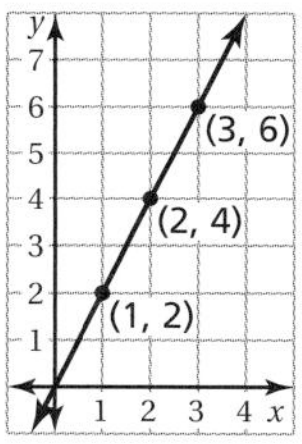

11.

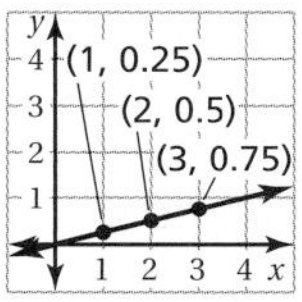

Section 4.4 Graphs *(continued)*

(pages 168 and 169)

13.

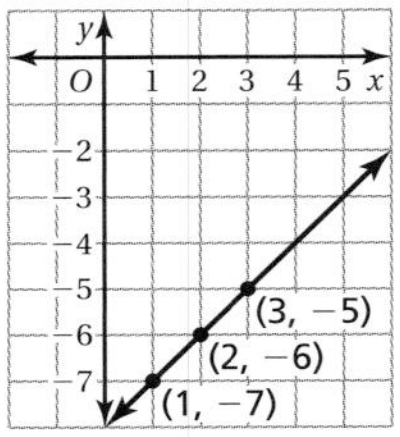

15.

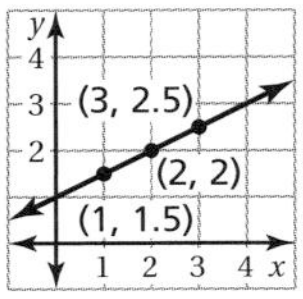

17.

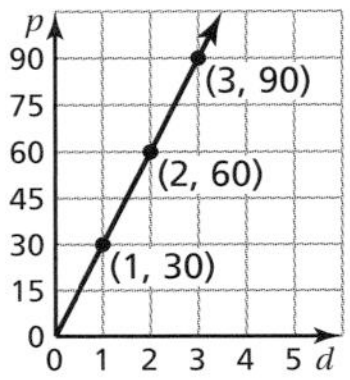

19. C

21. Part (c) asks for the sale price, not the discount.

23. 17

25. −2

27. 19

29. C

Section 4.5 Analyzing Graphs

(pages 174 and 175)

1. A function is called a linear function if its graph is a line.

3.

Radius, *r*	1	2	3	4
Diameter, *d*	2	4	6	8

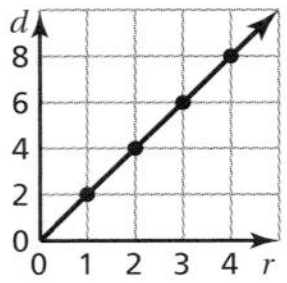

$d = 2r$

5. yes; The graph is a line.

7. no; The graph is *not* a line.

9. no; The graph is *not* a line.

11. yes; The graph is a line.

13. no; The graph is *not* a line.

15.

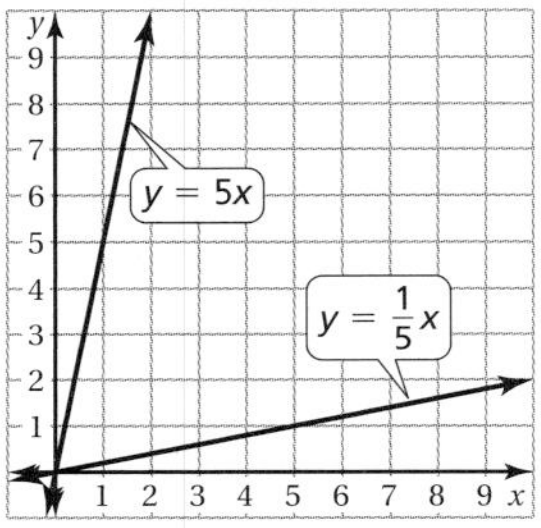

$y = 5x$; 5 is greater than $\frac{1}{5}$.

17.

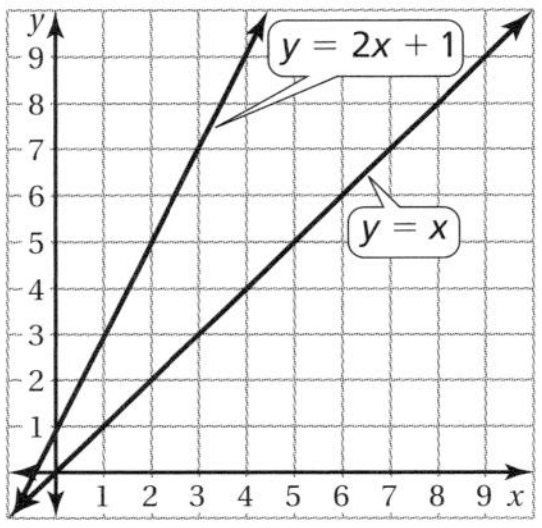

$y = 2x + 1$; 2 is greater than 1.

19.

Figure, *x*	**Area, *y***
1	1
2	2
3	4
4	8

no; The graph is *not* a line.

21. Airport 2; about 100 flights; *Sample answer:* From the graph, Airport 2 has about 350 flights each day and Airport 1 has about 250 flights each day.

23. $\frac{11}{25}$

25. 0.802

27. C

Section 4.6 Arithmetic Sequences

(pages 180 and 181)

1. Find the difference between a term and the previous term.

3.

Time (min)	1	2	3	4
Distance (in.)	10	20	30	40

5.

Year	0	1	2	3	4
Spoons	75	80	85	90	95

7. −60, −90, −120

9. $4\frac{2}{3}, 5\frac{1}{3}, 6$

11. 5.1, 6.9, 8.7

13. $y = n - 6$

15. $y = \frac{1}{2}n$

17. $y = -10n$

19. The difference between a term and its previous term is -1, not 1. So, the common difference is -1.

21. 45, 60, 75

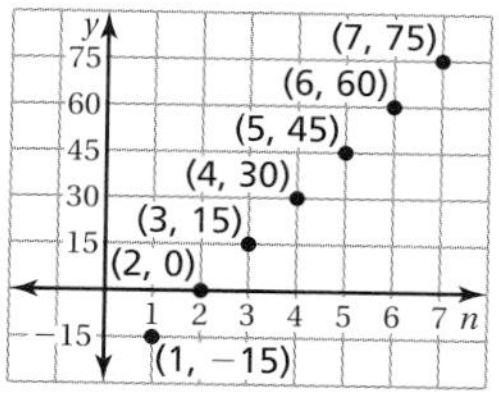

23. $-9, -11, -13$

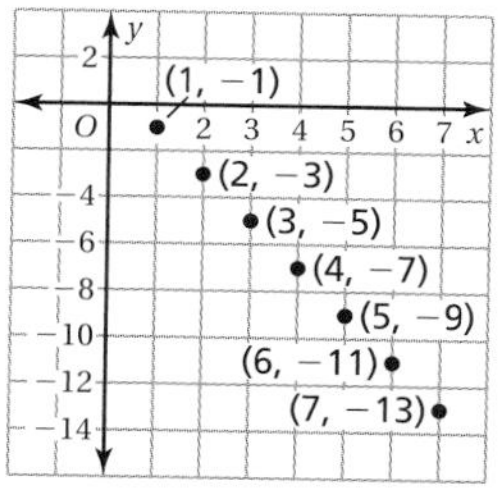

25. Make a table with position $n = 1$ corresponding to speed $s = 62$.

27. **a.** $y = 5n$

b. Substitute the number of minutes in a day, 1440, for n.

c. Substitute the number of minutes in a year, 525,600, for n.

29. (1, 8), (2, 4), (3, 0), (4, -4)

Section 4.7 Geometric Sequences
(pages 186 and 187)

Selected Answers

1. The ratio of a term to the previous term is the same for every pair of consecutive terms in a geometric sequence.

3. 1, -3, 5, -7, . . . does not belong, because it is not a geometric sequence.

5.

t	Principal and Interest	Annual Interest	Balance at End of Year
1	\$100.00	\$7.00	\$107.00
2	\$107.00	\$7.49	\$114.49
3	\$114.49	\$8.01	\$122.50
4	\$122.50	\$8.58	\$131.08

7. $-112, 224, -448$

9. $-\frac{3}{5}, -\frac{3}{25}, -\frac{3}{125}$

11. 40.5, 60.75, 91.125

13. arithmetic

15. neither

17. geometric

19. geometric

21. arithmetic

23. $\frac{1}{36}, \frac{1}{216}, \frac{1}{1296}$

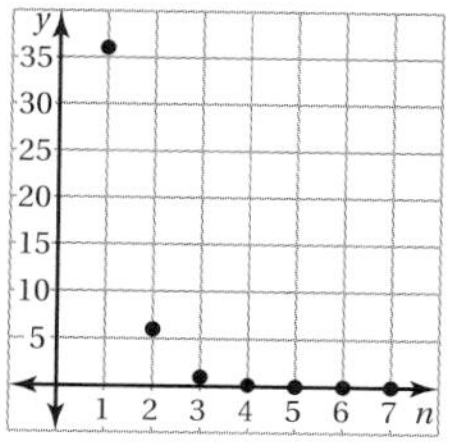

25. $-\frac{1}{8}, \frac{1}{16}, -\frac{1}{32}$

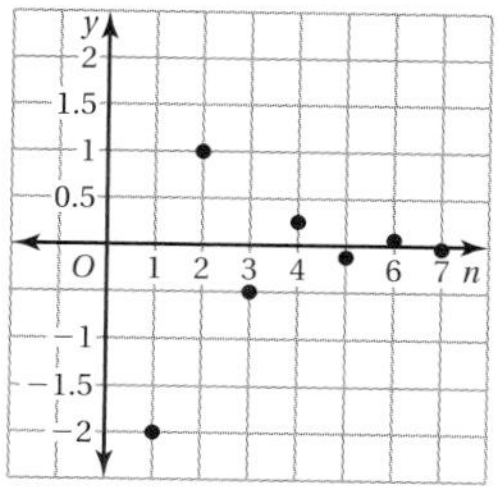

27. **a.** 12, 24, 48

b. 9, 12, 15

29. For part (b), rewrite the terms of the sequence using exponents to show the repeated multiplication.

31. $\frac{3}{4}$

33. $\frac{3}{4}$

Section 5.1 Ratios and Rates
(pages 202 and 203)

1. It has a denominator of 1.

3. *Sample answer:* A basketball player runs 10 feet down the court in 2 seconds.

5. \$0.10 per fl oz

7. \$72

9. 840 MB

11. $\frac{5}{9}$

13. $\frac{7}{3}$

15. $\frac{4}{3}$

17. 60 mi/h

19. \$2.40 per lb

21. 54 words per min

23. 90 calories per serving

25. 4.5 servings per package

27. 4.8 MB per min

29. **a.** It costs \$122 for 4 tickets.
    b. \$30.50 per ticket
    c. \$305

31. The 9-pack is the best buy at \$2.55 per container.

33. Try searching for "fire hydrant colors."

35 and 37.

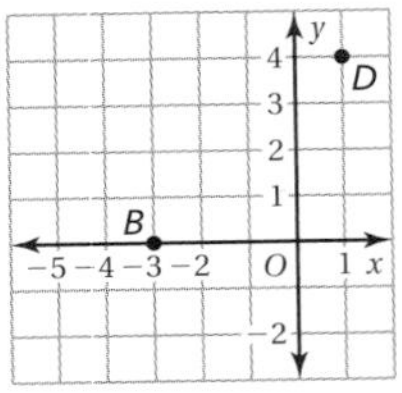

Section 5.2 Slope
(pages 208 and 209)

1. yes; Slope is the rate of change of a line.

3. 5; A ramp with a slope of 5 increases 5 units vertically for every 1 unit horizontally. A ramp with a slope of $\frac{1}{5}$ increases 1 unit vertically for every 5 units horizontally.

5. $\frac{3}{2}$

7. 1

9. $\frac{4}{5}$

11.

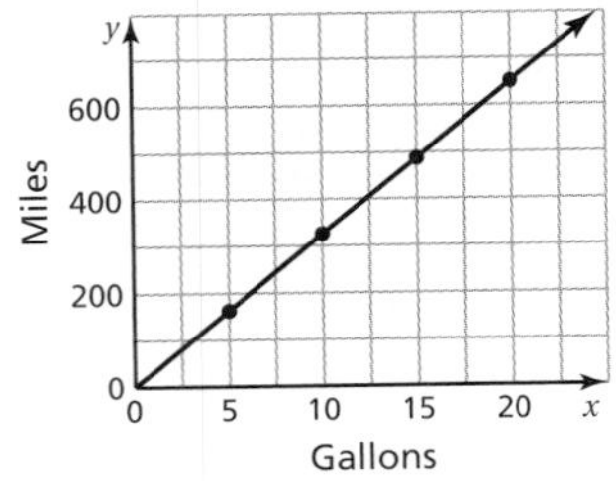

slope = 32.5

13.

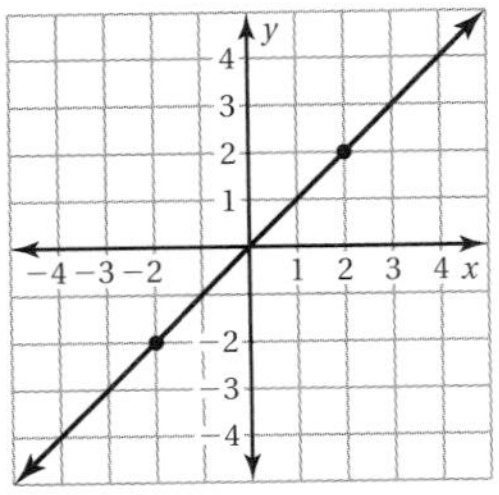

slope = 1

15. The change in y should be in the numerator. The change in x should be in the denominator.

Slope $= \frac{5}{4}$

17. **a.**

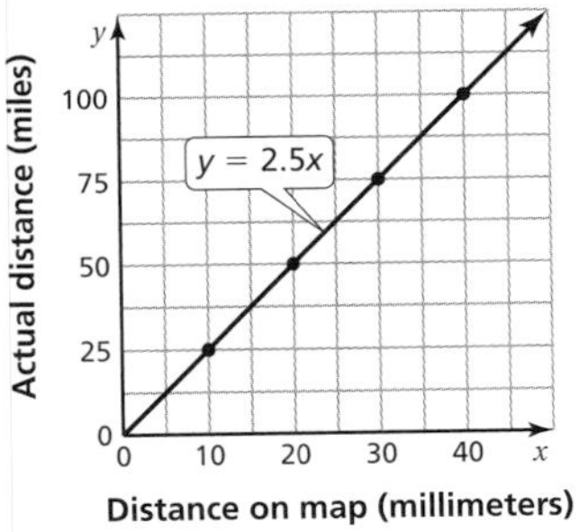

b. 2.5; Every millimeter represents 2.5 miles.

c. 120 mi

d. 90 mm

19. $y = 6$

21. $<$

23. $-\frac{4}{5}$

25. 3

Section 5.3 Proportions
(pages 216 and 217)

1. Both ratios are equal.
3. *Sample answer:* $\frac{6}{10}, \frac{12}{20}$
5. yes
7. no
9. yes
11. no
13. yes
15. yes
17. yes
19. no
21. yes
23. yes; Both can do 45 sit-ups per minute.
25. yes
27. yes
29. yes; They are both $\frac{4}{5}$.
31. **a.** Pitcher 3 **b.** Pitcher 2 and Pitcher 4
33. **a.** no

    b. *Sample answer:* If the collection has 50 quarters and 30 dimes, when 10 of each coin are added, the new ratio of quarters to dimes is 3 : 2.
35. -13
37. -18
39. D

Section 5.4 Writing Proportions
(pages 222 and 223)

1. You can use the columns or the rows of the table to write a proportion.
3. *Sample answer:* $\frac{x}{12} = \frac{5}{6}$; $x = 10$
5. $\frac{x}{50} = \frac{78}{100}$
7. $\frac{x}{150} = \frac{96}{100}$
9. $\frac{n \text{ winners}}{85 \text{ entries}} = \frac{34 \text{ winners}}{170 \text{ entries}}$
11. $\frac{100 \text{ meters}}{x \text{ seconds}} = \frac{200 \text{ meters}}{22.4 \text{ seconds}}$
13. $\frac{\$24}{3 \text{ shirts}} = \frac{c}{7 \text{ shirts}}$
15. $\frac{5 \text{ 6th grade swimmers}}{16 \text{ swimmers}} = \frac{s \text{ 6th grade swimmers}}{80 \text{ swimmers}}$
17. $y = 16$
19. $c = 24$
21. $g = 14$
23. $\frac{1}{200} = \frac{19.5}{x}$; Dimensions for the model are in the numerators and the corresponding dimensions for the actual space shuttle are in the denominators.
25. Draw a diagram of the given information.
27. $x = 9$
29. $x = 140$

Section 5.5 Solving Proportions
(pages 228 and 229)

1. mental math; Multiplication Property of Equality; Cross Products Property

3. yes; Both cross products give the equation $3x = 60$.

5. $h = 80$ 7. $n = 15$ 9. $y = 7\frac{1}{3}$ 11. $k = 5.6$

13. $n = 10$ 15. $d = 5.76$ 17. $m = 20$ 19. $d = 15$

21. $k = 5.4$ 23. 108 pens 25. $x = 1.5$ 27. $k = 4$

29. $\frac{2.5}{x} = \frac{1}{0.26}$; about 0.65

31. true; Both cross products give the equation $3a = 2b$.

33. 15.5 lb

35. **a.** about 14 min 34 sec

b. *Sample answer:* 5.84 sec

37. 4 bags 39. $x = 80$ 41. $n = -120$

Section 6.1 Quadrilaterals
(pages 244 and 245)

1. Match the characteristics of the quadrilateral to the most specific definition of a quadrilateral.

3. rhombus 5. trapezoid 7. kite 9. 130° 11. 90°, 90° 13. 75°

15. **a.** parallelogram **b.** 45°, 45°, 135°

17. A concave quadrilateral will have one interior angle that measures greater than 180°.

19. yes 21. no

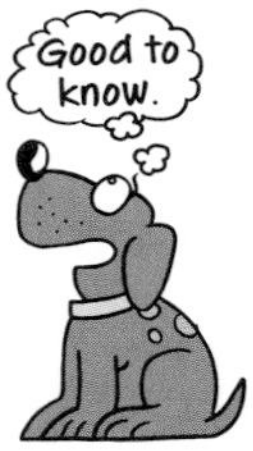

Section 6.2 Identifying Similar Figures
(pages 250 and 251)

1. They are congruent.

3. *Sample answer:* A photograph of size 3" × 5" and another photograph of size 6" × 10"

5. $\angle A$ and $\angle W$, $\angle B$ and $\angle X$, $\angle C$ and $\angle Y$, $\angle D$ and $\angle Z$;
Side AB and Side WX, Side BC and Side XY, Side CD and Side YZ, Side AD and Side WZ

7. 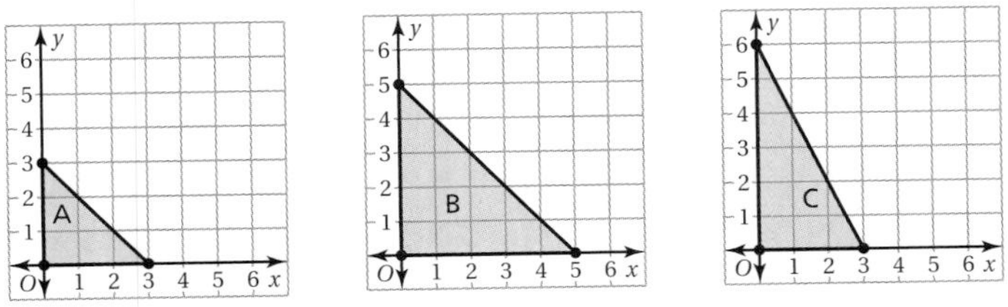

A and B; Corresponding side lengths are proportional and corresponding angles are congruent.

9. similar; Corresponding angles are congruent. Because $\frac{4}{6} = \frac{6}{9} = \frac{8}{12}$, the corresponding side lengths are proportional.

11. no

13. 48°

15. 42°

17. Simplify the ratios of length to width for each photo to see if any of the photos are similar.

19. yes; One could be a trapezoid and the other could be a parallelogram.

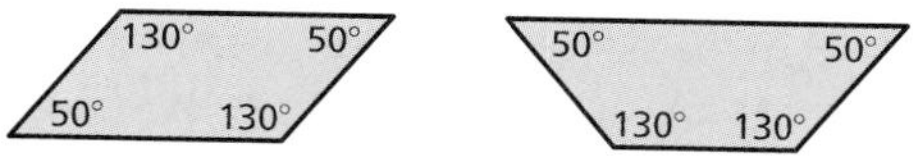

21. a. yes

b. yes; It represents the fact that the sides are proportional because you can split the isosceles triangles into smaller right triangles that will be similar.

23. $\frac{16}{81}$

25. $\frac{49}{16}$

27. B

Section 6.3 Perimeters and Areas of Similar Figures

(pages 256 and 257)

1. The ratio of the perimeters is equal to the ratio of the corresponding side lengths.

3. Because the ratio of the corresponding side lengths is $\frac{1}{2}$, the ratio of the areas is equal to $\left(\frac{1}{2}\right)^2$. To find the area, solve the proportion $\frac{30}{x} = \frac{1}{4}$ to get $x = 120$ square inches.

5. $\frac{5}{8}; \frac{25}{64}$

7. $\frac{14}{9}; \frac{196}{81}$

9. perimeter triples

11. Area is 16 times larger.

13. 45 in.

15. false; $\frac{\text{Area of } \triangle ABC}{\text{Area of } \triangle DEF} = \left(\frac{AB}{DE}\right)^2$

17. a. 400 times greater; The ratio of the corresponding lengths is $\frac{120 \text{ in.}}{6 \text{ in.}} = \frac{20}{1}$. So, the ratio of the areas is $\left(\frac{20}{1}\right)^2 = \frac{400}{1}$.

b. 180,000 in.^{2}

c. 1250 ft^2

19. $\frac{3}{4}$

21. 512, 2048, 8192

23. $\frac{11}{3}, -\frac{11}{9}, \frac{11}{27}$

Section 6.4 Finding Unknown Measures in Similar Figures

(pages 264 and 265)

1. You can set up a proportion and solve for the unknown measure.

3. 15

5. 14.4

7. 8.4

9. 35 ft

11. 108 yd

13. 3 times

15. 12.5 bottles

17. 31.75

19. 3.88

21. 41.63

Section 6.5 Scale Drawings
(pages 270 and 271)

1. A scale is the ratio that compares the measurements of the drawing or model with the actual measurements. A scale factor is a scale without any units.

3. Convert one of the lengths into the same units as the other length. Then, form the scale and simplify.

5. 10 ft by 10 ft

7. 112.5%

9. 50 mi

11. 110 mi

13. 15 in.

15. 21.6 yd

17. The 5 cm should be in the numerator.

$$\frac{1 \text{ cm}}{20 \text{ m}} = \frac{5 \text{ cm}}{x \text{ m}}$$

$$x = 100 \text{ m}$$

19. 2.4 cm; 1 cm : 10 mm

21. **a.** *Answer should include, but is not limited to:* Make sure words and picture match the product.
    b. Answers will vary.

23. Find the size of the object that would represent the model of the Sun.

25 and 27.

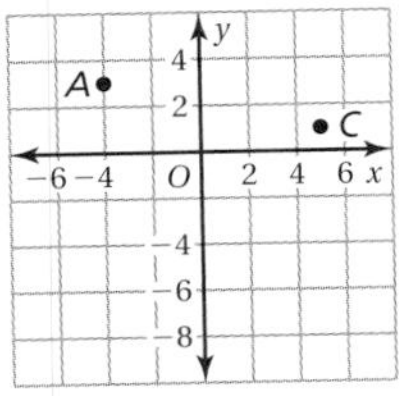

29. A

Section 7.1 Translations
(pages 286 and 287)

1. A

3. yes; Translate the letters T and O to the end.

5. no

7. yes

9. no

11. $A'(-3, 0)$, $B'(1, 0)$, $C'(1, -5)$, $D'(-3, -5)$

13.

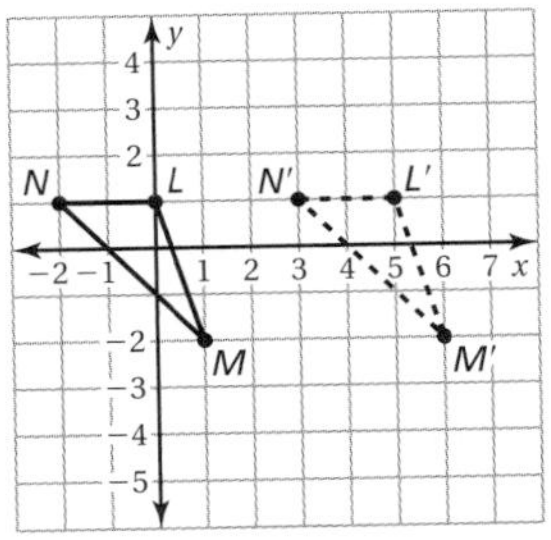

15.

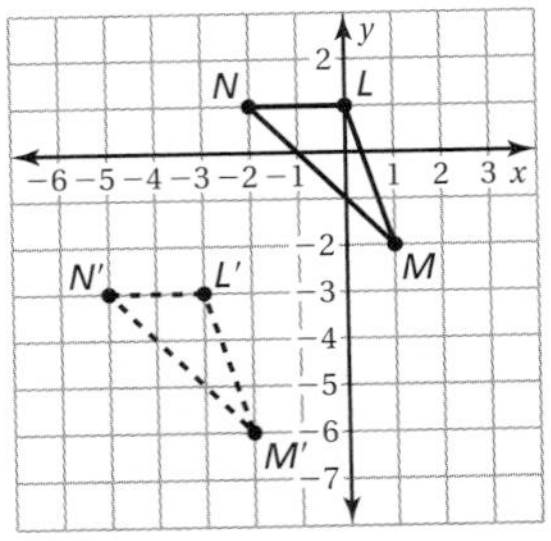

17. 2 units left and 2 units up

19. 6 units right and 3 units down

21. **a.** 5 units right and 1 unit up
    b. no; It would hit the island.
    c. 4 units right and 4 units up

23. If you are doing more than 10 moves and have not moved the knight to g5, you might want to start over.

25. no

27. yes

Section 7.2 Reflections
(pages 292 and 293)

1. The third one because it is not a reflection.

3. Quadrant IV

5. yes

7. no

9. no

11. $M'(-2, -1)$, $N'(0, -3)$, $P'(2, -2)$

13. $D'(-2, 1)$, $E'(0, 1)$, $F'(-2, 5)$, $G'(0, 5)$

15. $T'(-4, -2)$, $U'(-4, 2)$, $V'(-6, -2)$

17. $J'(-2, 2)$, $K'(-7, 4)$, $L'(-9, -2)$, $M'(-3, -1)$

19. x-axis

21. y-axis

23.

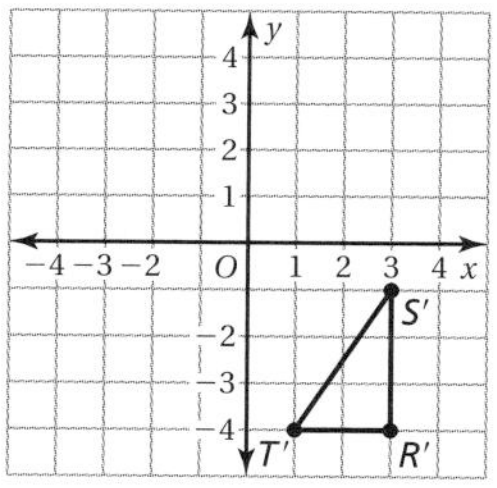

25. the first one; The left side of the face is a mirror image of the right side.

27.

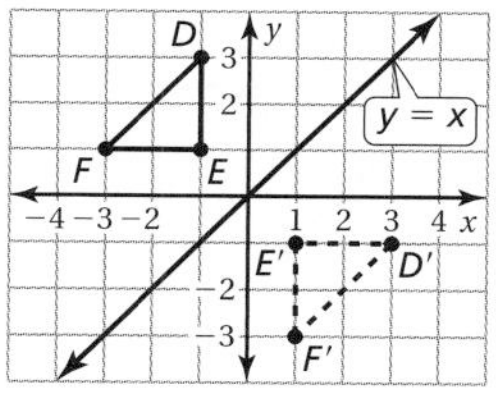

The x-coordinate and y-coordinate for each point are switched in the image.

29. straight

31. acute

Section 7.3 Rotations
(pages 300 and 301)

1. **a.** reflection **b.** rotation **c.** translation

3. Quadrant I

5. Quadrant III

7. no

9. yes; 180° clockwise or counterclockwise

11. It only needs to rotate 90° to produce an identical image.

13. $A'(-1, -4)$, $B'(-4, -3)$, $C'(-4, -1)$, $D'(-1, -2)$

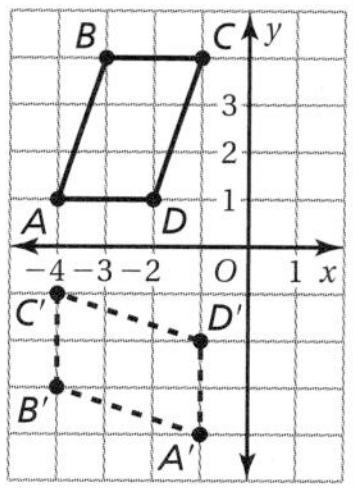

15. $A'(4, -1)$, $B'(3, -4)$, $C'(1, -4)$, $D'(2, -1)$

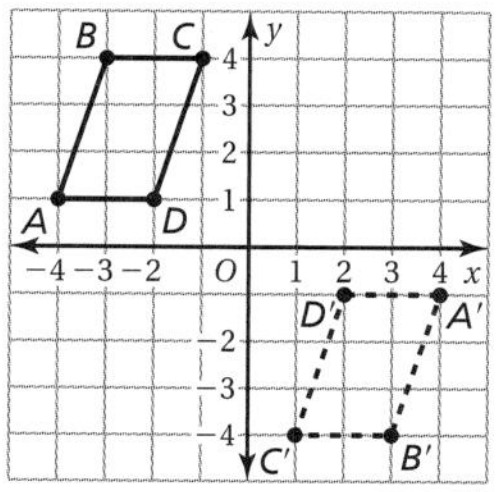

17. because both ways will produce the same image

19. Use *Guess, Check, and Revise* to solve this problem.

21. triangular prism

23. C

Section 7.4 Dilations
(pages 306 and 307)

1. A dilation changes the size of a figure. The image is similar, not congruent, to the original figure.

3. The middle red figure is not a dilation of the blue figure because it is a translation of the blue figure. The left red figure is a reduction of the blue figure and the right red figure is an enlargement of the blue figure.

5. 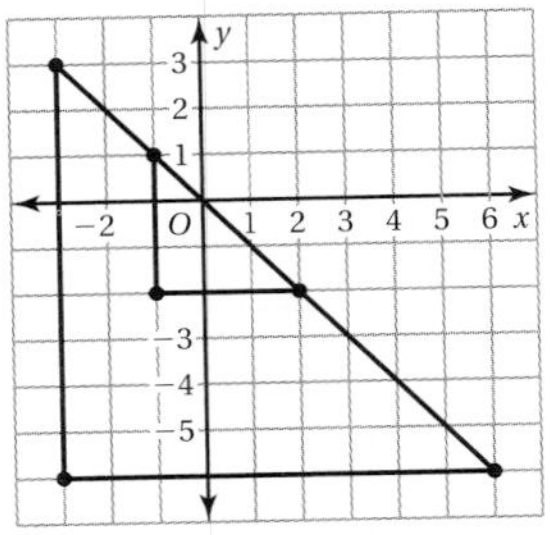

The triangles are similar.

7. yes

9. no

11. yes

13. 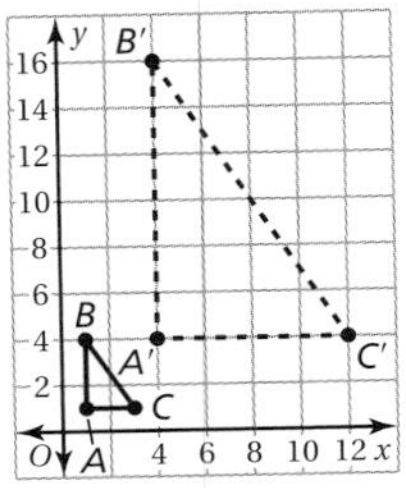

enlargement

15. 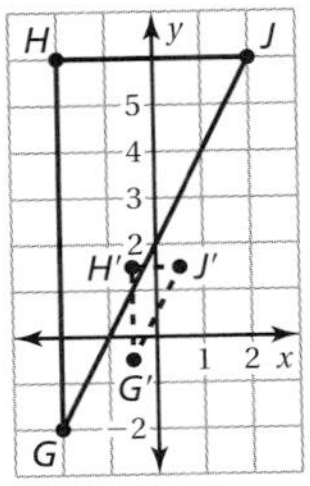

reduction

17. 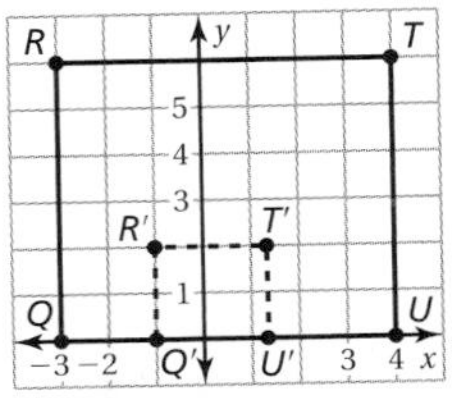

reduction

19. Each coordinate was multiplied by 2 instead of divided by 2. The coordinates should be $A'(1, 2.5)$, $B'(1, 0)$, and $C'(2, 0)$.

21. reduction; $\frac{1}{4}$

23. Use the ratio $\frac{\text{Area of dilated rectangle}}{\text{Area of original triangle}}$ to compare the areas.

25. $\frac{3}{2}$; Multiply the two scale factors.

27. 4

29. -1

Section 8.1 Surface Areas of Rectangular Prisms

(pages 322 and 323)

1. a three-dimensional figure with six rectangular sides

3.

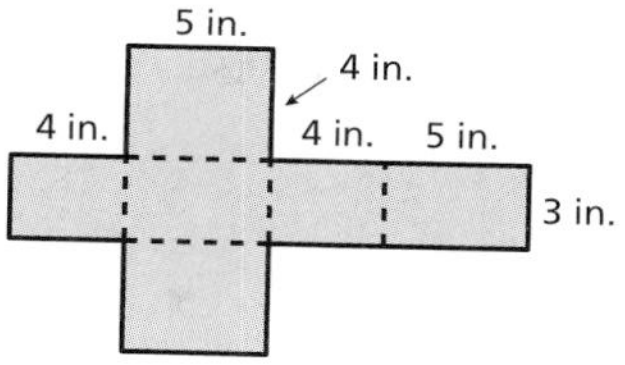

5. 6 units^2; 54 units^2; 9 times greater

7. 58 m^2

9. 141.8 mm^2

11. 49.2 yd^2

13. 9 times greater

15. **a.** 0.125 pint

b. 1.125 pints

17. **a.** 64 **b.** $24s^2$

c. 4 times greater

19. 3:7; 9:49

21. B

Section 8.2 Volumes of Rectangular Prisms

(pages 328 and 329)

1. cubic

3. 2 cubic units; 16 cubic units; 8 times greater

5. 224 m^3

7. 343 cm^3

9. 15 cm^3

11. Volume is measured in cubic units not square units; 21 in.^{3}

13. shorter dispenser; It has the larger volume.

15. 3 times greater

17. How many times greater is the new sandbox than the old sandbox?

19. **a.** The volume doubles; The surface area increases but does not double.

b. The volume triples; The surface area increases but does not triple.

c. The volume is 8 times greater; The surface area is 4 times greater.

d. The volume is 27 times greater; The surface area is 9 times greater.

21. translation

23. B

Section 8.3 Surface Areas of Cylinders

(pages 336 and 337)

1. $2\pi rh$

3. $36\pi \approx 113.0$ cm^2

5.

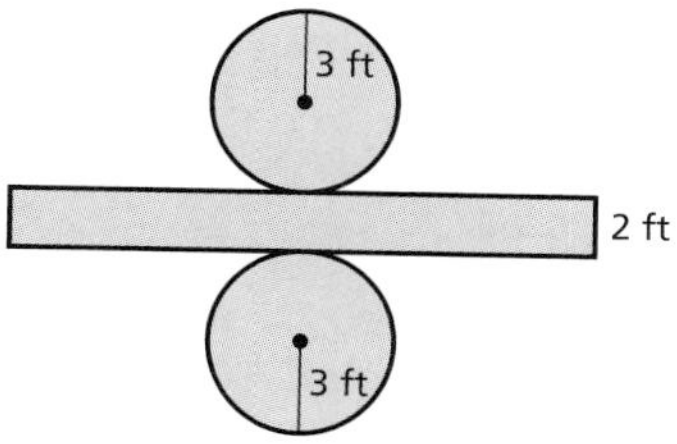

$30\pi \approx 94.2$ ft^2

7.

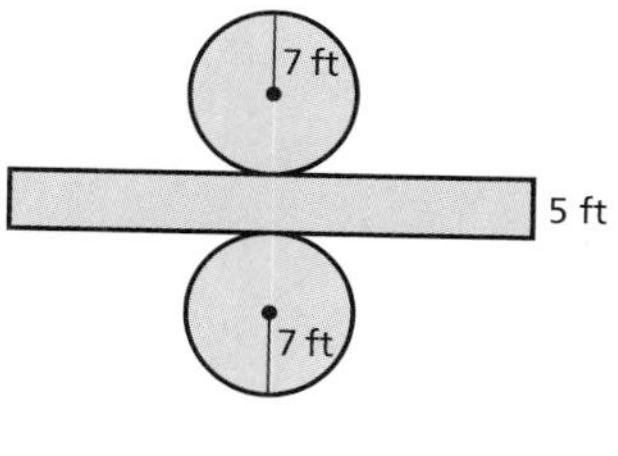

$168\pi \approx 527.5$ ft^2

9. $156\pi \approx 489.8$ ft^2

11. $120\pi \approx 376.8$ ft^2

13. $28\pi \approx 87.9$ m^2

Section 8.3 Surface Areas of Cylinders *(continued)*

(pages 336 and 337)

15. The error is that only the lateral surface area is found. The areas of the bases should be added;

$S = 2\pi r^2 + 2\pi rh$
$= 2\pi(6)^2 + 2\pi(6)(11)$
$= 72\pi + 132\pi$
$= 204\pi \text{ ft}^2$

17. The surface area of the cylinder with the height of 8.5 inches is greater than the surface area of the cylinder with the height of 11 inches.

19. After removing the wedge, is there any new surface area added?

21. 117

23. 56.52

Section 8.4 Volumes of Cylinders

(pages 342 and 343)

1. How much does it take to cover the cylinder?; $170\pi \approx 533.8 \text{ cm}^2$; $300\pi \approx 942 \text{ cm}^3$

3. $486\pi \approx 1526.0 \text{ ft}^3$

5. $245\pi \approx 769.3 \text{ ft}^3$

7. $90\pi \approx 282.6 \text{ mm}^3$

9. $63\pi \approx 197.8 \text{ in.}^3$

11. $256\pi \approx 803.8 \text{ cm}^3$

13. $\frac{125}{8\pi} \approx 5 \text{ ft}$

15. $\frac{240}{\pi} \approx 76 \text{ cm}$

17. Divide the volume of one round bale by the volume of one square bale.

19. $8325 - 729\pi \approx 6036 \text{ m}^3$

21. $y = 8$

23. $x = 5$

25. B

Section 9.1 Exponents

(pages 358 and 359)

1. An exponent describes the number of times the base is used as a factor. A power is the entire expression (base and exponent). A power tells you the value of the factor and the number of factors. No, the two cannot be used interchangeably.

3. 3^4

5. $\left(-\frac{1}{2}\right)^3$

7. $\pi^3 x^4$

9. $8^4 b^3$

11. 25

13. 1

15. $-\frac{1}{8}$

17. The bases are different. The base in Exercise 14 is $\frac{1}{2}$ and the base in Exercise 15 is $-\frac{1}{2}$.

19. $3^3 \cdot 5^2$

21. $12 \cdot \left(\frac{7}{10}\right)^3$; 4.116 in.

23. 189

25. 5

27. 2

29. **a.** about 99.95 g **b.** 99.95%

31. Commutative Property of Multiplication

33. Identity Property of Multiplication

Section 9.2 Zero and Negative Exponents
(pages 364 and 365)

1. no; Any nonzero base with an exponent of zero is always 1.

3. $5^{-5}, 5^0, 5^4$ **5.** $<$ **7.** $>$ **9.** 1 **11.** 144

13. 1 **15.** 50 **17.** 10,000,000 grains of sand

19. $\frac{6}{y^4}$ **21.** $\frac{81}{c^4}$ **23.** $4x^6$ **25.** $\frac{n^3}{m^2}$ **27.** 100 mm

29. 1,000,000 nanometers **31.** **a.** 10^{-9} m **b.** equal to

33. If $a = 0$, then $0^n = 0$. Because you cannot divide by 0, the expression $\frac{1}{0}$ is undefined.

35.

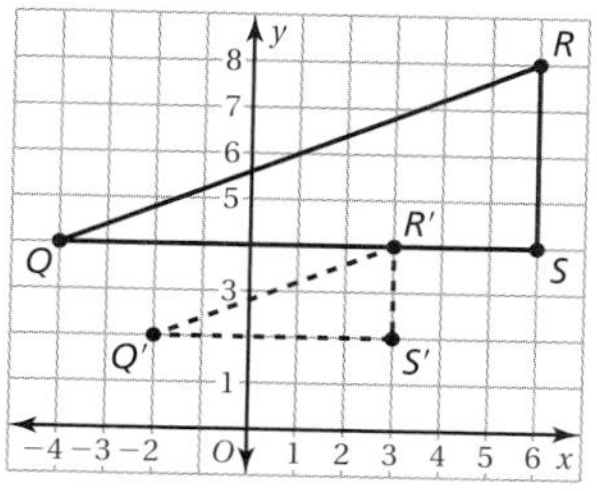

reduction

Section 9.3 Reading Scientific Notation
(pages 372 and 373)

1. Scientific notation uses a factor greater than or equal to 1 but less than 10 multiplied by a power of 10. A number in standard form is written out with all the zeros and place values included.

3. 0.00015 m **5.** 20,000 mm^3

7. yes; The factor is greater than or equal to 1 and less than 10. The power of 10 has an integer exponent.

9. no; The factor is greater than 10.

11. yes; The factor is greater than or equal to 1 and less than 10. The power of 10 has an integer exponent.

13. no; The factor is less than 1. **15.** 70,000,000

17. 500 **19.** 0.000044 **21.** 1,660,000,000 **23.** 9,725,000

25. **a.** 810,000,000 platelets
b. 1,350,000,000,000 platelets

27. **a.** Bellatrix
b. Betelgeuse

29. 5×10^{12} km^2

31. Be sure to convert some of the speeds so that they all have the same units.

33. 3^3y^3 **35.** C

Selected Answers

Section 9.4 Writing Scientific Notation
(pages 378 and 379)

1. If the number is greater than or equal to 10, the exponent will be positive. If the number is less than 1 and greater than 0, the exponent will be negative.

3. 2.1×10^{-3}

5. 3.21×10^{8}

7. 4×10^{-5}

9. 4.56×10^{10}

11. 8.4×10^{5}

13. 72.5 is not less than 10. The decimal point needs to move one more place to the left. 7.25×10^{7}

15. 6.09×10^{-5}, 6.78×10^{-5}, 6.8×10^{-5}

17. 4.8×10^{-8}, 4.8×10^{-6}, 4.8×10^{-5}

19. 6.88×10^{-23}, 5.78×10^{23}, 5.82×10^{23}

21. 4.01×10^{7} m

23. $680,\ 6.8 \times 10^{3},\ \frac{68{,}500}{10}$

25. $6.25 \times 10^{-3},\ 6.3\%,\ 0.625,\ 6\frac{1}{4}$

27. 1.99×10^{9} watts

29. carat; Because 1 carat $= 1.2 \times 10^{23}$ atomic mass units and 1 milligram $= 6.02 \times 10^{20}$ atomic mass units, and $1.2 \times 10^{23} > 6.02 \times 10^{20}$.

31. 16

33. -49

Section 9.5 Square Roots
(pages 384 and 385)

1. There is no integer that equals 50 when squared.

3. finding the square root; *Sample answer:* $3^2 = 9$ and $\sqrt{9} = \sqrt{3^2} = 3$

5. 2 ft

7. 7 in.

9. 11 and -11

11. 13 and -13

13. -12

15. 7

17. ± 14

19. 14

21. 38

23. 3

25. 6 ft

27. 2 cm

29. 8 in.

31. a; because $\sqrt{c} = \sqrt{a^2} = a$

33. The distance d that the hammer falls is 400 feet.

35. similar; Corresponding angles are congruent and corresponding sides are proportional.

37. C

Section 10.1 Histograms
(pages 401–403)

1. The *Test Scores* graph is a histogram because the number of students (frequency) achieving the test scores are shown in intervals of the same size (20).

3. No bar is shown on that interval.

5.

Number	Tally	Total
6	I	1
7	II	2
8	III	3
9	~~IIII~~ II	7
10	II	2

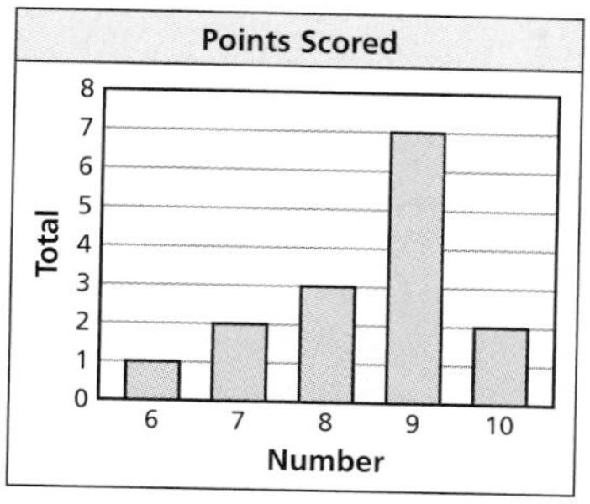

7.

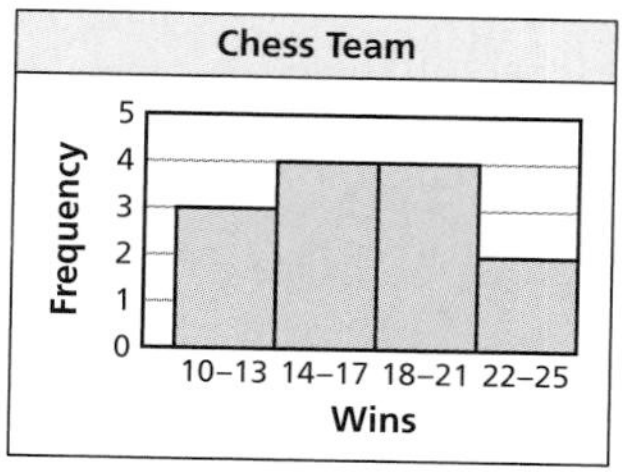

9. There should not be space between the bars of the histogram.

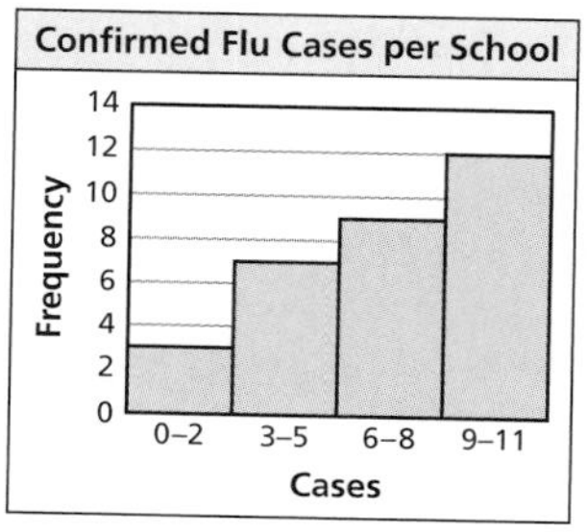

11. The frequency is the number of songs not the percent of songs. The statement should be "12 of the songs took 5–8 seconds to download."

13. Pennsylvania; You can see from the intervals and frequencies that Pennsylvania counties are greater in area, which makes up for it having fewer counties.

15. **a.** yes; The stem-and-leaf plot shows that 10 pounds is a data value.

b. no; Both displays show that 11 residents produced between 20 and 29 pounds of garbage.

17. Begin by ordering the data.

19. 27

21. 51.2

Section 10.2 Fundamental Counting Principle
(pages 408 and 409)

1. An event M has m possible outcomes and event N has n possible outcomes. The total number of outcomes of event M followed by event N is $m \times n$.

3. 125,000

5. 8

7. 20

9. 60

11. The possible outcomes of each question should be multiplied, not added. The correct answer is $2 \times 2 \times 2 \times 2 \times 2 = 32$.

13. Use *Solve a Simpler Problem* to solve this.

15. 2.56×10^9

17. 3.204×10^{11}

Section 10.3 Theoretical Probability
(pages 416 and 417)

1. There is a 50% chance you will get a favorable outcome.

3. Spinner 4; The other three spinners are fair.

5. $\frac{1}{6}$ or about 16.7%

7. $\frac{1}{2}$ or 50%

9. 0 or 0%

11. 9 chips

13. not fair, your friend

15. $\frac{1}{44}$ or about 2.3%

17. **a.** $\frac{4}{9}$ or about 44.4% **b.** 5 males

19. There are 2 combinations for each.

21. $\frac{1}{4}$

23. $-\frac{21}{40}$

25. C

Section 10.4 Experimental Probability
(pages 422 and 423)

1. Perform an experiment several times. Count how often the event occurs and divide by the number of trials.

3. $\frac{2}{5}$ or 40% **5.** $\frac{2}{5}$ or 40% **7.** $\frac{7}{50}$ or 14% **9.** $\frac{21}{25}$ or 84% **11.** $\frac{17}{50}$ or 34%

13. The theoretical probability was found, not the experimental probability. $P(4) = \frac{11}{50}$

15. 45 tiles

17. a. $\frac{5}{9}$ or about 55.6%

b. 40 songs

19 and 21. Make a list of the different sums and the number of times each was rolled. Use a double bar graph to compare the probabilities.

23. *Sample answer:* Roll two number cubes 50 times and find each product. Record how many times the product is at least 12. Divide this number by 50 to find the experimental probability.

25. $x = 5$

27. $x = 24$

Section 10.5 Compound Events
(pages 429–431)

1. What is the probability of choosing a 1 and then a blue chip?; $\frac{1}{10}$; $\frac{1}{15}$

3. *Sample answer:* choosing two marbles from a bag

5. $\frac{1}{8}$ **7.** $\frac{3}{8}$ **9.** $\frac{1}{42}$ **11.** $\frac{2}{21}$

13. The two events are dependent, so the probability of the second event is $\frac{1}{3}$.

$P(\text{red and green}) = \frac{1}{4} \cdot \frac{1}{3} = \frac{1}{12}$

15. $\frac{1}{28}$ **17.** $\frac{1}{36}$ **19.** 1

21. $\frac{1}{12}$ **23.** $\frac{7}{8}$

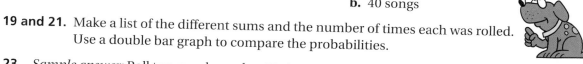

25. Use the strategy *Working Backward* to solve this problem.

27. 1 **29.** 64

Key Vocabulary Index

Mathematical terms are best understood when you see them used and defined *in context*. This index lists where you will find key vocabulary. A full glossary is available in your Record and Practice Journal and at *BigIdeasMath.com*.

Student Index

This student-friendly index will help you find vocabulary, key ideas, and concepts. It is easily accessible and designed to be a reference for you whether you are looking for a definition, a real-life application, or help with avoiding common errors.

Student Index

F

G

H

I

K

L

M

Q

R

Student Index

K

Strand		Focus
Number and Number Sense	–	Whole Number Concepts
Computation and Estimation	–	Whole Number Operations
Measurement	–	Instruments and Attributes
Geometry	–	Plane Figures
Probability and Statistics	–	Data Collection and Display
Patterns, Functions, and Algebra	–	Attributes and Patterning

1

Strand		Focus
Number and Number Sense	–	Place Value and Fraction Concepts
Computation and Estimation	–	Whole Number Operations
Measurement	–	Time and Nonstandard Measurement
Geometry	–	Characteristics of Plane Figures
Probability and Statistics	–	Data Collection and Interpretation
Patterns, Functions, and Algebra	–	Patterning and Equivalence

2

Strand		Focus
Number and Number Sense	–	Place Value, Number Patterns, and Fraction Concepts
Computation and Estimation	–	Number Relationships and Operations
Measurement	–	Money, Linear Measurement, Weight/Mass, and Volume
Geometry	–	Symmetry and Plane and Solid Figures
Probability and Statistics	–	Applications of Data
Patterns, Functions, and Algebra	–	Patterning and Numerical Sentences

3

Number and Number Sense	– Place Value and Fractions
Computation and Estimation	– Computation and Fraction Operations
Measurement	– U.S. Customary and Metric Units, Area and Perimeter, and Time
Geometry	– Properties and Congruence Characteristics of Plane and Solid Figures
Probability and Statistics	– Applications of Data and Chance
Patterns, Functions, and Algebra	– Patterns and Property Concepts

4

Number and Number Sense	– Place Value, Fractions, and Decimals
Computation and Estimation	– Factors and Multiples, and Fraction and Decimal Operations
Measurement	– Equivalence within U.S. Customary and Metric Systems
Geometry	– Representations and Polygons
Probability and Statistics	– Outcomes and Data
Patterns, Functions, and Algebra	– Geometric Patterns, Equality, and Properties

5

Number and Number Sense	– Prime and Composite Numbers and Rounding Decimals
Computation and Estimation	– Multistep Applications and Order of Operations
Measurement	– Perimeter, Area, Volume, and Equivalent Measures
Geometry	– Classification and Subdividing
Probability and Statistics	– Outcomes and Measures of Center
Patterns, Functions, and Algebra	– Equations and Properties

Mathematics Reference Sheet

Conversions

U.S. Customary
1 foot = 12 inches
1 yard = 3 feet
1 mile = 5280 feet
1 acre ≈ 43,560 square feet
1 cup = 8 fluid ounces
1 pint = 2 cups
1 quart = 2 pints
1 gallon = 4 quarts
1 gallon = 231 cubic inches
1 pound = 16 ounces
1 ton = 2000 pounds
1 cubic foot ≈ 7.5 gallons

U.S. Customary to Metric
1 inch ≈ 2.54 centimeters
1 foot ≈ 0.3 meter
1 mile ≈ 1.6 kilometers
1 quart ≈ 0.95 liter
1 gallon ≈ 3.79 liters
1 cup ≈ 237 milliliters
1 pound ≈ 0.45 kilogram
1 ounce ≈ 28.3 grams
1 gallon ≈ 3785 cubic centimeters

Time
1 minute = 60 seconds
1 hour = 60 minutes
1 hour = 3600 seconds
1 year = 52 weeks

Temperature
$C = \frac{5}{9}(F - 32)$

$F = \frac{9}{5}C + 32$

Metric
1 centimeter = 10 millimeters
1 meter = 100 centimeters
1 kilometer = 1000 meters
1 liter = 1000 milliliters
1 kiloliter = 1000 liters
1 milliliter = 1 cubic centimeter
1 liter = 1000 cubic centimeters
1 cubic millimeter = 0.001 milliliter
1 gram = 1000 milligrams
1 kilogram = 1000 grams

Metric to U.S. Customary
1 centimeter ≈ 0.39 inch
1 meter ≈ 3.28 feet
1 kilometer ≈ 0.6 mile
1 liter ≈ 1.06 quarts
1 liter ≈ 0.26 gallon
1 kilogram ≈ 2.2 pounds
1 gram ≈ 0.035 ounce
1 cubic meter ≈ 264 gallon

Surface Area and Volume

Rectangular Prism
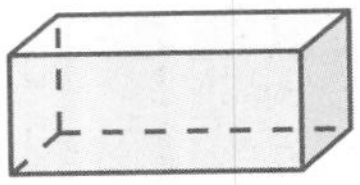

$S = 2\ell w + 2\ell h + 2wh$

$V = \ell wh$

Cylinder

$S = 2\pi r^2 + 2\pi rh$

$V = Bh$

Rules of Exponents

Zero Exponents

$a^0 = 1$, where $a \neq 0$

Negative Exponents

$a^{-n} = \frac{1}{a^n}$, where $a \neq 0$

Slope

$\text{slope} = \frac{\text{change in } y}{\text{change in } x}$

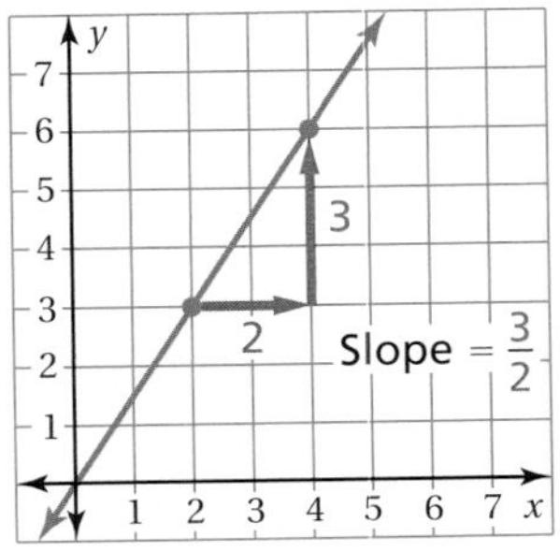